The Quantum Physics of Atomic Frequency Standards

Recent Developments

The Quantum Physics of Atomic Frequency Standards

Recent Developments

Jacques Vanier
Université de Montréal, Montréal, Canada

Cipriana Tomescu
Université de Montréal, Montréal, Canada

CRC Press
Taylor & Francis Group
Boca Raton London New York

CRC Press is an imprint of the
Taylor & Francis Group, an **informa** business

CRC Press
Taylor & Francis Group
6000 Broken Sound Parkway NW, Suite 300
Boca Raton, FL 33487-2742

First issued in paperback 2019

SBN-13: 978-1-4665-7695-7 (hbk)
ISBN-13: 978-1-138-89455-6 (pbk)

Visit the Taylor & Francis Web site at
http://www.taylorandfrancis.com

and the CRC Press Web site at
http://www.crcpress.com

Contents

Preface

Volumes 1 and 2 of *The Quantum Physics of Atomic Frequency Standards*, henceforth referred to as *QPAFS* (1989), were written in the 1980s and were published in 1989. They covered, in some detail, work done up to 1987 on the development of atomic frequency standards. The text included a description of their development at that time, as well as a description of the research on the physics supporting that development. Since that time, the field has remained a very active part of the research program of many national laboratories and institutes. Work has remained intensive in many sectors connected to the refinement of classical frequency standards based on atoms such as rubidium (Rb), caesium (Cs), hydrogen (H), and selected ions in the microwave range, while new projects were started in connection to the realization of stable and accurate frequency standards in the optical range.

For example, intensive studies were made on the use of lasers in the optical pumping and cooling of Rb and Cs as well as on the development of a new type of standard based on the quantum-mechanical phenomenon called coherent population trapping (CPT). Regarding Cs and Rb, laser cooling of atoms has made possible the realization of an old dream in which a small blob of atoms, cooled in the microkelvin range, is projected upward at a slow speed in the gravitational field of the earth and the atoms fall back like water droplets in a fountain. In their path, the atoms are made to pass through a microwave cavity, and upon falling back after having spent their kinetic energy, pass through the same cavity, mimicking, with a single cavity, the classical double-arm Ramsey cavity approach. The system is called *atomic fountain*. Its advantage over the classical approach resides in the reduction of the width of the resonance hyperfine line by a factor of the order of 100 relative to that observed in the room temperature approach. The resulting line width is of the order of 1 Hz. Work has also continued on the development of smaller H masers, in particular in the development of passive devices and in the use of a new smaller so-called *magnetron cavity*. The advent of the solid-state laser in the form of the conventional edge-emitting type (GaAs) and vertical structure (VCSEL) has opened the door to a new approach in optical pumping for implementing smaller and more performing Rb and Cs cell frequency standards.

Since the 1990s, laser cooling has been studied extensively and aside from providing a means for realizing the fountain clock mentioned above, it has allowed the realization of clocks based on microwave transitions in ions such as mercury (Hg^+), barium (Ba^+), strontium (Sr^+), and ytterbium (Yb^+) confined within an electromagnetic trap.

On the other hand, intense work has been carried out in several laboratories in extending the work done at microwave frequencies to the optical frequency range. The gain in that approach relies mainly on the increase in the frequency of the atomic transitions involved, which provides for a line width similar to that obtained in the microwave range a resonance quality factor millions of times larger. Laser cooling has been applied successfully to such atoms as mercury (Hg), ytterbium (Yb), and strontium (Sr) stored in optical lattice traps in order to reduce their thermal motion. Laser cooling has also been used in the mono-ion trap to implement optical frequency standards. In that case, a single ion, say Sr^+ or Yb^+, is maintained in a Paul or Penning trap and its motion within the trap is damped by laser cooling. Clocks at optical frequencies have been implemented as laboratory units with unsurpassed accuracy and frequency stability reaching the 10^{-16} to 10^{-18} range. In both cases, the clock frequency is derived from a transition between the ground S state of the atom and an excited metastable state with a lifetime of the order of 1 second or more leading to a very narrow resonance line. The clock transition is detected by means of monitoring changes in the fluorescence level created by the cooling radiation when the clock transition is excited.

The large gap in frequency between the microwave and the optical range has always been an roadblock in the use of optical frequencies in various applications such as frequency standards or still high precision spectroscopy and fundamental research. The reason is mainly due to the fact that gaps between available optical frequencies for the realization of clocks are very large. It is extremely difficult to connect those frequencies to the microwave range. This connection is required because most of the applications are in the low frequency range of the spectrum and, furthermore, because the SI (International System of Units) definition of the second is based on a microwave hyperfine transition in Cs, in the X band. We have given in Volume 2 of *QPAFS* examples of the conventional method used to make that connection. That method comprises frequency- and phase-locking together banks of lasers with appropriate heterodyning in several steps in order to interconnect various optical frequencies to reach finally the microwave range. The connection has to be done over a large number of steps and involves tremendous investment of space and time to finally measure what very often happens to be just a single frequency. Such a task has been reduced considerably by the invention of the so-called *optical comb*, which comprises locking the repetition rate of a femtosecond laser to a stable atomic frequency standard of high spectral purity, such as an H maser referenced in frequency to a primary Cs atomic clock. When observed by means of a nonlinear optical fibre, the resulting laser spectrum consists of a spectrum of sharp lines, themselves called the *teeth of the comb*, which covers a frequency range of the order of 1 octave. Frequencies over a broad range are then measured essentially in a single step on an optical table, resulting in a considerable reduction in work and size as compared to the previous heterodyning technique, which required entire rooms filled with lasers.

This volume covers those subjects in some detail. It is divided into five chapters. Chapter 1 is an introduction, presenting a review of recent developments made on the improvement of conventional atomic frequency standards described in the two volumes of *QPAFS*. It highlights the main limitations of those frequency standards and the physical basis of those limitations and outlines the progress made during the last 25 years. Chapter 2 is a description of recent advances in atomic physics, theory and applications, that opened new avenues. Chapter 3 is concerned with research and development done in the development of new microwave frequency standards. Chapter 4 describes research and development done in the optical range to implement optical frequency standards based on new results in atomic physics as described in Chapter 2. Chapter 5 summarizes the results in frequency stability and accuracy achieved with those new frequency standards and outlines selected applications. A short reflection is included giving some insight into future work.

Such a text cannot be written without significant help from experts in the field. We wish to recognize the contribution and collaboration of many scientists. In particular, we wish to recognize the invaluable help of André Clairon, who has read the whole manuscript and helped in improving its exactness and completeness. We also show our gratitude to the following scientists who helped us through their encouragement, supplied original figures or material, and contributed by means of comments on various sections of the text: C. Affolderbach, A. Bauch, S. Bize, J. Camparo, C. Cohen-Tannoudji, E. De Clercq, A. Godone, D. Goujon, S. Guérandel, P. Laurent, T. Lee, S. Micalizio, G. Mileti, J. Morel, W.D. Phillips, P. Rochat, P. Thomann, R.F.C. Vessot, and S. Weyers.

Jacques Vanier

and

Cipriana Tomescu
University of Montreal

Introduction

This book is about recent developments in the field of atomic frequency standards, developments that took place after the publication in 1989 of the first two volumes with the same title. Atomic frequency standards are systems providing an electrical signal at a cardinal frequency of, say, 10 MHz, a signal generated usually by a quartz crystal oscillator locked in phase or in frequency to a quantum transition inside an atom. The atom is selected for its properties such as easy detection of the particular quantum transition chosen and relative independence of its frequency of the environment. In early work, those conditions limited development around hydrogen and alkali atoms, which have transitions in the microwave range and could be manipulated easily as beams or atomic vapour with the techniques available at that time. Progress in the development of lasers and their stabilization extended that work to the optical range. A major task encountered in the early development of microwave standards has always been the elimination of Doppler effect. Atoms at room temperature travel at speeds of several hundred metres per second and, consequently, Doppler effect causes frequency shifts and line broadening of the resonance signal. This effect is generally eliminated by means of various storage techniques based on Dicke effect, or still beam techniques using the Ramsey double-arm cavity approach. These techniques are not well adapted to optical frequencies because of the shorter wavelengths involved. However, progress in the understanding of interactions between atoms and electromagnetic interactions has provided new means of reducing the velocity of atoms and reducing, if not eliminating, the constraints introduced by Doppler effect.

An atomic frequency standard that is operated continuously becomes an atomic clock. The operation is essentially a process of integration and the date set as the constant of integration provides the basis for implementing a timescale. This is the origin of atomic timescales, in particular the one maintained by the International Bureau of Weights and Measures. Various systems in operation have their own timescale, for example, the global positioning system (GPS) of the United States, the Russian Glonass system, the Chinese Beidou system, and the European Galileo systems under development, all playing an important role in navigation on or near the surface of the earth.

Although time is central to physics and is used in our day-to-day life, it is a concept that is difficult to grasp, let alone to define. We use it without questioning its origin and its exact nature. It is basic in physics for describing the dynamics of systems and ensembles of systems by means of equations that model the evolution of *objects* forming our universe. The concept is used as such without questioning much its exact nature and origin. In Newtonian mechanics, objects evolve in space and their behaviour is described by means of differential equations and functions of time and space. Both space and time are independent and in common language they are said to be absolute. In that context, time is not a function of space and space is not a function of time. However, in attempts to relate mechanics and electromagnetism by space and time transformations, a difficulty arose. This is due to the finiteness and invariability of the speed of light, made explicit in Maxwell's equations, whatever the motion of the frame of reference in which it is generated and measured. In this context, with Einstein, Poincaré, Lorentz, Minkowski, and others, time and space become entangled and functions of each other. There is no such thing as an absolute space in which objects evolve in an absolute time framework, both independent of each other. Time and space form a single four-dimensional framework and cannot be treated independently. This concept forms the basis of the theory of relativity. This theory has been shown to be valid through multiple experiments and verifications to a level that raises its validity to a high degree. It should be pointed out that the most accurate verifications were done with atomic clocks, the instruments that are the content of this book. There is another question also often raised regarding the nature of time: Could it be discrete? If so what would be the size of its smallest quantity, the time quantum? Could it be that Planck's time is the smallest time entity? This is a totally unknown subject and appears to be a roadblock to in the development of a sustainable quantum theory that includes the concepts elaborated in the theory of general relativity.

Although we may feel somewhat uncomfortable in the context of such questions, time remains the most basic concept in physics, is fundamental, and is the quantity that is measured with the greatest precision. Current atomic clocks can commonly keep time to an accuracy of 1 s in a million years, or in other words are stable to better than 1 μs in a year. For example, the timescale generated by the GPS satellites for navigation, based on atomic frequency standards on satellites and on ground, is stable after appropriate processing and filtering to about 1 ns/day. On the other hand, on the basis of our inability to measure time by astronomical means with such accuracy, it was decided in 1967 to replace the astronomical definition of the second by one in terms of a particular atomic hyperfine transition in the Cs atom. The frequency of that transition is set at 9,192,631,770 Hz. Furthermore, since now the speed of light is defined exactly as 299,792,458 m/s, providing at the same time a definition of the metre, the mechanical units of the SI become essentially determined by the basic time unit, the second. The concept of unifying all SI units in terms of a single quantity goes further due to the Josephson effect phenomenon, which relates voltage to frequency in a most fundamental expression, $2e/\hbar$, involving only fundamental constants. This is the subject that will be described in Chapter 5.

From this discussion, it is evident that time plays a most important role in physics and technology and the realization of the highest accuracy and precision of the SI unit, the second, has remained one of the most active preoccupations of several laboratories and institutes over the past 50 years. Starting with tremendous improvements in the realization of the second within the microwave range, work has extended to the optical range with proven increase in frequency stability and accuracy by several orders of magnitude. These achievements were possible mainly through a better understanding of the interactions between electromagnetic radiation and atoms, providing a means of altering the properties of atoms. This book is about those improvements that have taken place mainly during the past 25 years, on the realization of stable and accurate frequency standards.

Authors

Jacques Vanier completed his undergraduate studies in physics at the University of Montreal, Québec, Canada, before moving to McGill University to undertake his graduate studies. During his career he has worked in various industries (Varian, Hewlett-Packard); taught physics; and carried out research at Laval University, Montreal, Québec, Canada, and has also been an active member of the National Research Council of Canada, in Ottawa, Ontario, Canada. His research work is oriented towards the understanding and the application of the quantum electronics phenomena

and he has been a consultant for several companies engaged in the development of atomic clocks. Jacques has also been very active on the academic circuit, giving lectures and presenting at numerous conferences in universities, national institutes, and summer schools around the world. He has written more than 120 journal articles and proceedings papers and is the author of review articles and books on masers, lasers, and atomic clocks. His book *The Quantum Physics of Atomic Frequency Standards*, written with C. Audoin, is recognized as a main reference in the field. He is the author of *The Universe: A Challenge to the Mind* published by Imperial College Press/ World Scientific. Jacques is a fellow of the Royal Society of Canada, the American Physical Society, and the Institute of Electrical and Electronic Engineers. He has received several awards for his contributions to the field of measurement science. He is currently an adjunct professor in the Physics Department, University of Montreal, Québec, Canada.

Cipriana Tomescu completed her studies in physics at the University of Bucharest, Romania, where she obtained her PhD degree.

From 1982 to 2004, she was a researcher at the National Institute of Laser Physics, Plasma and Radiation, Bucharest, Romania. In the early years of her employment, she participated in the construction of H masers used by the Bucharest Observatory, the Institute of Metrology, and the Faculty of Physics. During the period 1996–2004, she was laboratory director. During the period 1992–2006, she also worked in various national laboratories, in particular,

Paris Observatory, LNE-SYRTE, France; Neuchâtel Observatory, Switzerland; and Communication Research Laboratory, Japan. At those locations, she contributed to the development of advanced state-of-the-art atomic frequency standards, such as Rb and Cs fountains using atom trapping techniques and laser atom cooling. From 2008 to 2012, she worked at the University of Liege, IPNAS, and at Gillam-Fei. She was responsible for the implementation of the first H maser realized in Belgium under

Plan Marshall: SKYWIN-TELECOM. She is the author of numerous publications in scientific journals and conference proceedings and she has been invited to make presentations at numerous symposia, universities, and national institutes. In 1985, she received the D. Hurmuzescu Prize of the Romanian Academy for work on the physics of the H maser. She is currently an invited researcher in the Physics Department of the University of Montreal, Québec, Canada.

1 Microwave Atomic Frequency Standards
Review and Recent Developments

At the end of the 1980s, atomic frequency standards reached a level of refinement that made it the envy of many other fields of physics. The accuracy of primary caesium (Cs) standards maintained in operation at national institutes reached a level better that 10^{-13} and the frequency stability of the hydrogen (H) maser in the medium term was better than 10^{-14}. These characteristics made possible the verification to great accuracy of basic physics predictions such as those resulting from the theory of relativity and made possible the maintenance of a timescale to an unsurpassed stability. It also opened the use of such devices in many applications. The time unit, the second, became the most accurate unit of the International System of Units (SI), with consequences for the implementation of other units such as the metre, the volt, and the ohm. On the other hand, Rb standards had reached a level of development that made them an excellent support of digital communication systems with improved reliability and also made them appropriate for navigation systems using satellites requiring medium frequency stability and small size.

There has been extensive research on the possibility of using other atoms as the basis for new types of frequency standards. However, those systems are still under study in laboratories; Cs, H, and Rb therefore remain the atoms at the heart of atomic frequency standards used at large either as references in basic research or in practical systems requiring precise and accurate timing. Although the Cs standard in its original beam implementation using magnetic state selection has been dethroned as the most accurate primary standard with the introduction of optical pumping and laser cooling, it still remains in many laboratories the work horse for implementing a local timescale, for confirming the accuracy of other standards, and, to a limited extent, for reliable reporting to the BIPM (Bureau International des Poids et Mesures) in the maintenance of the second.

In this chapter, we recall the physical construction and the characteristics of those frequency standards based on Cs, H, and Rb, as well as of some selected other types of microwave frequency standards, which still show promise regarding possible specific applications. We examine the physics at the heart of the operation of those standards and behind their limitations relative to size, accuracy, and frequency stability. We also see that those limitations were overcome to some extent, showing that, with some imagination, improvements could still be made on instruments that had already attained a very high level of maturity.

1.1 CLASSICAL ATOMIC FREQUENCY STANDARDS

We usually group Cs beam frequency standards, H masers, and optically pumped Rb standards under the terminology "classical atomic frequency standards." In the following paragraphs, we review their physical construction and recall the essential theoretical results that were developed in parallel with their implementation. Theoretical investigations were required for an understanding of the various phenomena causing biases observed when evaluated relative to accuracy and frequency stability. The reader will find in Volumes 1 and 2 of *The Quantum Physics of Atomic Frequency Standards (QPAFS)* a detailed description of the operation of such standards and a description of the basic physics involved. In the following sections, we recall the main concepts behind their operation in order to simplify reading of subsequent sections, in which we discuss recent progress in understanding the physics involved. We then present new analysis and realizations that have resulted in better understanding of their operation, greater accuracy, better frequency stability, and in some instances reduction in size and weight.

1.1.1 Cs Beam Frequency Standard

A frequency standard using Cs and the separate oscillatory field approach proposed by Ramsey (Ramsey 1950) was implemented as early as 1955 (Essen and Parry 1955). Intense laboratory and industrial development followed (see, e.g., McCoubrey 1996). Development showed great success and soon after the construction of Cs primary standards in several laboratories, the frequency of the Cs ground state hyperfine transition was adopted for the definition of the second by the Conférence Générale des Poids et Mesures (CGPM, 1967–1968). The frequency adopted was 9,192,631,770 Hz. It was the best number obtained by means of precise astronomical measurements by which the Cs hyperfine frequency was determined relative to the second, whose formal definition at the time of measurement was the ephemeris second based on astronomical observation (Markowitz et al. 1958). That choice has remained till date (2015).

Why was Cs selected for providing the basis of the definition of the second? First, the choice of the Cs atom in the implementation of a frequency standard has resulted from the considerable accumulation of knowledge on that atom over the years and from the several advantages that it provides over other candidates. In particular, Cs has a single stable isotope, ^{133}Cs, and is relatively abundant in nature. Its melting point is 28.4°C. Its vapour pressure is such that it is possible to implement a rather intense atomic beam from an oven at a relatively low temperature of the order of 425 to 500 K. Its ionization energy is low, 3.9 eV, making it easy to detect by conventional procedures such as ionization with a hot wire detector and ion counting. Finally, Cs has a ground state hyperfine frequency falling in the X band, a microwave region that has known extensive development, which makes possible atom–microwave interaction by means of small structures such as cavities whose dimensions are in the centimetre range.

The Cs atom has a nuclear spin $I = 7/2$ and has a single s electron outside closed electronic shells. Its ground state consists of two hyperfine levels $F = 3$ and $F = 4$

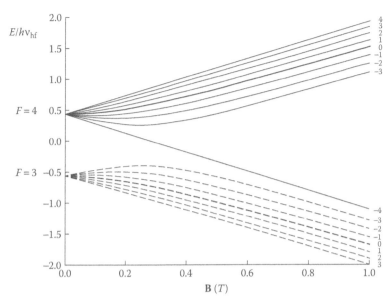

FIGURE 1.1   Ground state energy level manifold of the caesium atom as a function of the magnetic induction **B** in tesla.

and in a low magnetic field the structure consists of two manifolds of 7 and 9 energy levels, respectively. This ground state is shown in Figure 1.1 as a function of the magnetic induction **B**.

1.1.1.1   Description of the Approach Using Magnetic State Selection

A conceptual diagram of the classical Cs beam frequency standard using magnetic state selection is shown in Figure 1.2 (Vanier and Audoin 2005).

A beam of Cs atoms is formed by proper collimation from an oven heated at a temperature of the order of 50–100°C depending on the intensity required. This beam is directed as to pass through a so-called Ramsey cavity that provides a region of electromagnetic interaction and excite transitions between the two ground state hyperfine levels $m_F = 0$ of the atoms. Magnets A and B are generally dipole magnets and create an intense inhomogeneous field in which atomic trajectories are deflected. They are called Stern–Gerlach selector magnets or filters. The deflection is caused by the interaction of the atom magnetic moment with the magnetic field gradient and by the tendency of atoms to seek states of low potential energy. Consequently, according to Figure 1.1, atoms having higher energy in high magnetic fields are deflected to regions of low magnetic field in order to lower their potential energy. Similarly, those atoms having lower energies at high magnetic fields seek regions of high magnetic field for the same reason. Selection is accomplished by means of magnet A whose orientation is such as to force atoms in level $F = 4$, $m_F = 0$ to pass through the Ramsey microwave cavity, and reach the second deflecting magnet B. Atoms in the other $F = 3$, $m_F = 0$ level, being deflected away from the $F = 4$, $m_F = 0$ atoms, are eliminated from the beam by appropriate collimation. The analysis of the

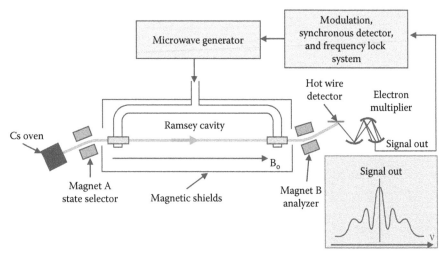

FIGURE 1.2 Simplified conceptual diagram of the Cs beam frequency standard using magnetic state selection. The inset shows the shape of the resonant signal observed when the frequency-lock loop is open and the microwave frequency is scanned slowly over the atomic hyperfine resonance. Although in the figure the magnetic induction is shown parallel to the beam direction, in practice it is very often made perpendicular to the beam. (Data from Vanier, J. and Audoin, C., *Metrologia*, 42, S31, 2005. Copyright Bureau International des Poids et Mesures. Reproduced by kind permission of IOP Publishing. All rights reserved.)

beam composition is done by the combination of magnet B, called the analyzer, and a hot wire ionizer followed by a counter usually assisted by an electron multiplier. In their transit through the Ramsey cavity, the atoms are submitted to an electromagnetic field of angular frequency ω in the two arms of the cavity called the interaction regions. In the first arm of the cavity, atoms are excited into a Rabi oscillation that puts them into a quantum superposition of the two hyperfine levels $F = 4$, $m_F = 0$ and $F = 3$, $m_F = 0$ of the ground state. We define τ, the time of transit of an atom at speed v inside that first arm of length l. The power fed into the cavity is adjusted to such a value as to make the electromagnetic radiation appear as a $\pi/2$ pulse, that is to say a microwave pulse that puts the atoms in an exact superposition state of the two hyperfine levels when they exit that first arm at the most probable speed. The atoms are subsequently left to drift unperturbed in the space within the double arm cavity. A uniform magnetic induction B_o provides an axis of quantization and the atoms remain in the same state. They then penetrate inside the second arm at distance L from the first arm. We call T the time of transit between the two arms of the cavity. If v is the speed of a given atom, then T is simply L/v and is affected the spread in v. In that second arm, the atoms are again submitted to a field of the same intensity and same frequency as in the first arm. Atoms having the same speed as in the first arm are, thus, submitted again to a $\pi/2$ pulse. The atoms at the exit of the second arm find themselves in the lower state $F = 3$, $m_F = 0$ and the transition is complete as if they had been submitted to a π pulse. If the frequency applied to the cavity is not exactly the resonance frequency of the atoms, the phase of the field in the second arm is not coherent with the phase of the magnetic moment of the entering atoms and interference takes place. The total effect of the radiation inside the two arms is smaller than that of a π pulse and the transition is not complete. The probability of the transition is not a maximum and upon detection some atoms appear as having made the transition to the lower state and some not. The actual state of affairs is analyzed by means of a second Stern–Gerlach filter that selects only atoms having made the transition and orients them towards the detector. That detector is made of a hot wire of tungsten or other material that ionizes the atoms reaching it. The resulting ion current is measured either directly (usual in laboratory standards) or by means of an electron multiplier (usual in commercial standards). If the frequency of the radiation fed to the cavity is swept over a certain range, a kind of interference pattern is observed at the output of the detector as shown in the inset of Figure 1.2. It is worth mentioning that the role of atoms in level $F = 4$, $m_F = 0$ and of atoms in level $F = 3$, $m_F = 0$ can be inverted without affecting the operation of the system.

This type of signal is called *Ramsey fringes* (Ramsey 1956). A complete calculation of the shape of that signal is done in *QPAFS*, Volume 2 (*QPAFS* 1989).

Since the central peak, being the resonance peak, is the signal of interest we concentrate on it. For a given atomic velocity v leading to a transit time τ in each arm of the cavity, the shape of the central fringe signal can be represented approximately by the following equation (for ease in reading, we import into the present text a few important equations from *QPAFS* 1989):

$$P(\tau) = \frac{1}{2}\sin^2 b\tau \left[1 + \cos\left(\Omega_o T + \phi\right)\right] \qquad |\Omega_o| \ll b \qquad (1.1)$$

where:
   τ is the time of interaction of the atom with the microwave field in each interaction region
   T is the time spent by the atom between the interaction regions
   φ is the phase difference that exists between the microwave fields in the two arms of the cavity, including the effect of asymmetries and cavity losses
   Ω_o is the difference between ω, the angular frequency of the microwave radiation in the Ramsey cavity, and ω_o, the resonance angular frequency of the atom:

$$\Omega_o = \omega - \omega_o \qquad (1.2)$$

The parameter b is the Rabi angular frequency in the interaction region and is a measure of the amplitude of the microwave induction B_{mw}. It is defined by the equation (our definition of b is different from that used by Ramsey (1956) by a factor of 2, consistent with the notation used in *QPAFS* 1989):

$$b = \frac{\mu_B}{\hbar} B_{mw} \qquad (1.3)$$

where:
   μ_B is the Bohr magneton
   $\hbar$ is the Planck's constant h over 2π

Assuming $\phi = 0$ or π, the full width at half maximum of the central fringe as given by Equation 1.1 is readily calculated as:

$$W = \frac{\pi}{T} \tag{1.4}$$

If the resonance signal was observed with a single cavity (Rabi resonance), its line width would be π/τ. With the double arm cavity the width of the signal is reduced by the factor, $L/l = T/\tau$, which could be very large in laboratory units designed with a large distance L between the two arms of the Ramsey cavity of individual lengths l. In those units, L/l may be in practice of the order of 100 or more.

We note that Equation 1.1 is valid under the assumption that $|\Omega_o| \ll b$. A better approximation to the central fringe is obtained by means of a first-order expansion of the full Ramsey fringe equation as developed in *QPAFS*, Volume 2 (*QPAFS* 1989):

$$P(\tau) = \frac{1}{2}\sin^2 b\tau \left[1 + \cos(\Omega_o T + \Phi) - \left(\frac{2\Omega_o}{b}\right)\tan\frac{1}{2}b\tau\sin(\Omega_o T + \Phi) \right] \tag{1.5}$$

Equation 1.1, however, is an excellent approximation of the central fringe shape and is used in most calculations concerned with frequency shifts introduced by various phenomena. The third term in Equation 1.5 introduces a correction that amounts in some cases to a few percent of the biases calculated and is sometimes used for better precision in the evaluation of various effects (Makdissi and de Clercq 2001).

The above calculation was done under the assumption that the atoms in the beam have all the same velocity and spend the same time in the two arms of the cavity. In practice, the beam is composed of atoms travelling at thermal velocities. In a gas, atomic velocities are spread according to a Maxwell distribution. However, in the case of a collimated beam and state selection by magnets, this distribution is greatly altered. If $f(\tau)$ is the resulting distribution of interaction times τ in each arm of the cavity, then an average of the probability P over this distribution must be made:

$$P = \int_0^\infty f(\tau)P(\tau)d\tau \tag{1.6}$$

where the following relations between speed v and interaction time distributions hold:

$$\int_0^\infty f(\tau)d\tau = \int_0^\infty p(v)dv = 1 \tag{1.7}$$

$$f(\tau) = \frac{1}{\tau^2}p\left(\frac{1}{v}\right) \tag{1.8}$$

The fringe pattern is smeared out to some extent by the velocity spread. It turns out, however, that the central fringe is not much affected by the averaging, if the velocity distribution is made sufficiently narrow as is done in some implementation (Becker 1976). A typical experimental result is shown in Figure 1.3 for two scan widths.

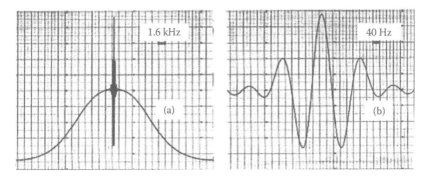

FIGURE 1.3 Ramsey fringes observed experimentally in the NRC Cs VI standard. (Data from Mungall, A.G. et al., *Metrologia*, 17, 123, 1981. Copyright Bureau International des Poids et Mesures. Reproduced by permission of IOP Publishing. All rights reserved.)

A practical implementation of a Cs standard consists of frequency locking to the atomic resonance the frequency of the quartz crystal oscillator that is used to generate the microwave radiation that induces the atomic transitions in the Ramsey cavity as illustrated in Figure 1.2. In that system, the frequency of the crystal oscillator is modulated at a low frequency with a modulation depth less than the line width of the central Ramsey fringe. The resulting signal obtained by means of synchronous detection is a discriminator pattern that is used in a feedback loop to lock the frequency of the crystal oscillator to the hyperfine resonance line.

As is evident from our earlier discussions, the atoms are relatively free in the beam. There are still present, however, some physical phenomena that cause small frequency shifts or biases. One of the main tasks in the implementation of a primary frequency standard of the type just described is the precise evaluation of those shifts or biases. It is only after such an evaluation that a given standard may be accepted as a representation of the SI unit, the second, which is the main goal in national standards laboratory implementations of such units.

Those shifts may be separated into three main groups: those intrinsic to atomic properties, those introduced in the detection of the resonance signal, and those introduced in the locking of the microwave generator to the resonance line. We outline the nature of these shifts. In Section 1.1.1.4 we show how progress was made in recent years in the accuracy of their evaluation and sometimes in their reduction.

1.1.1.2 Review of Frequency Shifts and Accuracy

1.1.1.2.1 Frequency Shifts (Biases) Associated with Physical Atomic Properties

1.1.1.2.1.1 Magnetic Field Shift In a low magnetic field, the shift of the resonant line of interest with the applied magnetic field is given by the equation (QPAFS 1989, Volume 1, Table 1.1.7):

$$\nu = \nu_{hf} + 427.45 \times 10^8 B_o^2 \tag{1.9}$$

where:

ν_{hf} is the unperturbed hyperfine frequency defined as 9,192,631,770 Hz
B_o is the value of the applied magnetic induction in Tesla

The field applied may be of the order of 50 to 100×10^{-7} Tesla (50 to 100 mG) and the displacement of the resonance peak is several parts in 10^{10}. This is the most important shift in the frequency standard and must be determined with an accuracy compatible with the accuracy desired in the final evaluation. It is obvious that field fluctuations must be minimized for reasons of frequency stability. This constraint forces the use of a rather stable current source supplying the device that creates the field. This device may be rods or a solenoid. On the other hand, efficient magnetic shielding is used to prevent environment field fluctuations from reaching the region of interaction, the Ramsey cavity. This is generally done by means of several layers of mu-metal or high permeability metallic cylinders surrounding the region of interaction.

1.1.1.2.1.2 Second-Order Doppler Effect   This shift originates from the time dilation phenomenon of special relativity. For an atom at speed v, the second-order Doppler effect frequency shift Δv_{D2} is given by the equation (*QPAFS* 1989, Volume 1)

$$\frac{\Delta v_{D2}}{v_{hf}} = -\frac{v^2}{2c^2} \tag{1.10}$$

where:
   c is the speed of light

In the beam, the velocities are spread over a relatively large range and this shift must be averaged over the velocity distribution. The average transition probability is given by Equation 1.6. Since the various shifts to be considered are expected to be small, we may consider them as independent of each other. When the second-order Doppler shift alone is taken into account, then the transition probability at the exit of the second selector magnet is given by:

$$P = \frac{1}{2} \int_0^\infty f(\tau) \sin^2 b\tau \left\{ 1 + \cos \left[\omega - \omega_o \left(1 - \frac{v^2}{2c^2} \right) \right] T \right\} d\tau \tag{1.11}$$

This expression may be used directly by proper adjustment and normalization to obtain the intensity of the beam reaching the detector. We may, thus, write the beam intensity as:

$$I = I_b + \frac{1}{2} I_o \int_0^\infty f(\tau) \sin^2 b\tau \left\{ 1 + \cos \left[\omega - \omega_o \left(1 - \frac{v^2}{2c^2} \right) \right] T \right\} d\tau \tag{1.12}$$

where:
   I_b is the background atomic flux reaching the detector, which, for example, is composed of atoms in a wrong state and do not contribute to the signal

The resonance frequency of the standard is identified with the maximum of the central Ramsey fringe signal, which is simply the maximum of the beam intensity I reaching the detector. The value of that frequency is obtained by differentiating Equation 1.12 with respect to ω (see Appendix 1.A). Hence, we obtain:

$$\frac{\omega_D - \omega_0}{\omega_0} = -\frac{\int_0^\infty v^2 T^2 f(\tau) \sin^2 b\tau \, d\tau}{2c^2 \int_0^\infty T^2 f(\tau) \sin^2 b\tau \, d\tau} \qquad (1.13)$$

where:
ω_D is the frequency of the maximum of the central Ramsey fringe

In order to evaluate the resulting shift, one needs to know the interaction time distribution function $f(\tau)$ or the speed distribution. In the 1960s, in the field of atomic beam resonance spectroscopy, a Maxwellian distribution of atomic velocities in the beam was assumed (Harrach 1966, 1967). With magnets as state selectors, the distribution may be very different from a Maxwell's distribution. In early developments of Cs primary standards, it was assumed that the distribution was Maxwellian, but with low and high speeds cutoff (Mungall 1971). It was then verified by numerical analysis that the velocity spectrum with its chosen width and cutoff frequencies, using Equation 1.6, reproduces the observed Ramsey fringe pattern to a good approximation. However, such an approach is rather empirical. We now outline more advanced methods for evaluating $f(\tau)$ and recent developments that have allowed the evaluation of velocity-dependent shifts to a very satisfying accuracy.

1.1.1.2.1.3   Black Body Radiation   This effect is caused by an interaction of the atoms with the oscillating electric field of the ambient thermal radiation. At an operating temperature of 300 K the shift using polarizability values reported in Table 1.1.8 of *QPAFS* (1989) is calculated to be -1.69×10^{-14} and varies as the fourth power of the absolute temperature (Itano et al. 1981). The effect was not important in the evaluation of Cs standards implemented before 1990, due to their limited accuracy and frequency stability. However, the shift was observed at room temperature in a Cs beam standard with greatly improved stability (Bauch and Schröder 1997). The measurement was made as a function of temperature over a range of more than 150°. The value found was 1.66×10^{-14} and was in good agreement with the theoretical prediction. The shift became important in the context of the accuracy reached with the Cs fountain to be described later. Furthermore, the polarizability value used in its theoretical evaluation was questionable in the light of new measurements (Micalizio et al. 2004). We return to this point in Chapter 3 dealing with the Cs fountain.

1.1.1.2.1.4   Spin–Exchange Frequency Shift   Collisions between atoms travelling at different velocities within the beam and with atoms in the background vapour pressure may cause an exchange of their electrons. This is called spin exchange and it creates a frequency shift proportional to the collision rate (*QPAFS* 1989). The collision rate is proportional to the relative velocities of the atoms and their collision cross section. The value of this cross section is not known for Cs at room temperature and the effect, although expected to be small, still needs to be evaluated. The effect is small in Cs beam standards operating at room temperature but as will be seen in the case of the Cs fountain it becomes important particularly when accuracy reaches the 10^{-16} level.

1.1.1.2.2 Shifts Introduced by the Resonance Detection System

1.1.1.2.2.1 Phase Shift between the Two Cavities If the phase ϕ between the fields in the two arms forming the Ramsey cavity differs from either 0 or π by a small amount ϕ, then the central fringe maximum or minimum is displaced. This phase shift may be caused by an asymmetry in the cavity construction or electrical losses in the waveguide creating a travelling wave within the structure.

For atoms at speed v travelling the distance L between the arms in time $T = L/v$, it is readily shown from Equation 1.1 that a phase shift ϕ between the fields of the two arms creates a frequency shift of the resonance line by the amount:

$$\Delta v_\phi = -\frac{\phi}{2\pi T} \tag{1.14}$$

For example, an asymmetry to the extent of 10^{-4} m between the lengths of the two arms of the Ramsey cavity may cause a frequency shift of the order of 10^{-13} depending on the electrical losses of the waveguide used. The frequency shift changes sign upon reversal of the velocity. It can thus be determined experimentally by reversing the direction of the beam. However, due to the presence of a phase shift that varies with position in the cavity, called generally distributed phase shift, and due to the fact that the beam does not necessarily retrace the same path upon reversal, the accuracy of determination of this shift is limited. In short commercial instruments, the frequency shift is larger since T is smaller than in laboratory standards. It may reach 1×10^{-12}.

We have provided in *QPAFS*, Volume 2 (*QPAFS* 1989), a detailed analysis of the effect of this shift. Let us recall the main points of that analysis. The equations to be used are Equations 1.1 and 1.6. Converting to beam intensity, we obtain:

$$I = I_b + \frac{1}{2}I_o \int_0^\infty f(\tau)\sin^2 b\tau \left\{1 + \cos\left[(\omega - \omega_o)T + \phi\right]\right\} d\tau \tag{1.15}$$

We emphasize the difference between this type of bias and that introduced by the second-order Doppler effect. The second-order Doppler shift affects directly the frequency of the atoms while the phase shift is introduced through the cavity and is considered as a step in the time evolution of the atoms. To obtain the frequency where beam intensity I is maximum, we need to do as in the case of second-order Doppler effect, that is, differentiate with respect to ω and set $\partial I/\partial\omega = 0$ (see Appendix 1.B). Using the relation $T = L/v$, the result is:

$$\omega_\phi - \omega_o = -\frac{\phi}{L}\frac{\displaystyle\int_0^\infty (1/v)f(\tau)\sin^2 b\tau\, d\tau}{\displaystyle\int_0^\infty \left(1/v^2\right)f(\tau)\sin^2 b\tau\, d\tau} \tag{1.16}$$

An important consideration follows from that analysis: the shift as measured from the maximum of the Ramsey pattern is a function of the velocity distribution and depends on the value of b, thus on the power fed to the cavity. This is the same thing as in the case of second order Doppler effect, as calculated through Equation 1.13. We now return to this point and see how recent advances in understanding the phase shift effect has resulted in its reduction by means of a new type of cavity.

1.1.1.2.2.2 Cavity Pulling The cavity tuning influences the position of the resonance maximum. The effect is small due to the fact that the cavity Q is low resulting in weak stimulated emission in the cavity since the number of atoms in interaction is small. In short, Cs beam frequency standards where the resonance is less selective, the effect may be significant. However, in laboratory standards where the cavity Q is intentionally made small and the atomic gain is low, this shift is generally under control and does not cause a problem. For example, for a cavity Q of 500, a line Q of 1.5×10^8 corresponding to a line width of 60 Hz as realized in the best standards shown in tables provided hereunder, a detuning of the cavity by 1 MHz would produce a fractional frequency shift of 6×10^{-15}. Full details of the calculation are given in *QPAFS*, Volume 2 (1989). On the other hand, it may be mentioned that cavity detuning may introduce another frequency shift when square wave modulation of the microwave interrogating frequency is used. The effect appears because the fields at the two frequencies resulting from the modulation may not have the same amplitude if the cavity is not tuned exactly to atomic resonance.

1.1.1.2.2.3 Bloch–Siegert Effect The microwave magnetic induction in the cavity may be thought of as linearly polarized radiation. A linearly polarized field may be decomposed into two counter-rotating fields. In the rotating frame approach, one component is seen as resonant with the atomic ensemble and the other rotating in counter direction is seen as having twice the resonance frequency. An elementary analysis shows that a frequency shift is introduced in the detection of the resonance frequency by this off resonance component. This shift is called the Bloch–Siegert effect (Bloch and Siegert 1940). It is proportional to the ratio l/L of the beam tube and in laboratory standards of large size it is of the order of 5×10^{-15}.

1.1.1.2.2.4 Majorana Transitions If the constant magnetic field along the atomic beam is inhomogeneous, transitions of random nature can be caused between m_F sublevels of the two manifolds $F = 3$ and $F = 4$. These are called Majorana transitions (Majorana 1932). It has been shown that these transitions can cause a shift of the resonance frequency of the central $\Delta m_F = 0$ transition (Ramsey 1956). Since in the classical approach permanent magnets are used for state selection and detection, it is possible that stray inhomogeneous fields created by those magnets excite Majorana transitions with a resultant frequency shift. This effect is absent in optically pumped beam tubes in which no selector magnets are used and where the magnetic field can be made very homogeneous all along the beam path.

1.1.1.2.2.5 Rabi and Ramsey Frequency Pulling This is an effect that is partly inherent to the atoms and partly introduced by the technique of detection of the resonance. A shift is introduced by the overlapping of the symmetrically situated field dependent Rabi pedestals with the central fringe of the $\Delta F = 1$, $\Delta m_F = 0$ resonance line (De Marchi et al. 1984; De Marchi 1987). When these pedestals have different amplitudes, which is the case for magnetic state selection, a small distortion of the central fringe is created by the tails of the field dependent Rabi pedestals, causing a frequency shift of the central fringe. Furthermore, the microwave field in the cavity may contain a small perpendicular component causing transitions $\Delta F = 1$, $\Delta m_F = \pm 1$ that are connected to the resonant transition of interest (the $\Delta m_F = 0$ transition) by a common energy level. These transitions may also distort the central fringe and cause a small frequency shift. This is called Ramsey pulling (Cutler et al. 1991). These shifts are a function of beam design and depend to some extent on the microwave power applied to detect the resonance. The effect is function of the applied magnetic field and is small when the resonance line is narrow. Consequently, these effects are much reduced in laboratory standards and in general are small. A considerable amount of theoretical analysis has been made on these effects (Shirley et al. 1995; Lee et al. 2003).

1.1.1.2.2.6 Microwave Leakage An undesired microwave field may be present all around the microwave cavity when microwave leakage occurs. The spurious field may originate from the cavity holes that let the atomic beam through or from unwanted small spacing between different parts of the cavity assembly causing microwave leaks. It may also originate from electrical feedthroughs. The atomic beam may then be subjected to a travelling wave in a place where no microwave field should be present. A Doppler frequency shift then occurs. A model of this effect has been presented, which accounts for the frequency shifts that may be observed (Boussert et al. 1998). Since the details of the field configuration are generally not known, it is rather difficult to evaluate the shift introduced and only careful design of the system avoiding leakage can minimize the effect.

1.1.1.2.2.7 Gravitational Effect According to the general theory of relativity, a clock rate is a function of the gravitational potential at the location of the clock. Consequently, the atomic Cs standard frequency is a function of its altitude in the earth's field. Since the laboratory Cs beam frequency standards used in the determination of the SI unit second are located at different altitudes, it is thus important to precisely determine the actual altitudes of these standards relative to the geoid and make the appropriate correction. This shift is small. In the earth's field, it varies with altitude h relative to the geoid as:

$$\frac{\Delta v_{gr}}{v} \cong \frac{gh}{c^2} \tag{1.17}$$

where:
   g is the acceleration due to gravity at the location of the clock
   c is the speed of light (Ashby et al. 2007)
   The altitude h is assumed to be small relative to the earth's radius

The fractional effect on the frequency is about 10^{-16} m^{-1}. The height above the geoid is difficult to determine to an accuracy of the order of 10 cm. This corresponds to an accuracy in clock frequency to the order of 10^{-17}. Consequently, since, as we will see in Chapter 4, it is possible that optical clock accuracies could reach a level of 10^{-18} soon, it should be possible to find a very useful application of atomic clocks in precision geodesy.

1.1.1.2.3   Offsets Introduced by the Electronic Servo System

1.1.1.2.3.1   Spectrum of the Microwave Radiation   Imperfection in the spectrum of the microwave radiation and in its modulation can cause frequency shifts (Audoin et al. 1978). The microwave radiation at 9.2 GHz is normally obtained by synthesis from a quartz crystal oscillator at a nominal frequency, say 10 MHz. The process generally creates sidebands at various frequencies and furthermore amplifies any spurious spectral components present in the spectrum of the quartz crystal oscillator. These sidebands create virtual transitions and cause small frequency shifts.

1.1.1.2.3.2   Frequency Shifts Introduced by the Modulation   When square wave frequency modulation of amplitude ω_m is used in the servo system to lock the microwave frequency to the resonance line, the system's role, by means of synchronous detection and feedback, is to make equal the amplitude of the signal detected on each side of the line at $+\omega_m$ and $-\omega_m$. If the central Ramsey fringe is asymmetrical, the servo system may lock to a frequency different from that corresponding to the maximum of the resonance line as given, for example, by Equation 1.13 including the Doppler shift (Mungall 1971). The difference in frequency depends on the distortion of the line and on the amplitude of the frequency modulation ω_m used. This is illustrated in Figure 1.4, where it is clearly seen that for a symmetrical line the central frequency detected by that means is the same as the frequency for maximum signal. This is not the case for an asymmetrical line for which the servo frequency ω_o' varies with amplitude of modulation ω_m and is not the same as the frequency ω_o for maximum signal.

The asymmetry of the Ramsey fringes may originate, for example, from frequency shifts that depend on velocity. The second-order Doppler effect causes such a

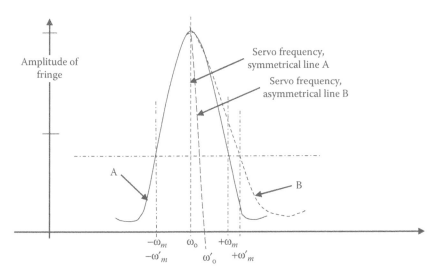

FIGURE 1.4   Illustration of the effect on the actual measured frequency of using square wave modulation for detecting an asymmetrical fringe. A is a symmetrical fringe while B is a fringe that is made asymmetrical by a bias such as second-order Doppler effect that depends on the velocity of the atoms.

velocity-dependent shift. In that case it is shown that the measured frequency called ω'_D is different from ω_D and is given by (see Appendix 1.C) (Audoin et al. 1974):

$$\frac{\omega'_D - \omega_o}{\omega_o} = \frac{\int_0^\infty v f(\tau)\sin^2 b\tau \sin \omega_m T d\tau}{2c^2 \int_0^\infty (1/v) f(\tau)\sin^2 b\tau \sin \omega_m T d\tau} \tag{1.18}$$

A similar expression is obtained for the shift introduced by a residual phase shift between the two arms of the Ramsey cavity:

$$\omega_\phi - \omega_o = -\frac{\phi}{L} \frac{\int_0^\infty f(\tau)\sin^2 b\tau \sin \omega_m T d\tau}{2c^2 \int_0^\infty (1/v) f(\tau)\sin^2 b\tau \sin \omega_m T d\tau} \tag{1.19}$$

Consequently, it is clearly seen from these expressions that the locking frequency of the servo system depends on both the amplitude of the microwave field, b, in the interaction regions and the amplitude of the frequency modulation, ω_m. In conclusion, extreme care needs to be taken in order to evaluate the appropriate corrections for the bias effects mentioned above that are velocity sensitive and distort the central Ramsey fringe. The reader is referred to *QPAFS*, Volume 2 (*QPAFS* 1989), for more details on this subject.

1.1.1.2.3.3 Frequency Shifts Related to Imperfections in Modulation and Demodulation
These shifts are related to distortion of the modulation and demodulation signals that are used in the creation of the error signal by the synchronous detection process (*QPAFS* 1989, Volume 2; Audoin 1992). Even harmonics in the spectrum cause frequency shifts. Distortion ratios less than 10^{-6} are preferable in order to make the effects negligible at the level of accuracy encountered in state-of-the-art standards.

1.1.1.2.3.4 Frequency Control Loop
Finite dc gain in the control loop and voltage offsets can cause frequency offsets in the frequency lock loop. In state-of-the-art designs, digital servo loops are used and such offsets are eliminated (Garvey 1982; Nakadan and Koga 1985; Rabian and Rochat 1988; Sing et al. 1990).

The size of the various offsets described above is summarized in Table 1.1 along with the present state-of-the-art accuracy in the determination of these offsets. The table is given without reference to particular systems implemented and is given solely as a guide to the reader making explicit the relative importance of a given shift and how accurately it can be determined in the best experimental conditions. At present, it appears that the biggest shift is the magnetic field offset. However, it is felt that the accuracy with which it is determined does not cause a major problem if care is taken in the design of the magnetic environment around the clock. The greatest cause of inaccuracy is probably still the cavity distributed phase shift limiting the accuracy to which phase asymmetry in the Ramsey cavity can be determined.

TABLE 1.1

Approximate Size of Biases or Offsets Present in Laboratory Cs Beam Frequency Standard[a]

	Typical Size in Laboratory Standards (parts in 10^{15})	Typical Smallest Evaluation Uncertainty Achieved (parts in 10^{15})
Magnetic field	>100,000	0.1
Second-order Doppler effect	Depends on construction $> \lvert -50 \rvert$	1
Black Body radiation	~20	0.3
Spin–exchange interactions	Unknown	Expected ≤ 1
Cavity phase shift	Depends on construction >100	1 to 10
Cavity pulling	~5 to 10	0.6
Bloch–Siegert effect	~1	Expected ≤ 0.3
Majorana transitions	~2	<1.3
Rabi and Ramsey pulling	<2	0.02
Microwave spectrum	<1	0.1
Electronics, modulation, demodulation, etc.	1	1
Microwave leakage	Depends on construction	<1
Gravitation	Depends on location	<0.1
Microwave spectrum	<1	0.1
Fluorescence light shift in optically pumped standards	<2	<0.5

Source: Vanier, J. and Audoin, C., *Metrologia*, 42, S31, 2005. Copyright Bureau International des Poids et Mesures. Reproduced by kind permission of IOP Publishing. All rights reserved.

[a] The uncertainty given is that achieved in best circumstances and is given as a reference to the accuracy that may be achieved in practice.

1.1.1.3 Frequency Stability of the Cs Beam Standard

The frequency stability of the Cs beam frequency standard depends on the averaging time, on factors such as modulation and frequency locking scheme and on the constancy of all the shifts enumerated above. In the so-called short-term region, where shot noise at the beam detection is important, the frequency stability is given approximately by (*QPAFS* 1989, Volume 2):

$$\sigma(\tau) = \frac{k'}{Q_l(S/N)\tau^{1/2}} \tag{1.20}$$

where:

Q_l is the atomic line Q

S/N is the signal-to-noise ratio essentially limited by shot noise at the detector

k' is a factor close to unity

The range of application of this equation depends on the servo loop, integrating filter type and bandwidth. As an example, in some well-designed laboratory standards using magnetic state selection, a frequency stability of $5 \times 10^{-12} \, \tau^{-1/2}$ over a range extending to 40 days has been measured, in general agreement with the above expression (Bauch et al. 1999). In the case of optically pumped standards, a better signal-to-noise ratio (S/N) can be obtained and a frequency stability better than that just mentioned by an order of magnitude can be realized ($3.5 \times 10^{-13} \, \tau^{-1/2}$) (Makdissi and de Clercq 2001).

The long-term frequency stability of the Cs beam frequency standards depends on the stability of the various frequency shifts and offsets enumerated above. Consequently, the frequency of a unit is dependent to a certain extent on its environment. Depending on construction type, temperature, humidity, atmospheric pressure, and magnetic field play a role to various degrees in determining long-term frequency stability. Temperature fluctuations appear to have the most important effect acting through some of the shifts enumerated above. In general, best results are obtained in a temperature controlled environment.

Fluctuations of unknown origins generally limit the frequency stability in the very long term. When the averaging time τ is increased, frequency stability, as given by Equation 1.20, improves and reaches a plateau called the flicker floor. The level of this flicker floor is generally a function of unknown parameters. In practice it is found that better quality in construction and design lowers this flicker floor to nearly undetectable levels.

Several national institutes and laboratories have been very active during the period 1970–1990, in developing Cs beam frequency standards using the classical approach. Those standards reached a high level of maturity. That stage was attained through intensive research and development, sophistication of the units, better understanding of the fundamental phenomena taking place and collaboration between the institutions. Table 1.2 is a compilation of the main characteristics of several selected laboratory units that have been developed during that period. They have played and, in some cases, still play an important role in the accuracy of TAI (Temps Atomique International) maintained by the BIPM. Most of them have been influential in the design of classical primary Cs standards implemented later.

1.1.1.4   Recent Accomplishments

During the years following 1990, there appeared to be a radical change regarding the study and development of laboratory atomic frequency standards, in particular, laboratory primary standards using the Cs atom. That state of affair occurred because of the refinement of stable solid-state laser diodes that became available with the proper wavelength and spectrum for efficient optical pumping of alkali atoms such as Cs. It was then possible to replace the selector and detector magnets in the classical Cs beam standard by using the technique of optical pumping providing extremely versatile approaches to state selection. That approach also avoided the problems caused by inhomogeneous magnetic fields that may be created by the state selector magnets as well as the problems introduced by the alteration of the velocity distribution of the atoms by the same selector magnets. With optical pumping for state selection, the velocity distribution is known analytically and the evaluation of frequency biases

TABLE 1.2
Characteristics of Selected Laboratory Primary Cs Frequency Standards Developed during the Period 1970–1987

	NRC (Canada) Cs V	NRC (Canada) Cs VIA & C	PTB (Germany) CS2	GOSSTRDT (USSR) MCs R 101	GOSSTRDT (USSR) MCs R 102	NIST (USA) NBS 6	CRL (Japan) Cs1	NRLM (Japan) NRLM II	NIM (China) Cs2
Distance between Ramsey cavities (m)	2.1	1	0.8	0.65	1	3.7	0.55	1	3.68
Microwave-magnetic field direction relative to the Cs beam	⊥	⊥	=	⊥	=	⊥	⊥	⊥	
State selector analyzers	2 poles	2 poles	Tandem: hexapole–quadrupole	2 poles	Hexapole	2 poles	Hexapole	2 poles	2 poles
Mean atom velocity (m/s)	250	200	93	170–220	220	195	110	300	
Line width (Hz)	60	100	60	130–200	110	26	100	150	
$\sigma_y(\tau)\,\tau^{-1/2}$	3×10^{-12}	3×10^{-12}	2.7×10^{-12}	3×10^{-12}	5×10^{-12}	2×10^{-12}	5×10^{-12}	$<8 \times 10^{-12}$	1.8×10^{-11}
Accuracy	1×10^{-13}	1×10^{-13}	2.2×10^{-14}	1×10^{-13}	5×10^{-14}	9×10^{-14}	1.1×10^{-13}	2.2×10^{-13}	4.1×10^{-13}
References	Mungall et al. 1973; Mungall and Costain, 1977	Mungall et al. 1981	Bauch et al. 1987	Abashev et al. 1983, 1987	Abashev et al. 1983, 1987	Lewis et al. 1981	Nagakiri et al. 1981, 1987	Koga et al. 1981; Nakadan and Koga 1982	Xiaoren 1981

Source: Vanier, J. and Audoin, C., *Metrologia*, 42, S31, 2005. Copyright Bureau International des Poids et Mesures. Reproduced by kind permission of IOP Publishing. All rights reserved.

that are functions of velocity can be computed more easily. Furthermore, those same laser diodes allowed the mechanical manipulation of atoms such as reducing their speed to such limits that it became possible to create small atomic ensembles of atoms, essentially small balls of the order of a cm or so in diameter, characterized by a very low temperature. The old dream of the atomic fountain clock, called in the early days as *Fallotron*, proposed by Zacharias in the 1950s, could be realized (see Forman 1985).

Nevertheless, some institutions continued the refinement of their classic Cs beam standards, sometimes even in parallel with developments in the new avenues, optical pumping state selection and atomic fountain just mentioned. Those refined frequency standards using the classical approach reached a level of accuracy such that it proved the high degree of understanding reached in the physics involved by those who persisted in improving them. It left no doubts on the quality of the work that was performed. Some of those standards, at the time of writing of this book, are still used in some cases in the implementation of the atomic timescale (TAI). They are also used as reference to check, within their own limits, the reliability and absolute accuracy of the standards of the new wave. However, it is clear that their contribution as a primary standard is limited by their accuracy. An examination of all the work published during that period makes evident that the goal was that of breaking the 10^{-14} barrier as far as accuracy is concerned. It was only achieved after extensive work.

The Cs and Rb fountains are now the workhorse in many primary standards laboratories and have reached a level of accuracy in the 10^{-15} and better. However, it is worth examining the physics behind the work that was done in the improvement of the classical Cs frequency standards using magnetic state selection since their understanding is essential to the success of the optical pumping and fountain approaches. We hence review in the following paragraphs some of those refinements that were accomplished during the last two decades and have provided the scientific community with some of the most reliable frequency standards.

As readily observed from Table 1.1, the largest frequency biases, which are observed in classical Cs beam standards using magnetic state selection and which need to be evaluated as accurately as possible, are the magnetic field and its homogeneity, the second-order Doppler effect, and the cavity phase shifts, either nominal between the two interaction regions or distributed within each region. Black Body radiation, cavity pulling, Majorana transitions, and Rabi–Ramsey pulling cause shifts much smaller, less than 10^{-14}, but have also been carefully re-examined in order to certify that they do not introduce inadvertently any important bias in the measurements.

1.1.1.4.1 *Magnetic Field Generation*

The magnetic field is applied to provide a quantization axis to the system. It is also required to separate the field-dependent transitions from the clock transition $F = 4$, $m_F = 0 - F = 3$, $m_F = 0$ in order to prevent overlapping of the field-dependent lines as much as possible. We mentioned above the effect to the tail of the neighbouring Rabi pedestal on the clock transition. If the amplitude of these transitions on each side of the clock transition is not identical, a distortion of the central fringe takes place and the peak of the signal is displaced. The size of the magnetic field required, then, depends on the accuracy desired. In general, the field is set to a value of the order of

50 to 100×10^{-7} T (50–100 mG). This gives a Zeeman frequency of the order of 25 to 50 kHz for transitions between field-dependent levels. It provides enough separation between the field dependent Rabi pedestals to guarantee little effect from overlap. Thus, the applied field is quite large. The displacement Δv_B of the central fringe of the Ramsey pattern is obtained from Equation 1.9 as:

$$\Delta v_B = 427.45 \times 10^8 B_0^2 \qquad (1.21)$$

where:
B is in Tesla

For 50×10^{-7} T (50 mG), this corresponds to a displacement of 1 Hz or 1×10^{-10}. Consequently, in order to reach fractional frequency stability of the order of 10^{-15}, the field must be stable to something like 10^{-5} in fractional value. This is very demanding on several aspects of the construction of such a device. This includes the stability of the current source driving the circuit generating the field. It also sets requirements on homogeneity of the field within the system that includes the Ramsey cavity, which is between 1 and 2 m long in laboratory standards. Finally, magnetic shielding of the region where the interaction takes place must be well-designed in order to avoid external field fluctuations that can affect the atomic beam resonance frequency. These problems have been addressed in various ways. In one type of design, the field is created by four metallic rods situated along the length of the standards (Mungall et al. 1973). In that case, the rods are spaced very accurately by means of a number of pyrex spacers interspaced with coils whose purpose is to excite $\Delta m = +/-1$ transitions and whose resonance frequencies v_z are in the kHz range. Such transitions, when excited, affect the amplitude of the clock transition, an effect seen on the amplitude of the detected signal. The field can then be determined exactly through the relation:

$$v_z = 349.86 \times 10^7 B \qquad (1.22)$$

That relation applies within the upper states manifold $F = 4$. In another design, the field is created by means of a long solenoid enveloping the whole Ramsey cavity with small end coils used for trimming the field at both ends (Bauch et al. 1996). The whole structure is enveloped in all cases in multilayer high permeability material such as mu-metal or moly-permalloy. In that approach, as mentioned in the article cited, one main cause of instability is temperature affecting the length of the solenoid support, causing a change in actual solenoid dimensions with a resulting fluctuation of the magnetic field. Temperature regulation is thus required.

Nevertheless, as reported in Table 1.1, it appears that the task of creating a field of sufficient homogeneity and stability for implementing a clock with accuracy in the 10^{-15} range is not insurmountable. Actually, even at this high magnetic field, an accuracy in magnetic field bias determination of 10^{-15} has been realized in Physikalisch-Technische Bundesanstalt (PTB) CS1 (Bauch et al. 2000a, 2000b). The question of the magnetic field bias is thus not a major problem in the construction of such a device although it must be implemented with great care, particularly regarding homogeneity. The main reason of this requirement is that what is measured by means of the Zeeman frequency of Δm_F transitions is the actual value of the field along the beam while the bias frequency of the clock transition is proportional to the average of the field squared. In general, $<B^2>$ is not equal to $<B>^2$. In practice, the effect may be evaluated from the position of the Ramsey resonance on top of the Rabi pedestal (Bauch et al. 1996).

It should, however, be recalled that the Ramsey cavity design approach introduces a desirable property regarding frequency shifts introduced by magnetic field inhomogeneities. For example, the magnetic field may be different in the two interaction regions and may be different from the field in the drift region. The resulting frequency shift of the central Ramsey fringe, for a drift region L long compared to the interaction region l, is the sum of the shifts multiplied by the ratio l/L. The actual shift is given by the equation (*QPAFS* 1989):

$$\frac{\omega - \omega_0}{\omega_0} = \frac{l}{L}\left(\frac{\omega_0' - \omega_0}{\omega_0} + \frac{\omega_0'' - \omega_0}{\omega_0}\right) \frac{\int_0^\infty \tau f(\tau)(1 - \cos b\tau)\sin b\tau \, d\tau}{\int_0^\infty b\tau^2 f(\tau)\sin^2 b\tau \, d\tau} \tag{1.23}$$

where:

ω_0' and ω_0'' are the resonance frequencies in the two interacting regions, respectively
ω_0 is the resonance frequency in the drift region

For a laboratory standard with a 1 m long cavity, a frequency shift of the order of 10^{-4} Hz may thus be reduced by a factor of 100 and made negligible.

1.1.1.4.2 Second-Order Doppler Effect

As is made explicit in Equation 1.18, the second-order Doppler shift is a function of the distribution of velocities in the detected beam. Because the second-order Doppler effect is velocity dependent, it makes the line asymmetrical and shifts the fringe line to the centre. According to the explanation given in Figure 1.4, if square wave modulation is used in the servo system, as is generally the case, the frequency measured is dependent on the modulation amplitude. This is shown by Equation 1.18 above and reviewed in some detail in *QPAFS*, Volume 2 (1989).

An important step in the evaluation of the actual frequency shift is, thus, the evaluation of the integrals included in that equation. For this, the velocity or interaction time distribution must be known. A major effort has been spent by several laboratories on the evaluation of such a distribution. It is well known that the velocity distribution at the exit of magnetic dipole selectors is not Maxwellian as was assumed by Harrach (Harrach 1966, 1967). In one type of approximation, cut-off velocities were introduced in the velocity spectrum assumed to be Maxwellian (Mungall 1971). This is a gross approximation for selector dipole magnets. Nevertheless, the approximation was checked by calculating numerically the actual Ramsey fringe shape with the assumed velocity spectrum, and approximate agreement was obtained with the experimental data. However, in such a procedure the cut-off velocities are chosen rather arbitrarily and are not a priori based on experimental data. This did not appear to be a proper avenue for calculating a bias affecting the accuracy of a primary standard. We now outline regarding how the question of determining the velocity distribution to evaluate carefully the frequency bias caused by the second-order Doppler effect as well as by cavity phase shift bias was addressed.

1.1.1.4.3   Cavity Phase Shift

According to Table 1.1, and as discussed above, the second most important bias is the phase shift that can exist between the Ramsey cavity arms. In general, this phase shift is evaluated by means of beam reversal. This can be done because of the unique property that the phase shift between the two arms of the cavity changes sign upon inversion of the beam direction. The measurement may be done by interchanging the Cs source and the detector. This is a major operation and depending on design may lead to opening the system to atmospheric pressure and re-evacuating the system after the interchange has been effectuated. In some cases, the instrument is constructed with both, source and detector, at both ends (Mungall et al. 1973; Bauch et al. 1996). Their position is then switched directly under vacuum by means of a rotating or sliding mechanism. In all cases, the question remains as to the exact retrace of the Cs beam within the cavity and consequently as to the reproducibility of the phase shift with that retrace. This question was examined carefully in *QPAFS* (1989). Let us recall the main conclusions.

The Ramsey cavity is normally implemented with an X band waveguide in a U shape. Two ways for the orientation of the waveguide relative to the direction of the beam are represented in Figure 1.5, which, for convenience, is reproduced from *QPAFS* (1989). The relative orientation dictates the direction of the static magnetic field applied in order to satisfy the quantum transition probability condition for $\Delta m_F = 0$ transitions. The radiation H-field in the cavity must be parallel to the static induction B_0 applied. In Figure 1.5a, the static field must be perpendicular to the beam direction and in Figure 1.5b it must be parallel. The ends of the U cavity are short circuits. If the material of the waveguide were a perfect conductor, the electromagnetic radiation within the cavity would be represented by a perfect standing wave. There would be no travelling waves in the waveguide required to feed energy losses in the walls of the guide. The position of the antinodes of the magnetic field would be well determined. In that case, if the Cs beam passes, say close to the short circuit (Figure 1.5a) or at the $\lambda_g/2$ antinode point from that short circuit (Figure 1.5b), the atoms would see the same phase of the field in both arms of the cavity. That phase would change sharply by π from one antinode to the other, but would be constant in the region of each antinode. All atoms in a narrow beam, a few mm in dimension, would see the same phase independently of their position within the beam and furthermore, they would see the same phase whatever is the actual position of the beam in traversing the cavity around an antinode. This is schematically represented in Figures 1.6a and b.

This is an ideal situation. In practice, the cavity is made of a material with electrical losses and the wave is attenuated all along its path. Furthermore, upon reflection at the short circuit its amplitude is smaller than the incoming wave because the reflection coefficient is less than unity. The standing wave ratio varies along the path traversed by the wave and its phase is continuously changing with distance travelled.

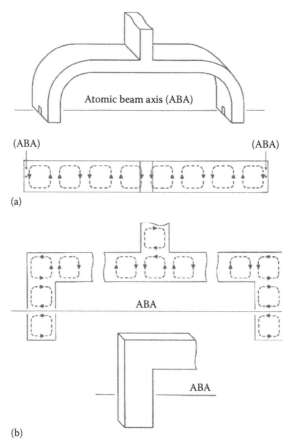

FIGURE 1.5 Schematic representation of the two usual implementations of the Ramsey cavity. In (a) ABA stands for atomic beam axis. The coupling to the cavity is made through an E-plane T-junction and the cavity is bent in the E-plane; (b) the coupling is done through an H-plane T-junction. The magnetic field is represented by dotted lines. In the implementations chosen, in (a) the atomic beam passes close to the end while in (b), it passes at a so-called anti-node $(1/2)\lambda_g$ from the short circuit end. (Data from Bauch, A. et al., *IEEE T. Instrum. Meas.*, IM-34, 136, 1985; Mungall, A.G. et al., *Metrologia*, 9, 113, 1973.)

This is illustrated in Figures 1.6c and d. Let us examine the phase of the microwave standing wave along the direction travelled called the z_g direction. If we call z'_g as the distance within the guide from the nearest antinode point identified by the number $p = 0, 1, 2 ...$, then the phase is given by:

$$\phi\left(z'_g\right)=\left(\frac{1}{2} p\alpha_g\lambda_g + \alpha_g z'_g + r_m\right)\tan\frac{2\pi z'_g}{\lambda_g} \tag{1.24}$$

The full calculation is given in *QPAFS*, Volume 2 (1989).

In order to understand the following discussion and development, it is important to know well the effect of each term in this expression and their size. The first term

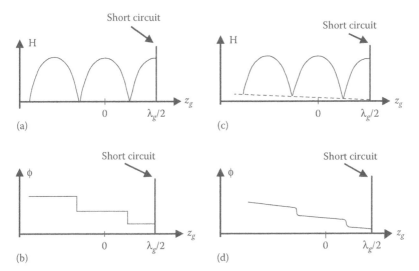

FIGURE 1.6 Schematic representations of the amplitude of the magnetic field H and the phase of H at the first antinode(s) near the end of the arms of the Ramsey cavity. (a) and (b): case where the walls of the waveguide are perfect conductors. (c) and (d): case where the walls of the waveguide have finite electrical conductivity. (Data from Bauch, A. et al., *IEEE T. Instrum. Meas.*, IM-34, 136, 1985.)

on the right represents absorption along the waveguide characterized by the absorption coefficient α_g due to the extra length travelled by the wave upon its reflection to reach the antinode point $(1/2)\,p\lambda_g$. The second term is of the same nature but represents the same effect as a function of distance close to an antinode with z'_g being a small distance from the antinode centre. The last term, r_m, represents the effect of the finite conductivity of the cavity termination on the actual reflection of the wave at that short circuit. The parameter r_m is the real part of the surface impedance of the short-circuit material normalized to the waveguide impedance. Hereunder, we provide the value of the various parameters for a copper waveguide WR 90

$$\alpha_g - 1.33 \times 10^{-2} \ \mathrm{m}^{-1}$$
$$r_m = 4.6 \times 10^{-5}$$
$$\lambda_g = 4.65 \ \mathrm{cm} \ (9{,}192{,}631{,}770 \ \mathrm{Hz})$$

We recall that p is an integer with value 0 at the short circuit and 1 at the first antinode close to the short circuit, which in some implementation is the point where the beam is oriented as in Figure 1.5b. Because of the electrical losses, two effects result. If the arms do not have exactly the same length, that is, the T feeding the cavity is not well centred, the phase is different in the two arms. As a first approximation, neglecting the small effect of reflection less than unity at the short circuit, r_m can be neglected relative to the other terms in Equation 1.24 and the phase shift ϕ between the two arms is calculated as:

$$\phi = \frac{2\pi\alpha_g L_0 \Delta L_0}{\lambda_g} \tag{1.25}$$

where:

L_o is the mean value of the length of each of the two arms

ΔL_o is the difference between their lengths caused by an inexact centring of the T feeding the cavity

For example, for $L_o = 15\ \lambda_g$ and an error in construction of $\Delta L_o = 10^{-4}$ m between the two arms length, we obtain $\phi = 1.25 \times 10^{-4}$ rad. The guide wavelength is 4.65 cm, and for an average speed of 200 m/s, using Equation 1.19, one obtains a fractional frequency shift of the Ramsey central fringe of the order of 6×10^{-13}, which is very large in the present context. In order to reach the 10^{-15} range accuracy, the phase shift should be known to the order of one microradian. This sets relatively rigid constraints on the fabrication of the cavity, the centring of the coupling T, and the measurement process itself.

On the other hand, as made evident by Equation 1.24, the phase of the magnetic field varies with distance within the arms. Thus, the phase seen by the atoms traversing the cavity is different depending on their exact transverse position within the beam. If the beam is large, several mm to cm depending on the collimation and the hole made in the cavity, different atoms will see a different phase shift.

According to Equation 1.19 the resulting frequency shift is a function of the sign of the difference in phase between the two arms of the cavity. In practice, this property is used to determine the value of the phase shift by measuring the frequency for reverse orientations of the atomic beam. As we have already discussed, a basic requirement is thus that the beam retraces exactly the same path in both directions. This is made evident from actual values of phase shift with distance as calculated from Equation 1.24. For small values of z'_g, we have approximately:

$$\phi_1\left(z'_g\right)=\frac{\pi\alpha_g pL_o z'_g}{\lambda_g}+\frac{2\pi\alpha_g}{\lambda_g}\left(z'_g\right)^2+\frac{2\pi r_m z'_g}{\lambda_g} \tag{1.26}$$

For convenience, the results are shown in Figure 1.7, partly replicated from *QPAFS* (1989) and for the value of the parameters given above for a WR 90 copper waveguide. The variation of the phase along the other axis x'_g perpendicular to the beam direction for the case of Figure 1.5b, that is in the direction of the larger transverse dimension, a, of the waveguide, is calculated in the same manner. Those results are also shown in Figure 1.7.

In practice, the measured gradient of the distributed phase shift may be as large as 10^{-4} rad/mm, slightly larger than that calculated above (as reported for PTB's CS2, Bauch et al. 1987). On the other hand, in a particular design, the precision of the retrace upon beam reversal and after great experimental care was found to be of the order of 0.13 mm (Bauch et al. 1993). This would thus correspond to a possible error of 1.3×10^{-5} rad in the evaluation of the phase shift. Using Equation 1.9, for a cavity structure 1 m long and a mean atomic beam speed of 100 m/s, this would correspond to a fractional frequency shift of the order of 2×10^{-14}. As can be seen, it thus appears that the phase shift, although great care is taken in evaluating it by means of beam reversal, limits the accuracy of the clock. Furthermore, the beam trajectory is affected by the earth's gravitational field (*QPAFS* 1989). A simple calculation shows

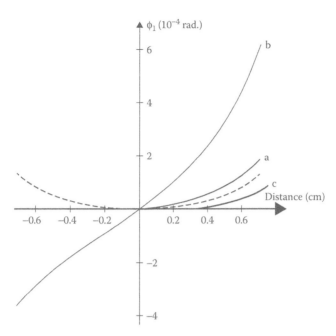

FIGURE 1.7 Variation of the phase as a function of the transverse direction across the beam. Curve b is for a beam traversing the waveguide at an antinode point situated at $\lambda_g/2$ from the short circuit while curve a is for the beam traversing the waveguide close to the short circuit. Curve c is for the ring cavity studied below. The dotted line curve is for direction x, perpendicular to the direction of propagation as explained in the text.

that for a 1 m long structure and even with an average atomic beam speed of 100 m/s, the atoms fall by 0.5 mm during the traversal of the cavity. This number varies with the speed of the individual atoms forming the beam. This effect of course creates a real challenge in mechanical design and adjustment to reproduce reliably the beam path upon reversal. With such an error in path reproducibility, the corresponding shift could be nearly 10^{-13} for the case above. For this reason, some systems using slow atoms have been constructed vertically. However, the particular approach used was abandoned due to other difficulties encountered in connection to several other detrimental effects in particular with magnetic shielding (A. Bauch 2012, pers. comm.).

As is readily observed in Equation 1.26, if the odd terms could be eliminated by means of a different cavity configuration, the phase shift could be reduced. This can be accomplished by means of a so-called ring cavity (De Marchi 1986). Such a cavity is shown in Figure 1.8.

In that configuration, the cavity resonant at 9.192 GHz is excited by means of a rectangular waveguide, and waves are generated symmetrically around the ring. A standing wave is thus excited in the structure. In the configuration shown, at the entrance through the tee, the wave separates into two waves travelling to the right and to the left with amplitudes b_1 and b_2, or since we are interested in the magnetic field components of the wave as H_{10} and H_{20}. We assume that the structure is characterized by a propagation vector $\gamma = \alpha + i\beta$, where α is the absorption

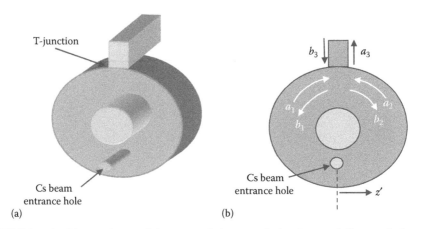

FIGURE 1.8 Ring cavity used in some of the recently implemented Cs atomic beam frequency standards. (a) 3D view, (b) Identification of the various field components.

coefficient and β the propagation constant $2\pi/\lambda_g$, characteristic of the waveguide. At a small distance z' from the central point where the beam enters the interaction region, we have:

$$H_1(z') = H_1 e^{-i\omega t - \gamma(l/2 + z')}$$

$$H_2(z') = H_2 e^{-i\omega t - \gamma(l/2 - z')}$$

(1.27)

With a ring having a circumference $l = (n + 1/2)\lambda_g$, where n is an integer, and assuming equal amplitude of the field ($b_1 = b_2$) at the entrance, we may write at the antinode where the beam passes through the cavity:

$$H(z') = H_0 e^{i\omega t - \gamma l/2}\left(e^{-\gamma z'} + e^{+\gamma z'}\right)$$

(1.28)

By using trigonometric relations (see Appendix 1.D), it is then shown that the phase of H close to the beam axis is given approximately by:

$$\phi = \alpha\beta z'^2$$

(1.29)

What are the advantages of such a cavity in comparison to the standard one using short circuits? This question can be addressed by comparing the result just obtained to that of Equation 1.26 for the standard approach using a cavity terminated with short circuits. Its first advantage is that the value of phase shift is independent of the antinode chosen for the interaction. The second advantage is that, since there is no short circuit used to terminate the cavity, there is no reflected wave and thus the phase shift does not depend on the attenuation caused by partial reflection at such a short circuit. For example, for a displacement of the beam along z' we have the following result for the phase shift, assuming the same characteristics as those used previously for the waveguide:

- Cavity with short circuits:
  - Beam passing at $\lambda_g/2$ from short circuit: $\phi = 2.09 \times 10^{-5}$ rad/mm
  - Beam passing close to the short circuit: $\phi = 4.87 \times 10^{-6}$ rad/mm
- Ring cavity: $\phi = 1.8 \times 10^{-6}$ rad/mm

The distributed phase shift thus appears to be reduced somewhat in the ring cavity compared to the standard short-circuit approach. However, asymmetries in the ring may have an effect on the phase shift and position of the antinode. Asymmetries can be of two types. A phase shift may be introduced in construction through a mechanical asymmetry, a tilt angle of the T-junction, for example. The resulting effect is an imbalance of the two counter-propagating waves excited in the cavity (De Marchi et al. 1988). In such a case, the asymmetry may cause a small displacement of the antinode relative to its position when the structure is symmetrical. It is evaluated that a tilt angle of 50 mrad can produce a frequency shift of the order of 3×10^{-15}. A similar effect may result from an asymmetry of the propagation constant γ in the two halves of the cavity. It is calculated that if the asymmetry, in terms of $\Delta\gamma/\gamma$, is less than 10^{-2}, the displacement of the antinode is then less than 0.2 mm, which is negligible in the present context.

The ring cavity was used in a few recent implementations of laboratory primary standards (Bauch et al. 1998, 1999a, 1999b). It was also used in some units where optical pumping was used for state selection (Shirley et al. 2001; Hasegawa et al. 2004). It was found that the distributed phase shift was reduced to some extent when compared to the short-circuit approach. In the case of PTB's CS1 refurbished with ring cavities, it was concluded that the phase gradient could be as large as 20 μrad/mm although using the analysis presented above the phase over the beam diameter of 3 mm should not vary more than approximately 4 μrad. It is concluded, however, that this is much better than the 94 μrad/mm expected in their case for the standard cavity using short-circuit terminations (Bauch et al. 1998). We note that in order to evaluate correctly the effect of the phase shift, we must evaluate it by means of Equation 1.19 and that the velocity distribution or interaction time distribution must be established for the particular device used. This is what we examine next.

1.1.1.1.1 Velocity Distribution Evaluation and Control

The evaluation of some of the biases just mentioned, as shown above by means of Equations 1.13 and 1.16 relies heavily on the exact determination of the velocity distribution at the exit of the state selector magnet. It is interesting to follow the evolution of the assumptions and technique of determination of this distribution over the years. As briefly described above, this evolution started from the use of a Maxwell distribution (Harrach 1967) and went to the use of a distribution altered by cut-offs at both low and high speeds made necessary by the use of selector magnets (Mungall 1971). This was followed by an experimental approach that uses a numerical analysis of the response of the beam standard to pulsed radiofrequency (RF) excitation (Hellwig et al. 1973). Another approach comprised analyzing the shape of the observed Ramsey fringes directly, which depends on the velocity distribution of the atoms in the beam. This was done numerically and the so-called moments of the velocity distribution were evaluated (Audoin et al. 1974). Finally, a most powerful approach using Fourier transform techniques to analyze the observed Ramsey fringes was introduced early in the development of laboratory standards (Kramer 1973; A. Bauch 2012, pers. comm.). The technique was used in various ways and has been an important factor in improving accuracy. It was outlined in *QPAFS*, Volume 2 (1989). Let us recall that technique and address the question in the more general context of the control of atomic velocities in laboratory Cs beam frequency standards.

The need of knowing accurately the velocity distribution, or the interaction time distribution, arises in the evaluation of both the second-order Doppler effect and the cavity phase shift. These are given by Equations 1.13 and 1.16. In those equations, it is observed that although the shifts are functions of velocity, they are rather small, being of the order of parts in 10^{13}. Consequently, although producing a measurable frequency shift, the effect on the shape of the Ramsey fringes themselves arising from those shifts is very small. The shape of the signal detected, as given by Equations 1.12 and 1.15 with the term $v^2/2c^2$ and ϕ neglected can then be used as an excellent representation of the Ramsey line shape, which remains a function of velocity through the function $f(\tau)$. In that context, those expressions become:

$$I = I_b + \frac{1}{2}I_0 \int_0^\infty f(\tau)\sin^2 b\tau \left[1 + \cos\left(\omega - \omega_0\right)T\right]d\tau \tag{1.30}$$

However, this equation can be rearranged. We remove the constant term I_b not contributing to the useful signal and hence obtain:

$$I = \frac{1}{2}I_0 \int_0^\infty f(\tau)\sin^2 b\tau\, d\tau + \frac{1}{2}I_0 \int_0^\infty f(\tau)\sin^2 b\tau \cos\left(\omega - \omega_0\right)d\tau \tag{1.31}$$

This equation relies on the approximation made earlier that $(\omega - \omega_0)$ is much smaller than $b\tau$. The first term is independent of ω and is simply the maximum of the Rabi pedestal. The second term is the Ramsey modulation of that pedestal. This expression contains the effect of the atoms' velocity distribution on the Ramsey fringes. The Rabi pedestal being broad with a flat top, a simple approach consists in looking only at the second term as a good representation of the effect of the velocity spectrum on the signal observed. It is readily seen that the expression is a cosine transform of the term $f(\tau)\sin^2 b\tau$. Consequently, with a knowledge of the Rabi frequency b in the cavity, $f(\tau)$ can in principle be obtained through an inverse Fourier transform of the measured Ramsey pattern (Kramer 1973; Daams 1974).

A particular approach in applying such a technique consists in setting $\omega = \omega_0$, and measuring the signal amplitude as a function of the field intensity b in the cavity (Boulanger 1986). Let us now recall the general idea. For $\omega = \omega_0$, the expression for the Ramsey fringe amplitude becomes:

$$I = I_0 \int_0^\infty f(\tau)\sin^2 b\tau\, d\tau \tag{1.32}$$

This can also be written as

$$I = \frac{1}{2}I_o - \frac{1}{2}I_o \int_0^\infty f(\tau)\cos 2b\tau\, d\tau \tag{1.33}$$

In that case, the second term is the cosine transform of $f(\tau)$ that we call $F(b)$. An experimental plot of I against b gives $F(b)$ and its inverse transform gives:

$$f(\tau) = \frac{4}{\pi} \int_0^\infty F(b)\cos 2b\tau\, db \tag{1.34}$$

This expression is evaluated by means of numerical analysis of the experimental results obtained for I. The difficulty in the techniques is the evaluation of b, a measure of the microwave field in the cavity. It requires a good knowledge of the cavity Q, its dimension, and the power fed into it. This is a difficult and not a precise exercise. In order to initiate the calculation, it is best to use an approximately known value of b, such as a value for which the signal is optimal. This value is given by $b_{opt} = (\pi/2)(v/l)$. The value of b is then altered by changing the power fed into the cavity and normalized to that value. A graph of the maximum of the central fringe is obtained as a function of b and the calculation of the Fourier transform of this result can be done. The subsequent exercise then consists in calculating the shape of the Ramsey patterns with the results obtained and in comparing the result to the shape obtained experimentally. The width of the patterns obtained is then an important parameter in concluding about the agreement of theoretical and experimental results. The exercise can be repeated until the agreement is satisfactory. The value of $f(\tau)$ can then be obtained through normalization by means of Equation 1.7. The technique was used successfully and provided clear information of the velocity spectrum. It was claimed for example that the second-order Doppler shift could be evaluated with accuracy better than a few parts in 10^{15}.

Another approach consists in using Equation 1.31, considering, as mentioned above, that the Ramsey fringes given by the second term are the cosine transform $R(\Omega_o)$ of $f(\tau)\sin^2 b\tau$ through the equation (Daams 1974):

$$R(\Omega_o) = \int_0^\infty f(\tau)\sin^2 b\tau \cos(\Omega_o \tau)\, d\tau \tag{1.35}$$

The function $f(\tau)\sin^2 b\tau$, called $F(\tau)$, can then be obtained from the data recorded as a function of Ω_o by means of an inverse transform (Shirley 1997):

$$F(\tau) = f(\tau)\sin^2 b\tau = \frac{4}{\pi} \int_0^\infty R(\Omega_o)\cos(\Omega_o \tau)\, d\Omega_o \tag{1.36}$$

As in the previous approach, b needs to be determined in order to isolate $f(\tau)$. This may be done as in the previous case using b at its optimal value, by making several measurements at various values of b relative to that value and averaging the results.

These techniques, although rather powerful, are somewhat tedious to implement, and are time consuming. Furthermore, since the exercise is done in several steps by adjusting parameters to obtain a velocity spectrum that reproduces the shape of the fringes observed, there is often doubt on the actual resulting accuracy. However, the techniques have been used at large and it appears that confidence in the accuracy achieved has reached a satisfying degree, permitting the evaluation of frequency shifts with accuracy better than 10^{-14}.

One obvious way to increase accuracy would be to reduce the width of the central fringe. This can be done by increasing the length L of the space between the arms of the cavity. However, there is a limit to that approach. As shown in Table 1.2, several Cs beam devices with various lengths have been constructed, with L varying from 0.55 m to 3.7 m. The accuracy reported for these devices varies from 2.2×10^{-14} to 4.1×10^{-13}. Actually, there is no direct relation between accuracy reported and length of the device. For example, NRC (National Research Council of Canada) reports an accuracy of 1×10^{-13} for a 2.1 m long Ramsey cavity while the NIST (National Institute of Standards and Technology) reports an accuracy of 9×10^{-14} for 3.7 m. However, the best accuracy reported is for an interaction region about 1 m long. The same kind of remark can be made about frequency stability. The standards have frequency stability between 2×10^{-12} and 1.8×10^{-11} at an averaging time of 1 s. We can observe that the line width is reduced considerably by increasing length, but the expected improvement on frequency stability does not follow. It is thus obvious that the approach of simply lengthening the Ramsey interaction region for reducing line width does not necessarily lead to a proportional improvement in accuracy and frequency stability. The improvement expected seems to be cancelled by some other effects. One obvious reason appears to be the fact that the Cs beam is never perfectly collimated and that increasing the length reduces the number of atoms being detected. This has a direct effect on S/N. Increasing the length of the interaction region also increases the demand on the quality of the magnetic field regarding homogeneity and stability, free of environmental fluctuations. Finally, the actual control of the position of the beam within the interaction regions is more difficult to achieve with long beams affecting accuracy through distributed cavity phase shifts.

Another way of reducing line width consists in reducing the speed of the atoms traversing the Ramsey cavity. The time spent by the atoms between the two arms of the cavity is lengthened in inverse proportion to the reduction in speed and the line width is effectively reduced according to Equation 1.4. That avenue was used early in the design of the laboratory standards at PTB, Germany, and perfected until recently (Becker 1976; Bauch 2005). The approach was optimized, for example, by using hexapole and quadrupole selector magnets in tandem. This combination uses velocity selective focussing properties of both magnets for forming a Cs beam of the order of 3 mm in diameter. It is then possible to produce a beam formed mainly of slow atoms by proper collimation. At the same time the group of selected velocities has a considerably smaller spread. A typical design using an hexapole magnet resulted in a beam with mean atomic speed of 72 m/s and a full width at half maximum of the distribution of 12 m/s (Bauch et al. 1996). This is to be compared to a mean speed of the order of 200 m/s and a spread of the order of 100 m/s for Cs VI at NRC using dipole magnets as state selectors. In comparison, PTB's CS3 beam looks

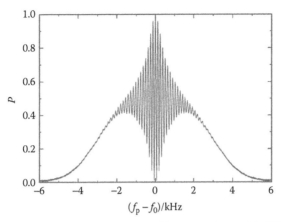

$(f_p - f_0)/\text{kHz}$

FIGURE 1.9 Cs beam clock transition Ramsey fringes as observed in PTB's CS1. Due to the use of hexapole selector magnets the number of visible fringes is increased when compared to the standard approach using selector dipole magnets (see Figure 1.3). (Courtesy of PTB, Germany.)

almost monokinetic. Such low speed and spread allow the observation of many more Ramsey fringes than in the case where single magnetic dipole selectors are used. It also has the obvious advantage of providing better control on shifts that are velocity dependent. This is due to the fact that the beam velocity distribution, in the case of multipole magnets used in tandem, is closer to a monokinetic beam and interferences take place over a larger spectrum of frequencies, the coherence of the atoms not being destroyed by a large spread in times of arrival at the second arm of the cavity due to a velocity spread. This is illustrated in Figure 1.9 in which Ramsey fringes are reported for the clock transition in CS1. The pattern can be compared to the fringes obtained in NRC's Cs VI using dipole magnets state selectors and shown in Figure 1.3.

As can be seen from Table 1.3, the results of the vertical Cs beam clocks were not up to expectations and actually the approach was abandoned because of other complications such as a reduced magnetic shielding (A. Bauch 2012, pers. comm.). It appears that in the actual device other shifts were present, such as those caused by microwave leakage in the region occupied by the cavity, and could not be controlled easily. They could be a major cause of discrepancy between expected and observed results. A photograph of Cs beam frequency standard CS 1 and 2 developed at PTB is shown in Figure 1.10. As can be observed from the photograph, such systems are not elementary.

1.1.1.4.5   Progress in the Evaluation of Some Other Frequency Shifts

Other various shifts enumerated above have also been addressed and progress has been made recently in their exact evaluation. The effect of Majorana transitions is difficult to identify among the other shifts present. However, changes in magnetic field and trimming by coils between the state selecting magnets and the cavity may be used to verify the presence of the undesirable effect to the level desired. The final conclusion remains at the level of an evaluation uncertainty of the order of 2×10^{-15}. Its actual value in a particular device is difficult to assess.

TABLE 1.3

Results Obtained at PTB upon Refinement of Classical Cs Beam Frequency Standards

	PTB (Germany) CS1	PTB (Germany) CS2	PTB (Germany) CS 3
Distance between the arms of the Ramsey cavity (m)	0.8	0.8	0.77 (vertical construction)
Microwave-magnetic field direction/beam	Parallel	Parallel	Parallel
State selector analyzers	Hexapole + quadrupole	Hexapole + quadrupole	Hexapole
Mean atom velocity (m/s)	93	93	72
Line width (Hz)	59	60	44
$\sigma_y(\tau)\,\tau^{+1/2}$	5×10^{-12}	4×10^{-12}	9×10^{-12}
Accuracy	7×10^{-15}	12×10^{-15}	1.4×10^{-14}
References	Bauch et al. 1998, 2003	Bauch et al. 2003	Bauch et al. 1996

Source: Vanier, J. and Audoin, C., *Metrologia*, 42, S31, 2005. Copyright Bureau International des Poids et Mesures. Reproduced by kind permission of IOP Publishing. All rights reserved.

FIGURE 1.10   PTB's CS 1 and 2 developed and improved over several decades. At the time of writing the units are still used as primary frequency standards providing input to the BIPM for maintaining atomic timescale. (Courtesy of PTB, Germany.)

Fortunately, in the case of state selection by means of optical pumping, no magnets are used and the effect should be considerably reduced if not eliminated completely. The question of Rabi and Ramsey shifts has also been addressed. The first effect, Rabi pulling, is caused by the asymmetry in the Rabi pedestals neighbours of the central Ramsey fringe. The shape of the central fringe is distorted by the slanted base line of those Rabi pedestal neighbours, which do not have the same intensity when magnet selectors are used. This effect is considerably reduced by operating the device at a high magnetic field. It is essentially absent in case optical pumping is used for state selection since the neighbouring Rabi pedestals have equal intensities (Audoin 1992). The actual value of the shift was calculated in some detail in *QPAFS*, Volume 2 (1989). The Ramsey shift is caused by the presence of $\Delta m = (+/-)$ 1 transitions which share a level in common with the transition responsible for the central Ramsey fringe, that is, the $|3,0\rangle \rightarrow |4,0\rangle$ transition (Cutler et al. 1991). These transitions are made possible because, in practice, the microwave magnetic field in the interaction regions has a small component orthogonal to the dc magnetic induction. The amplitude of those transitions is a function of the microwave power and magnetic field, and the effect depends directly on the population difference of levels $|4,1\rangle$ and $|4,-1\rangle$ (Fisher 2001). The effect is small in long laboratory standards. Its evaluation in a given unit led to a value of 0 with an uncertainty of 3×10^{-15} (Bauch et al. 1998). The actual theoretical original analysis (Cutler et al. 1991) has been re-addressed (Fischer 2001; Lee et al. 2003) and the subject is still under investigation.

1.1.2 HYDROGEN MASER

The hydrogen maser invented by N. Ramsey in 1959 (Goldenberg et al. 1960) is one of the most stable microwave atomic frequency standards at least in its active configuration. Its concept originated in part from independent fundamental studies made on attempts to increase the travel time of Cs atoms by means of a storage approach between the arms of the Ramsey cavity in the device described in Section 1.1.1 (Goldenberg et al. 1961). It also relies on fundamental concepts elaborated and results obtained in studies made earlier on the properties of stored hydrogen, including stimulated emission, coherence, spin–exchange interactions, and relaxation in general (Dicke 1953, 1954; Wittke and Dicke 1956).

We review the present state-of-the-art of this most useful frequency standard in its active as well as in its passive configuration. Much has been written on that device particularly in Volume 2 of *QPAFS* (1989). However, we limit ourselves to the essentials, summarizing its theory in a somewhat different approach as that used up until now. We also outline the most recent results obtained by several laboratories engaged in making the device as stable as possible. We then describe new approaches in implementing the device to make it smaller, using either dielectric loading of the cavity or using a so-called magnetron type cavity design.

1.1.2.1 Active Hydrogen Maser

A conceptual schematic diagram of the classical hydrogen maser is shown in Figure 1.11 with the ground state energy levels of the H atom as an inset. The energy levels manifold is created by the magnetic interaction of the unpaired electron with the nucleus consisting of a single proton. This is one of the simplest atomic structure. The interaction creates two hyperfine energy levels $F = 0$ and $F = 1$. The energy separation of these levels corresponds to a frequency v_{hf} equal to 1420.405 MHz, in the L band of the microwave spectrum. A magnetic field applied to the system removes the residual degeneracy and causes a splitting of the $F = 1$ state into three

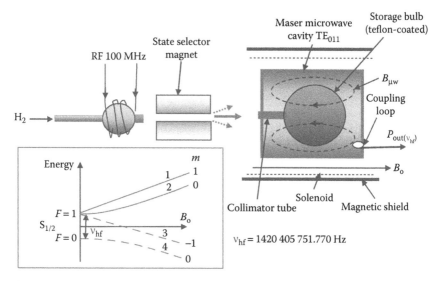

FIGURE 1.11 Simplified conceptual diagram of a hydrogen maser. The inset shows the ground state energy levels of hydrogen.

levels identified as $m_F = 1, 0, -1$. The maser operates on the transition $F = 1, m_F = 0$ to $F = 0, m_F = 0$. The dependence of the resulting frequency on the magnetic induction is given by:

$$\nu = \nu_{hf} + 1399.08 \times 10^7 B_o^2 \qquad (1.37)$$

where:

B_o is the magnetic induction expressed in Tesla

This is the clock frequency, which is also perturbed by some other small effects which will be made explicit hereunder.

In Figure 1.11, molecular hydrogen is first introduced in a dissociator consisting of a glass enclosure of the order of 5 cm in diameter and placed inside the tank coil of an oscillator. The hydrogen pressure inside that dissociator is of the order of 0.1 Torr. The frequency of the oscillator is about 100 MHz and the power delivered to the molecular hydrogen gas is of the order of a few watts. The dissociation of the molecules takes place with a relatively high efficiency and a beam of atomic hydrogen is formed at the exit of the glass enclosure generally through a small multihole collimator. The beam is directed along the axis of a hexapole magnet which has the property of deflecting atoms in the $F = 1, m_F = 0$ and 1 towards the symmetry axis of the magnet and deflecting the other atoms away from the axis. The hexapole magnet acts as a lens and the geometry of the system is such as to focus the atoms in the two states mentioned into the entrance of a storage bulb placed inside a microwave cavity resonating at the hyperfine frequency, 1420.405 MHz. For reasons of dimensional stability, the cavity resonator is generally made of fused silica or of a low thermal expansion material such as Cervit or Zerodur. It is coated internally with a metallic silver film, operates in the TE_{011} mode, and has an unloaded Q of the order of 50,000. In some cases, for reasons of simplicity, the cavity is made of a metal such as aluminium. In that case, the cavity resonant frequency is very sensitive to temperature and needs to be stabilized electronically. The storage bulb is made of fused silica in order to reduce microwave losses and preserve the high quality factor of the cavity. The inner surface of the storage bulb is coated with a substance that prevents recombination of the atoms into molecular hydrogen and relaxation of the atoms in the ground state upon collision with the surface. Teflon™ has been found very efficient in this regard. In a bulb having a 15 cm diameter, the lifetime of an atom in one particular level, experiencing thousands of collisions with the surface, may be of the order of 1 s.

The principle of operation is as follows. Assuming the presence of a microwave field in the cavity, those atoms that have entered the storage bulb in the $F = 1, m_F = 0$ level emit their energy at 1420.405 MHz through the process of stimulated emission of radiation. The radiation emitted adds in phase to the existing radiation, a process that results in amplification. The energy given by the atoms is partly dissipated in the walls of the cavity and partly delivered to external circuitry via the coupling loop. Atoms in the $F = 1, m_F = 0$ level are continuously replenished by the input beam, and if the microwave losses are small and the relaxation time sufficiently long, a continuous oscillation occurs. This situation results in an active maser. In the other case where microwave losses are too large, no continuous oscillations are present. However, it is still possible to observe the stimulated emission phenomenon through appropriate passive amplification-detection techniques. This is called the passive maser approach.

The solenoid shown in the figure creates a magnetic field, B_0, that provides an axis of quantization for the atomic ensemble. The clock frequency originates from $\Delta m_F = 0$ transitions and quantum mechanical selection rules require that the dc magnetic field and the microwave magnetic field be parallel. This is the situation shown in Figure 1.11. The selector magnet, storage bulb, and cavity are maintained under a vacuum better than 10^{-7} Torr created by VacIon™ or getter type pumps. The storage bulb and cavity are placed inside a set of concentric magnetic shields to reduce the influence of environmental fluctuations of the earth's magnetic field. As we discuss in subsequent paragraphs, the output frequency of the device is sensitive to a small extent to the cavity tuning which is a function of its dimension. For this reason, the temperature of the ensemble is generally regulated to a high degree.

The output power of the maser is of the order of 10^{-13} to 10^{-14} W and has a nominal frequency of 1420.405 MHz which, in general, is not readily usable. For that reason, this signal is normally processed by means of a digital system, phase locking a quartz crystal oscillator at a nominal frequency of 10 MHz to the maser signal. A typical phase-locked-loop system is shown in Figure 1.12.

The signal is detected by means of heterodyning techniques. The so-called reference frequency used in such a detection scheme is generated from a quartz crystal oscillator multiplied to a frequency, say 1400 MHz, resulting in a beat frequency of 20.405 MHz called the intermediate frequency (if). The beat signal obtained may be heterodyned down to dc in a few stages. In the final stage, the signal, containing information on the relative phase of the crystal oscillator and the maser output

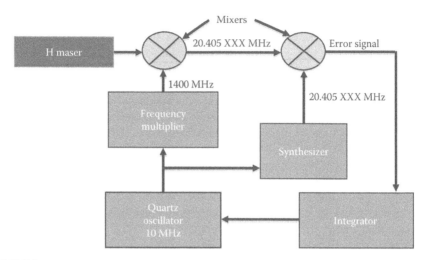

FIGURE 1.12   Block diagram of a typical analogue phase-lock-loop system used with the hydrogen maser. In the present case, the frequency has been set at 20 MHz. For simplicity in construction, the second mixing at 20.405 MHz is generally done by means of several intermediate steps.

signal, is used to lock in phase the crystal oscillator to the maser signal output as shown in the block diagram.

1.1.2.1.1   Basic Theory of Operation

Various types of analysis describing the theory of operation of the H maser are of course possible. The following derivation is given as an illustration of the H maser theory in a mathematical context slightly different from those already published, using conditions on the phase of the generated radiation as the main element for permitting continuous oscillation (Vanier 2002).

1.1.2.1.1.1   Rate Equations   In this analysis, the ensemble is represented by the density matrix ρ. The diagonal elements of this matrix represent the fractional population of each level of the atomic ensemble in question. The trace is equal to unity since it represents the total fractional population. The off diagonal elements represent the coherence existing in the ensemble. In the case of the H maser, the atoms in the storage bulb are all in the ground state $S_{1/2}$ and the equilibrium density matrix is:

$$\rho = \begin{pmatrix} \rho_{11} & 0 & 0 & 0 \\ 0 & \rho_{22} & 0 & \rho_{42} \\ 0 & 0 & \rho_{33} & 0 \\ 0 & \rho_{24} & 0 & \rho_{44} \end{pmatrix} \tag{1.38}$$

The levels are numbered from high to low energy as in Figure 1.11. It is assumed that the only coherence existing is that corresponding to transitions from level 2 to level 4. This is due to the fact that the cavity is tuned to the corresponding frequency and the microwave radiation existing in the cavity is resonant with only that transition. Furthermore, it has the appropriate orientation with the dc magnetic field for exciting so-called $\Delta m = 0$ transitions. This coherence oscillates at the frequency of the micro-wave field. As we discuss hereunder, an atomic magnetization is associated with that coherence and oscillates at the same frequency. This magnetization acts as a source term for the microwave field as made explicit in Maxwell's equations. It is the energy contained in this oscillating magnetization that is extracted from the ensemble and coupled out through the cavity coupling loop. We now develop these concepts in a formal mathematical approach.

In general, the various perturbations are decoupled from each other and can be assumed to act independently. In the case of the H maser, the evolution of the density matrix elements can be written as:

$$\frac{d\rho}{dt} = \left(\frac{d\rho}{dt}\right)_{flow} + \left(\frac{d\rho}{dt}\right)_{wall} + \left(\frac{d\rho}{dt}\right)_{se} + \left(\frac{d\rho}{dt}\right)_{rad} \tag{1.39}$$

where the subscripts flow, wall, se, rad, refer respectively to atomic flux in and out of the storage bulb, collisions with that internal wall of the storage bulb, spin–exchange interactions between H atoms and effect of the microwave radiation field. The flow term takes into account the input flux of atoms into the storage bulb and the escape of the atoms through the entrance hole and is given by:

$$\left(\frac{d\rho}{dt}\right)_{flow} = \begin{pmatrix} \dfrac{I_1}{N} & 0 & 0 & 0 \\ 0 & \dfrac{I_2}{N} & 0 & 0 \\ 0 & 0 & 0 & 0 \\ 0 & 0 & 0 & 0 \end{pmatrix} - \gamma_b \rho \tag{1.40}$$

where:
I_1 and I_2 are the input flux in atoms per second in levels 1 and 2, respectively
N is the total number of atoms in the storage bulb

In general, the selector magnet geometry makes $I_1 = I_2$. The parameter γ_b character-izes the rate of escape of the atoms from the bulb and can be calculated from the bulb geometry. The exit hole (collimator tube) is usually made such as to make γ_b equal to about 2 s^{-1}. In equilibrium, the population of the levels is constant and $(d/dt)\rho = 0$. Defining $I_t = I_1 + I_2$, we have:

$$\frac{I_t}{N} = \gamma_b \left(\rho_{11}^0 + \rho_{22}^0 + \rho_{33}^0 + \rho_{44}^0\right) = \gamma_b \tag{1.41}$$

The trace of ρ is equal to 1, since it represents the total fractional population. In a stor-age bulb, coated with a substance such as Teflon™, the atoms have a long lifetime in a given state. Physical mechanisms that limit the lifetime of an H atom as an indepen-dent entity are absorption by the surface and recombination to form an H molecule.

In both cases, the atom is totally lost to the emission process. It is as if the atom was quitting the storage bulb. In some cases, the atom may bounce back from the surface in another state. Furthermore, in some weak collisions, the atom may stay in the same m_F state, but its oscillating magnetic moment may suffer from a small phase shift. This last mechanism creates, on an average through accumulation of phase shifts during the emission time, a small frequency shift of the signal delivered by the atom. These atomic phenomena are rather complex and the relative importance of each mechanism is not entirely known. In order to simplify the analysis, we assume that the collisions with the surface cause uniform relaxation and we assume that both the coherence and the population relaxation rates are equal. The phase shift introduced by the collisions is taken into account in a phenomenological way by adding an imaginary term to the coherence relaxation rate. The phenomenon is represented by the general equations:

$$\left(\frac{d\rho_{ii}}{dt}\right)_{wall} = -\gamma_w \rho_{ii} \tag{1.42}$$

$$\left(\frac{d\rho_{24}}{dt}\right)_{wall} = -\gamma_w \rho_{24} + i\Omega_w \rho_{24} \tag{1.43}$$

where:
  γ_w is the wall relaxation rate
  Ω_w is the so-called wall shift

In a collision between two H atoms, there is a probability that an interchange in spin coordinates takes place. This phenomenon has been studied in detail in *QPAFS*, Volumes 1 and 2 (1989), and we hence outline only the main conclusion. The collision process is characterized by a cross section σ_{se} and the resulting relaxation rate is:

$$\gamma_{se} = n\bar{v}_r \sigma_{ex} \tag{1.44}$$

where:
  n is the atomic density
  $\bar{v}_r$ is the relative velocity of the atoms

The collision also introduces a small phase shift of the atomic magnetic moment that results in a small average frequency shift represented by the rate γ_{se}^{λ}. A detailed calculation shows that the density matrix elements are affected by spin–exchange collisions as follows:

$$\left[\frac{d(\rho_{22} - \rho_{44})}{dt}\right]_{se} = -\gamma_{se}\left(\rho_{22} - \rho_{44}\right) \tag{1.45}$$

$$\left(\frac{d\rho_{24}}{dt}\right)_{se} = -\frac{\gamma_{se}}{2}\rho_{24} - i\frac{\gamma_{se}^{\lambda}}{2}\rho_{24}\left(\rho_{22} - \rho_{44}\right) \tag{1.46}$$

As is readily seen, the frequency shift introduced by spin–exchange collisions is proportional to the difference in fractional population of levels 2 and 4.

The dynamic behaviour of the density matrix under the influence of the microwave field is obtained from Liouville's equation:

$$\frac{d\rho_{ij}}{dt} = \left(\frac{1}{i\hbar}\right)\sum_k \left(\mathcal{H}_{ik}\rho_{kj} - \rho_{ik}\mathcal{H}_{kj}\right) \qquad (1.47)$$

where:
$\mathcal{H}_{ik}$ is the atom-field interaction Hamiltonian

The microwave field is assumed to have the form:

$$\vec{B}_{rf}(r) = \vec{k}B_z(r)\cos(\omega t + \phi) \qquad (1.48)$$

where ω is the angular frequency of the field in the cavity and ϕ is its phase. We are considering only the z component of the field since only this component has the appropriate orientation to excite the $\Delta m_F = 0$ transition. The Hamiltonian of this perturbation is:

$$\mathcal{H}_{24} = -\frac{1}{2}\mu_B g_j B_z(r)\cos\left(\omega t + \phi\right) \qquad (1.49)$$

where:
μ_B is Bohr magneton
g_j is the spin splitting factor equal to 2

We define the Rabi angular frequency, b, as in the case of the Cs standard described above:

$$b = \frac{\mu_B B_z(r)}{\hbar} \qquad (1.50)$$

We expand the cosine in Equation 1.49 in exponential terms and use the rotating wave approximation as in magnetic resonance. We obtain:

$$\mathcal{H}_{24} = -\frac{1}{2}\hbar\left(be^{-i\phi}\right)e^{-i\omega t} \qquad (1.51)$$

We assume for the off diagonal elements of the density matrix a solution of the form:

$$\rho_{24} = \delta_{24}e^{-i\omega t} \qquad (1.52)$$

We define fractional population difference between the levels of interest as:

$$\Delta = \rho_{22} - \rho_{44} \qquad (1.53)$$

We expand Equation 1.47 and use the previous relations. We obtain two equations describing the behaviour of the atomic ensemble under the simultaneous effect of the microwave field and of the relaxation processes:

$$\dot{\Delta} + \gamma_1 \Delta = \frac{I_2}{N_t} - 2b\,\mathrm{Im}\,e^{-i\phi}\delta_{42} \tag{1.54}$$

$$\dot{\delta}_{24} + \left[\gamma_2 - i\left(\omega - \omega_{24} + \Omega_w + \frac{\gamma_{se}^\lambda}{2}\Delta \right) \right]\delta_{24} = -\frac{i}{2}be^{-i\phi}\Delta \tag{1.55}$$

where γ_1 and γ_2 are the population and coherence total relaxation rates defined as:

$$\gamma_1 = \gamma_b + \gamma_w + \gamma_{se} \tag{1.56}$$

$$\gamma_2 = \gamma_b + \gamma_w + \frac{\gamma_{se}}{2} \tag{1.57}$$

1.1.2.1.1.2 Field Equation   As mentioned earlier, the coherence in the ensemble creates an oscillating magnetization. Its expectation value is given by:

$$\langle M_z \rangle = Tr\left(\rho M_{op} \right) \tag{1.58}$$

where:

M_{op} is the equivalent quantum mechanical operator of the classical magnetization M
Tr is the sum of the diagonal elements of the matrix resulting from the product of ρ by M_{op}

The result is:

$$\langle M_z \rangle dv = -\frac{1}{2}n\mu_B \left(\rho_{24} + \rho_{42} \right)dv \tag{1.59}$$

This classical magnetization is coupled to the rf field through Maxwell's equations. In a cavity, the relation is (*QPAFS* 1989, Volume 1):

$$\ddot{\vec{H}}(r,t) + \left(\frac{\omega_c}{Q_L} \right)\dot{\vec{H}}(r,t) + \omega_c^2 \vec{H}(r,t) = \vec{H}_c(r)\int_{V_c} \vec{H}_c(r) \bullet \ddot{\vec{M}}(r,t)dv \tag{1.60}$$

where:

ω_c is the cavity angular resonance frequency
Q_L is the cavity loaded quality factor
V_c is the cavity volume
$\vec{H}_c(r)$ is the orthonormal cavity field mode
$\vec{M}(r,t)$ is the magnetization calculated above

We write *H* and *M* in complex form:

$$\vec{H}(r,t) = \left[H^{+*}(r)e^{-i\omega t} + H^+(r)e^{i\omega t} \right]\vec{z} \tag{1.61}$$

$$\vec{M}(r,t) = \left[M^{+*}(r)e^{-i\omega t} + M^+(r)e^{i\omega t} \right]\vec{z} \tag{1.62}$$

where:

H^+ and M^+ are complex amplitudes of the field and of the magnetization, respectively

We replace these expressions in Equation 1.60, and keeping only the resonant component as in the rotating wave approximation used before, we obtain using Equation 1.59:

$$\left|\vec{H}\right|e^{-i\phi} = \frac{-iQ_L}{1+2iQ_L\left(\Delta\omega_c/\omega\right)}\vec{H}_c(r)\int_{V_c}\vec{H}_c(r)\cdot\left(\frac{1}{2}\right)n\mu_B\left(\delta_{24}^r + i\delta_{24}^i\right)dv \qquad (1.63)$$

where the off diagonal density matrix element has been written explicitly in complex form. Algebraic manipulations show that the phase of the field is given by:

$$\phi = \frac{\pi}{2} + \tan^{-1}2Q_L\frac{\Delta\omega_c}{\omega} - \tan^{-1}\frac{\delta_{24}^i}{\delta_{24}^r} \qquad (1.64)$$

making explicit the phase quadrature, $\pi/2$, between the field and the magnetization.

1.1.2.1.2 Oscillation Condition

The other unknown in Equations 1.54 and 1.55 is the Rabi frequency, b. The energy given by the atomic ensemble is lost in the cavity resonator walls and in the external coupling loop. The power dissipated in the cavity is given by (Collin 1991):

$$P_{diss} = \frac{\omega}{Q_L}\mu_0\int_{V_c}\overline{H}(r,t)\,dv \qquad (1.65)$$

where the bar over H means an average over time. Using Equation 1.50, this expression can be written in terms of the Rabi angular frequency b as:

$$P_{diss} = \frac{1}{2}\frac{N\hbar\omega}{k}\mu_0\langle b\rangle^2_{bulb} \qquad (1.66)$$

where k is defined as:

$$k = \frac{NQ_L\eta'\mu_B^2\mu_0}{\hbar V_{bulb}} \qquad (1.67)$$

The filling factor is defined as:

$$\eta = \frac{\langle H_z(r)\rangle^2_{bulb}}{\langle H^2(r)\rangle_c} \qquad (1.68)$$

$$\eta' = \frac{V_{bulb}\langle H_z(r)\rangle^2_{bulb}}{V_c\langle H^2(r)\rangle_c}$$

On the other hand, the power given by the atoms can also be obtained from Equation 1.66 by realizing that the value of H as given by Equation 1.61 can be written in terms of the magnetization by means of Equation 1.63. Assuming that $2Q_L(\Delta\omega_c/\omega)$ is much smaller than 1, which is always the case of the oscillating H maser, one obtains:

$$P_{at} = \frac{1}{2}N\hbar\omega k \left|2\delta_{24}\right|^2 \tag{1.69}$$

The energy given by the ensemble compensates for the losses in the cavity. Equating Equations 1.66 and 1.69, we obtain:

$$\langle b \rangle = 2k\left|\delta_{24}\right| \tag{1.70}$$

This result is used with Equation 1.64 to evaluate the terms containing the phase ϕ in Equations 1.54 and 1.55. Hence, we obtain:

$$2b\,\mathrm{Im}\,e^{-i\phi}\delta_{42} = 4k\cos\psi\left|\delta_{24}\right|^2 \tag{1.71}$$

$$-\frac{i}{2}be^{-i\phi} = -ke^{-i\psi}\delta_{24} \tag{1.72}$$

where ψ stands for:

$$\psi = \tan^{-1}2Q_L\frac{\Delta\omega_c}{\omega} \tag{1.73}$$

and is a measure of the cavity detuning effect on the behaviour of the maser. Using these relations, Equations 1.54 and 1.55 become:

$$\dot{\delta}_{24} + \left[\gamma_2 - i\left(\omega - \omega_{24} + \Omega_w + \frac{\gamma_{se}^{\lambda}}{2}\Delta\right)\right]\delta_{24} = ke^{-i\psi}\delta_{24}\Delta \tag{1.74}$$

$$\dot{\Delta} + \gamma_1\Delta = \frac{I_2}{N} - 4k\cos\psi\left|\delta_{24}\right|^2 \tag{1.75}$$

These last two equations describe the dynamical behaviour of the H maser in a self-consistent approach relative to energy and phase.

In the case of self-sustained continuous oscillations, that is a stationary situation, the derivatives in Equations 1.74 and 1.75 are set equal to zero. We separate the real and imaginary parts of these equations and assume the cavity to be tuned, making $\psi = 0$. We obtain:

$$\left|2\delta_{24}\right|^2 = \frac{I_2}{kN} - \frac{\gamma_1}{k}\Delta \tag{1.76}$$

$$\Delta = \frac{\gamma_2}{k} \tag{1.77}$$

and using Equation 1.69, the power given by the ensemble is:

$$P_{at} = \frac{1}{2}\hbar\omega N\left(\frac{I_2}{N} - \frac{\gamma_1\gamma_2}{k}\right) \tag{1.78}$$

We define the following terms which have been used extensively in the past in the description of the maser operation (Kleppner et al. 1965):

$$P_c = \frac{1}{2}\hbar\omega\frac{\hbar V_c \gamma_b^2}{Q_L \eta \mu_B^2 \mu_o} \tag{1.79}$$

$$I_{th} = \frac{\hbar V_c \gamma_b^2}{Q_L \eta \mu_B^2 \mu_o} \tag{1.80}$$

$$q = \frac{\hbar V_c \bar{V}_r \sigma_{ex}}{2Q_L \eta \mu_B^2 \mu_o V_b}\frac{\gamma_b + \gamma_w}{\gamma_b}\left(\frac{I_t}{I_2}\right) \tag{1.81}$$

The parameter, q, is called the oscillation parameter and I_{th} is the threshold atomic flux that would be required to obtain self-sustained oscillation if spin–exchange interactions did not exist. Using these definitions, we obtain:

$$P_{at} = P_c \left[-2\left(\frac{I_2}{I_{th}}\right)^2 q^2 + (1-3q)\left(\frac{I_2}{I_{th}}\right) - 1 \right] \tag{1.82}$$

The power coupled out of the cavity through the coupling loop is given by:

$$P_{out} = P_{at}\left(\frac{\beta}{1+\beta}\right) \tag{1.83}$$

where β is the coupling factor of the cavity, which is assumed to be undercoupled as is normally the case.

Equation 1.82 is used to describe the oscillation condition of the H maser. It was obtained here through a mathematical approach slightly different from that used in the past, by first setting an arbitrary phase for the field and then calculating this phase by means of the field equation. The amplitude of the field was then determined in a self-consistent approach. The method results in two time-dependent equations describing the maser behaviour. A steady-state solution was then obtained by setting time derivatives equal to zero.

Equation 1.82 is represented graphically in Figure 1.13 for various values of the oscillating parameter q making explicit the minimum and maximum fluxes allowed for continuous oscillation. In an oscillating maser, the power output must be positive. This sets conditions on the parameter q and the flux I_2. It is readily shown that for continuous oscillation we must have:

$$q \le 0.172 \tag{1.84}$$

From a practical point of view, this can also be expressed in a simple manner by realizing that most of the parameters in Equation 1.81 are known constants, except for γ_b, γ_w, and I_t/I_2, which are fixed by design of the device and quality of the Teflon^{TM} coating. Unless a double selection magnet utilizing adiabatic fast passage (AFP) is used for selection of a single state in the beam, we have $I_t/I_2 = 2$. Furthermore, the wall

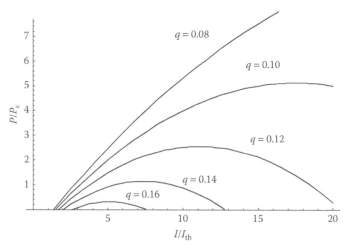

FIGURE 1.13 Hydrogen maser power output as a function of beam intensity for various quality factors.

relaxation rate, γ_w, may vary, depending on the quality of the Teflon™ wall coating and the size of the storage bulb. Finally, it is a standard practice to attempt to obtain a line quality factor Q_l of the order of a few times 10^9. This requires some control on the escape rate γ_b of the bulb. We may set $\gamma_b = 1 \text{ s}^{-1}$ and, from experience, we may have with an average bulb size and good coating, $\gamma_w = 1 \text{ s}^{-1}$ ($T_w = 1 \text{ s}$). The line Q is then 2.2×10^9. The oscillation condition then requires that:

$$Q_L \eta' > 10^4 \tag{1.85}$$

where the filling factor η' as defined by Equation 1.68 is given by:

$$\eta' = \eta \frac{V_b}{V_c} \tag{1.86}$$

For a cylindrical cavity, η' may be as high as 0.45 (see *QPAFS* 1989, Volume 2). This means that in such a case the loaded cavity Q must be larger than ~22,000 for oscillation.

Figure 1.14 is a typical result obtained on a maser developed in a research program at Laval University, Canada, in the 1980s (Vanier et al. 1984). In that particular case the maser bulb was relatively large, 16 cm in diameter, and the bulb entrance hole was made of 12 small tubes of commercial Teflon™. The bulb itself was coated with three layers of Teflon™ FEP 120 type. This construction resulted in a quality factor q smaller than 0.1 and made possible continuous oscillation. The large aperture of the bulb collimator allowed the entrance into the bulb of atoms on the whole cross section of the H beam making possible the operation of the maser over a large range of source pressures. The long time constant of the arrangement allowed operation of the maser for beam intensities varying by a factor larger than 10 while the line Q varied by a factor of 3. Concepts developed in the construction of that maser, such

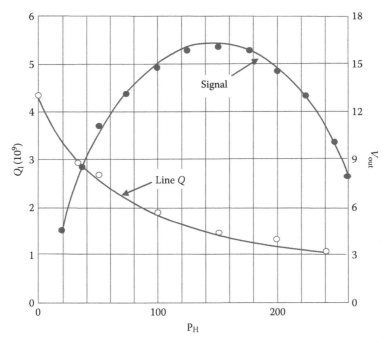

FIGURE 1.14 Typical result obtained on an experimental hydrogen maser showing the variation of output power and line Q against atomic flux.

as general structure, H pressure control, cavity design, and temperature control were used at the NRC to construct masers that are used at present with primary Cs beam clocks in maintaining the Canadian atomic timescale (Morris 1990).

1.1.2.1.3 Frequency Shifts

1.1.2.1.3.1 Magnetic Field Shift   As in the Cs beam standard, a very important frequency shift is that caused by the magnetic field required to provide an axis of quantization to the ensemble of atoms. This shift was discussed above. With sufficient magnetic shielding, masers can normally be operated in rather low fields, less than 10^{-7} T (1 mG). In such a case, the magnetic field fractional frequency shift is less than 2×10^{-12}. Since the field can be determined very accurately by means of excitation of $\Delta m = \pm 1$ transitions between Zeeman levels, this shift is generally known to be of a fractional accuracy of 10^{-15}. Consequently, the magnetic field does not affect the accuracy of the maser to that level. Its effect may be seen, however, on the maser frequency stability if it is not constant with time. This is the reason why multilayers of magnetic shields are required to prevent environmental fluctuations of the magnetic field to reach the ensemble of atoms in the cavity. On the other hand, the current source driving the solenoid must have a stability that is compatible with the frequency stability desired. This current stability is directly calculated from Equation 1.37 and the solenoid characteristics.

Other important frequency shifts affecting the output of the maser are contained in the imaginary term of Equation 1.74. Assuming ψ to be small, the equation may be written as:

$$\left\{ \gamma_2 - k\Delta + i \left[\omega - \omega_{24} + \Omega_w + \left(\frac{\gamma_{se}^\lambda}{2} \right) \Delta + \psi k\Delta \right] \right\} \delta_{24} = 0 \qquad (1.87)$$

In continuous oscillation, δ_{24} is different from zero and both the real and imaginary parts must be set equal to zero to satisfy that equation. From the imaginary part, we thus have:

$$-\left\{ \left[\omega - \omega_{24} + \Omega_w + \left(\frac{\gamma_{se}^\lambda}{2} \right) \Delta \right] + \psi k\Delta \right\} = 0 \qquad (1.88)$$

It is readily observed that the maser output frequency is shifted by various small perturbations that alter the transition frequency ω_{24}, which we now examine.

1.1.2.1.3.2   The Wall Shift, Ω_w   As mentioned above, the hydrogen atom, upon collision with the Teflon™ surface experiences a small phase shift. On the average, this is reflected on the output frequency by a small frequency shift. This frequency shift is proportional to the collision rate and thus for spherical bulbs to the inverse of the bulb diameter. For a 15-cm-diameter bulb, coated with Teflon™, this shift is negative and of the order of a few parts in 10^{11}, a number that depends on the type of Teflon™ used and the quality of the coating. Unfortunately, it is difficult to reproduce this shift from bulb to bulb to better than 10%, giving the maser a frequency accuracy of about 1 to 2×10^{-12} (Vanier et al. 1975).

1.1.2.1.3.3   The Spin–Exchange Shift, $\gamma_{se}^\lambda \Delta$, and the Cavity Pulling, $\psi k \Delta$ It is observed that both these shifts are proportional to the population difference Δ. Using Equation 1.77, these shifts can be written as:

$$\Delta\omega_{se} + \Delta\omega_{cp} = \left[-\frac{\gamma_{se}^\lambda}{2} + \psi k \right] \frac{\gamma_2}{k} \qquad (1.89)$$

showing that the two frequency shifts are proportional to the total line width $(1/\pi) \gamma_2$. It is readily shown that the cavity pulling is equal to:

$$\Delta\omega_{cp} = \frac{Q_{cL}}{Q_{at}} \Delta\omega_c \qquad (1.90)$$

Since the spin–exchange frequency shift is also proportional to the total line width, a simple technique for tuning the cavity consists in altering γ_2 in Equation 1.89 by varying the beam flux which causes a change of the density within the bulb volume. By tuning the resonator so that the frequency of the maser becomes independent of the beam flux, the cavity detuning is such as to cause a shift opposite and exactly equal to the spin–exchange shift given by:

$$\Delta\omega_{se} = -\frac{\bar{v}_r \hbar \lambda \gamma_2}{4 Q_L \eta' \mu_B^2 \mu_o} \qquad (1.91)$$

where λ is the spin–exchange frequency shift cross section (Vanier and Vessot 1964). This shift is of the order of a few parts in 10^{-11}. In practice it is possible to tune the cavity such as to make the residual uncertainty in the maser frequency due to cavity detuning less than 1×10^{-14}.

1.1.2.1.3.4 Second-Order Doppler Shift Another shift that is not included in the previous calculation is that due to a relativistic effect caused by time dilation and called the second-order Doppler shift. It is present as in the case of the Cs beam standard. It is given by:

$$\Delta\omega_D = -\frac{3}{2}\frac{k_B T}{Mc^2}\omega_{\text{hf}} \qquad (1.92)$$

In H at 40°C, this shift is -4.31×10^{-11} and with a temperature determination of the storage cell better than 0.1 K the uncertainty in the frequency is less than 2×10^{-14}.

1.1.2.1.3.5 Magnetic Field Inhomogeneity Finally, the maser frequency may be affected by other small frequency shifts connected to inhomogeneities in the applied magnetic field. These shifts depend on the construction geometry of the maser and are generally quite small in well-built H masers. They do not affect the frequency stability of the maser to a significant extent (*QPAFS* 1989, Volume 2).

A typical large size active H maser developed at Université de Liège is shown in Figure 1.15a (Mandache et al. 2012). As in many other laboratories, that breadboard

(a)   (b)

FIGURE 1.15 (a) Large size H maser recently developed at Université de Liège. The maser was built as a demonstration tool within a program of developments on atomic frequency standards. (b) Small H maser developed at the SAO Harvard by RFC Vessot (shown in the figure) and his team. (Courtesy of R.F.C. Vessot; Data from Mandache, C. et al., *Appl. Phys. B: Lasers Opt.*, 107, 675, 2012.)

type maser was constructed as an experimental tool within a program of develop-ment on atomic frequency standards. It provided essential information and experi-ence for the development of other units. In the study if its characteristics, an elegant technique was developed for the determination of the oscillation parameter q using the variable cavity Q approach (Mandache et al. 2012). Figure 1.15b shows a small maser developed at Smithsonian Astrophysical Observatory, Cambridge, MA (R.F.C. Vessot 2014, pers. comm.).

1.1.2.2 Passive Hydrogen Maser

The component that plays an important role in determining the size of the H maser is the cavity resonator. The size of this resonator can be reduced by operating it in a mode different from the conventional TE_{011} mode. The resonator can also be loaded with dielectric material. We examine these options under the section outlining recent pro-gresses on size reduction. However, in some circumstances, the cavity Q is not likely to be high enough to achieve continuous oscillation. Consequently, in order to relax the requirements on the quality factor of the cavity, it is possible to operate the maser in the so-called passive mode. The maser is then used essentially as an amplifier (Vuylsteke 1960; Siegman 1964, 1971). The system may be operated with two coupling loops, one being used to inject a microwave signal at the hyperfine frequency, while the other is used to detect the amplified signal. Another approach consists in using a so-called microwave circulator with a single coupling loop as shown in Figure 1.16. Here, one observes the amplified power reflected from the cavity.

1.1.2.2.1 Theory of Operation

For passive operation, microwave energy close to the hyperfine frequency is injected inside the cavity through the coupling loop. This energy creates a field in the cavity which causes stimulated emission of the H atoms. The process creates coherence and

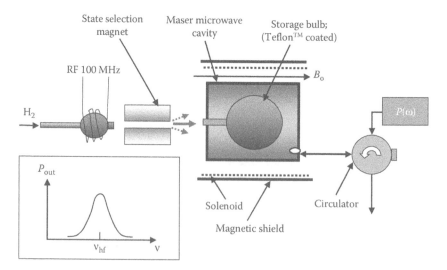

FIGURE 1.16 Simplified conceptual block diagram of a reflection-type passive hydrogen maser.

an oscillating magnetization in the ensemble. Since the upper level ($F = 1$, $M_F = 0$) is more populated than the lower one ($F = 0$, $m_F = 0$), power is emitted by the ensemble. The Rabi angular frequency, b, now includes the contribution from a source external to the cavity as well as the contribution from the created magnetization. The interaction Hamiltonian given by Equation 1.51 is now written as:

$$\mathcal{H}_{24} = -\frac{1}{2}\hbar\left(b_i e^{-i\phi_i} + b_e e^{-i\phi_e}\right)e^{-i\omega t} \tag{1.93}$$

where the subscripts i and e refer to internal and external, respectively. We make an analysis very similar to that made in the case of the active maser. There are now two source terms in the field equation: the injected field and the magnetization. One observes readily that these two source terms are separated. The field created by the magnetization leads to feedback on the atomic ensemble, an effect which can be analyzed in the same way as in the active maser. Equations 1.74 and 1.75 become:

$$\dot{\delta}_{24} + \left\{\gamma_2 - k\Delta - i\left[\omega - \omega_{24} + \Omega_w + k\psi\Delta + \left(\frac{\gamma_{se}^\lambda}{2}\right)\Delta\right]\right\}\delta_{24} = -i\frac{b_e}{2}e^{-i\phi_e}\Delta \tag{1.94}$$

$$\dot{\Delta} + \gamma_1\Delta = \frac{I_2}{N} - 4k\cos\psi|\delta_{24}|^2 - 2b_e\,\mathrm{Im}\,e^{-i\phi_e}\delta_{42} \tag{1.95}$$

These equations describe the dynamic behaviour of the H maser submitted to a field injected in the cavity. A solution of these equations may be obtained numerically. However, the general behaviour of the passive H maser may be obtained by means of simple approximations. We assume steady-state operation and set all time derivatives equal to zero. The coherence is obtained directly from Equation 1.94 as:

$$\delta_{24} = \frac{-i\left(b_e/2\right)e^{-i\phi_e}\Delta}{\left(\gamma_2 - k\Delta\right) - i\left[\omega - \omega_{24} + \Omega_w + \left(\gamma_{se}^\lambda/2\right)\Delta + k\psi\Delta\right]} \tag{1.96}$$

We are interested in the gain of the maser that we define as:

$$G = \left|\frac{b_i e^{-i\phi_i} + b_e e^{-i\phi_e}}{b_e e^{-i\phi_e}}\right| \tag{1.97}$$

The internal field part is produced by the induced coherence as given by Equation 1.74 and using solution (1.96) for δ_{24}, the gain can be written as:

$$G = \left|1 + \frac{k\Delta e^{-i\psi}}{\left(\gamma_2 - k\Delta\right) - i\left[\omega - \omega_{hf} + \Omega_w + \left(\gamma_{se}^\lambda/2\right)\Delta + k\psi\Delta\right]}\right| \tag{1.98}$$

where Δ is a solution of Equations 1.94 and 1.95. We can obtain some insight into the behaviour of the maser by examining a special case. We assume that the injected field is weak and that the gain factor is small. In such a case, Δ obtained from Equation 1.95 is:

$$\Delta = \frac{I_2}{N\gamma_1} \tag{1.99}$$

Using definitions 1.67 and 1.80, we obtain:

$$\frac{k\Delta}{\gamma_2} = \frac{I_2}{I_{th}} \frac{\gamma_b^2}{\gamma_1\gamma_2} = I_{eff} \frac{\gamma_b^2}{\gamma_1\gamma_2} \tag{1.100}$$

Assuming that the cavity is exactly tuned to the atomic resonance, Equation 1.98 becomes:

$$G = \left| 1 + \frac{I_{eff}}{1 - I_{eff} - i\Omega_{24}} \right| \tag{1.101}$$

where

$$\Omega_{24} = \left[\omega - \omega_{24} + \Omega_w + \left(\frac{\gamma_{se}^{\lambda}}{2} \right) \Delta \right] \gamma_2 \tag{1.102}$$

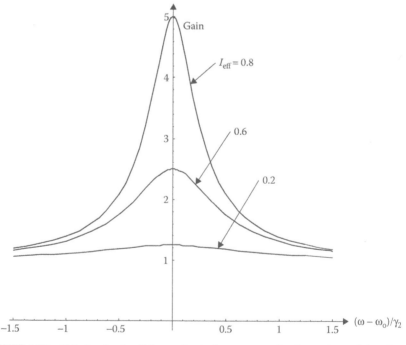

FIGURE 1.17 Calculated gain of the passive hydrogen maser for three values of the effective hydrogen beam intensity.

Equation 1.101 is plotted in Figure 1.17 for three values of the normalized flux I_{eff} as a function of the angular frequency ω normalized to γ_2. For $I_{eff} = 1$, the gain becomes infinite and in principle the system becomes an active maser. However, this is not exact since saturation takes place at high gain and the threshold flux is somewhat larger than I_{eff}. The gain given by Equation 1.101 represents the ratio of the cavity field with the atoms present to the field without the atoms. In practice, a system such as the one shown in Figure 1.16 may be used to detect the amplified signal.

In that case the signal is observed as the power reflected by the cavity through the circulator. We examine the particular case where the cavity is matched when there are no atoms in the cavity. In such a case there is no energy reflected by the cavity in the absence of atoms. The situation is the same if the frequency is tuned well off the atomic resonance. The reflection cavity amplitude gain, G_{rc} is then given directly by the cavity reflection coefficient ρ (*QPAFS* 1989, Volume 1):

$$G_{rc}(\omega) = |\rho| = \frac{1}{1 - |Q_m|/Q_L} \tag{1.103}$$

where:

Q_L is the cavity loaded quality factor without the contribution of the atomic ensemble
Q_m, called the magnetic Q, is the quality factor considering only the losses or the gain of the atomic ensemble

This Q is negative since the atomic ensemble emits energy under the influence of the applied field. In the case where there is sufficient gain in the atomic ensemble, we may have $Q_m = -Q_L$ and the reflection coefficient is infinite. The system hence becomes an oscillator.

The power given by the atoms is calculated from Equations 1.69 and 1.96. Using the definition of the quality factor as the energy stored over the power lost, which in the present case becomes negative since it is power emitted, the gain is readily calculated as:

$$G_{rc}(\omega) = \left| \frac{1}{1 - (1/I_{eff})(1 + \Omega_{24}^2)} \right| \tag{1.104}$$

In such a case, in the absence of atoms, the gain is zero since the cavity resonator is matched and no energy is reflected from the cavity.

1.1.2.2.2   *Frequency Shifts*

In such a device the output signal frequency is always equal to the input frequency. However, the output signal shows a maximum when $\Omega_{24} = 0$ and a servo system may be used to lock the frequency of the input generator to this maximum. The frequency of this maximum is given by Equation 1.102 and is thus displaced by the wall shift, Ω_w, the spin–exchange shift, $(\gamma'_{se}/2)\Delta$, and the cavity pulling. If the gain is very low, $k \ll 1$, there is little feedback of the atoms on the field inside the cavity and the term $k \psi \Delta$ may be neglected. The effect of a cavity detuning may then be obtained from Equation 1.63. The maximum of the gain is displaced by a small quantity proportional to the square of the ratio of the cavity Q to the line Q:

$$\Delta\omega_{cp} = \left(\frac{Q_{cL}}{Q_{at}}\right)^2 \Delta\omega_c \qquad \text{(low gain)} \qquad (1.105)$$

In the case of high gain there is strong feedback of the atoms on the cavity field and the term $k\,\psi\,\Delta$ in Equation 1.98 cannot be neglected. The frequency of the maximum of amplification is displaced by the quantity given by Equation 1.90 as in the case of the active maser. The effect of the wall shift is the same as in the case of the active maser. However, since the cavity is tuned for maximum power output, the spin–exchange frequency shift is not compensated by an opposite cavity detuning pulling effect as is the case of the active maser when the cavity is tuned at a frequency which makes the maser frequency independent of beam flux.

1.1.2.2.3   Practical Implementation

The passive H maser can be implemented in practice in a very similar manner as the active H maser (Walls and Howe 1978). However, as mentioned above, it can operate with a low cavity Q and consequently the physical system can be constructed using a cavity operating in a mode different from the TE_{011}. We now examine a few approaches, in particular one approach using a so-called magnetron cavity that can be made much smaller than the standard TE_{011} type.

Since the maser is passive, there is no signal with a given phase to compare a local oscillator to and to lock it to the atomic transition. Consequently, the phase-lock-loop described earlier in the case of the active maser cannot be used. Therefore, a frequency lock-loop as in the case of the Cs beam frequency standard must be used. A simple system illustrating such a servo is shown in Figure 1.18.

The reader will find an analysis of such a servo system using either slow or fast modulation in *QPAFS*, Volume 2 (1989). It should be emphasized, as shown above, that cavity frequency pulling exists in the passive mode configuration of the H maser. Therefore, the cavity frequency must be stabilized. It is possible to reduce considerably this cavity pulling and at the same time lock the frequency of the rf

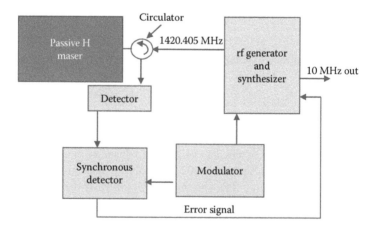

FIGURE 1.18   Block diagram of a frequency-lock-loop that can be used to implement a passive hydrogen maser frequency standard.

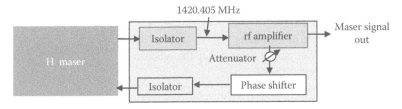

FIGURE 1.19 Basic block diagram of a system using an external feedback loop to enhance the cavity Q in order for the maser to satisfy the oscillation condition. The section containing the electronic feedback components needs to be controlled in temperature to avoid any change in the feedback loop parameters.

generator to the atomic transition by using fast frequency modulation. The technique uses the approach developed by Pound to stabilize a microwave generator on a cavity by means of frequency discrimination utilizing an intermediate frequency modulation (Pound 1946). This is usually called a Pound stabilizer. It is possible to use the technique in reverse to frequency lock a cavity to a generator which itself is locked to the atomic resonance (Busca and Brandenberger 1979; Lesage et al. 1979) using a single modulation frequency. This approach is very effective and is described in *QPAFS*, Volume 2 (1989).

The previous discussion emphasized the approach where the Q of the cavity is too low to satisfy the oscillation condition, that is, $q < 0.172$. However, it is possible to increase the cavity Q by using external positive feedback. This is done, for example, by using a cavity with two coupling loops, the first one extracting energy to the cavity and the second one feeding back this energy into the cavity after amplification with a low noise amplifier (Wang 1980). A block diagram of a typical system is shown in Figure 1.19.

In such an approach, a phase shifter is required in the feedback loop in order to compensate for any phase shift that may be created within various components in the loop and cause a frequency shift. This phase shift creates a perturbation in the resonance frequency of the cavity. It is important to temperature stabilize all components of the feedback loop, amplifier, attenuator, phase shifter, transmission lines, and isolators since they are essentially part of the cavity. Even with such stabilization, it is possible that the phase of the signal injected in the cavity varies with time because of aging of components, for example. Cavity pulling is then present and an autotuning system of the cavity may be required.

1.1.2.3 Frequency Stability of the Hydrogen Maser

In well-designed active H masers, the frequency stability for averaging times in the range 10 s $<\tau < 1000$ s is mostly limited by internal white frequency noise. The frequency stability expressed in terms of the Allan variance is given by (*QPAFS* 1989):

$$\sigma_y(\tau) \cong \left(\frac{kT_c}{2P_o}\right)^{1/2}\frac{1}{Q_1\tau^{1/2}} \qquad (1.106)$$

where:

k is Boltzmann constant
T_c is the temperature of the cavity
P_o is the maser power output delivered to the receiver
Q_1 is the maser atomic line Q
τ is the averaging time interval

In the best performing active masers, a stability of $2.2 \times 10^{-14} \tau^{-1/2} for\ 30\,s < \tau < 6000\,s$ in a 6 Hz bandwidth is obtained in agreement with the above expression (Vessot et al. 1977, 1988).

In the short-term region, say for averaging times below 10 s, the frequency stability of the active maser is limited by additive noise originating from the cavity and the first stage of mixing and detection of the maser signal:

$$\sigma_y(\tau) \cong \left(\frac{F\,kT_c\omega_R\,Q_{\text{ext}}}{2P_b\omega_o^2 Q_{\text{cL}}} \right)^{1/2} \frac{1}{\tau} \tag{1.107}$$

where:

F is the noise figure of the receiver
ω_R is its bandwidth
Q_{ext} is the cavity external Q
Q_{cL} is the cavity loaded Q
P_b is the power delivered by the H beam in the cavity

The frequency stability is generally of the order of a few $\times 10^{-13}$ for $\tau = 1$ s.

In the case of the passive maser, the short-term frequency stability is a function of the type of system used to lock the frequency of the external generator to the maximum of emission. This stability is given by an equation similar to Equation 1.106 but multiplied by a factor K larger than unity. With the *S/N* available in an actual device, the frequency stability is of the order of $1 \times 10^{-12} \tau^{-1/2}$ in the same range of averaging time intervals as in the case of the active maser.

Several laboratories have developed passive and active H masers. Some of the pioneers have been the Smythsonian Astrophysical Observatory (SAO) of Harvard University in Cambridge, MA; Vremya in Nizhny Novgorod, Russia; Sigma Tau Co., Gaithersburg, MD; and Spectratime, Switzerland.

The frequency stability of an SAO VLG 11 maser reflects the high quality of its design (Vessot et al. 1984; Vessot 2005). The stability is a few times 10^{-13} at an averaging time of 1 s and reaches 5×10^{-16} at an averaging time of 10^4 s. No cavity auto tuning system is used and a slow drift of the order of 6×10^{-15} day is reported.

A design overview of an active H maser developed at Spectratime, Switzerland, originally for the ACES (Atomic Clocks Ensemble in Space) mission is shown in Figure 1.20. In that design, the medium-term stability relies on an automatic cavity tuning (ACT) based on a carrier free fm sideband interrogation system described in *QPAFS*, Volume 2 (*QPAFS* 1989). Its frequency stability is given in Table 1.4. Its frequency drift is less than 2×10^{-16}/day.

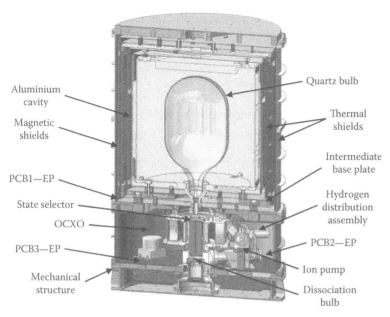

Aluminium cavity

Magnetic shields

PCB1—EP

State selector

OCXO

PCB3—EP

Mechanical structure

Quartz bulb

Thermal shields

Intermediate base plate

Hydrogen distribution assembly

PCB2—EP

Ion pump

Dissociation bulb

FIGURE 1.20   3D view of the Spectratime space active hydrogen maser design. (Courtesy of Spectratime, Switzerland.)

TABLE 1.4

Frequency Stability of a Space H Maser Developed at Spectratime, Switzerland

Averaging Time	Allan Variance
1 s	9.8×10^{-14}
10 s	2.8×10^{-14}
100 s	7.0×10^{-15}
1,000 s	2.2×10^{-15}
10,000 s	1.1×10^{-15}

Source: Goujon, D. et al., Development of the space active hydrogen maser for the ACES mission. In *Proceedings of the European Forum on Time and Frequency* 17-02, 2010; Goujon, D., *Personal Communication*, 2014.

The kind of stability reported above is achieved only by extremely careful design of the cavity. In the case of the SAO maser, the cavity is made of low thermal expansion material called Cervit and maintained in place under as little stress as possible by means of so-called Bellville springs, as well as under extremely stable temperature control. The Vremya maser cavity (Demidov et al. 2012) is made of *Sitall* (a crystalline glass-ceramic with ultra low coefficient of thermal expansion of $0 \pm 1.5 \times 10^{-7}/°C$ in the temperature range $-60°C$ to $60°C$). In that case, the origin of the observed drift could be relaxation of the bulk material of the cavity, which is not stabilized by an automatic cavity tuning. In fact, with a line Q of 1.4×10^9 (1 Hz linewidth) and a cavity Q of 40,000, the cavity must stay tuned to within 0.05Hz for a maser frequency stability of 10^{-15}. With an axial tuning rate of 10^7 Hz/cm, the length of the cavity must stay stable within about 5×10^{-9} cm, which is a dimension encountered at the atomic level.

It is thus not surprising to observe drifts in an H maser whose cavity is made of assembled individual parts in mechanical contacts with each other and relaxing with time under local efforts. It is extremely difficult to construct a cylinder and end plates that fit together exactly. This can be done only using optical polishing techniques for contact between cavity parts. Even then, it is possible that those contacts, because of the presence of asperities, relax with time. On the other hand, the origin of the long-term drift as a time dependent wall shift should not be discarded.

The cavity slow drift with time can only be compensated by retuning the cavity. As was mentioned above several times, there have been several proposals and attempts to construct systems that can automatically tune the cavity to the hydrogen emitting frequency without disturbing the maser signal. They can be classified in two groups. One consists in modulating the line Q of the maser emission line and in adjusting the cavity frequency to a value that makes the maser output frequency independent of the line Q. One method consists in changing the H beam intensity to alter the pressure in the storage bulb. This procedure alters the line Q through spin–exchange interactions. It is called the spin–exchange tuning technique. For best results, an external reference as stable as the maser to be tuned should be used. The method can also be used by means of broadening the resonance line through the application of a magnetic field gradient (Vanier and Vessot 1966; Vanier 1969). The maser is operated at high beam flux to allow the broadening to take place while the maser is kept in oscillation. These techniques are delicate in the sense that they act directly on the maser signal. A method that avoids that problem consists in acting directly on the cavity. The cavity can be tuned by means of an external interrogating radiation applied to the cavity itself. The interrogating signal is modulated in frequency and the reflection of the cavity is analyzed. In order not to perturb the maser signal, the interrogation signal is modulated in frequency, square wave, by say half the width of the cavity resonance (~15 kHz) with the carrier suppressed. However, a more direct approach consists in modulating the cavity resonance itself at high frequency (Gaigerov and Elkin 1968; Peters 1984; *QPAFS* 1989). For a tuned cavity the atomic resonance line is not affected, but for a detuned cavity the maser signal is modulated in amplitude. The detuning is thus detected directly on the maser output signal. The modulation of the maser signal is detected and analyzed through synchronous detection to create an error signal that is fed back to the cavity to keep it tuned. This type of automatic cavity tuning system was used in an actual maser, and an outstanding frequency stability of the order of 3×10^{-16} at an averaging time $>10^5$ s was reported. No drift larger than 10^{-16}/day was observed (Demidov et al. 2012).

It should be mentioned that the wall shift may vary with time (Morris 1990). The clear identification of such an effect, which could be at the level of parts in 10^{16}/day or smaller, requires rather long-term measurements. It appears that its existence may depend very much on the actual technique of depositing the Teflon™

on the inner surface of the storage bulb, the actual source of the Teflon™ and its quality. No conclusion can be drawn at this time on its exact value and its presence is rather difficult to evaluate in a particular maser due to the possible presence of other effects of the same order of magnitude. It should be mentioned that degradation with time of the resonance line quality factor Q_l has been observed in most of the masers constructed (Bernier and Busca 1990; Morris 1990). The effect in some cases is such as to prevent the maser from oscillating after a given time of functioning. In such a case the storage bulb needs to be recoated. However, no direct relation has been found between this degradation of line Q and long-term drift. At this time it does not appear that systematic work is done in order to elucidate those two parasitic effects, masers being used in situations for which the range of frequency averaging times required falls where their stability is optimum. If need is required they can be recalibrated against a Cs fountain.

1.1.2.4   State of the Art of Recent Developments and Realizations

We have just described some interesting results on classical standard H masers that were refined and improved from earlier models. Our review was limited to the refinements of the classical maser approach. However, parallel research and developments have taken place on the possibility of making drastic changes in construction to realize an instrument with greater frequency stability by orders of magnitude or, on the other hand, reduce its size considerably. We will outline now some of the most important research and development that have taken place recently in that direction. Often, due to the great complexity involved in the systems developed, they remained at the level of basic laboratory research instruments and did not lead to application directly usable in the field. They nevertheless provided some interesting insight into possible directions for improvements.

1.1.2.4.1   Improvement of Accuracy and Frequency Stability

Concerning improvement of the short-term stability of the active maser, a quick look at Equation 1.106 shows us that the choice we can make is rather limited. We can act on the noise figure, the power output of the maser and the temperature. Generally an amplifier with a low noise figure, less than 2 dB, is used as the first stage of the receiver. Consequently, not much can be done about that parameter. Similarly, maser power output cannot be increased freely. Figures 1.13 and 1.14 show that the power output cannot be increased arbitrarily by increasing atom flux because spin–exchange interactions cause relaxation and the maser power output goes through a maximum and then decreases at larger atomic fluxes. The range of oscillation is limited. If more power is desired the bulb time constant could be decreased allowing an increase in atom flux. However, in that case the line width of the transition is increased affecting the line quality factor. This has a direct effect on the frequency stability in the range where noise within the line width is important as shown by Equation 1.106. It appears that the best way to improve short-term frequency stability is by decreasing the temperature T_c of the maser region of emission. An improvement in frequency stability of several orders of magnitude can thus be predicted at very low temperatures. However, this direction is a function of the possibility of finding a storage bulb surface material that still works at low temperature. Several studies have been made with the goal of finding such a surface coating that would still allow long storage times at cryogenic temperatures. Coatings such as CH_4 at 30 K (Vessot et al. 1979), Ne at 9 K (Crampton et al. 1984), and He at temperatures below 0.5 K (Hardy and Morrow 1981) have been found to provide a possible solution. The most attractive approach appeared to be that last proposal using a superfluid helium film. However, although it is most interesting in connection to the basic physics involved, the complexity of the supporting systems required, such as dilution refrigeration, has limited the development to laboratory projects. A system developed in a joint project between Harvard University Center for Astrophysics, Lyman Laboratory at Harvard University, and the Massachusetts Institute of Technology is shown in Figure 1.21.

In view of its rather interesting exotic construction, we will discuss that system. One main objective of the project was to operate the maser in the active mode. That was achieved and a signal output was obtained at the level of 5×10^{-13} W. The main physical characteristics of that maser are as follows. The cavity, operating in the TE_{011} mode, is made of a large pure single crystal sapphire cylinder whose hollow central part serves as storage "bulb." End Teflon™ septums are used for limiting the motion of atoms to a region where the microwave field has the proper orientation. The sapphire cylinder, 17 cm long and 10 cm in diameter, was machined out of a single crystal and was silvered on its outside. The dielectric losses of sapphire are considerably reduced at low temperature and at 0.5 K the Q of that cavity, operating in the TE_{011} mode, was 27,000 with a coupling coefficient of 0.33. Assuming a filling factor of the order of 0.4, this amply satisfies the oscillation condition of Equation 1.85. Atomic hydrogen was produced in a room temperature discharge as in the standard classical maser, but after dissociation the hydrogen gas was allowed to drift to a collimator inside a piece of Teflon™ tubing cooled progressively to about 10 K, forming a low velocity H beam. The beam was directed to a short hexapole magnet state selector and atoms in the states $F = 1$, $m_F = 0, 1$ at a temperature of 10 K were then directed to the combined storage bulb-sapphire cavity. It is assumed that due to the limited number of collisions in the Teflon™ tube little relaxation occurs in their travel and the state of the atoms is not much altered at the entrance of the cavity acting as the storage container. The arrangement, rather complex because of the extremely low temperature involved, was cooled to 0.5 K by means of dilution refrigeration within a vacuum enclosure cooled itself at 77 K. Helium was admitted in the system at a low rate and, being in its superfluid state at 0.5 K, coated all internal surfaces including the inner surface of the hollow sapphire cavity.

This idea of using a surface coating made of a helium film originated from experiments done at the University of Amsterdam which showed the possibility of maintaining a moderate density of atomic hydrogen in a space whose containing surface was coated with superfluid helium (Silvera and Walraven 1980). Other groups (Cline et al. 1980; Morrow et al. 1981; Wallsworth et al. 1986) applied successfully the approach to the storage of hydrogen atoms for the purpose of implementing an H maser. The maser shown in Figure 1.21 is one of those approaches. A similar technique, but using a more compact cavity, was developed by Hardy and Whitehead at University of British Columbia, Canada (Hardy and Whitehead 1981). It is described in *QPAFS*, Volume 2 (1989).

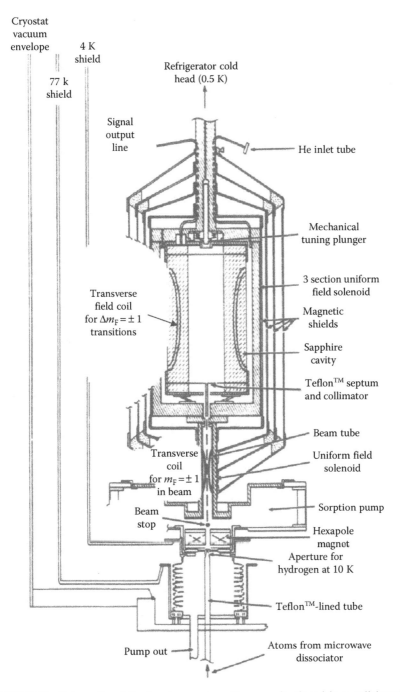

FIGURE 1.21 Schematic of the low temperature H maser developed in a collaboration project between CfA, Harvard University and MIT in the late 1980s. (Data from Vessot, R.F.C., *Metrologia*, 42, S80, 2005. Copyright Bureau International des Poids et Mesures. Reproduced by permission of IOP Publishing. All rights reserved.)

One other advantage of using such a low temperature is the decrease of the cross section in spin–exchange collisions. As is reported in *QPAFS* (1989), the cross section σ_{ex} at a temperature of 0.5 K is considerably reduced to a few $\times 10^{-16}$ cm^2 compared to 2.8×10^{-15} cm^2 at room temperature. Since the oscillation parameter q is a function of the product $< v_r > \sigma_{ex}$, it can readily be seen that the oscillation parameter q is reduced by a factor of the order of 10^3 at such a low temperature making maser oscillation much easier. This conclusion is reinforced by the fact that the atoms travelling at low speed make less frequent collisions with the walls of the helium coated storage sapphire cavity, reducing furthermore wall relaxation with a direct effect on the decrease of q.

In a complex system as the one described, it should be realized that phenomena not observed in the classical room temperature maser could be present. A most important one is the presence in the system of the injected helium as a gas in equilibrium with the superfluid film coating all surfaces. One first effect of the presence of that gas at the pressure required to provide a superfluid film is collisions that reduce the mean free path of the H atoms to about 5 cm. Those collisions may not only affect the H beam free motion but also produce relaxation and a frequency shift. However, the ionization energy of H is 13.6 eV and that of He is 24.6 eV. Because of this high ionization energy, their interaction is small, as discussed in *QPAFS*, Volume 1 (1989). Nevertheless, the interaction is still present even at the low temperature involved. A buffer gas frequency shift is thus present and adds to the shift caused by wall collisions. The predicted mean free path and frequency shift of H in He including wall collisions effect is shown in Figure 1.22 as a function of temperature. One observes that there is a minimum in the predicted frequency shift at a temperature of about 0.5 K. However, this prediction relies on the quality of the superfluid He film coating the internal surface of the sapphire cavity. If this film is not saturated, that is to say thin, the H atoms upon collisions are influenced by the potential of the more polarizable substrate seen by the atoms directly through the He film. In practice it was found that the shift measured at a temperature of 0.493 K was of the order of several Hz and decreased with the amount of He injected into the system (Vessot et al. 1986). This appears to confirm that the film increases in thickness with increasing He pressure. However, that shift is of the order of a few Hz, an order of magnitude larger than that expected. Furthermore, the line Q was found to be of the order of 10^9 much smaller than the predicted value of about 3×10^{10} (Vessot et al. 1986). Consequently, more experimental data is required before a final conclusion can be reached on the possibility of that type of cryogenic H maser in fulfilling the goals set forth in its construction.

From the development just described, the cryogenic approach to the construction of an H maser could in principle lead to greater frequency stability although accuracy would be influenced by the ability to evaluate the combined wall-buffer gas shift mentioned above. As of 2015, no further results than those reported above have been made public. It appears that the complexity of the system is such that it requires much more work. In view of the success reached by means of other approaches, using Cs and Rb with laser cooling that we will examine in Chapter 2, it is not clear at this stage that further development of a practical standard based on that cryogenic approach is still of interest to the community although the physics involved is most interesting.

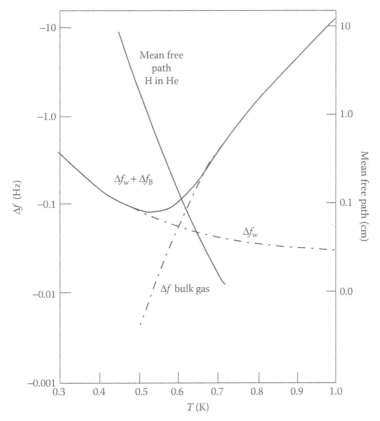

FIGURE 1.22 Predicted frequency shift and mean free path in the cryogenic H maser as a function of temperature. (Data from Vessot, R.F.C., *Metrologia*, 42, S80, 2005.)

In the previous paragraphs, we have limited the analysis to the question of frequency stability. How about accuracy? The main frequency shift that needs to be addressed is the wall shift. In the standard room temperature maser it is of the order of parts in 10^{11}. It can only be determined independently from a primary standard by comparison between two similar masers but equipped with storage bulbs having a different diameter (Vanier and Vessot 1964, 1970; Morris 1971, 1990; Vanier et al. 1975; Vanier and Larouche 1978). The reproducibility of the coating film of Teflon™ is not expected to be much better than a few percent thus a few parts in 10^{13}. Consequently, it does not appear that the maser may become a primary standard competing even with the room temperature Cs beam frequency standard. Furthermore, it has been found that the Teflon™ surface may change with time or age, creating a time dependent wall shift (Bernier and Busca 1990; Morris 1990). This effect has not yet been determined accurately. In the cryogenic maser, using a superfluid He film as coating of the storage container, the "wall shift" adding to the inherent buffer gas shift is of the order of parts in 10^9. Such a shift is 10^6 times larger than the accuracy reached in the classical Cs beam approach and it does not appear that it can be determined with better accuracy. In the Cs fountain approach, it is now possible to reach accuracy of the order of a few parts in 10^{16}. Consequently, at this stage, due to the uncertainty in the wall shift effect, it appears that the H maser may play a role mainly as an outstandingly stable source of radiation over a range of averaging time extending to 10^5 s. Nevertheless, it can be used as an accurate reference at the level of ~10^{-12} accuracy without calibration with an external primary Cs standard.

1.1.2.4.2 Reduction in Size

Since the first realization of the H maser as a practical instrument, the question of its size has always been a subject of interest particularly for space applications (Beard et al. 2002). As was just outlined, it is one of the most stable atomic microwave frequency standards and its use was often suggested in applications where maximum frequency stability was required. However, due to its size and weight, aside from two special rocket launches containing H masers (Vessot et al. 1980; Demidov et al. 2012), use of classical H masers equipped with a TE_{011} cavity was essentially limited to ground-level applications.

As was made evident by the previous analysis, the dimension of the maser is essentially controlled by the size of the cavity. The frequency of the ground state hyperfine transition of H is 1420 MHz corresponding to a wavelength of 21 cm. The cylindrical TE_{011} cavity resonant at that frequency has a dimension of about 27–28 cm in length and diameter. The majority of existing active masers have been designed with this type of cavity. The cavity controls the size of the maser by limiting considerably the freedom of reducing the size of the whole ensemble including vacuum tank, solenoid, magnetic shields, thermal control, and insulation. Nevertheless, relatively small size active H masers were realized by perfecting the supporting necessary equipment resulting in a weight of 260 kg (e.g., the VLG 10, 11, and 12 of the SAO shown in Figure 1.15b). One unit designed especially for space application had a weight of 90 kg and, as mentioned above, was flown on a rocket as a gravity probe (Vessot et al. 1980). Other masers with reduced size were also developed using a TE_{011} elongated cavity approach (Peters and Washburn 1984; Goujon et al. 2010; VREMYA-CH 2012). However, they remained somewhat heavy, a characteristic that still limited their use to applications at ground level. With recent development in navigation by means of satellites, such as the European Galileo system, the use of H masers as primary time standards on satellites was considered as a valuable alternative to the passive Cs and Rb standard used in other satellite navigation systems such as the USA-GPS and the Russian Glonass. A typical arbitrary requirement for such an application is a weight of the order of 15 kg and this appeared to require a drastic change in the maser design.

We have outlined in *QPAFS* (1989) how new cavity designs could be used to reduce dimensions. These designs are based on a rather limited number of possibilities, forced by the quantum condition that the direction of the rf magnetic field in the cavity must be parallel to the applied dc magnetic field and that the oscillation condition, $q < 0.172$, be satisfied. This condition can also be expressed very approximately as requiring that the product of the loaded Q of the cavity with the filling factor η' be larger than 10^4 as dictated by Equation 1.85 ($\eta'Q_c > 10,000$). Furthermore, because H atoms travel freely within the maser storage bulb or container, the rf magnetic vector must keep the same phase in space in order to avoid interferences that would reduce the filling factor and by the same fact increase the oscillation parameter q to a value such as to inhibit continuous oscillation. It appears that three promising cavity designs can be used to reduce at the same time size and weight. One is a cavity operating in the TE_{111} mode. One other is a TE_{011} cavity loaded with a dielectric material. A last one is a totally different design incorporating symmetrical metallic plates, enveloping the storage bulb within essentially a TE_{011} configuration. This approach leads to an elegant design called a magnetron cavity. We will examine these three approaches.

1.1.2.4.2.1 TE_{111} Mode Cavity The TE_{111} mode requires a cylindrical cavity of the order of 16 cm in diameter for being resonant at the H ground state hyperfine frequency. It is thus relatively small compared to the TE_{011} cavity, which has dimensions of the order of 27–28 cm. However, that mode, although very satisfactory for a situation where a buffer gas restricts the motion of the atom to a small region of space during the time of emission, as in the passive optically pumped Rb frequency standard, is characterized by a rather unfavourable distribution of rf field for its use in an H maser. The microwave magnetic field is zero on the symmetry axis of the cylinder, increases in intensity towards the cylinder walls, but has a phase shift of 180° in opposite regions of the central plane of symmetry of the cylinder. Such a configuration requires the presence of a so-called septum that divides the cavity in two symmetrical sections of uniform phase. This divides the atoms into two sub-ensembles restricting the motion of the atoms to separate sections where they are exposed to a field of the same phase thus avoiding interferences (Mattison et al. 1976). The actual configuration is shown in Figure 1.23 with a septum in place.

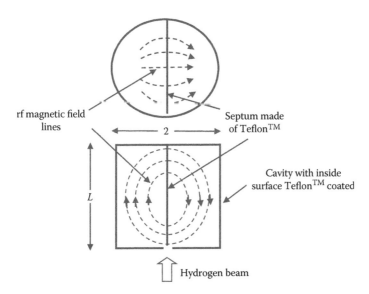

FIGURE 1.23 TE_{111} microwave cavity adapted for use in a Hydrogen maser.

The two sections on each side of the septum act as independent storage bulbs, each fed by approximately half of the incoming H beam. Such a cavity is characterized by a resonant wavelength given by (Wang et al. 2000):

$$v_{res} = c\sqrt{\frac{1}{4L^2} + \left(\frac{1}{3.41\,R}\right)^2} \qquad (1.108)$$

Such a cavity was constructed and mounted on a test maser (Mattison et al. 1976). The cavity, made of copper, had a length L of 18 cm and a diameter $D = 2R$ of 16 cm. This is about 40% smaller than the TE_{011} cavity. The loaded Q of the cavity was 26,000 with a coupling coefficient β of 0.1. The calculated filling factor was $\eta' = 0.42$, giving a $\eta'Q$ of the order of 11,000. The maser could be set into oscillation and its line quality factor was measured as $Q_l = 0.8 \times 10^9$. The storage containers escape rate was adjusted to 1 s^{-1} and a measurement of the decay rate of stimulated emission gave a total relaxation time of 0.34 s. The wall relaxation rate is thus calculated as being of the order of 2 s^{-1}. The low value of the measured line Q is due to the contribution of spin–exchange collisions, ~2.5 s^{-1}, when the maser is in its active mode with a large beam flux. Such numbers are comparable to those obtained in storage bulbs with dimensions similar to those of the two halves formed by the cavity and its septum, although the smaller dimensions of the half cylinders may be responsible for a larger collision rate and a relaxation rate greater than in a standard spherical storage bulb.

It is worth repeating that the problem in all designs and construction of a smaller cavity as that shown above is the practical fulfilment of the oscillation conditions: $\eta'Q$ is required to be larger than 10^4 leading approximately to the requirement of having a loaded cavity Q of the order of 20,000 or more. In practice, this condition, for various reasons of material characteristics and quality of construction, is not always attained. In that case, the maser has to be operated in a passive mode or with cavity Q enhancement introduced externally as described earlier. This last approach has been used at the Shanghai observatory in China in the construction of an H maser with a TE_{111} cavity similar to the one just described (Wang et al. 2000).

1.1.2.4.2.2 Dielectrically Loaded Cavity   Another approach consists in introducing a dielectric material inside the cavity to reduce its size. There are two ways of using that approach. The first one consists in introducing a relatively thin cylinder of dielectric material inside a standard TE_{011} cavity (Peters et al. 1987; Gaygorov et al. 1991; Busca et al. 1993). The second one consists simply in constructing the cavity with a cylinder of dielectric material, silver painted on its outside as in the case of the cryogenic maser described above. In both cases, the evacuated interior surface of the dielectric cylinders is used directly as the storage container or bulb. Substantial size reduction is realized, but a compromise must be made between various requirements connected to the fulfilling of the oscillation condition if it is desired to construct an active maser. For example, the size of the storage container, a cylinder, must be sufficiently large in order to make the filling factor large enough to allow oscillation. This is also conditional to the realization of a high cavity Q, which depends on the dielectric losses of the dielectric material and the quantity of the material

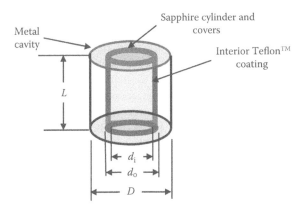

FIGURE 1.24 Schematic diagram of an H maser cavity loaded with a cylinder of dielectric material. In the present case sapphire is used as dielectric material.

exposed to the microwave electric field in the cavity. It is thus important to choose a proper dielectric material with a high dielectric constant but with little microwave losses. We will first examine the case of the cavity loaded with a thin cylinder of dielectric material. The general design of such an approach is shown in Figure 1.24.

We first examine the case where sapphire is chosen as dielectric material. It is assumed that the cavity operates in the TE_{011} mode. Sapphire has a room temperature dielectric constant of about 9.4 and its losses are characterized by a loss tangent (tan δ) equal to ~1.5×10^{-5}, a rather small value. The first question is to decide on its size and where to situate such a cylinder within the cavity. However, since the size of the bulb is an important parameter, its diameter and length must be as large as possible to make the filing factor large of the order of 0.4 or more. A procedure described by Bernier (1994) consists first in setting the diameter of the cavity desired with a fixed arbitrarily chosen length of 160 mm. This fixes at the same time the length of the sapphire cylinder and of the hydrogen container (bulb). One then chooses the inner diameter of the sapphire cylinder. This determines the volume of the hydrogen container. The outside diameter of the sapphire cylinder is then adjusted to make the whole ensemble resonant at the hyperfine frequency of H, 1.420 GHz. The thickness of the cylinder sets at the same time the dielectric losses. The cavity Q is then calculated. A computer program (Sphicopoulos et al. 1984) is used to make the calculation and, after several attempts with various starting dimensions of cavity and sapphire cylinder inner diameters, a choice can be made that satisfies dimension requirements for the application desired. At the same time, the calculation verifies oscillation conditions relative to the filling factor and the cavity Q computed. A typical result obtained is (Bernier 1994):

- $D = 20$ cm; $L = 14$ cm; $d_i = 12.5$ cm; $d_o = 13.7$ cm
- e (thickness of sapphire cylinder) $= 6.0$ mm
- Hydrogen storage container volume (bulb) $V_B = 1.7$ litre
- Cavity volume $V_c = 4.40$ litres
- Cavity $Q = 47,000$

This is indeed an excellent choice of dimensions making the volume of the cavity much smaller than a full size TE_{011} cavity while the hydrogen storage volume is very similar to that of a standard storage bulb 15 cm in diameter. The volume of the cavity is about five times smaller than a TE_{011} cavity without dielectric loading for the same resonant frequency. Furthermore, the oscillation conditions are fully satisfied through the high resulting cavity Q and large storage volume.

A system using that approach was constructed and tested at the Observatoire de Neuchâtel in Switzerland (Jornod et al. 2003). The maser worked according to expectation. The loaded cavity Q_L was 35,000 while the line quality factor Q_l was 1.5×10^9. This gives a cavity frequency pulling factor of 2.3×10^{-5}. This means that for obtaining a fractional frequency stability of 10^{-15} the cavity must stay tuned at 1.42 GHz to a precision better than 0.1 Hz (Jornod et al. 2003). As mentioned earlier, when transformed into equivalent dimensional conditions this corresponds to a precision of the order of atomic dimensions. With a complex construction as that described this is extremely difficult to achieve. Furthermore, it is found that the dielectric constant of the sapphire material is a function of temperature establishing for the cavity a temperature coefficient of about 40 KHz/K. For the conditions just described, these characteristics exert a stabilization burden on the temperature control of the ensemble that is impossible to accomplish in practice. Consequently an ACT, similar to those already mentioned, is required to compensate for the limited capacity of practical temperature controls. In the present case, the fast cavity auto tuning concept has been used (Audoin 1981, 1982; see *QPAFS* 1989, Volume 2; Weber et al. 2007). The measured frequency stability was of the order of 2×10^{-13} at 1 s averaging time and 7×10^{-15} at 100 s averaging time. With ACT measurements showed that at 1000 s averaging time the maser frequency stability was improved by an order of magnitude (2.3×10^{-15}) as compared to the free running operation (2×10^{-14}).

The sapphire loaded maser was constructed with the objective of being part of a proposed European Space Agency project in the early development of the project "Atomic Clock Ensemble in Space" (ACES). The ensemble was to include such a maser and a laser-cooled Cs atomic beam clock. We shall describe later that Cs beam clock. The H maser would provide frequency stability in the short term while the Cs clock would maintain stability in the long term. The goal is to do fundamental research in space in micro gravity in such subjects as general relativity, time and frequency metrology, and very long base line interferometry.

Similar approaches, using sapphire cylinders inside a TE_{011} cavity to construct small H masers dedicated to space applications such as navigation systems have been studied by other laboratories (Gaygorov et al. 1991; Morikawa et al. 2000; Ito et al. 2002, 2004; Hartnett et al. 2004). In the last case, the cavity had a diameter equal to its length, 160 mm, and the sapphire cylinder had an inner diameter of 72 mm and a thickness of 7.8 mm. A calculation showed that $\eta'Q$ was about 32,000 for that configuration. That maser operated also in the active mode with a frequency stability reaching a few in 10^{15} at $\tau = 5 \times 10^3$ s. Calculations were also performed in order to optimize the parameters. It was found for example that with a somewhat smaller and thicker sapphire cylinder $\eta'Q$ could be as high as 49,000, fulfilling easily the oscillation condition. However, that makes the size of the H container of the order 0.2 l, which is rather small and may be a handicap regarding wall collision rate affecting relaxation and line quality factor.

Quartz can also be used as inner loading of a TE_{011} cavity. Unfortunately, dielectric losses are larger than those of sapphire and do not allow substantial reduction without reducing the cavity Q to such a value that maser self-oscillation is no longer possible. In one example (Peters et al. 1987), maser operation could be realized with a storage bulb inside a rather thin quartz cylinder that reduced the cavity size to 218 mm in diameter and 303 mm in length. The cavity Q was 38,000 and maser self oscillation could be realized.

The other approach consists in building the cavity of a block of dielectric material as in the case of the cryogenic maser described above. An empty cylinder is cut of a piece of dielectric material. The inside of the cylinder is used as the storage container of the H ensemble after proper coating with Teflon™. The outside of the cylinder is coated with silver paint. This is shown in Figure 1.25. In that case, a rather heavy dielectric loading results with substantial reduction in size since the dielectric material overlaps with the region where the electric field is a maximum. However, this is accompanied by a substantial increase in losses and a reduction in cavity Q even in the case where low loss Al_2O_3 was used (Howe et al. 1979; Walls 1987; Yahyabey et al. 1989). The Q of the cavity of a unit built of that material was around 1.7×10^3 (Mattison et al. 1979). The losses are too large to allow fulfilment of the oscillation condition. In the case of the cryogenic maser the sapphire cavity dimension was 17 cm long by 10 cm in diameter. At low temperature, dielectric losses of single crystal sapphire are reduced considerably and an unloaded Q as high as 3.4×10^4 was achieved allowing maser oscillation as mentioned earlier (Vessot et al. 1986).

It should be realized that, in general, as the volume of the cavity is reduced, the volume of the H storage space is also necessarily reduced in size. There may not be a loss in the quality factor q of the maser since it is proportional to the ratio V_c/V_b but the number of atoms forming the ensemble is considerably reduced which may

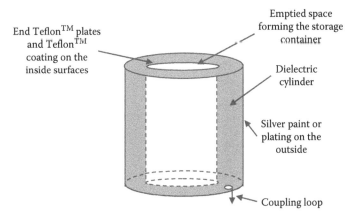

FIGURE 1.25 TE_{011} cavity made of a cylinder of dielectric material coated externally with a high electrical conductivity material such as silver.

raise the threshold flux. Another important parameter is the wall relaxation rate γ_w. This relaxation rate is proportional to the collision rate of the atoms with the surface, which is larger for a smaller storage container. This has a direct effect on the threshold flux I_{th}. Consequently, oscillation may require larger fluxes of atoms than in the case of larger storage bulbs. Furthermore, the escape rate through the entrance hole or tube to the storage space may have to be increased; otherwise the value of the quality factor q may be too large to allow oscillation because of the increase in γ_w.

1.1.2.4.2.3 Magnetron Mode Design The design of a magnetron cavity is shown in Figure 1.26. It consists of a cylindrical cavity with internal metallic plates placed in such a way as to create a magnetic field configuration very similar to that inside a TE_{011} cavity.

There has been a relatively long history of development behind such a concept. It started from a design in which internal metallic electrodes were glued directly on the storage bulb and evolved to the structure shown in the figure. In very loose terms, we may say that the system operates in the following way (Peters 1978a, 1978b). The plates inside the container surrounding the storage bulb act as a one turn coil of inductance L in series with capacitances C representing the gaps between the plates.

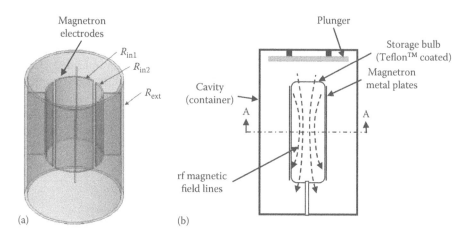

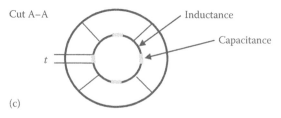

FIGURE 1.26 Magnetron cavity used in the development of small hydrogen masers. (a) Perspective view without storage bulb and plunger. (b) Horizontal view showing the position of the storage bulb and plunger. (c) Cut A–A showing vertical cut view. The possibility of placing a dielectric between the gaps of the magnetron plates is shown.

The resonance frequency of such a circuit is then grossly $(LC)^{-1/2}$ (Appendix 1.E). The rf magnetic field within the structure may be represented by a superposition of En_{11} modes. Rough calculations show that the filling factor of such a setup is expected to be rather large (>0.5) and its Q to be in the 10,000 range. Its dimensions can be adjusted to the requirements of the application, but a minimum dimension of the storage bulb must be preserved in order to maintain a wall relaxation rate that allows a line Q of the order of 10^9 or more; otherwise the maser itself with its sharp resonance hyperfine line, for which it was conceived, loses its interest.

An exact analytical calculation of the important parameters involved in the construction of a maser for such a complex structure is not possible. However, approximate solutions, leading to computer evaluation, are possible (Sphicopoulos et al. 1984; Sphicopoulos 1986; Belyaev and Savin 1987; Wang 1989; COMSOL Multiphysics Program [http://www.comsol.com/products]). A typical design could result in:

$$\nu_o = 1420 \text{ MHz} \qquad Q_{uc} = 9000 \qquad \eta' = 0.5$$

The temperature coefficient of such a structure may be of the order of 14 kHz/K (Wang et al. 1979). Such a maser does not oscillate by itself since the Q of the setup is only of the order of 9000 which makes $\eta'Q$ smaller than the required 10,000. However, it can be operated in the passive mode and several laboratories have adopted that approach. In particular, it is worth mentioning the work of the Institute of Electronic Measurements, KVARZ (Demidov et al. 1999) that has resulted in a passive maser operating on the basis of the magnetron cavity. The cavity is 200 mm long and has a diameter of 120 mm. Cavity Q is 10,000 and filling factor η' is 0.5. The maser gain is of the order of 6–8 dB. The complete system including automatic cavity tuning had a frequency stability of $8 \times 10^{-13} \times \tau^{-1/2}$ for 1 s $< \tau <$ 1000 s and showed fluctuations of the order of a few in 10^{-14} for averaging times of 1 hour and 1 day. Such an approach has also been used in the development of an H maser adapted for space environment (Wang 1989; Mattioni et al. 2002; Berthoud et al. 2003; Busca et al. 2003; Droz et al. 2006, 2009; Wang et al. 2006; Rochat et al. 2007; Belloni et al. 2010, 2011) or simply for reducing its size (Lin et al. 2001). Passive H masers of the type just described are presently used in the Galileo navigation system. Ground tests have been reported on those passive masers. Stability better than 10^{-12} at 1 s was observed with an averaging time dependence of $\tau^{-1/2}$ up to 10^5 s, a flicker floor of 4×10^{-15}, and a drift less than 10^{-15}/day (Wang et al. 2013).

1.1.3 OPTICALLY PUMPED RB FREQUENCY STANDARDS

1.1.3.1 General Description

The Rb standard is the third so-called "classical" atomic frequency standard that has known extensive laboratory research and industrial development. Its interest lies mainly in its small size, coupled with good medium term frequency stability. The standard operates between the field independent energy levels $F = 2$, $m_F = 0$ and $F = 1$, $m_F = 0$ of the ground state of ^{87}Rb and the frequency of the transition is 6.834 GHz making possible the use of a relatively small cavity to provide the exciting microwave field. This approach is generally called the double resonance technique. Optical pumping is used in order to obtain an imbalance in the population of these two levels within the hyperfine ground state, rather than magnetic state selection as in the previous two cases described above. This is done in a closed cell, containing the ^{87}Rb vapour and a chemically inert gas, placed inside the cavity resonating at 6.8 GHz. This gas acts as a buffer that reduces considerably the motion of the atoms through collisions, without affecting much the internal state of the atoms or in other words without causing much relaxation. The reduction of motion prevents the atoms from colliding with the walls of the containing cell during interaction with the microwave field, collisions that would cause relaxation. In some cavity configuration, it also prevents the atom from experiencing a change in the exciting rf magnetic field inside the cavity. This is in effect an extension of the method suggested by Dicke for reducing Doppler effect (Dicke 1953). The system is illustrated in Figure 1.27.

In one approach, optical pumping is done by means of radiation emitted by a small ^{87}Rb lamp filtered by a cell containing ^{85}Rb as illustrated in the figure. The spectral lines of that isotope are displaced from those of ^{87}Rb by a fortuitous desirable amount. Absorption in that filter cell is such that the line corresponding to the transition from $S_{1/2}$, $F = 2$ to the P state is absorbed. The filter thus provides at its output mainly the resonance radiation corresponding to the transition from the state $S_{1/2}$, $F = 1$ to the P state. The output spectrum of the filter is thus favourable to create in the ^{87}Rb resonance cell in the cavity an inversion of population through optical pumping. In another type of implementation, the filter action is integrated directly in the resonance cell through the use of a proper mixture of isotopes, both in the lamp and in the cell. We will examine later the case when a solid-state laser is used rather

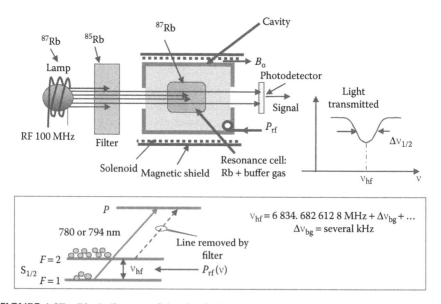

FIGURE 1.27 Block diagram of the classical optically pumped Rb standard optical package. The inset shows the type of transition involved in the optical pumping process. The separated hyperfine filter approach is that illustrated.

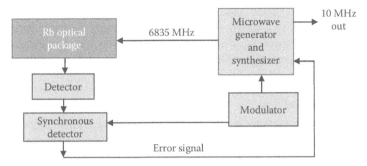

FIGURE 1.28 Block diagram of the classical optically pumped Rb standard making explicit the frequency lock-loop approach generally used to implement a frequency standard. The whole system can be made digital with automatic frequency locking.

than a spectral lamp. Since the atoms are pumped into the other hyperfine level by the optical radiation, the resonance cell becomes transparent. If a microwave signal at the proper frequency is applied to the microwave cavity, transitions are excited among the ground state levels of the ensemble of atoms and the cell returns to a state of absorption of the incident optical radiation. A photodetector is used to detect the level of transmitted radiation and the resulting signal is used, through proper frequency modulation and synchronous detection, to lock in frequency the crystal oscillator that is used to generate the microwave radiation fed to the cavity. This is a frequency-lock-loop, as in the case of the Cs standard. It is shown in Figure 1.28

1.1.3.2 State-of-the-Art Development

The theory of operation of this classical passive atomic frequency standard has received much attention in the past and was given in detail in (*QPAFS* 1989). It has been one of the first atomic frequency standards, with the Cs beam standard, to be studied in the laboratory and developed in industry. Its basic theory of operation is well-known. The interest on that standard over half a century has been maintained by the fact that it is a small device when compared to H maser and Cs beam standards described above and by the fact that it provides a source of radiation that is rather stable in frequency relative to its size. Its frequency stability is orders of magnitude better than that of best quartz oscillators of the same volume, that is, of the order of a litre or so. Although its frequency needs to be calibrated against a primary standard, such characteristics as size, weight, and good frequency stability have made it the preferred standard in many applications requiring moderate timing precision.

However, there is essentially no limit in the demand for size reduction and frequency stability improvement in applications such as those developed today, extending from basic research to communication and navigation systems. Consequently, in view of the simplicity of the device, interest has remained very high in attempting to still reduce its size, volume, and weight, and improve its frequency stability. Furthermore, due to the interesting physics involved in its functioning and the elegance of the system, it has remained in a sense one of the cherished laboratory devices to address the validity of various ideas and theories derived in the atomic physics field. Also, the fact that the optical pumping spectral lamp could now be replaced by newly developed solid-state lasers with narrow spectral line widths has raised the interest to a totally new level. We will study that new approach in Chapter 2. For the moment, we will examine what can be done for addressing the question of size and frequency stability and what has been proposed in the standard approach using a spectral lamp for optical pumping.

1.1.3.2.1 Question of Size

The size of the unit is essentially dictated by the optical package that includes, as shown in Figure 1.27, the lamp, filter cell, cavity and resonance cell, photodetector, solenoid, and magnetic shields. However, as is readily seen in the figure, the cavity controls the size of that package. In early development, a TE_{011} cavity was used. That cavity is relatively large and the smaller resonance cell is usually attached inside the cavity at one end. It is fabricated of glass with a volume much smaller than the cavity itself, and the atoms are submitted to a microwave field of nearly constant phase. Unfortunately, due its diameter of the order of 5 cm and a similar length, much space is wasted.

However, in that type of atomic frequency standard, due to the presence of the buffer gas that restricts the motion of the atoms to a small region during microwave interaction, the requirement of constant phase throughout the cell is not essential. This is a fortunate situation and for this reason, another type of cavity working in a mode, such as TE_{111}, described earlier in the case of the H maser, can be used directly without septum. This approach allows a noticeable reduction of the size of the optical package and has been used at large in many implementations. A relatively recent study of that type of cavity has resulted in a cavity design 24 cc in volume (diameter of ~24 mm) and an overall optical package having a total volume of 160 cc. The frequency stability of $1.3 \times 10^{-11} \tau^{-1/2}$ for 10^{-1} s $< \tau < 10^4$ s, obtained with a frequency standard based on the use of such a cavity appears quite attractive (Koyama et al. 1995).

However, the size of the system can still be reduced by means of introduction of a dielectric material such as Al_2O_3 inside the cavity. As mentioned in the case of the H maser, such a material has a dielectric constant of about 10 and allows a substantial reduction of dimensions. It should also be mentioned that since this type of frequency standard is passive, the cavity Q does not play as important a role as in the case of masers. Operation with a Q of a few hundred is entirely satisfactory since the loss in Q can be easily compensated by means of an increase in exciting microwave power fed to the cavity. In a particular case, a cavity volume of 9.7 cc was achieved resulting in an optical package of 25 cc (Koyama et al. 2000). In operation, the atomic resonance signal has a line width of 550 Hz (line $Q \sim 1.2 \times 10^7$) and is characterized by an S/N of the order of 58 dB. The frequency stability of a stand-alone complete unit was $7 \times 10^{-11} \tau^{-1/2}$. In the long term, a residual frequency drift of the order of $3 \times 10^{-11}/$ month was observed. It is thus concluded that a drastic reduction in size could be achieved and the resulting frequency stability remained at a level acceptable for some specific applications.

A reduction in size can also be accomplished by using, as in the case of the H maser, a magnetron type cavity. In the present application, however, since the characteristics of the cavity relative to quality factor q is not critical, a slightly altered design can be used with added loading with a dielectric material (Huang et al. 2001). Such a design is shown schematically in Figure 1.29. In that particular case, the

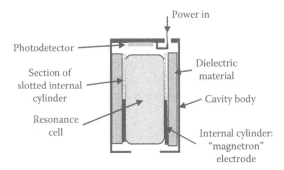

FIGURE 1.29 Magnetron type cavity used in the implementation of a passive optically pumped Rb frequency standard. The internal electrode is slotted as in the magnetron cavity approach in the H maser, but on the upper part only. (Data from Xia et al. 2006; see also Bandi, T. et al., Laser-pumped high-performance compact gas-cell Rb standard with <3 × 10⁻¹³ τ⁻¹/² stability. In *Proceedings of the European Forum on Time and Frequency* 494, 2012.)

volume of such a cavity was 9.5 cc with a Q of about 400. The results of a preliminary test of short-term frequency stability were compatible with the other approaches ($\sigma(\tau) = 3 \times 10^{-11}\, \tau^{-1/2}$) (Hu et al. 2007).

This magnetron cavity approach has received a fair amount of attention recently (Stefanucci et al. 2012). The structure studied is similar to that shown in Figure 1.26 with six electrodes (Appendix 1.E). Aside from the gain in size, another advantage resides in the homogeneity of the microwave field in the cavity which is very much like the field of a TE_{011} cavity in its central part. However, in a magnetron cavity, the field is almost constant over the whole cell that occupies the volume between the internal electrodes. In the structure proposed, the radius was 36 mm diameter, allowing the use of a cell with a 25 mm diameter. The unloaded Q of the cavity was 488. The volume of the whole resonator was less than 0.045 dm³. As pointed out by the authors, this has to be compared to the TE_{011} cavity used in the classical approach which has a volume of 0.14 dm³. A frequency standard was implemented using that cavity and laser optical pumping providing a very high S/N ratio as will be shown later (Stefanucci et al. 2012). The measured frequency stability of that ensemble was

$$\sigma(\tau) = 2.4 \times 10^{-13}\, \tau^{-1/2}$$

for $1\ \text{s} < \tau < \sim 100\ \text{s}$ in agreement with the stability estimated from the S/N of the observed double resonance signal. The approach was perfected through the use of printed circuit technology. The resonator was made of planar conductive electrodes structures printed on dielectric structures and stacked on an axial direction. This structure was made in order to accommodate a micro-fabricated cell produced also by means of so-called micro-technology (Violetti et al. 2012, 2014).

1.1.3.2.2 Short-Term Frequency Stability

In order to examine recent progress in frequency stability of the optically pumped Rb standard, we must differentiate between short, medium, and long averaging times because of the presence of various distinct phenomena that take place and that influence the atomic ensemble in very different ways. These regions of averaging times are generally differentiated by the type of noise or fluctuations that are predominant although this differentiation is not sharp and the regions tend to overlap.

The short-term region is usually affected by shot noise. It is controlled by the S/N and the atomic resonance line width. In the case of sine wave frequency modulation, the spectral density of fractional frequency fluctuations is given by ($QPAFS$ 1989):

$$S_{y,t}(f) = \frac{3^3 \pi^2}{2^8} \frac{e_M^2}{(S/N)Q_l^2} \tag{1.109}$$

where:

e_M is the amplitude of the error signal
Q_l is the resonance line quality factor
S/N is the signal-to-noise ratio observed at the synchronous detector

The resulting time domain frequency stability for sine wave frequency modulation is:

$$\sigma(\tau) = \frac{0.16}{(S/N)^{1/2} Q_l \tau^{1/2}} \tag{1.110}$$

There is not much difference between various types of frequency modulation and we keep this expression as representing a good reference in order to evaluate the frequency stability that is possible in various practical implementations. With the kind of S/N ratio (55 dB) and atomic line Q (10^7) mentioned above, we expect a short-term frequency stability of the order of a few in 10^{11} as observed experimentally for averaging times of 1 s to ~10^4 s. Above that averaging time other effects than shot or thermal noise predominate. We will see later that laser optical pumping makes possible larger signal contrasts allowing a larger S/N ratio. The observed short-term frequency stability is then improved considerably as was reported above for the standard developed by Stefanucci et al. (2012).

1.1.3.2.3 Medium- and Long-Term Stability

The main problem regarding the medium-term frequency stability of the optically pumped Rb standard is the number of biases present that may change with time. Any change in those biases is bound to affect the observed frequency and is perceived as a random perturbation. It is often said in a humoristic way that the standard is somewhat like a can of worms in which everything can move and that it is surprising that it works so well. Some of those perturbations may originate from environmental fluctuations influencing internal parameters that are intrinsic to the operation of the standard and can change with time. These include for example: magnetic field bias that can change if magnetic shielding is not sufficient, buffer gas bias that is sensitive to temperature of the cell and atmospheric pressure, light shift bias that is a function of pumping lamp spectrum and intensity, cavity pulling that is temperature sensitive, power shift that is a function of inhomogeneous broadening of the resonance line, spin–exchange frequency shift that is a function of temperature, and probably other factors of unknown exact nature such as interaction and chemical reaction of Rb with the resonance cell walls and possibly the buffer gas. These biases were studied in some detail in *QPAFS*, Volume 2 (1989). We will not review them here in detail but will examine those that have been addressed recently. We must be conscious that such effects as those enumerated may extend on the long term and that it is not possible off hand to separate medium- and long-term frequency stability without studying, at least in an elementary way, the origins of those fluctuations.

1.1.3.2.3.1 Cavity Pulling The error is often made that since the Rb standard is a passive device, not working on the basis of stimulated emission and not emitting radiation as in a maser, cavity frequency pulling may be a function only of the square of the ratio of the quality factors of the cavity and resonance line, $(Q_c/Q_l)^2$, as calculated above for the case of the Cs standard. Consequently the effect of stimulated emission is often neglected. However, even though the population difference between the hyperfine levels is the parameter detected by means of the optical radiation transmitted by the cell, stimulated emission within the atomic ensemble is not totally negligible under normal operation. The density in the cell is usually raised to such a level as to provide a good *S/N*. In such a case, the factor α that was introduced earlier and which may be understood as a gain factor measuring essentially stimulated emission cannot be neglected. The cavity pulling needs to be represented by the expression (*QPAFS* 1989):

$$\Delta v_{clock} = \frac{Q_c}{Q_l} \frac{\alpha}{1+S} \Delta v_{cav} = P_{pop} \Delta v_{cav} \qquad (1.111)$$

where α is given by:

$$\alpha = kQ_cT_1T_2N_n \qquad (1.112)$$

where:

T_1 and T_2 are, respectively, the population and coherence relaxation times
N_n is the normalized population in the upper hyperfine state
S is the saturation factor
Q_c is the cavity quality factor
The factor k is a term that is a function of the construction of the optical package
and fundamental constants:

$$k = \frac{\mu_0\mu_B}{\hbar V_{cell}} \eta' \qquad (1.113)$$

in which the various terms have been defined previously. P_{pop} is called the pulling factor, characterizing population difference between the hyperfine states. For a proper *S/N*, α is normally made equal to 10^{-2} and S is set at about 2. The Q of the cavity may be of the order of 500 while that of the resonance line is of the order of $1 - 1.3 \times 10^7$. The pulling factor is then equal to about 4×10^{-5}. In such a case, for a fractional frequency stability of 10^{-14}, the cavity needs to be stable to ~200 Hz. The temperature coefficient of a metallic cavity operating in the TE_{111} mode may be as high as 213 KHz/degree (Huang et al. 2001). Consequently, in such a case the temperature of the cavity needs to be stabilized to better than 10^{-3} degree. Temperature control to such a precision is not easy to achieve in practice. This is particularly very difficult in environmental situations where temperature may fluctuate by large amounts in rooms where temperature is not controlled and in satellites where the temperature of the base plate to which the clock is attached fluctuates by several degrees (Camparo et al. 2005a, 2012). In the design shown in Figure 1.29, the introduction of a dielectric material to reduce the size of the magnetron cavity has the further advantage of reducing the temperature coefficient of the ensemble by nearly a factor of 10. This is due to the fact that the dielectric material has a temperature dependent dielectric coefficient opposite to that of the metallic cavity. This relaxes the demand on the temperature control of the unit (Huang et al. 2001).

It is concluded that although cavity pulling is small in devices such as the optically pumped Rb frequency standard, it is not totally negligible. Care must be taken in evaluating the performance of a particular standard relative to the actual cause of instability particularly in environmental situations where large temperature fluctuations are present.

1.1.3.2.3.2 Buffer Gas Shift   The buffer gas used in the cell has a double purpose: augmenting diffusion time and thus reducing collisions with the cell walls, a phenomenon that causes relaxation and broadens the resonance line, and introducing a mechanism that quenches the scattered or random fluorescence radiation that would cause relaxation in the ensemble of Rb atoms through random optical pumping. Nitrogen at a pressure of the order of 10 Torr is found to be very efficient in relation to that last mechanism and is generally used. Its fractional frequency pressure coefficient is 80.0×10^{-9}/ Torr. However, it is also characterized by a rather large linear positive fractional frequency temperature coefficient of 79.3×10^{-12}/Torr/degree. For this reason it is usually mixed with another gas with a negative temperature coefficient. Argon has been the most often used gas. Its fractional frequency pressure coefficient is -9.4×10^{-9}/Torr and temperature coefficient is $-53.5 - 10^{-12}$/Torr/degree. A ratio of pressures P_A/P_N of the order of 1.5 reduces considerably these linear temperature coefficients by cancellation.

However, positive quadratic temperature dependence remains. In principle, the temperature could be adjusted to the optimal point, which is the maximum frequency of the frequency dependence on temperature (Vanier et al. 1982). In practice, this is difficult and the temperature of that maximum may fall outside the desired temperature of operation for maximum contrast, for example. Consequently, normally a small residual linear temperature coefficient remains that we have to live with. For example, in a typical case, the residual fractional frequency temperature coefficient could be 10^{-11}/degree requiring a temperature control of 10^{-3} degree to achieve in the mean term a stability of 10^{-14} (Affolderbach et al. 2006). One notes that this residual temperature dependence is comparable to the cavity pulling effect examined above. In some cases, depending on which side of the maximum of the buffer gas temperature dependent residual frequency shift the cell is operated, both effects could cancel to some extent. Since the actual residual coefficient of the cell depends very strongly on the cell filling process, which is usually rather imprecise, it is possible that this effect may be a reason why some particular units behave differently depending on the quality of reproduction of the particular fabrication process.

1.1.3.2.3.3   Spin Exchange   Rb atoms in the cell are in motion and collide with each other. Their rate of collision is independent of the presence of the buffer gas and is a function of the temperature (their velocity) and their density. During collisions, the electrons being identical particles, the so-called phenomenon of spin exchange takes place. As described earlier in the case of the H maser, this effect introduces a relaxation phenomenon that broadens the resonance line and shifts the resonance frequency of the ensemble of atoms by a small amount. In the case of ^{87}Rb, the relaxation is characterized by rates equal to (*QPAFS* 1989):

$$\gamma_1 = n\bar{v}_r\sigma, \qquad \text{population relaxation rate} \qquad (1.114)$$

$$\gamma_2 = \frac{5}{8}n\bar{v}_r\sigma, \qquad \text{coherence relaxation rate} \qquad (1.115)$$

In the Rb standard, this phenomenon is usually detected through a broadening of the resonance line caused by the loss of coherence. However, the effect is accompanied by a frequency shift that is usually neglected being assumed negligible relative to the other shifts that we are presently examining. The total effect can be obtained from the rate equations of the density matrix elements outlined in *QPAFS*, Volume 2 (1989). Considering the atomic system as a three-level system, one obtains a frequency shift (Micalizio et al. 2006):

$$\Delta\nu_{se} = \frac{1}{4}n\bar{v}_r\lambda_{se}\Delta \qquad (1.116)$$

where:
  λ_{se} is the spin–exchange frequency shift cross section
  Δ is the fractional population difference between the two levels involved in the clock transition

The phenomenon has been addressed in some detail. The cross section λ_{se} is computed as 6.0×10^{-15} cm^2. We take the temperature of operation as 60°C. At that temperature, a change of +1 degree changes the Rb density by 8% and depending on the scheme of optical pumping, from level $F = 1$ or level $F = 2$, the fractional frequency shift dependence on temperature is:

$$\frac{\delta\left(\Delta\nu_{se}/\nu_o\right)}{\delta T} = 1.3\times10^{-11}/\text{K}, \qquad \text{for pumping from } F=2 \qquad (1.117)$$

and

$$\frac{\delta\left(\Delta\nu_{se}/\nu_o\right)}{\delta T} = 0.8 \times 10^{-11}/\text{K}, \qquad \text{for pumping from } F=1 \qquad (1.118)$$

Although the absolute shift has no importance on the clock quality itself, since it needs to be calibrated anyway, its variation with temperature is important for medium- and long-term frequency stability. One may point out that a change in temperature by one degree is relatively large since temperature is controlled to probably better than 0.01 degrees. However, even a change of that magnitude (0.01 degrees) would cause a frequency shift of the order of 10^{-13}, a factor of 10 larger than the actual long-term frequency stability desired. We will leave this discussion for the moment and come back to it below when we will examine the behaviour of actual clocks.

1.1.3.2.3.4 Light Shift   The light shift is probably the most problematic shift of all those that have been identified. It is normally seen as a frequency shift that is proportional to the radiation intensity. It is usually written as (*QPAFS* 1989):

$$\Delta \nu_l = \alpha_l I_o \qquad (1.119)$$

where:
   I_o is the photon flux at the entrance of the resonance cell
   α_l is the light shift coefficient which is a function of several fundamental parameters such as pumping light line width and shape, overlap of lamp emission lines and cell absorption lines, filter effectiveness, and several other parameters

It is general practice to plot the cell resonance frequency as a function of light intensity for various adjustable parameters such as filter cell temperature in the separated filter approach and adjust the filter temperature that makes the cell resonance frequency independent of light intensity. In the case of the integrated filter approach, it appears that the best setting is the light intensity for which the cell is independent of temperature. One then relies on the stability with time of all the parameters that affect either the spectrum of the incident light and its intensity. If we examine carefully Equation 1.119 above, we see readily that the shift is a function not only of the light intensity, but of the constancy of α_l with time. We may differentiate Equation 1.119 in respect to time and obtain:

$$\frac{d(\Delta \nu_l)}{dt} = I_o \frac{d\alpha_l}{dt} + \alpha_l \frac{dI_o}{dt} \qquad (1.120)$$

We realize that the shift may be a function of time if anything changes either in the lamp, the filter cell or the resonance cell, changes affecting either α_l or I_o. What may change effectively with time is temperature that has a direct effect on the Rb density. On the other hand, the Rb density itself may change through interactions of unknown origins with the walls of the glass enclosures of the lamp, filter, and resonance cell. This is a very complex situation since the only parameter that can be measured in the completed unit is the light intensity at the photodetector and the frequency of the resonance cell. This, furthermore, has to be done over very long periods of time, in order to address the question properly.

   It should be pointed out that the long-term stability of Rb clocks has most of the time been assumed to be represented by a drift of the order of parts in 10^{11}/month. Furthermore, it was found that some units were more stable than the average clock. The shift could be negative or positive and until recently, its origin led essentially to speculations without successful convincing experiments.

Recently, Camparo and colleagues have addressed the question experimentally in a rather serious way in order to possibly identify the origin of that drift unambiguously (Camparo 2005; Camparo et al. 2012). The authors had access to data accumulated over many years on clocks in earth artificial satellites. A typical result is shown in Figure 1.30. Although clocks from different satellites behaved in a slightly different way, the general behaviour was essentially the same for all. The data for that standard in a GPS satellite covers a period of nearly three years. Data from a number of other clocks of a different construction and in Milstar satellites forming a different system showed similar time dependence. That behaviour was also observed on ground in a laboratory unit.

The most striking feature about the data is that the light intensity and frequency of most clocks under operation showed an exponential behaviour with time, a tendency towards pseudo-equilibrium in detected light intensity emitted by the lamp and measured output frequency. The behaviour for the clock fractional frequency $y(t)$ could be characterized by an empirical equation of the form:

$$y(t) = Ae^{-\gamma_w t} + Be^{-\gamma_e t} + Dt \qquad (1.121)$$

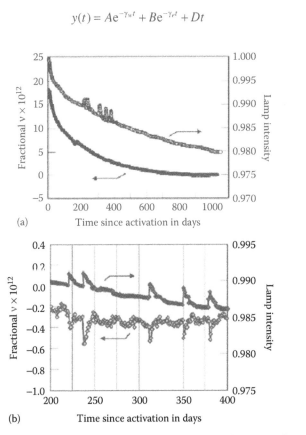

(a)

(b)

FIGURE 1.30 Results of the behaviour of one Rb clock in GPS satellite SVN-54. Graph (b) is an expanded view of a section of Graph (a). (Reproduced with permission from Camparo et al. 2005. Copyright 2005 IEEE.)

where γ_w^{-1} and γ_e^{-1} are time constants of the processes taking place (not to be confused with relaxation rates generally used in the study of rate equations). The time constant γ_w^{-1} may be of the order of a month or so, while γ_e^{-1} may be of the order of a year or so. The constants A and B determining the importance of the process in action vary drastically from clock to clock. The last term appears as a residual linear frequency drift that varies from clock to clock. It is of the order of parts in 10^{-14}/day but in some clocks appears to be much less, almost not measurable. What can be concluded of that behaviour? There is not really a clear answer to that question. However, some general conclusions can be drawn and some possible answers in terms of the analysis made earlier relative to the frequency shifts present in the system can be proposed.

First, it is readily observed that there is a drastic change in the lamp intensity (decrease) within the first few months of operation. It is observed from Figure 1.30a that the clock frequency in the very long term also varies in the same direction. This overall process towards some equilibrium has been called "equilibration" (Camparo et al. 2005). The first conclusion would be that according to the previous analysis, particularly Equation 1.120, there is a light shift present that is proportional to light intensity and has the same sign in its dependence on time. The correlation appears to be very strong. However, Figure 1.30b, which shows a higher resolution of the clock frequency variation over a shorter period, contradicts this conclusion. That graph shows that the lamp changes its intensity randomly on a short timescale, somewhat like in short bursts. This may happen sometime with sudden redistribution, like bubbling, of Rb within the lamp glass enclosure. Such a process would result in a sudden increase in Rb density with an increase in light intensity. From the figure it is observed that the clock frequency decreases. This contradicts the conclusion that could be reached from Figure 1.30a. This conclusion is also supported by data obtained with a clock on another satellite for which it was found that, for some unknown reason the light intensity increased with time, but the clock frequency still decreased over the same period. A light shift effect is thus eliminated as the source of the frequency change during the period in which a static state is searched by the system. It is thus not clear what changes are taking place that causes the observed frequency shift, uncorrelated to light intensity. It is possible that the lamp spectrum is altered at the same time as intensity. It is not clear, however, how this process could explain the overall behaviour of the intensity and frequency with time. Although the rate and time constant is different from one clock to another, one important aspect is that the changes have the same sign except maybe in the very long term. Although the light shift cannot be totally eliminated in some cases, as an added mechanism altering the rate and time constant to reach equilibrium, it is possible that the main effect is a reaction of Rb with glass or buffer gas itself in all components, lamp, filter, and resonance cell. For example, qualitative experiments show that in resonance cells that have not been well-cleaned, a large variation in frequency with time is observed over a long period. Furthermore, it is not possible to uniquely control the density of Rb in a cell by altering only the temperature of the cell where a Rb film has been deposited. It seems that a slow adsorption of Rb within the glass takes place and the new vapour pressure equilibrium expected when temperature is altered is not dictated only by temperature but by the actual content of Rb on or within the glass surface. There is a delay observed in the reaching of equilibrium in all cases and these results are usually not reproducible. It appears that the Rb density within a cell is a function not only of temperature but of its history, from its construction time to its present existence. The process must also take place in the lamp. Consequently, we may expect that the Rb density may have an effect on resonance frequency through various mechanisms. It is also possible that the buffer gas itself reacts with the Rb film within the cell, although very slightly, physically by adsorption or by chemical reaction. Both, adsorption or reaction would alter the buffer gas pressure in the cell, which results in a frequency shift.

We have also identified above a mechanism for which the atom frequency in the resonance cell is directly proportional to density: it is spin exchange. The shift is relatively large. It varies by about 10^{-11} for a change in density of 8% corresponding to an equivalent change in temperature of 1 K. The change in frequency observed in Figure 1.30a above is of that order of magnitude. Such changes in density are very possible through the migration of Rb and the equilibration with the glass surface with time. On the other hand, this kind of change in the lamp ampoule may very well take place at a greater rate since it operates at a higher temperature. This would have consequences on light intensity and spectrum shape through line reversal, for example. It does not appear that we can identify absolutely which mechanism is the most important. It is very possible that many mechanisms are active at the same time and age at different rates. Some of them may compensate each other and such a process could explain the very long-term behaviour of the frequency standard in various environments.

As is seen from the previous discussion, the classical optically pumped Rb standard still raises great interest today. Improvements have been made in its construction through the design of a smaller cavity and some refinement through the use of digital electronics in the frequency-lock-loop system just mentioned. It does not appear, however, that the question of the origin of the long-term frequency stability has been resolved. It may be that the complexity of the problem is such that no easy solution with long-term drift can be expected. We have also mentioned in this chapter the possibility of using a laser rather than a spectral lamp to invert the populations in such a standard. In the next chapter concerning new developments in atomic physics in general, we will present a rather complete analysis of that approach as well as develop a new approach using coherent population trapping (CPT) to implement a small passive Rb or Cs passive frequency standard without the use of a microwave cavity (Cyr et al. 1993; Levi et al. 1997; Vanier et al. 1998). We will also describe a recent development using a pulse interrogation scheme that duplicates in the time domain Ramsey's approach of space separated interaction regions used in the realization of primary Cs standards and described above (Godone et al. 2004, 2006; Micalizio et al. 2008, 2012, 2013). The question of long-term drift could possibly be re-addressed using those techniques. In such cases, with lasers well-stabilized in frequency, only one element, the cell, would be questionable regarding Rb and buffer gas reactions.

In *QPAFS* (1989), we have also described the functioning of an optically pumped maser (Vanier 1968). Except for the use a larger microwave cavity with a higher quality factor, it is based essentially on the same optical pumping scheme described for the implementation of the passive standard. We will describe later a new approach for implementing a Rb or a Cs maser using the phenomenon of coherent population trapping (Vanier et al. 1998; Godone et al. 1999). Such a maser does not require a cavity with a high quality factor since it does not have a threshold of oscillation. Consequently, it can be made much smaller than the classical Rb maser. We will describe it in Chapter 3.

1.2   OTHER ATOMIC MICROWAVE FREQUENCY STANDARDS

A legitimate question about atomic frequency standards just described is what criteria were used to select those elements, H, Cs, and Rb, as the heart of atomic frequency standards. There are several reasons for that, but an important one was the available technology when the choice was made. In the middle of the twentieth century, microwave technology had just been made available as a by-product of the development of Radar making available rf sources in the microwave range. Furthermore, H and alkali atoms had been extensively studied by means of beam techniques and much was known about their internal properties and spectroscopy (Ramsey 1956). In particular, the magnetic properties of their ground state had been studied, the energy separation between the hyperfine levels falling in the microwave frequency range. Furthermore, magnetic properties of that ground state were such that the spacing between two of the hyperfine levels were independent of the applied magnetic field in first order. This is an important property, which using modest magnetic shielding makes such an atomic transition suitable for the implementation of a standard nearly independent of fluctuations of the environmental magnetic field. Another property, which was conclusive at that time, was the relative easiness of measuring the intensity of a flux of atoms through ionizing and counting of the ions.

The choice of Cs relied in great part on that technique because of its low ionization energy (3.894 eV). Furthermore, a single isotope, ^{133}Cs, exists in nature, avoiding the requirement of purifying it as is required in the case of Rb, which contains two isotopes in its natural form. These properties coupled with the fact that the Cs hyperfine frequency falls in the X band of the microwave spectrum and the existence of an important vapour density at a temperature of the order of 100°C are probably the most fundamental reasons for the extensive study that was made at that time to implement a frequency standard with that atom. It turned out that the choice, based mainly on those properties, was appropriate. Because of that intense research and development and the practical device that resulted, coupled with the fact that it is much easier and more accurate to implement in practice the SI unit, the second, from electromagnetic radiation rather than from astronomical observations, the hyperfine frequency of that atom was chosen as the basis for the definition of the second in 1967. This choice appears to be excellent since, in 2015, we still live with that definition. However, as we will see, this situation may change, due mainly to outstanding new atomic physics research, which has resulted in a better understanding of the interaction of atoms with optical radiation. That new knowledge has resulted in a quasi-perfect control of the mechanical properties of atoms such as their velocity and movement in space. This has led to new approaches to the development of frequency standards that can be characterized by accuracies that are orders of magnitude larger than those achieved with the classical approach described above.

The other element, H, basic in this universe, is certainly very attractive with its simple internal structure. Its hyperfine frequency falls in the low microwave range, 1.4 GHz giving a 21 cm wavelength. Microwave cavities at that frequency, in their fundamental mode, are rather large but in principle do not cause a major problem, although not as attractive as Cs with its 9.2 GHz hyperfine frequency or 3 cm wavelength. The high ionization potential of H of the order of 13.6 eV, however, prevents its use in a scheme similar to that of Cs, that is to say detection by means of ionization. It has thus been used for the development of a maser, as described above, where stimulated emission is used and microwave radiation is detected as signal rather than atom counting. Because of the presence of the wall shift, the accuracy of the H maser was not comparable to that of the Cs standard and it could not be used as a primary standard. However, in the light of the kind of frequency stability obtained, the H maser has remained a workhorse mainly in ground applications where an ultrastable reference with high spectral purity is required. Nevertheless, new devices with smaller cavities of the magnetron type have appeared allowing their use in satellite navigation systems (Rochat et al. 2007; Waller et al. 2010).

The choice of Rb as the atom for implementation of a frequency standard is rather special. This choice is really based on the fortuitous coincidence of some lines of the D_1 and D_2 spectrum of the two isotopes ^{85}Rb and ^{87}Rb. That coincidence allowed very efficient optical pumping in a cell containing one isotope by using a spectral lamp filled with the same isotope and a filter using the other isotope. The resulting resonance signal obtained in the type of construction described above is very large providing an excellent S/N. Furthermore, a cavity at the resonance frequency of 6.8 GHz may be made very small by choosing the proper mode of operation. This coupled to easiness in construction led to a reduction of size of the standard without affecting much its frequency stability. The interest in the standard has remained high for more than 50 years and due to successful development in its supporting electronics, making it very reliable, it has found applications in such fields as satellite navigation systems and communication systems at large.

Nevertheless, work has continued during all those years on new approaches to implement frequency standards using other elements. The reader is referred to *QPAFS*, Volume 2 (1989) for a detailed account of the state of the study on the use of other elements at the end of 1980s. We will only recall here those investigations that led subsequently to some interesting developments or still opened avenues for the use of other atoms and techniques. Some were studied in their ionized state and allowed a totally different approach in their storage, manipulation and technique for implementing a standard.

1.2.1 ^{199}HG$^+$ ION FREQUENCY STANDARD

1.2.1.1   General Description

In its singly ionized state, the isotope mercury, ^{199}Hg$^+$, appears to be a most interesting element. Its ground state is an $S_{1/2}$ state with a spin $I = 1/2$ as in the case of H but with a hyperfine frequency of 40.5 GHz. The difficulty of using ions resides in their manipulation in space and in the required inversion of their population, since beams and state selector magnets cannot be used as in the case of neutral atoms

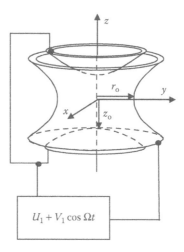

FIGURE 1.31   Paul trap using hyperbolic-shaped electrodes.

such as H and Cs. Furthermore, containers with wall coatings cannot be used since the ions would react violently with the surface of such containers. Consequently, one technique used comprises storing the ions in a so-called electromagnetic trap. Combinations of magnetic, electric, and rf fields have been used for such a purpose. A structure using only static electric fields cannot be used because it leads to an unstable situation. The use of combined static magnetic and electric fields leads to stable trapping and is called a Penning trap. On the other hand, the use of combined dc and rf electric fields leads also to stable trapping. It is called a Paul trap and is the one that has been used most extensively in various shapes in the field of atomic frequency standards (Paul et al. 1958; Paul 1990). A particular configuration of such a trap is shown in Figure 1.31.

The potential inside such a trap, assuming electrodes having hyperbolic cross sections with dimensions r_0 and z_0 as indicated in the figure and with the geometrical condition $r_0^2 = 2z_0^2$, may be written in cylindrical coordinates as:

$$V(r,z)=\frac{V}{2r_0}\left(r^2-2z^2\right)$$
(1.122)

If the potential V contains a dc and an oscillating component of frequency Ω, such as:

$$V = U_1 + V_1 \cos\Omega t$$
(1.123)

the centre of the trap acts as region of minimum potential energy and, under certain conditions, charged particles may be trapped or stored for a long time. In the case where $U_1 = V_1$, the potential is said to be spherical and calculations show that the potential energy of an ion in such a trap is given by:

$$E_p=\frac{1}{2}\frac{e^2V_1^2}{M\Omega^2 r_0^4}\left(r_{av}^2+z_{av}^2\right)$$
(1.124)

where:

    e is the electronic charge

    V_1 is the amplitude of the applied rf field of frequency Ω

    M is the mass of the ion

    r_{av} and z_{av} are the average position of the ion in its motion

The ion is forced into a harmonic motion, called micromotion, at the frequency Ω of the driving field. Depending on its initial energy the ion travels also around the centre of the trap, a motion called macro- or secular motion. It is standard practice to define the term:

$$\omega = \frac{eV_1}{M\Omega r_o^2} \tag{1.125}$$

as the frequency of the macromotion. The motion of an ion in the trap looks somewhat like a Lissajou figure.

One of the first successful developments in the use of ions for developing a microwave frequency standard was based on the mentioned ion, $^{199}Hg^+$, and the use of such a trap. The inversion of population is done in a similar way as in the case of the Rb cell standard, by means of optical pumping with radiation emitted by an isotopic ^{202}Hg lamp, whose emission line at 194.2 nm coincides fortuitously with the transition from state $S_{1/2}$, $F = 1$ to state $P_{1/2}$ of $^{199}Hg^+$. In early studies, the atomic ensemble was exposed simultaneously to optical pumping and microwave radiation at the hyperfine transition in the same region of space. A line width of 8.8 Hz was observed leading to a frequency stability $\sigma(\tau) = 3.6 \times 10^{-11}\,\tau^{-1/2}$ (Jardino et al. 1981). A narrower line was obtained by cooling the ion cloud with a low pressure buffer gas, such as He, and using a pulse approach, that is turning the optical pumping radiation off during interaction with the microwave radiation. The approach led to a narrower resonance line and improved frequency stability. For example, a line width of 0.85 Hz was measured in that approach leading to a line Q of 4.8×10^{10} (Cutler et al. 1983). In a similar type of arrangement, a line width of 1.6 Hz was observed leading to a frequency stability of $\sigma(\tau) = 4.4 \times 10^{-12}\,\tau^{-1/2}$ for 20 s $< t <$ 320 s (Prestage et al. 1987; Tanaka et al. 2003).

1.2.1.2 Frequency Shifts

In those early studies, various parameters affecting the frequency of the standard were studied and their effect on accuracy and frequency stability was evaluated. In general, as we have seen in the case of frequency standards described previously, a frequency shift does not affect the quality of a standard if it can be evaluated accurately and is stable with time. Three of the shifts identified in the case of ion storage in a Paul trap appeared to be large enough to be possible sources of inaccuracy and frequency instability. Those are (a) the applied magnetic field, (b) the light shift, (c) second-order Doppler effect. Schemes and techniques have been found recently that address at least to a satisfactory level the questions raised by those shifts. We will examine them.

1.2.1.2.1 Applied Magnetic Field

The dependence of the resonance frequency on the dc induction B required to quantize the system and give an axis of orientation is given by (*QPAFS* 1989):

$$\nu = \nu_{hf} + 97.00 \times 10^8 B^2 \tag{1.126}$$

where:

   B is in Tesla

For an induction of 10 µT (100 mGauss), the shift is 2.4×10^{-11}. Although this appears to be a large shift, the value of the field can be determined accurately with the help of Zeeman transitions within the $F = 1$ states. It also turns out that in the trap described, the atomic ensemble occupies a rather small volume when compared to the classical standards described previously and field homogeneity is not a problem. On the other hand, the field must be stable with time. This is usually accomplished by means of well-designed current sources driving a solenoid. Furthermore, the field can be lowered to some extent because the resonance line width is rather narrow (~1 Hz) and the effect of neighbouring transitions is not a problem. A reduction in field reduces the demand on field stability. Consequently, the magnetic field, if well-controlled, does not introduce either inaccuracy or frequency instability that would inhibit the stored ions approach to implement a frequency standard of high frequency stability and accuracy.

1.2.1.2.2   Light Shift

We have addressed the question of the light shift to some extent in *QPAFS*, Volume 2 (1989) for the case of Hg$^+$ and have calculated a fractional frequency shift of about 6×10^{-12} in standard operating conditions. This is not negligible and its stability with time and its precise evaluation are a concern. The lamp used is excited at a level of several watts rf, is hot and emits UV radiation that affect the glass enclosure generally made of quartz. It is found in practice that Hg lamps age and light intensity tend to vary with the period of operation. The light shift can be studied by measuring the resonance frequency with pumping light intensity and its value can be determined in principle by means of an extrapolation to zero light intensity. However, it remains uncertain to a certain extent and may have an effect on long-term frequency stability as well on accuracy. As mentioned above, that question was addressed in early development stages by means of a pulse technique. The pumping light is turned off during the application of the microwave radiation that excites the ground state hyperfine transition. Since the technique avoids the presence of radiation during the period of application of the microwave radiation, there is no broadening present from optical pumping and in principle no light shift. As mentioned above, outstanding results were obtained relative to line Q and frequency stability in the short term. Frequency stability measurements in reference to an H maser showed random fluctuations of the order of 4 to 8×10^{-15} for periods of days or so (Cutler et al. 1987).

1.2.1.2.3   Second-Order Doppler Effect

As in the case of Cs and H, the second-order Doppler effect is a major concern. In the present case, the line is extremely narrow and there is no concern relative to symmetry. However, the ions temperature is rather high after dissociation and their speed within the trap is also affected by their motion caused by the trapping electric fields.

As mentioned above, the motion of the ions caused by the applied radio-frequency field at a frequency Ω, of the order of 250–500 kHz, is called microscopic while the motion caused by their evolution of the ions around the centre of the trap is called macroscopic or secular motion. The fields increase with distance from the centre of the trap and the energy of the ions increases also. As we will see later, the motion of a single ion can be reduced considerably by means of laser cooling and set to the middle of the trap where the intensity of the fields is zero. In that case the motion of the ion is minimized and the second-order Doppler effect becomes negligible. However, in the microwave standard described here, the trap is loaded with something like 10^6 ions and these ions having all the same charge repel each other. They form a cloud. Assuming that the space charge adjusts itself to produce a gradient of the total potential equal to zero, it is possible to calculate the various parameters characterizing that cloud (Dehmelt 1967). The density is uniform and equal to:

$$n = \frac{3\varepsilon_o M \omega^2}{e^2} \tag{1.127}$$

Assuming a trap with $r_o = 19$ mm, $V_1 = 250$ V, $\Omega = 2\,\pi \times 250$ rad/s and Hg ions with $M = 199$, one obtains $n = 1.54 \times 10^{13}$ ions m^{-3} and $\omega = 2\pi \times 33.7 \times 10^3$ rad/s. For a total number of ions $N = 10^6$, the radius of the cloud is $r_c = 2.5$ mm. This justifies the assumption made on the smallness of the cloud in the discussion related to the applied magnetic field. The frequency shift caused by second-order Doppler effect is given by:

$$\frac{\Delta v}{v} = -\frac{1}{2}\frac{v^2}{c^2} \tag{1.128}$$

where v is the ion velocity. In terms of average energy $<E>$ this can be written as:

$$\frac{\Delta v}{v} = -\frac{<E>}{Mc^2} \tag{1.129}$$

and the shift is given by

$$\left(\frac{\Delta v}{v}\right)_{2ndD} = -\frac{3}{10c^2}\left(\frac{N\omega e^2}{4\pi\varepsilon_o M}\right)^{2/3} \tag{1.130}$$

The potential energy is of the order of a tenth of the trap potential or of the order of 1 eV and the fractional frequency shift is of the order of a few in 10^{12}. The size of this shift can be reduced considerably by reducing the speed of the ions. This can be done by laser cooling mentioned above and to be described later. This can also be done by means of exchange of energy with a low pressure buffer gas such as He. The technique was applied successfully with Hg$^+$ as mentioned above. However, another small frequency shift is introduced, caused by the effect of ion–buffer gas collisions on the hyperfine splitting. At the pressure usually set, this shift is of the order of 10^{-14}.

One of the main practical goals in the implementation of such a frequency standard is to reach a frequency stability in the short term comparable to that of the classical frequency standards described above. For square wave frequency modulation, the frequency stability is given by (*QPAFS* 1989, Volume 2):

$$\sigma^2(\tau) = \frac{8}{27} \frac{1}{Q_i^2} \frac{1}{C^2 I_m} \frac{1}{\tau}$$

(1.131)

where C is the contrast or fractional signal amplitude I_M relative to the background I_m. As pointed out earlier, the technique allows the realization of long storage times with resulting line Q's in the $10^{10}-10^{11}$. The other factor on which the experimenter may have some control is the contrast. An increase of such a factor can be realized by increasing the number of ions contributing to the resonance signal. However, this contributes to the space charge resulting in an increase in size of the volume of the ion cloud. The ions at the periphery are exposed to a larger potential and have more energy. Consequently, the second-order Doppler shift increases and a greater demand is made on the stability of the ions density. Nevertheless, cooling by a buffer gas tends to reduce this effect to some extent.

1.2.1.3 Linear Trap

Another very successful approach consists in altering the hyperbolic trap described above into a linear trap as shown in Figure 1.32 (Prestage et al. 1989, 1990). The properties of such a trap can be calculated in a way similar to the case of the hyperbolic Paul trap. The ions are distributed along the length of the trap in its centre and for the same total number they do not experience electric fields as strong as in the case of the spherical ion cloud in a Paul trap of hyperbolic geometry. The second-order Doppler frequency shift is calculated as:

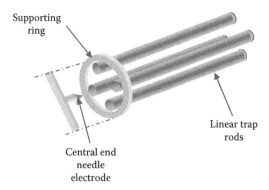

FIGURE 1.32 One end view of a linear trap formed of four rods with appropriate alternate polarities. The needle end cap for retaining the ions between the rods is shown. The construction of the other end is similar and contains a source of atoms and ionization electrodes. (Data from Prestage, D. et al., *J. Appl. Phys.*, 66, 1013, 1989; Prestage, J.D. et al., *IEEE Trans. Ultrason. Ferroelectron. Freq. Control*, 37, 535, 1990.)

$$\left(\frac{\Delta v}{v}\right) = \left(\frac{e^2}{8\pi\varepsilon_{o}mc^2}\right)\frac{N}{L} \tag{1.132}$$

where N/L is recognized as the linear density. When this result is compared to the case of the spherical ion cloud case, it is found that, for the same second-order Doppler shift, the linear trap can store a much larger number of ions given by:

$$N_{lin} = \frac{3}{5}\frac{L}{r_c}N_{sphe} \tag{1.133}$$

For example for $L = 75$ mm and $r_c = 2.5$ mm, the linear trap can store 18 times more ions than the spherical trap for the same second-order Doppler shift. Such a trap has received much attention and has been perfected by the introduction of sections as shown schematically in Figure 1.33 (Prestage et al. 1993, 1995, 2008; Burt et al. 2007).

In that arrangement, the ions ensemble can be trapped and shuttled back and forth between region A and B by altering the bias between the two set of trapping rods. Ions are produced in section A where they are also optically pumped. The ions are then shuttled to region B where they are exposed to microwave radiation. They are then shuttled back to region A where they are optically pumped and fluorescence is detected verifying if the condition of resonance with the microwave was satisfied in section B. The technique thus avoids completely the presence of optical pumping radiation during the atoms-microwave interaction, avoiding the light shift and optical broadening effect. Such a trap using a time domain Ramsey interrogation scheme for the detection of the resonance gave a frequency stability of $6.5 \times 10^{-14}\, \tau^{-1/2}$ up to nearly 10^5 s where a frequency stability of 5×10^{-16} was measured (Tjoelker et al. 1995). Further improvement was realized through the use of a multipole structure

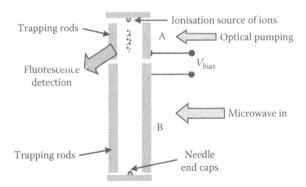

FIGURE 1.33 RF trap with two sections allowing a transfer of the ions ensemble from one section where optical pumping is done to another section where the resonance signal is detected. (Data from Prestage, J.D. et al., Improved linear ion trap physics package. In *Proceedings of the IEEE International Frequency Control Symposium* 144, 1993; Prestage, J.D. et al., Progress report on the improved linear ion trap physics package. In *Proceedings of the IEEE International Frequency Control Symposium* 82, 1995.)

(12 poles) in the region where the ions are interacting with the microwave radiation. However, the pumping region is kept as a quadrupole structure type allowing space for optical pumping and detection. In such a configuration, it is shown that the total number of ions can be increased considerably (up to 10^7 ions) and that changes in number of ions influence the second-order Doppler shift by a factor less than 10 times the same changes in the quadrupole trap. Finally, studies were also made on the use of different buffer gases for cooling. The intent was to find a buffer gas efficient in cooling but causing a smaller collision frequency shift than helium. Neon was found to have a collision shift of 8.5×10^{-9}/Torr, a factor of 2.5 less than helium (Chung et al. 2004). A record line Q of 5×10^{12} was observed using time domain Ramsey separated pulses with a cycle time of 12 s resulting in a frequency stability of $5 \times 10^{-14}\, \tau^{-1/2}$ (Tjoelker et al. 1995; Burt et al. 2008a and b). The size of such a clock has been reduced to a volume of the order of 3 l, while keeping a frequency stability of $1 - 2 \times 10^{-13}\, \tau^{-1/2}$ and 10^{-15} at one day (Prestage et al. 2007, 2008). It should be mentioned that this is comparable to the frequency stability obtained with H masers of much larger size.

1.2.2   OTHER IONS IN A PAUL TRAP

Other ions have also been proposed and studied for implementation of a microwave frequency standard using a Paul trap. Those are for example $^9Be^+$, $^{135}Ba^+$, $^{137}Ba^+$, $^{171}Yb^+$, $^{173}Yb^+$, $^{87}Sr^+$, and several others (see *QPAFS* Table 1.1.3b, Volume 1) (1989). The advantages of those ions over Hg^+ resides in the lower microwave frequency of the ground state transitions and in the longer wavelength required for state selection by means of optical pumping, making possible the use of lasers. Those properties lead to somewhat easier experimentation and implementation.

In particular, considerable work has been made on the ion $^{171}Yb^+$. An energy level diagram for this ion is shown in Figure 1.34. The main difference from an alkali

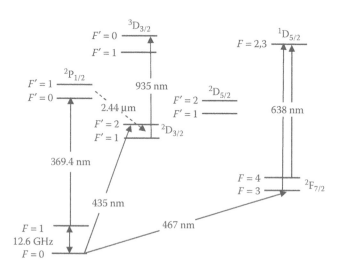

FIGURE 1.34   Lower energy levels manifold of $^{171}Yb^+$ ion of interest in the implementation of a frequency standard.

atom resides in the existence of D and F states lower than the P state. Since these states are metastable and ions falling into them have long lifetimes, they act as traps. Depending on the application, that property may be undesirable.

In general, optical pumping using isotopic filtering or a spectral lamp with a different isotope as in the case of ^{87}Rb and ^{199}Hg$^+$ cannot be done. Lasers at the proper wavelength need to be used. Furthermore, as in the case of Hg$^+$, an ensemble needs to be cooled in order to reduce the second-order Doppler effect. Fortunately, efficient lasers frequency doubling techniques exist with the proper wavelength to effectuate laser cooling for several ions. We will introduce those techniques of laser optical pumping and cooling in Chapter 2.

1.2.2.1 ^{171}Yb$^+$ and ^{173}Yb$^+$ Ion Microwave Frequency Standards

As an example, we will at this point outline immediately the results obtained with the ion ^{171}Yb$^+$ for implementing a microwave frequency standard at 12.6 GHz. Proposal for using this ion were made early in the development of ion standards (Blatt et al. 1983). Precise measurements were also done on the determination of its ground state hyperfine frequency splitting (Tamm et al. 1995). Actually, the ion can be used to implement a standard either in the optical range or in the microwave range (Barwood et al. 1988; Blatt et al. 1988; Webster et al. 2001). It was also studied for implementing a single ion optical frequency standard (Gill et al. 2008; Tamm et al. 2008).

In the implementation of a microwave frequency standard at 12.6 GHz, optical pumping is done at 369 nm as shown in the figure in a cycling transition. Since a fraction of ions decay to the metastable D$_{3/2}$ state, they need to be returned to the cycling transition otherwise they accumulate in that state and are lost for the double resonance signal. This ground state resonance signal is detected on the fluorescence at 369 nm. In a particular case, radiation at that wavelength is generated from the doubling of the frequency of a titanium-doped sapphire laser while the pumping of ions from the D$_{3/2}$ metastable state is done with a solid-state laser (Fisk et al. 1993; Sellars et al. 1995). In those experiments, the trapping system used was a linear trap similar to that described earlier and the ions were cooled by means of a buffer gas such as He. The ions were confined to a volume of the order of 10 mm long by 1.5 mm in diameter containing about 10^4 ions. In later experiments the ions were laser-cooled (to be described in Chapter 3) below 1 K (Warrington et al. 1999). It was claimed that with such cooling, the second-order Doppler shift was reduced by a factor of 180 when compared to the case where ions were cooled by means of a buffer gas. In order to avoid systematic shifts due to the optical radiation, the technique of pulsed optical pumping and cooling was used in a Ramsey type of double resonance in the time domain. Using a Ramsey pulse separation of 10 s and a cycle time of 13 s, a frequency stability $\sigma_y(\tau) = 5 \times 10^{-14} \, \tau^{-1/2}$ is expected while fractional frequency uncertainty of 4×10^{-15} is projected (Dubé et al. 2005; Park et al. 2007; King et al. 2012). It appears that these characteristics are nearly comparable to those of the Cs fountain to be described in detail later. We shall come back on those results later after the introduction of laser optical pumping and laser cooling.

^{173}Yb$^+$ is another ion that could be used for implementing a microwave atomic frequency standard. However, it appears to be less attractive than ^{171}Yb$^+$ since, its nuclear spin I being 3/2, it has a more complex ground state energy level structure. It was studied by Münch et al. (1987) in a rf Paul trap using a light buffer gas for cooling. Optical pumping was done with a dye laser at 369.5 nm and hyperfine resonance was detected by monitoring the fluorescence. Its frequency was determined as:

$$\nu_{hs}\left({}^{173}\text{Yb}^{+}\right) = 10,491,720,239.550(0.093) \text{ Hz}$$

1.2.2.2 ^{201}Hg^{+} Ion Microwave Frequency Standard

Two stable odd isotopes of mercury with singly ionized hyperfine structure are suitable for implementing a microwave clock: ^{199}Hg^{+} and ^{201}Hg^{+}. We have described above in some detail the progress made in the development of a frequency standard based on the isotope ^{199}Hg^{+}. It was reported that best results were obtained with a linear Paul trap. With that architecture, short-term stability ~$1 - 2 \times 10^{-13}$ at 1 s averaging time, reaching 10^{-15} at 1 day averaging time was demonstrated. A ground version of that clock that displays performance of $5 \times 10^{-14} \tau^{-1/2}$ and drift of 2×10^{-17}/day was also developed (Prestage et al. 2005; Burt et al. 2008a).

A standard using ^{201}Hg^{+} in a Penning trap was proposed early by Wineland et al. (1981). In such a trap, as mentioned earlier, a large magnetic field is required and this forces the use of transitions between field dependent levels. Fortunately, at high field, some of those transitions become dependent quadratically on the magnetic induction. For example, one transition that could be used is the ($F = 2$, $m_F = 0$) → ($F = 1$, $m_F = 1$) ground state transition at a magnetic induction B_o of 0.29 T with a frequency of 25.9 GHz. However, magnetic fields of that size cause usually experimental difficulties and it is desirable to use the so-called 0-0 transition of the ground state. That last approach has been proposed by Burt et al. (2008b) in a linear trap similar to the one described earlier in the case of ^{199}Hg^{+}.

The lower energy level manifolds of interest are shown in Figure 1.35. It is noted that the nuclear spin of that ion is negative and the Zeeman energy levels position is inverted from that of the isotope ^{199}Hg^{+}. Optical pumping of the ^{201}Hg^{+} isotope can be done efficiently with a ^{198}Hg^{+} lamp. Actually due to a good overlap of the optical resonance lines, optical pumping can be done rather efficiently from state S$_{1/2}$, $F = 2$ of the ground state to the excited state P$_{1/2}$, $F = 2$. Furthermore, the population of the $F = 1$ level of the ground state can be concentrated into $m_F = 0$ level by means of appropriate mixing of states $m_F = \pm 1$ with those of the ground state levels $F = 2$ at frequencies near 29.95 GHz.

Such a clock was investigated at Jet Propulsion Laboratory, Pasadena, CA (Taghavi-Larigani et al. 2009). According to the authors, such a clock has potential accuracy and stability exceeding 1 part in 10^{15}. The unperturbed frequency of the ^{2}S$_{1/2}$, $F = 1$, $m_F = 0$ to $F = 2$, $m_F = 0$ clock transition was measured as 29.9543658211(2) GHz.

A great interest in ^{201}Hg^{+} resides in its simultaneous use with the isotope ^{199}Hg^{+} in the same trap. Since the two isotopes would be submitted to the same perturbation and bias, an accurate measurement of their relative hyperfine splitting would be possible. It would then be possible to examine the possibility of the constancy of fundamental constants, which determine their hyperfine structure (Burt et al. 2008a and b; Flambaum and Berengut 2008).

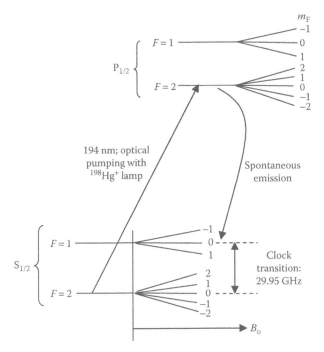

FIGURE 1.35   Lower manifolds of energy levels of the isotopes ^{201}Hg$^+$.

1.3   ON THE LIMITS OF CLASSICAL MICROWAVE ATOMIC FREQUENCY STANDARDS

We have described room temperature microwave atomic frequency standards in detail in the two volumes of *QPAFS* (1989). In the present text, we have outlined developments using the standard approach that were made to those frequency standards after publication of those volumes. We have shown what the limits were. At this stage, we would like to examine further and in a general way the physics behind those limitations.

In all implementations, two important parameters were the quality factor, $Q_l = v/\Delta v$, of the resonance line used for the clock transition and its S/N in its detection. These parameters were determined by contributions from several physical phenomena and played the basic role in the determination of frequency stability. On the other hand, accuracy was a function of the possibility of identifying and evaluating biases that affected the frequency of the atomic resonance chosen as the clock transition. The main question thus regards the possibility of doing better on those parameters to improve both frequency stability and accuracy.

In all cases, we were concerned with Doppler effect either linear or quadratic. Linear Doppler effect produces a large broadening of spectral lines that reduces considerably the line Q. We managed to get rid of that broadening by means of various schemes such as using Dicke effect in maintaining the atoms in a field of constant phase either in a box with non-relaxing walls, a storage trap such as the Paul trap and even a buffer gas. We could also avoid linear Doppler effect in using an atomic beam and reducing line width by means of Ramsey's double interrogation scheme either in space or in time. Due to mechanical constraints, the techniques were essentially limited to microwave. Even in the case where the first-order Doppler broadening could be eliminated, the second-order effect, dependent on the square of the atomic velocities, was present at a level that reduced considerably the accuracy of a given implementation.

During the last few decades work was done at addressing those questions and attempting to reduce if not eliminating those effects. Much work was done using laser optical pumping and other techniques such as sophisticated atom trapping. However, parallel work was done in the field of atomic physics at reducing the temperature. Similarly, transitions with optical frequencies were investigated as possible clock transitions. The first approach, atom cooling, would reduce linear Doppler effect. When used in combination with traps cooling would cancel it completely and at the same time reduce second-order Doppler effect to a negligible value. The second approach, the use of clock transitions at a higher frequency, would result in a larger line Q when those transitions have narrow line widths. The question regarding S/N and biases has to be evaluated for each choice of techniques used and is addressed in each individual implementation. The rest of this book is dedicated to an outline of the physics involved in materializing those ideas.

APPENDIX 1.A: FORMULA FOR SECOND-ORDER DOPPLER SHIFT

An atom travelling at velocity v experiences time dilation in its reference frame. An observer in the laboratory frame sees its frequency altered by the quantity:

$$\Delta \Omega_{D2} = -\frac{v^2}{2c^2} \tag{1.A.1}$$

where:

   c is the speed of light

In the classical Cs frequency standard, this atom is part of a beam with a velocity spread that gives rise to a distribution of interacting times $f(\tau)$ that is a function of the type of state selector used. The central Ramsey fringe is thus affected by this frequency shift and is written as:

$$I = I_b + \frac{1}{2} I_o \int_0^\infty f(\tau) \sin^2 b\tau \left\{ 1 + \cos\left[\omega - \omega_o \left(1 - \frac{v^2}{2c^2} \right) \right] T \right\} d\tau \tag{1.A.2}$$

In practice, it is important to know the exact effect of this frequency shift on the fringe. We can see readily that it distorts its shape and shifts its maximum. The frequency of the maximum can be obtained by differentiating Equation 1.A.2 relative to ω. This is done by first expanding the cosine part in series up to second order:

$$I = I_b + I_o \int_0^\infty f(\tau) \sin^2 b\tau \, d\tau - \frac{1}{4} I_o \int_0^\infty f(\tau) \sin^2 b\tau \left[\omega - \omega_o \left(1 - \frac{v^2}{2c^2} \right) T \right] d\tau \tag{1.A.3}$$

This procedure is valid since the frequency shift is very small. Differentiating and setting $dI/d\omega = 0$, we obtain readily the position of the maximum of the fringe as:

$$\frac{\omega_{max} - \omega_o}{\omega_o} = \frac{\int_0^\infty v^2 T^2 f(\tau) \sin^2 b\tau d\tau}{2c^2 \int_0^\infty T^2 f(\tau) \sin^2 b\tau d\tau} \qquad (1.A.4)$$

APPENDIX 1.B: PHASE SHIFT BETWEEN THE ARMS OF RAMSEY CAVITY

The effect of the phase shift ϕ between the arms of the cavity on the shape and frequency of the maximum of the central Ramsey fringe, when different from 0 or π, can be calculated from:

$$I = I_b + \frac{1}{2} I_o \int_0^\infty f(\tau) \sin^2 b\tau \left\{ 1 + \cos\left[(\omega - \omega_o)T + \phi \right] \right\} d\tau \qquad (1.B.1)$$

We expand the cosine term in series as above, in the case of the Doppler shift. This again is valid since the frequency shift associated with the phase shift is very small. We differentiate the resulting expression with respect to ω, equate the result to 0 and obtain:

$$\omega_\phi - \omega_o = -\frac{\phi}{L} \frac{\int_0^\infty (1/v) f(\tau) \sin^2 b\tau d\tau}{\int_0^\infty \left(1/v^2 \right) f(\tau) \sin^2 b\tau d\tau} \qquad (1.B.2)$$

APPENDIX 1.C: SQUARE WAVE FREQUENCY MODULATION AND FREQUENCY SHIFTS

In most practical implementations, the Ramsey central fringes is detected by means of a low frequency square wave modulation of the microwave frequency feeding the Ramsey cavity. We define ω_m as the depth of the frequency modulation of the microwave radiation. The value of ω_m is set approximately equal to one-half the width of the fringe as shown in Figure 1.C.1. The value of the signal on each side of the Ramsey fringe at $-\omega_m$ and $+\omega_m$ is then compared by means of a synchronous detector and the resulting signal is used to lock the microwave generator to the frequency that makes those signals equal. It is readily observed that the central frequency is dependent on the shape of the fringe and on possible distortions that are introduced by the frequency shifts that depend on the interaction time of the atoms.

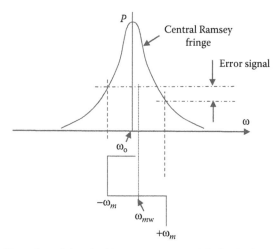

FIGURE 1.C.1   Detection of signal with square wave modulation in the case of the Cs beam frequency standard.

Let us assume that the frequency of a given atom is shifted by the amount $\Delta(\tau)$. We can then write the transition probability as:

$$P = \frac{1}{2}\int_0^\infty f(\tau)\left\{1+\cos\left[\omega_0 - \Delta(\tau) - \omega\right]T\right\}\sin^2 b\tau\, d\tau \qquad (1.C.1)$$

The cosine term can be transformed to:

$$\cos\left[\omega_0 - \omega - \Delta(\tau)\right] = \cos\left(\omega_0 - \omega\right)T\cos\Delta(\tau)T + \sin\left(\omega_0 - \omega\right)T\sin\Delta(\tau)T \qquad (1.C.2)$$

The term $\Delta(\tau)$ is small and we can write $\cos\Delta(\tau)\,T \sim 1$ and $\sin\Delta(\tau)T \sim \Delta(\tau)T$. We then have:

$$P = \frac{1}{2}\int_0^\infty f(\tau)\sin^2 b\tau\, d\tau$$

$$+\frac{1}{2}\int_0^\infty f(\tau)\left[\cos\left(\omega_0 - \omega\right)T + \sin\left(\omega_0 - \omega\right)T\Delta(\tau)T\right]\sin^2 b\tau\, d\tau \qquad (1.C.3)$$

This probability can be evaluated at the two frequencies $\omega = (\omega_1 - \omega_m)$ and $\omega = (\omega_1 + \omega_m)$, giving P_- and P_+ for the two cases. Making $P_- = P_+$, a condition fulfilled by the servo system and using trigonometric relations cosine and sine of summations of angles, we obtain the frequency shift as:

$$\left(\omega_0 - \omega_\Delta\right) = \frac{\displaystyle\int_0^\infty \Delta(\tau)Tf(\tau)\sin\omega_m\tau\sin^2 b\tau\, d\tau}{\displaystyle\int_0^\infty f(\tau)T\sin\omega_m\tau\sin^2 b\tau\, d\tau} \qquad (1.C.4)$$

where:

ω_A is the shifted measured frequency of the atoms

This expression is an approximation but is valid for any small shifts $\Delta(\tau)$ that are introduced in the system by physical phenomena such as second-order Doppler effect and light shift that will be examined later when optical pumping is used for state selection. However, it cannot be used in the case when a phase shift exists between the two arms of the cavity. In that case, a calculation similar to that just done gives for the frequency displaced by the cavity phase shift ϕ:

$$\left(\omega_o - \omega_\phi\right) = -\frac{\phi}{L}\frac{\int_0^\infty f(\tau)\sin\omega_m\tau\sin^2 b\tau\, d\tau}{\int_0^\infty \frac{1}{v}f(\tau)T\,\sin\omega_m\tau\sin^2 b\tau\, d\tau} \tag{1.C.5}$$

where we recall the $T = L/v$, and $\tau = l/v$.

APPENDIX 1.D: RING CAVITY PHASE SHIFT

In the text, we have introduced the concept of a ring cavity that in principle reduces the variation of phase shift with position across the Cs beam. We wish to derive the expression for this variation. Referring to Figure 1.8, the field around point $z = 0$ is the sum of the two components b_1 and b_2 injected at point A of the cavity. If we assume a propagation coefficient:

$$\gamma = \alpha + i\beta \tag{1.D.1}$$

where the propagation vector β is given by:

$$\beta = \frac{2\pi}{\lambda_g} \tag{1.D.2}$$

we may write the amplitude of the field as the sum of the two components H_1 and H_2

$$H = H_1 + H_2 = H_{10}e^{i\omega t}e^{\gamma(l/2+z')} + H_{20}e^{i\omega t}e^{\gamma(l/2-z')} \tag{1.D.3}$$

We may assume that $b_1 = b_2$ or $H_{10} = H_{20} = H_0$ at the entrance of the cavity and the expression becomes:

$$H = H_0 e^{i\omega t + \gamma l/2}\left(e^{\gamma z'} + e^{-\gamma z'}\right) \tag{1.D.4}$$

This can be written as:

$$H = 2H_0 e^{i\omega t + \gamma l/2}\cosh\left(\gamma z'\right) \tag{1.D.5}$$

Expanding in series to second order the cosh term and neglecting terms small relative to 1, we obtain:

$$H = 2H_0 e^{i\omega t + \gamma l/2}\left(1 + i\alpha\beta z'^2\right) \tag{1.D.6}$$

giving a phase shift $\alpha\beta\,z'^2$ at position z'.

APPENDIX 1.E: MAGNETRON CAVITY

The structure shown in Figures 1.26 and 1.E.1, called here "magnetron cavity," is also known as a loop gap resonator and was used in several fields of research in the domain of magnetic resonance (Hardy and Whitehead 1981; Fronzcis and Hyde 1982; Rinard and Eaton 2005). It is extremely versatile and its shape is well-adapted to applications in the field of atomic frequency standards in which a cell is used for containing the ensemble of atoms, such as in the case of H masers (Peters 1978b; Belayev and Savin 1987) and double resonance Rb clocks (Stefanucci et al. 2012).

The field within the electrodes, sometimes qualified as a TE_{011} like mode, is rather homogeneous and, in the application considered here, the cell containing the ensemble of atoms fills normally the space between those electrodes. Those electrodes can be thought of as inductances and the gaps between them as capacitances. Using the notation outlined in Figure 1.E.1, the resonance frequency of the structure is given by (Stefanucci et al. 2012):

$$v_r = \frac{1}{2\pi}\sqrt{\frac{n}{\pi r^2 \varepsilon \mu}}\frac{t}{w}\sqrt{1 + \frac{r^2}{R^2 - (r+w)^2}}\sqrt{\frac{1}{1 + 2.5(t/w)}} \tag{1.E.1}$$

The first term represents the resonance of the equivalent LC circuit, the second term the effect of the shield, and the third term the effect of fringing. This expression neglects the presence of the storage cell and the supports holding the electrodes. It is accurate to the order of a few percent (Stefanucci et al. 2012). It can be used as a first approximation for the determination of the various dimensions for a given frequency. A program, such as COMSOL Multiphysics software, is needed to adjust final dimensions and determine the quality of the field in the structure.

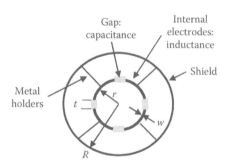

FIGURE 1.E.1 Cross section view of the magnetron cavity showing the electrodes and the shield, providing the notation used in the text.

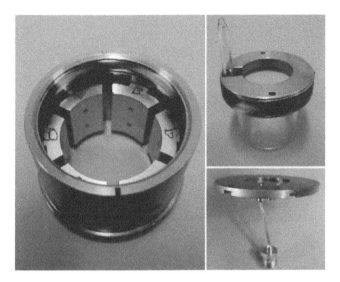

FIGURE 1.E.2 Photograph of the magnetron cavity implemented and used in an optically pumped Rb frequency standard. (Reproduced with permission from Stefanucci, C. et al., *Rev. Sci. Instr.*, 83, 104706, 2012. Copyright 2012 by the American Institute of Physics.)

For application in the case of the H maser, an important parameter is the filling factor given by Equation 1.68. In the case of a magnetron cavity, the magnetic field has the same phase in the whole volume enclosed by the electrodes and consequently the structure is well-adapted for that particular application. On the other hand, in the case of the Rb clock, the atoms movement is restricted by a buffer gas and the appropriate factor for qualifying the field in such a case is (*QPAFS* 1989):

$$\eta = \frac{\left\langle H_z^2(r) \right\rangle_{\text{cell}}}{\left\langle H^2(r) \right\rangle_{\text{cavity}}} \tag{1.E.2}$$

However, since the cell fills essentially the whole space between the electrodes, the numerator and denominator can both be considered as the average over the cell. In that case, the filling factor becomes really a measure of the quality of the field in the appropriate direction (parallel to the applied static magnetic field) and is called an orientation factor. In the case studied, this factor is very high (0.88) (Stefanucci et al. 2012). Figure 1.E.2 is a photograph of an actual prototype constructed with six electrodes, showing the cell sealed to the top tuning attachment and the coupling loop attached to the bottom section.

2 Recent Advances in Atomic Physics That Have Impact on Atomic Frequency Standards Development

Over the past several decades, the field of atomic physics has seen outstanding developments, which have had an important impact on the development of atomic frequency standards. This is mainly due to improvements in laser technology, which has had a direct effect on the technique of optical pumping and made possible the cooling and trapping of ensembles of atoms to sub-Kelvin (sub-K) temperatures. The laser as such was developed in the 1960s. In early stages of its development, however, the use of the laser in the study of interactions between electromagnetic radiation and atoms was limited by the very restricted number of wavelengths available with the existing lasers and the rather narrow tunability of the instruments developed. The so-called dye laser extended the range of possible studies, but such an instrument is rather cumbersome and its size and complexity restricted its use to limited laboratory applications. The advent of room temperature solid-state diode lasers, which could be fabricated to emit wavelengths in resonance with many transitions in alkali atoms such as Cs and Rb, for example, as well as in ions such as Sr^+ and Yb^+, among many others, changed the whole picture. The power emitted by such lasers may be several tens of milliwatts. The spectrum of the radiation emitted is relatively narrow and the frequency can be adjusted over a relatively broad range by means of doping the substrate with selected elements at the time of fabrication. Fine frequency tuning of the device can also be done by means of the temperature of the substrate and its driving current. This property allows the exciting of transitions in selected atoms from one single ground-state hyperfine level to an excited state. This makes possible nearly complete inversion of population in ground states through optical pumping. Furthermore, as will be seen below, the exchange of momentum between a photon and an atom makes possible alteration of atomic velocities leading to so-called laser cooling of atoms. These two possibilities by themselves have totally altered the landscape of research in the field of atomic frequency standards leading to new realizations in both the microwave and the optical ranges. We will review selected subjects and recent advances in atomic physics that have had a specific impact on the development of atomic frequency standards. For a broader overview of such advances in atomic physics, the reader is referred to the excellent review on the subject by Cohen-Tannoudji and Guéry-Odelin (2011).

2.1   SOLID-STATE DIODE LASER

In the following paragraphs, we will give an elementary description of the solid-state semiconductor diode laser. The basic idea behind the operation of such a laser is the emission of a photon when an electron recombines with a hole in the transition region of a pn junction. The emission of such a photon takes place in the so-called direct band gap semiconductors. This is not the case of indirect band gap semiconductor materials such as Ge and Si used at large in solid-state electronics. However, the condition exists with combined elements of the so-called III–V columns of the table of elements such as Ga and As forming a gallium–arsenide semiconductor. In indirect band gap semiconductors such as Si and Ge, recombination of electrons with holes injected across the transition region leads to the creation of phonons rather than photons, producing heat rather than electromagnetic radiation.

2.1.1   BASIC PRINCIPLE OF OPERATION OF A LASER DIODE

We have outlined the basic quantum theory behind laser operation in *QPAFS*, Volume 2 (1989). The reader is referred to that volume for details on the theory of stimulated emission of radiation within an optical cavity. The basic idea is similar in the case of diode lasers. As illustrated in Figure 2.1, a pn junction is made by the joining of two identical semiconductor materials doped with different atoms known as impurities, and identified as acceptors (p) and donors (n) of electrons. When the doping is large, the conduction and valence bands adjust themselves in such a way that the Fermi level resides in both the conduction band in the n type material and the valence band in the p type material. The highly doped substrate is called a degenerate material.

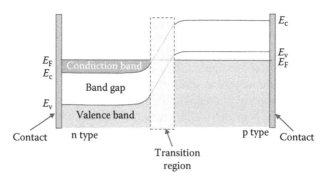

FIGURE 2.1   Schematic representation of the energy band structure of a pn junction for highly doped (degenerate) semiconductor materials. The Fermi level is in the conduction band for the n type region and in the valence band for the p type region.

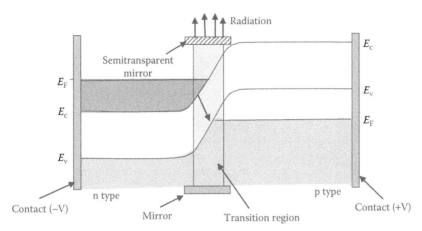

FIGURE 2.2 Schematic illustration of the effect of applying a direct voltage on a pn junction formed with highly doped material.

In thermal equilibrium, when no voltage is applied to the junction, the Fermi level is constant through the device. When a direct voltage is applied to the material by means of contacts, + on the p side and – on the n side, the Fermi level is displaced in each region and there is injection of holes and electrons from the respective p and n substrates into the transition region. This effectively creates an inversion of population in the transition region as illustrated in Figure 2.2.

An electron injected in the transition region tends to recombine with a hole and emit in a direct process a photon whose energy corresponds to the band gap energy. This is different from the process that takes place in a standard semiconductor pn junction diode, in which the junction is thin and electrons are injected directly in the bulk p material where they recombine with holes, producing heat. In the present case, the recombination creates a photon, like in the case of spontaneous emission. If mirrors are placed appropriately forming an optical cavity enclosing the junction as shown in Figure 2.2, such a photon is reflected and may stimulate another recombination process, producing a photon in phase and with the same polarization as the stimulating one, like in the process of stimulated emission discussed earlier. We thus have an amplifying medium, and laser action may take place if threshold is reached, that is, if the gain is high enough and compensates for the losses in the medium. As in the case of gas lasers, one mirror is made semitransparent and energy is coupled out of the device. The mirrors are usually created by cleavage at the edges of the substrate, creating parallel reflecting facets. Below the threshold, the process of recombination leads to incoherent radiation, which is the characteristic of light-emitting diodes known as LED. The physical construction of such a device may look as shown in Figure 2.3. It is generally known as a Fabry–Pérot (FP) semiconductor laser. The radiation is maintained in the active region, acting as a guide through the use of materials of different indexes of refraction (index guided). The emitted beam is slightly divergent because of the short distance between the mirrors and the narrow width of the transition region.

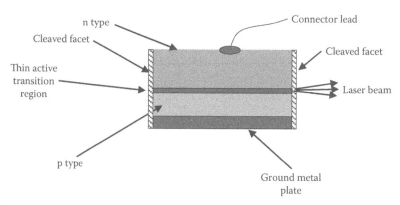

FIGURE 2.3 Typical construction of a conventional Fabry–Pérot semiconductor laser diode using cleaved facet of the substrate as the mirrors (dimensions are not to scale). The output laser beam is somewhat divergent due to the thinness (μm) and limited length (mm) of the active region from which it originates.

The beam generally also shows an asymmetry in orthogonal directions due to the flat geometry of the active region.

The condition of oscillation of the semiconductor laser may be derived as follows. The probability of occupation of level E_n by an electron in the conduction band is given by Fermi–Dirac statistics:

$$f(E_n) = \left[1 + \exp\left(\frac{E_n - \bar{\mu}_n}{kT}\right)\right]^{-1} \tag{2.1}$$

Similarly, for a hole in the valence band, the probability of occupation of level E_p is:

$$f(E_p) = \left[1 + \exp\left(\frac{E_p - \bar{\mu}_p}{kT}\right)\right]^{-1} \tag{2.2}$$

where:
   $\bar{\mu}_n$ and $\bar{\mu}_p$ are called the quasi Fermi levels since they are displaced by the applied voltage

We define ρ_v the radiation density at frequency v, W_{cv} the probability of a transition, and C a constant containing the line width, a characteristic of the transition. The rate of photon absorption dN_a/dt is proportional to the transition probability W_{cv}, the occupation probability f_v of a level in the valence band, the probability f_c that a level is free in the conduction band, and ρ_v the radiation density in the junction:

$$\frac{dN_a}{dt} = CW_{vc} f_v \left(1 - f_c\right)\rho_v \tag{2.3}$$

while the rate of emission following the same line of thought is:

$$\frac{dN_e}{dt} = CW_{cv}f_c(1-f_v)\rho_v \tag{2.4}$$

We have net gain and continuous coherent radiation emission if:

$$\frac{dN_e}{dt} > \frac{dN_a}{dt} \tag{2.5}$$

As in stimulated emission, we have $W_{cv} = W_{vc}$ and the oscillation condition is thus:

$$f_c(1-f_v) > f_v(1-f_v) \tag{2.6}$$

that is:

$$f_c > f_v \tag{2.7}$$

or in other words:

$$f(E_n) > f(E_p) \tag{2.8}$$

This means that the probability of occupation must be larger in E_n than in E_v leading to an inversion of population. Consequently, using Equations 2.1 and 2.2, we have:

$$E_n - \bar{\mu}_n < E_p - \bar{\mu}_p \tag{2.9}$$

which leads to:

$$\bar{\mu}_n - \bar{\mu}_p > E_n - E_p \tag{2.10}$$

Since we have:

$$\bar{\mu}_n - \bar{\mu}_p = eV_a \tag{2.11}$$

where V_a is the voltage applied to the junction, it is thus required that:

$$eV_a > h\nu \tag{2.12}$$

This is the condition of laser oscillation. This is a very elementary description of the semiconductor operation; for a more complete analysis, the reader is referred to books on fundamentals of laser diodes (e.g., Chow and Koch 2011).

2.1.2 BASIC CHARACTERISTICS OF THE SEMICONDUCTOR LASER DIODE

In their early development, laser diodes were inefficient, required much driving power, and could be operated essentially only in pulse mode. Heterojunction diodes were developed in which a thin layer of a smaller band gap semiconductor is introduced in the transition region between the p and n substrate. The wavelength emitted depends primarily on the band gap of the material chosen and on its doping. With Al doping, it is

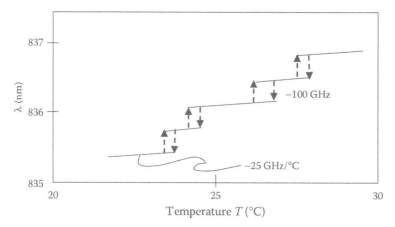

FIGURE 2.4 Typical (idealized) characteristics, λ versus *T*, of AlGaAs laser diode. (Based on data from Hitachi Application Note HLN500.)

possible to obtain a wavelength output in the range of 750–850 nm. These wavelengths fall most interestingly in the range of the wavelength used in optical pumping of Rb and Cs. Unfortunately, depending on the size of the junction, various modes of oscillation may be excited with slightly different wavelengths. This is due to the fact that the gain of the active region extends over a rather broad range of wavelengths and wavelength selection is done largely on the basis of device gain. Although such a laser may operate at a single frequency, a shift in gain due to a change in temperature may cause mode jumps, temperature having an effect on both the device index of refraction and length. This may be undesirable in applications where the frequency needs to be well determined and set accurately in order to interact with a single pair of levels in absorption experiments, such as in optical pumping and atom cooling. On the other hand, the diode current tends to heat up the junction leading to temperature changes of approximately 0.06–0.2 K/mA, depending on construction, causing simultaneously a frequency shift of 1.5–6 GHz/mA. This characteristic can be used, of course, for fine-tuning in well-behaved regions between mode hops. A particular characteristic (somewhat idealized) of a laser diode emission wavelength is illustrated in Figure 2.4 as a function of temperature. It is observed that the wavelength varies in steps and these abrupt steps are the mode jumps mentioned above. These jumps may be rather disturbing since the mode transition may fall at the wavelength desired and that, furthermore, hysteresis may cause the emission wavelength to be irreproducible.

2.1.3 Types of Laser Diodes

The problem just mentioned may be resolved by etching a diffraction grating in the active region of the device, providing active internal feedback. These devices are called distributed feedback (DFB) lasers. A schematic of the arrangement is shown in Figure 2.5.

    In such a construction, as its name indicates, the radiation created in the active region is reflected back and forth by the grating creating DFB with appropriate phase

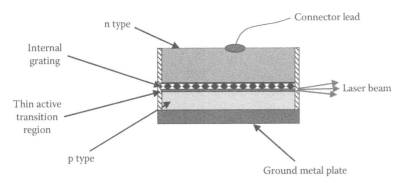

FIGURE 2.5   Schematic representation of a DFB laser diode.

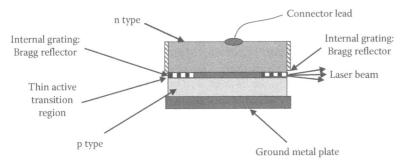

FIGURE 2.6   Schematic representation of a distributed Bragg reflector semiconductor laser diode (DBR).

for particular wavelengths. One mode is re-enforced by constructive interferences at the expense of other modes whose excitation are inhibited by destructive interference. A single wavelength is delivered at the output. In another type of laser, the grating is etched near the cavity ends as in Figure 2.6. These regions are then inactive regarding gain and act as wavelength selective mirrors. They are called Bragg reflectors. Selectivity of wavelength forces the device to emit a single wavelength with efficient suppression of side modes. The spectral width of the radiation emitted is at least an order of magnitude (~MHz) smaller than in the case of standard FP diode lasers (~100 MHz).

Another interesting type of semiconductor laser is constructed in such a way that the radiation emission is perpendicular to the transition or active region. It is called a vertical cavity surface emitting laser or VCSEL and is constructed also with internal structures (Bragg reflectors) acting as selective reflectors. They are small, consume very little power, and emit radiation as single modes at the milliwatts level. A schematic of its construction is shown in Figure 2.7.

In solid-state semiconductor lasers, the spectral width of the radiation emitted, although narrow relative to some atom spectral lines is generally large enough as to cause undesirable effects in several applications. As mentioned above, the wavelength of the emitted radiation is very sensitive to the driving current and the temperature of operation. It is generally required to control rigidly both the temperature of operation and the driving current in order to stabilize grossly the frequency of the emitted radiation.

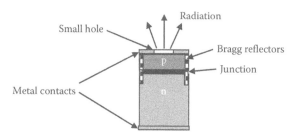

FIGURE 2.7   Schematic diagram of a VCSEL diode.

An efficient method of reducing the spectral width consists in placing the laser diode in an external optical cavity (FP) with a high finesse, the pn junction diode acting then as an amplifier. The frequency of the emitted radiation is determined by the optical cavity. The width of that radiation is then determined by the finesse of the optical cavity, with a width of the order of 100 kHz easily obtained. The line width and mode suppression can also be accomplished by means of a single element that reflects the radiation into the laser itself. The radiation acts as selective feedback. On the other hand, the frequency can be stabilized by locking it to an atomic absorption line. We shall outline these various stabilization techniques in Section 2.2.

2.1.4   OTHER TYPES OF LASERS USED IN SPECIAL SITUATIONS

A laser with variable wavelength is an extremely desirable tool in spectroscopy and is required in many instances in the field of frequency standards. The semiconductor lasers described above have addressed that question to some extent but as mentioned earlier have very limited range of tunability. In many occasions, the wavelength required in a particular application is not available with those lasers. We have mentioned in *QPAFS*, Volume 2, (1989) that the existence of the so-called dye lasers solve partly that problem. In that system, a chemical (dye) is diluted in alcohol or other solvent and pumped with an Ar^+ or other ion laser. The system can be tuned to emit a limited range of wavelength. However, such systems are rather complex, cumbersome, and delicate to operate. A new type of laser that addressed that question was realized in the 1980s and made commercially available in the 1990s (Moulton 1986). It is the Ti^{3+}-Al_2O_3 or so-called Ti-sapphire laser. The radiation originates from quantum transitions between vibration states of titanium ions imbedded as doping atoms in a sapphire crystal. Its gain is very broad, providing an emission spectrum from 650 to 1100 nm with a maximum at ~800 nm. With such a spectral breadth, the amplification gain is low and it requires powerful pumping. This is realized as in the case of the dye laser with an Ar^+ laser or other types of laser. However, the system is much simpler and more versatile than the dye laser and can be operated either continuously or in pulsed mode in the femtosecond range. It is extensively used in the field of atomic frequency standards particularly in the implementation of optical frequency standards and in the realization of the so-called optical comb used to measure optical frequencies in terms of microwave frequencies.

As will be seen later, laser cooling may require wavelengths that fall outside the spectrum of available lasers. For example, the Hg atom, which offers promising properties for the implementation of an optical frequency standard, can be cooled using an S–P transition at a wavelength of 254 nm. A coherent radiation field at that wavelength can only be reached by means of doubling or quadrupling the frequency of a longer wavelength laser. This process requires a rather powerful source of radiation. A wavelength in the range of 1015 nm can be generated with sufficient power by means of a so-called Yb-doped YAG crystal laser (*Versadisk*). Optical pumping of the laser is done with an array of diode lasers at 938 nm and 100 W. The crystal assembly is water cooled. The laser delivers up to 8 W at 1015 nm. For obtaining the required wavelength for cooling Hg atoms, the laser frequency is doubled twice. A power output of 600 mW at the desired frequency of 254 nm has been obtained with such an approach (Petersen et al. 2007; Mandache et al. 2008).

2.2 CONTROL OF WAVELENGTH AND SPECTRAL WIDTH OF LASER DIODES

2.2.1 LINE WIDTH REDUCTION

The characteristics just mentioned, frequency instability, line width, and mode hops may make laser diodes totally inappropriate to accomplish certain tasks required in optical pumping and laser cooling, subjects that will be described in Sections 2.3 and 2.5. Consequently, several techniques, either optical or electronic, have been devised to narrow the emission line width of those diode lasers and stabilize their frequency. Excellent reviews of these techniques have been published (de Labachelerie et al. 1992; Ohtsu et al. 1993). We will examine here the most important ones that are extensively used in the implementation of atomic frequency standards.

2.2.1.1 Simple Optical Feedback

It is possible to force a multimode laser diode to oscillate on a single mode by providing optical feedback from a small mirror or glass plate perpendicular to the radiation beam and placed close to one of the laser facets. The idea is to force the laser operation in a cavity mode created by the external mirror rather than in modes created by the cleavage surface of the semiconductor substrate. The limitations to the technique are that it is not efficient with all lasers and that it still requires some sort of frequency reference to stabilize the laser frequency. Furthermore, it can be difficult to achieve a long-term stable performance with such systems (Kanada and Nawata 1979; Goldberg et al. 1982; Saito et al. 1982).

2.2.1.2 Extended Cavity Approach

In a typical arrangement used in that approach, the system uses a frequency selective element, such as a grating, that reflects light in directions that are a function of the wavelength and redirects it back to the laser itself. Such an arrangement is called *extended cavity laser* (ECL) since it essentially places the laser-amplifying medium inside a cavity formed with one cleavage facet of the laser and an outside reflector. The grating is thus used as frequency discriminator. The arrangement reduces laser threshold current, narrows the line width, and provides broad tunability. Two optical configurations are generally used: the Littrow and the Littman–Metcalf configuration. They are illustrated in Figure 2.8.

Littrow design

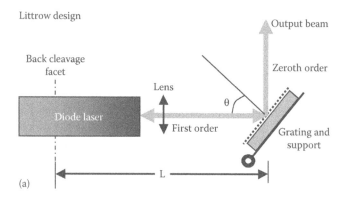

(a)

Littman–Metcalf design

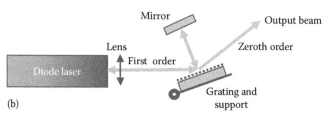

(b)

FIGURE 2.8   Schematic representation of the (a) Littrow and (b) the Littman–Metcalf configurations for the implementation of an ECL diode used in the stabilization of the frequency of such a laser.

In the Littrow configuration, the grating is aligned in such a way as to reflect back directly into the laser the first-order diffracted beam. The grating placed in front of the diode laser, acting as a frequency discriminator, feeds back into the laser the first diffraction order, and couples out part of the reflected light. In this configuration, the laser wavelength discrimination obeys the Bragg reflection condition as a function of the angle of incidence, θ:

$$\lambda = 2d \sin \theta \qquad (2.13)$$

where:
   d is the grating's line spacing
   θ is defined in Figure 2.8

On the other hand, the length of the extended cavity formed by the grating and the diode back reflecting face must be such as to satisfy the relation:

$$L = m \frac{\lambda}{2} \qquad (2.14)$$

where:
   m is an integer, satisfying the condition for providing a standing wave in the cavity

The zeroth-order reflection is used as the output beam. In a typical setup, with a grating of 1200 lines/mm and θ ~ 30°, one obtains $(d\lambda/d\theta) = 1.4$ nm/mrad. Tuning of the laser redirects the output beam by a value of $2(d\lambda/d\theta)^{-1} \approx 1.4$ mrad/nm since wavelength selection and discrimination are done simultaneously with the same element.

In the Littman–Metcalf configuration, the output beam from the laser diode is aligned at grazing incidence with the grating. The first-order diffracted beam is sent to a mirror (retroreflector) which reflects the beam back on itself. This reflected beam then hits the grating and the diffracted beam then couples back into the laser. Tuning in this case is achieved by varying the angle of the retroreflector, which changes the wavelength of the radiation sent back to the laser as optical feedback. With a mirror attached to the grating mount at a distance of 15 mm, a transverse displacement of the output laser beam of $(dx/d\lambda) = 18$ μm/nm results, which may be undesirable in some situations. The output is the zeroth-order beam reflected off the grating (Baillard et al. 2006).

The spectral line width for such configurations is strongly dependent on the feedback and also on the ratio between external cavity and diode cavity lengths. Line widths of the order of 100 kHz are achievable. Over long periods, various noise sources (acoustic wave propagation, air movements, driving current instabilities, mechanical and thermal drifts) contribute to increase the line width. The line width can also be temporally affected by phase fluctuations of the reflected light induced by mechanical vibrations of the external reflector (de Labachelerie and Cerez 1985). As pointed out above, the useful output beam direction is affected by tuning of the arrangement, which may be an inconvenience when the system is used as part of a complex arrangement with mirrors for reorienting the beam towards a collimator at the entrance of the desired experimental setup.

Another configuration employs a low loss interference filter as wavelength selector rather than a diffraction grating as used in the set ups described previously (Baillard et al. 2006). The external cavity is then formed of an external mirror with the interference filter placed between that mirror and the diode laser. The whole setup acts as a high finesse discriminator. A schematic of the arrangement is shown in Figure 2.9. The integration of an interference filter rather than a grating allows a linear design

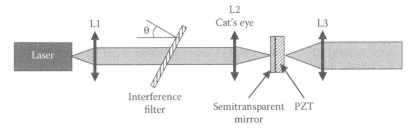

FIGURE 2.9 Schematic diagram of an external cavity laser configuration using an interference filter for wavelength selection. The various components are identified as: (L1) collimating lens 3–5 mm focal length; (L2) lens forming a so-called *cat's eye*, ~18 mm focal length; (L3) lens providing a collimated output beam with same focal length as L2; (PZT) piezoelectric transducer used for adjusting the position of the mirror, that is the length of the cavity. (Data from Baillard, X. et al., *Opt. Commun.*, 266, 609, 2006.)

of the setup, an arrangement that reduces the sensitivity of wavelength and external cavity feedback to misalignment. The feedback and wavelength selection functions are separated and both can be optimized independently leading to an increased tunability of the laser.

Wavelength discrimination of the interference filter is based on multiple reflections within its dielectric coatings and behaves as a thin FP etalon with effective index of refraction n_{ref}. The transmitted wavelength is given by:

$$\lambda = \lambda_{max}\sqrt{1 - \frac{\sin^2 \theta}{n_{ref}^2}} \qquad (2.15)$$

where:
  θ is the angle of incidence on the arrangement
  λ_{max} is the wavelength at normal incidence

It is possible to have filters with 0.3 nm bandpass and 90% transmission. With a λ_{max} of 853 nm, $n_{eff} \sim 2$, and the nominal wavelength at 6°, one obtains $d\lambda/d\theta = -23$ pm/mrad. This is much smaller than in the Littrow setup. This characteristic makes the setup less sensitive to vibrations. The horizontal displacement of the output beam with wavelength tuning in that setup, as reported by Baillard et al. (2006), is $dx/d\lambda = 8$ µm/nm, due to the thickness of the interference filter glass substrate, a large improvement over the Littrow setup.

2.2.1.3 Feedback from High-Q Optical Cavities

The essence of the optical feedback methods is that the line width is reduced by increasing the quality factor Q of the laser's resonator. The simplest implementation of the optical method for spectral narrowing is just to reflect back to the laser a small fraction of its output power. Using this technique, one can reduce the line width by more than a factor 1000 achieving line width of a few kilohertz (Dahmani et al. 1987; Laurent et al. 1989; Li and Telle 1989).

2.2.1.4 Electrical Feedback

Spectral line width reduction can also be accomplished by means of electrical feedback derived from the laser spectrum as measured with a FP interferometer (Ohtsu et al. 1985). In that implementation, the minimum value obtained was 330 kHz for an InGaAsP laser at 1.5 µm, which was 15 times narrower than that of a free-running laser.

2.2.1.5 Other Approaches

Kozuma et al. (1992) proposed an optical feedback method to control a semiconductor laser frequency by utilizing velocity-selective optical pumping (VSOP) and polarization spectroscopy of a rubidium (^{87}Rb) atomic vapour. In this way, the laser field spectral line width was reduced to 20 times less than that of a free running laser. Another approach to reduce the laser emission line width consists in placing the laser diode inside a relatively long cavity, with some means of wavelength selection (Fleming and Mooradian 1981; de Labachelerie 1988). This technique narrows the laser line width and provides a means of tuning the laser over a given range. The disadvantage is that it requires high quality antireflection coatings on the laser diode and a stable mechanical structure for the external cavity.

2.2.1.6   Locking the Laser to an Ultra-Stable Cavity

The approaches just described, which provide a significant improvement of a semiconductor laser spectrum, have nevertheless limitations. In some key applications, ultra-stable laser radiation with an ultra narrow spectrum is required. Those applications can be optical frequency standards (Ludlow et al. 2008; Rosenband et al. 2008), tests of relativity (Müller et al. 2007), transfer of optical stable frequencies by fibre networks (Jiang et al. 2008; Williams et al. 2008), and gravitational wave detection (Danzmann and Rudiger 2003; Acernese et al. 2006; Waldman 2006). These research topics have stimulated new approaches in the design of FP reference cavities which are used to stabilize semiconductor lasers. We will outline briefly some of those approaches.

One way to improve the spectral performance of stabilized lasers is to reduce vibration sensitivity by carefully designing the cavity geometry and its mounting. Several groups have proposed and successfully implemented low vibration sensitivity cavities (Nazarova et al. 2006; Ludlow et al. 2007; Webster et al. 2008). A second important issue is the reduction of thermal noise in cavity elements (Numata et al. 2004; Millo et al. 2009; Dawkins et al. 2010).

We will describe one typical setup assembled at Systèmes de Référence Temps-Espace (SYRTE), Paris, France. The core component is a FP cavity assembly designed to be highly immune to environmental perturbations such as temperature fluctuations and vibrations. The FP cavity is made of two high-quality mirrors (one flat and one concave with a 500 mm radius of curvature) attached with optical contact quality to the end of a 100 mm spacer made from ultra-low expansion material (ULE), which is very insensitive to environmental temperature fluctuations. Fused silica or ULE can be used as mirror substrate. The cavity is housed inside two nested vacuum enclosures, as shown in Figure 2.10. The outer vacuum surrounds an inner vacuum chamber which is actively temperature controlled via a three-wire

FIGURE 2.10   Photograph of the ultra-stable optical cavity used at LNE-SYRTE before final assembly. (Courtesy of SYRTE, France.)

temperature measurement by means of four series thermistors and feedback to four Peltier elements also in series. Additional passive isolation to temperature fluctuations is accomplished by means of two polished gold-coated aluminium shields. The windows used on the inner enclosure are made from BK7 and transmit the laser beam while blocking much of the thermal radiation above.

The other potential cause of fluctuations is the deformation induced by mechanical accelerations of the cavity. In order to minimize the effects of residual vibration, a vibration isolation system is used. The cavity is mounted in a vertical configuration (Dawkins et al. 2010) such that vertical vibrations at a centrally located mounting point cause approximately equal and opposite strain in the top and bottom half of the cavity structure (Taylor et al. 1995). The cavity can also be mounted in horizontal configuration (Millo et al. 2009).

In one setup, the system is mounted horizontally (cavity, vacuum chamber on an optical table) and is supported by an active vibration isolation platform. The cavity itself is supported under vacuum with four 2 mm² Viton pads 0.7 mm thick. In another setup, the cavity is mounted vertically and is isolated from vibration using a passive isolation table; it is supported under vacuum with the same Viton pads used for the horizontal setup. Air flow, acoustic noise, and large temperature fluctuations are strongly filtered by containing the whole system in a thermo-acoustic isolation box. The vacuum chamber temperature is actively stabilized at ~22°C.

For locking the laser onto one of the ultra-stable cavity resonances in order to obtain ultra-stable radiation, a Pound–Drever–Hall stabilization method is used (Drever et al. 1983). In that technique, the original Pound stabilizer used to frequency lock a klystron oscillator to a microwave cavity is adapted to the optical range. The microwave cavity is replaced by the high finesse optical cavity. Such a system is shown in Figure 2.11.

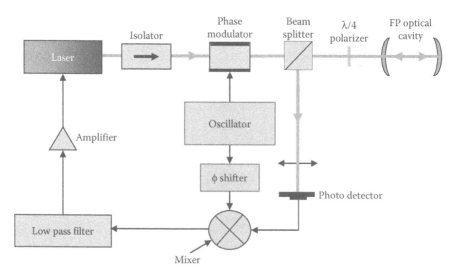

FIGURE 2.11 Block diagram of a Pound–Drever–Hall stabilizer used in the optical range to stabilize a laser on a high finesse cavity.

The finesse of the FP can be determined by means of the ringing of energy in the optical cavity under pulse excitation. Using that technique, a finesse of about 850,000 was measured for the particular cavity constructed by the Laboratoire National de Métrologie et d'Essais (LNE)-SYRTE group (Dawkins et al. 2010). With a free-spectral range of 1.5 GHz, the measured finesse corresponds to a cavity line width of 1.8 kHz. Using the setup just described, the frequency of a 1062.6 nm Yb-doped fibre laser was stabilized to the high finesse cavity. Using a link to LNE-SYRTE's primary frequency standards via an optical frequency comb (to be described in Section 4.6) the stabilized laser radiation signal produced was found to be stable to the 10^{-15} level and its frequency could be referenced directly to the primary frequency time standards, the Cs fountain.

2.2.1.6.1   Note on Laser Noise

Radiation emitted by a laser is subject to the presence of noise like all oscillators such as masers studied earlier. In the case of lasers, spontaneous emission taking place in the cavity is added into the laser coherent radiation mode. This is different from masers in the microwave range, in which spontaneous emission between hyperfine levels is totally negligible. It is the physical process that limits the line width of the laser radiation beam. We define the line width of a single mode laser as the full width at half maximum (FWHM) of the optical spectrum. Each spontaneous emission event adds energy with random phase to the optical radiation field. Some of the radiation created by the spontaneous emission process propagates along the same direction as that of the stimulated emission radiation. It cannot be separated from it. The main consequence of the spontaneous emission noise is to give the laser output radiation a finite spectral width (Schawlow and Townes 1958). The line shape is Lorentzian and the line width is given by (Schawlow–Townes formula):

$$\Delta v_{\text{laser}} = \frac{4\pi h v \left(\Delta v_c\right)^2}{P_{\text{out}}} \tag{2.16}$$

where:

P_{out} is the laser power output

Δv_c is the laser resonator bandwidth (half width at half maximum, HWHM)

Another type of noise that plays an important role in the use of a laser, for the purpose of optical pumping in particular, is intensity noise. It is usually characterized by means of the symbol RIN, for *relative intensity noise*, defined as

$$\text{RIN} = \frac{N_{\text{L}} + N_{\text{q}} + N_{\text{th}}}{P} \tag{2.17}$$

where:

N_{L} is the laser noise

N_{q} is shot noise

N_{th} is thermal noise

P is the laser average power

We will refer on occasion to these definitions.

2.2.2 LASER FREQUENCY STABILIZATION USING AN ATOMIC RESONANCE LINE

In the previous discussion, we outlined various ways of controlling the frequency of a laser essentially by means of mechanical devices such as mirrors and gratings. We have shown that the laser radiation line width could be reduced considerably and its frequency stabilized to a great extent. However, the laser frequency is determined by the tuning of the mechanical device used, in occurrence a FP cavity. This frequency is thus set arbitrarily by the actual construction of the mechanical device. In many cases, it is desirable to use the laser in experiments where its frequency must be set and maintained tuned to a given atomic resonance frequency. For example, in laser optical pumping and laser cooling the frequency of the laser must be tuned exactly to the frequency of a given quantum transition or close to such transition. Furthermore, that frequency must be kept tuned exactly to that frequency. This cannot be done with a mechanical device such as a FP cavity since its frequency is determined primarily by construction. We will now examine approaches to satisfy that requirement.

2.2.2.1 Locking the Laser Frequency to Linear Optical Absorption

Several microwave atomic frequency standards described in Chapter 1, for example, optical pumping for creating population inversion or still simply for detecting the resonance line. However, in such techniques, a typical situation may lead to a significant light shift of the microwave hyperfine transition used as the clock reference frequency. This shift may be as large as 5×10^{-11}/MHz of laser detuning. This sets the requirement that the laser frequency be stabilized to ~2 kHz to achieve a clock frequency stability of 10^{-13} over long-term periods. This is not a trivial requirement since it requires the frequency stability of the laser used to be of the order of 10^{-12} or better over the same long-term periods. This cannot be done by means of mechanical devices such as those described earlier.

The most common and simplest approach consists in locking the laser frequency to the optical absorption line observed in the same resonance cell as that used in the frequency standard. Such a system is shown in Figure 2.12.

However, these spectral absorption lines have rather large widths (700–1000 MHz), being broadened by Doppler effect and buffer gas collisions in the application mentioned. Consequently, the approach requires a tight frequency lock to a few parts in 10^6 of the atomic absorption line width. The locking of the laser frequency is generally done by low frequency modulation of the laser frequency through the driving current and synchronous detection. The error signal developed is applied directly to the driving current source, as illustrated in Figure 2.12.

In general, laser frequency stability is evaluated over short-term periods below 1000 s. This is a range where measurements can be made easily and information on the characteristics of the frequency lock loop efficiency can be obtained rapidly. For example, Tsuchida et al. (1982) evaluated from the error signal developed in the servo loop a laser frequency stability of the order of $10^{-11} \tau^{-1/2}$ for 10^{-2} s $< \tau < 100$ s, using stabilization on the D_2 linear absorption line of ^{85}Rb in a cell without buffer gas. In such a case, the optical absorption line width is Doppler broadened and is of the order of 500 MHz, which may limit the efficiency of the locking system. At averaging times longer than 100 s, laser frequency stability appeared to degrade rapidly.

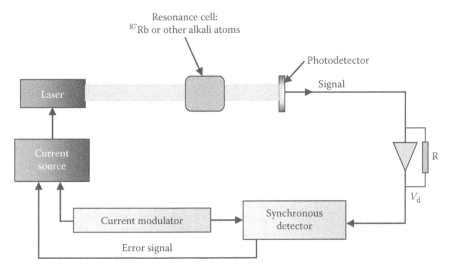

FIGURE 2.12   Block diagram of a typical servo system used to lock the frequency of a semiconductor laser diode to a linear absorption resonance in an Rb cell. The system may be used for optical pumping in a laser-pumped passive Rb standard.

Similar results were reported by Ohtsu et al. (1985) and Hashimoto and Ohtsu (1987), using the D_2 absorption line of ^{87}Rb as the reference. It should be mentioned that in those cases laser frequency stability was inferred from the error signal in the servo loop and results regarding frequency stability may have been optimistic. In experiments of the same nature in which lasers were frequency stabilized on an absorption line, two such lasers were measured and frequency stability was evaluated to be $4.2 \times 10^{-10} \, \tau^{-1/2}$ for the range 0.2–40 s and $\sim 10^{-11} \, \tau$ for longer periods, probably caused by drift (Barwood et al. 1988).

2.2.2.2   Locking the Laser Frequency to Saturated Absorption

In later experiments, Barwood et al. (1991) used saturated absorption technique and a laser whose line width was narrowed by optical feedback. Saturated absorption is the phenomenon observed in a cell when the atoms are excited by two radiation beams of the same frequency superimposed and travelling in opposite directions. The two radiation beams interact with the same atoms whose component of velocity in the direction of the beams creates a Doppler shift smaller than the natural line width of the transition excited. A frequency stability measured directly from the beat frequency between two such stabilized lasers gave $1.9 \times 10^{-11} \, \tau^{-1/2}$ (Barwood et al. 1991). A laser frequency stability of 4×10^{-12} was observed at an averaging time of 10 s corresponding to fluctuations of the order of 1.5 kHz, degrading slightly above that averaging time. In measurements made over several months, however, the laser frequency could be reproduced only to about 44 kHz. A typical setup for laser stabilization to saturated absorption is given in Figure 2.13. For a detailed discussion of the physics of the saturated absorption phenomenon, the reader is referred to Section 8.10.2 of *QPAFS*, Volume 2 (1989).

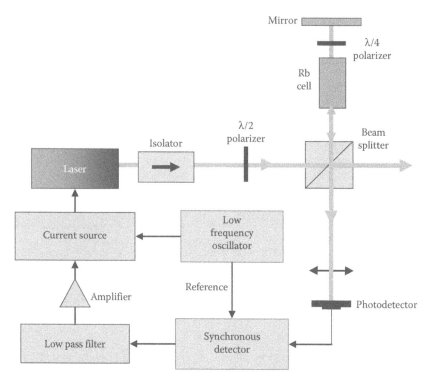

FIGURE 2.13 Block diagram of a typical servo system used for locking a laser frequency to saturated absorption. The system is similar to the linear absorption system described previously aside for the laser radiation beam that passes twice in the resonance cell, creating saturated absorption for those atoms in the path.

In another type of approach, Beverini used the dichroic property of a Cs vapour in a magnetic field for stabilizing the frequency of an external cavity diode laser system in the Littman configuration (Beverini et al. 2001). Although the width of the absorption line used for locking is still Doppler broadened, the technique, called DAVLL for dichroic-atomic-vapour-laser-lock, provides a much larger signal than that obtained by means of saturated absorption and is particularly robust against mechanical perturbations. It also offers a series of advantages requiring fewer optical and optoelectronic components and no frequency modulation of the diode laser current.

More recently, Affolderbach (Affolderbach and Mileti 2003, 2005; Affolderbach et al. 2004) has reported measurements on extended-cavity diode lasers (ECDL), stabilized to Rb reference cells by means of saturated absorption. They have measured the frequency stability of the laser itself. Their results are reproduced in Figure 2.14.

It is readily observed that the laser stabilized on Doppler broadened line gives a frequency stability of the order of 10^{-10} in the range of 10,000 s reflecting frequency fluctuations of the order of 375 kHz. Following the discussion made above relative to the possible presence of a light shift of the order of 5×10^{-11}/MHz laser detuning in optically pumped passive Rb standards, it is quite evident that such laser frequency fluctuations could be a handicap on long-term frequency stability of the clock.

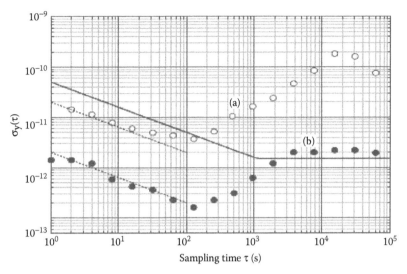

FIGURE 2.14 Allan standard deviation of lasers stabilized (a) on linear absorption (open circles) and (b) saturated absorption (filled circles). Dashed lines correspond to short-term stabilities of $2 \cdot 10^{-11} \tau^{-1/2}$ and $2 \cdot 10^{-12} \tau^{-12}$ for each scheme, respectively. The solid line illustrates typical stability required to avoid important light shift effects as discussed in the text. (Reproduced from Affolderbach, C. and Mileti, G., *Rev. Sci. Instrum.*, 76, 073108, 2005. With permission.)

2.3 LASER OPTICAL PUMPING

We have addressed the question of laser optical pumping in Volume 2 of *QPAFS* (1989) and have outlined the general characteristics of such an approach. To simplify reading of the detailed analysis concerned with microwave standards using laser optical pumping presented in Chapter 3, we will recall the main steps in the analysis in the present volume.

The three-level system used for the analysis is shown in Figure 2.15. It is a quite general case and may be applied to a particular atom by making explicit the hyperfine and Zeeman structure of the ground state and the fine and hyperfine structure of the excited state.

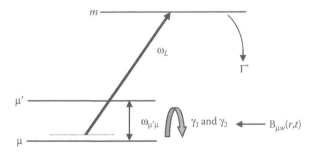

FIGURE 2.15 Three-level model used in the present analysis for double resonance in an alkali-type atom using coherent radiation at both microwave and optical frequencies. In practice, the laser is tuned close to either of the two transitions from μ or μ' to m.

In this model, the laser may be tuned to either of the ground-state levels. In the analysis, the laser spectrum is assumed to be broad enough to cover the possible splitting of each hyperfine level caused by an applied magnetic field, but much narrower than the ground-state hyperfine splitting. In microwave atomic frequency standards we are interested in the resonance signal of the field-independent transition F, $m_F = 0$ to F', $m_F = 0$. This may be taken into account at the end of the analysis by means of considerations on the number of atoms contributing to the resonance signal as compared to those contributing to the total light absorption. (Some material in Section 2.3 is reproduced with appropriate alterations from Vanier and Mandache [2007], with the kind permission of Springer-Verlag+Business Media, Berlin.)

2.3.1 RATE EQUATIONS

The dynamical behaviour of the matrix elements representing the population of the energy levels, ρ_{ii}, and the coherence, ρ_{ij}, existing in the system, is obtained as usual from Liouville's equation:

$$\frac{d}{dt}\rho_{jk} = (i\hbar)^{-1}\sum_{i}(\mathcal{H}_{ji}\rho_{ik} - \rho_{ji}\mathcal{H}_{ik})$$

(2.18)

where:

$\mathcal{H}$ is the interaction Hamiltonian of the various fields applied, in occurrence the laser and the microwave radiation fields

In the analysis, we use an approach slightly different from that used in *QPAFS* (1989). We keep all terms corresponding to the laser interaction with both levels of the ground state. The approach allows the possibility of examining the effect of pumping from one level or the other and leads to a more realistic evaluation of the light shift. The optical radiation field emitted by the laser is written as:

$$E(\omega_L,t,z) = E_{oL}e_\lambda \cos(\omega_L t + k \cdot r)$$

(2.19)

where:

E_{oL} is the radiation field amplitude
e_λ is the polarization vector
ω_L is the angular frequency
k is the propagation vector

We have a single laser radiation field applied. It is identified as subscript 1 and defines the optical Rabi angular frequencies $\omega_{1m\mu'}$ and $\omega_{1m\mu}$ as:

$$\omega_{1m\mu'} = \left(\frac{E_{oL}}{\hbar}\right)\langle\mu'|er\cdot e_\lambda|m\rangle = \left(\frac{E_{o1}}{\hbar}\right)d_{\mu'm}$$

(2.20)

$$\omega_{1m\mu} = \left(\frac{E_{oL}}{\hbar}\right)\langle\mu|er\cdot e_\lambda|m\rangle = \left(\frac{E_{o1}}{\hbar}\right)d_{\mu m}$$

(2.21)

where:

$\hbar$ is the Plank's constant over 2π

The terms within brackets are the so-called dipole matrix elements $d_{\mu m}$ and $d_{\mu'm}$ of the two possible transitions. We will study later the case where two laser radiation fields are present leading to interesting effects such as population trapping in a single state. We assume that a microwave field is present and has the form:

$$B_m(r,t) = zB_1\cos(\omega_M t) \tag{2.22}$$

where:

B_1 is the microwave field amplitude
ω_M is its angular frequency

This field could be created inside a cavity through excitation of a particular mode by means of an external oscillator. The Rabi angular frequency associated with this field is defined as:

$$\omega_{1g} = \frac{\mu_B\mu_o H_z(r)}{\hbar} \tag{2.23}$$

where:

μ_B is Bohr magneton
μ_o is the permeability of free space

For simplifying the calculation, we assume that the microwave field is homogeneous over the resonance cell. This assumption has no fundamental consequences on the calculation.

We evaluate the various terms in Equation 2.18 and obtain a set of rate equations for the populations and the coherence of the atomic ensemble. Relaxation in the ground state characterized by γ_1 and γ_2 and decay from the excited state characterized by the rate Γ^*, which includes spontaneous emission and relaxation by buffer gas collisions, are introduced in a phenomenological way. In fact, the ensemble of atoms is in close contact with the vacuum radiation field. This random radiation field has effects on the behaviour of the atomic ensemble in its excited state, causing a frequency shift observed at the optical pumping frequency (Lamb shift) and spontaneous emission taken care of by the term Γ^*. In view of the low laser light intensities used and the large decay rate from the excited P state, the fractional population of the excited state is always very small. This is taken into account in the equations for the ground-state populations, by assuming that the total population is the sum of the populations of the two ground-state energy levels. The rate equations are:

$$\frac{d\rho_{\mu'\mu'}}{dt} = -\omega_{1g}\,\mathrm{Im}\,\rho_{\mu\mu'}e^{-i\omega_M t} + \omega_{1m\mu'}\,\mathrm{Im}\,\rho_{\mu'm}e^{-i\omega_L t} + \frac{\Gamma^*}{2}\rho_{mm} - \gamma_1\left(\rho_{\mu'\mu'} - \frac{1}{2}\right) \tag{2.24}$$

$$\frac{d\rho_{mm}}{dt} = -\omega_{1m\mu}\,\mathrm{Im}\,\rho_{\mu m}e^{-i\omega_L t} - \omega_{1m\mu'}\,\mathrm{Im}\,\rho_{\mu'm}e^{-i\omega_L t} - \Gamma^*\rho_{mm} \tag{2.25}$$

$$\frac{d\rho_{\mu'm}}{dt} = i\omega_{m\mu'}\rho_{\mu'm} + i\frac{\omega_{1\mu'm}}{2}e^{i\omega_L t}\left(\rho_{mm} - \rho_{\mu'\mu'}\right)$$

$$+ i\frac{\omega_{1g}}{2}e^{-i\omega_M t}\rho_{\mu m} - i\frac{\omega_{1\mu m}}{2}e^{i\omega_L t}\rho_{\mu'\mu} - \frac{\Gamma^*}{2}\rho_{\mu'm} \tag{2.26}$$

$$\frac{d\rho_{\mu\mu'}}{dt} = i\omega_{\mu'\mu}\rho_{\mu\mu'} + \frac{i\omega_{1g}}{2}e^{i\omega_M t}\left(\rho_{\mu'\mu'} - \rho_{\mu\mu}\right)$$

$$+ i\frac{\omega_{1\mu m}}{2}e^{i\omega_L t}\rho_{m\mu'} - i\frac{\omega_{1m\mu'}}{2}e^{-i\omega_L t}\rho_{\mu m} - \gamma_2\rho_{\mu\mu'} \tag{2.27}$$

$$\frac{d\rho_{\mu m}}{dt} = i\omega_{m\mu}\rho_{\mu m} + i\frac{\omega_{1\mu m}}{2}e^{i\omega_L t}\left(\rho_{mm} - \rho_{\mu\mu}\right)$$

$$+ i\frac{\omega_{1g}}{2}e^{i\omega_M t}\rho_{\mu'm} - i\frac{\omega_{1\mu'm}}{2}e^{i\omega_L t}\rho_{\mu\mu'} - \frac{\Gamma^*}{2}\rho_{\mu m} \tag{2.28}$$

$$\frac{d\rho_{\mu\mu}}{dt} = +\omega_{1g}\,\mathrm{Im}\,\rho_{\mu\mu'}e^{-i\omega_M t} + \omega_{1m\mu}\,\mathrm{Im}\,\rho_{\mu m}e^{-i\omega_L t} + \frac{\Gamma^*}{2}\rho_{mm} - \gamma_1\left(\rho_{\mu\mu} - \frac{1}{2}\right) \tag{2.29}$$

$$\rho_{\mu'\mu'} + \rho_{\mu\mu} = 1 \tag{2.30}$$

These equations contain all information on the response of the atomic ensemble to the two electromagnetic fields applied, microwave and optical, taking into account the various relaxation mechanisms that are present in the ensemble. In practice, one is interested in determining the frequency at which resonance takes place between the microwave radiation at frequency ω_M and the atoms at frequency $\omega_{\mu'\mu}$ of the ground-state. The two ground state energy levels μ and μ' are hyperfine levels. There are various ways of detecting that resonance. One may detect the effect of transitions excited between those levels either on the fluorescent light resulting from the optical pumping or on the transmitted light. The first method in the case of dense medium as those encountered in Cs and Rb resonance cells is not appropriate. The reason is that generally a buffer gas such as nitrogen is used to quench the scattered or fluorescence radiation in order to avoid undesirable random optical pumping of the dense atomic ensemble by that radiation. Consequently, in most applications using closed cells, the resonance effect of the microwave radiation with the ground-state hyperfine transition is detected by means of its effect on the transmitted radiation.

2.3.2 FIELD EQUATION AND COHERENCE

The physical parameter looked for is the intensity of the radiation at the exit of the resonance cell as detected by a photodetector. Authors have taken various approaches in evaluating theoretically the effect of absorption on the transmitted radiation (Vanier and Audoin 1989; Mileti 1995; Scully and Zubairy 1999; Godone et al. 2002). All those approaches are equivalent. We choose the method developed in Godone et al. (2002). The technique is straightforward and gives rise to a first-order differential equation for the field amplitude at any point within the cell as a function of parameters that can be calculated from a solution of the set of Equations 2.24 through 2.30.

The local electric field E is a function of distance z within the cell and is related to the electric polarization P at that point by the equation (*QPAFS* 1989, Vol. 1):

$$\frac{\partial^2 E}{\partial z^2} - \varepsilon_0 \mu_0 \frac{\partial^2 E}{\partial t^2} = \mu_0 \frac{\partial^2 P}{\partial t^2} \qquad (2.31)$$

where:

ε_0 is the vacuum dielectric constant
μ_0 is the permeability of free space

The Rabi frequencies for both optical transitions and the electric field are related by Equations 2.20 and 2.21. We assume that the laser is tuned close to the transition μ to m and only atoms in level μ contribute to the absorption. This assumption is made possible because the laser spectrum is generally much narrower than the hyperfine splitting (MHz for the laser compared to GHz for the transition). However, the effect of the other transitions is kept in the equations since, even though it is off resonance, it causes a light shift through virtual transitions. Equation 2.31 can be transformed readily into (Godone et al. 2002):

$$\frac{\partial \omega_{1\mu m}}{\partial z} = \alpha \operatorname{Im} \delta_{\mu m} \qquad (2.32)$$

where $\delta_{\mu m}$ is the complex amplitude of the optical coherence $\rho_{\mu m}$ to be calculated below and

$$\alpha = \left(\frac{\omega_{m\mu}}{c \varepsilon_0 \hbar} d_{\mu m}^2 \right) n \qquad (m^{-1}s^{-1}) \qquad (2.33)$$

is the absorption coefficient, with c being the speed of light and n the atomic density.

The problem, thus, consists in evaluating $\delta_{\mu m}$ from Equations 2.24 through 2.30, and integrating Equation 2.32 for obtaining the Rabi frequency proportional to the electric field at any point within the resonance cell and particularly at the exit, where $z = L$, L being the length of the cell. From the information obtained, it is then possible in principle to draw conclusions regarding background radiation, contrast, and line width as a function of several parameters such as light intensity, atomic density, and microwave power.

In the following analysis we assume for the moment that the intensity of the laser radiation is constant in the radial direction. We assume that the coherences excited by the laser and the microwave radiation are given by the expressions:

$$\rho_{\mu\mu'}(z,t) = \delta_{\mu\mu'}(z,t)e^{i\omega_M t} \qquad (2.34)$$

$$\rho_{\mu m}(z,t) = \delta_{\mu m}(z,t)e^{i\omega_L t} + \varepsilon_{\mu m}(z,t)e^{i(\omega_L + \omega_M)t} \qquad (2.35)$$

$$\rho_{\mu'm}(z,t) = \delta_{\mu'm}(z,t)e^{i\omega_L t} + \varepsilon_{\mu'm}(z,t)e^{i(\omega_L - \omega_M)t} \qquad (2.36)$$

The calculation of $\delta_{\mu m}$ required in Equation 2.32 involves some algebra but is straight-forward, as described in *QPAFS*, Volume 2 (1989). In the analysis, we assume the following values for various parameters, values that are typical of experimental conditions:

$$\gamma_1 \sim \gamma_2 \sim 500 \;\; (\text{s}^{-1}), \qquad\qquad \Gamma^* = 2\times10^9 \;\; (\text{s}^{-1})$$

The Rabi frequencies are variable but are of the order of:

$$\omega_{1g} \sim 1 \text{ to } 5\times10^3 \;\; (\text{s}^{-1}), \qquad\qquad \omega_{1\mu n} = \omega_{1\mu'm} \sim 1 \text{ to } 5\times10^6 (\text{s}^{-1})$$

In view of the various decay and relaxation rates, the values just chosen make all optical coherences very small relative to the ground-state coherence. Furthermore, at laser intensities used in the present application and as mentioned above, the fractional population of the excited state ρ_{mm} is of the order of 10^{-6}. These considerations make possible a number of approximations valid over a broad range of microwave and optical Rabi frequencies. We replace Equations 2.34 through 2.36 into Equations 2.24 through 2.30 and we look for a stationary solution of each of those equations. We find that the ground-state coherence created by the microwave field is given by:

$$\delta_{\mu\mu'} = i\frac{(b/2)\left[\gamma_2 + \left(\Gamma_{p\mu} + \Gamma_{p\mu'}\right)/2\right]}{\left[\gamma_2 + \left(\Gamma_{p\mu} + \Gamma_{p\mu'}\right)/2\right]^2 + \left[\omega_M + \left(\Delta\omega_{l\mu} - \Delta\omega_{l\mu'}\right) - \omega_{\mu'\mu}\right]^2}\Delta \qquad (2.37)$$

where Δ, the population difference between the two ground-state levels, is calculated as:

$$\Delta = \frac{(1/2)\left(\Gamma_{p\mu} - \Gamma_{p\mu'}\right)}{\left[\gamma_2 + \left(\Gamma_{p\mu} + \Gamma_{p\mu'}\right)/2\right] + \dfrac{b^2\left[\gamma_2 + \left(\Gamma_{p\mu} + \Gamma_{p\mu'}\right)/2\right]}{\left[\gamma_2 + \left(\Gamma_{p\mu} + \Gamma_{p\mu'}\right)/2\right]^2 + \left[\omega_M + \left(\Delta\omega_{l\mu} - \Delta\omega_{l\mu'}\right) - \omega_{\mu'\mu}\right]^2}} \qquad (2.38)$$

On the other hand, the optical coherence $\delta_{\mu m}$ is a solution of the equation:

$$\delta_{\mu m} = -i\frac{\omega_{1\mu n}/2}{\left[\left(\Gamma^*/2\right) + i\left(\omega_L - \omega_{m\mu}\right)\right]}\rho_{\mu\mu}$$

$$+ \frac{\left(\omega_{1g}/2\right)^2}{\left[\left(\Gamma^*/2\right) + i\left(\omega_L - \omega_{m\mu}\right)\right]\left[\left(\Gamma^*/2\right) + i\left(\omega_L - \omega_M - \omega_{m\mu'}\right)\right]}\delta_{\mu m} \qquad (2.39)$$

$$- \frac{\left(\omega_{1\mu n}/2\right)\left(b/2\right)}{\left[\left(\Gamma^*/2\right) + i\left(\omega_L - \omega_{m\mu}\right)\right]\left[\left(\Gamma^*/2\right) + i\left(\omega_L - \omega_M - \omega_{m\mu'}\right)\right]}\delta_{\mu'\mu}$$

A similar expression is obtained for $\delta_{\mu'm}$:

$$\delta_{\mu'm} = -i \frac{\omega_{1\mu'm}/2}{\left[\left(\Gamma^*/2\right)+i\left(\omega_L-\omega_{m\mu'}\right)\right]}\rho_{\mu'\mu'}$$

$$-\frac{\left(\omega_{1g}/2\right)^2}{\left[\left(\Gamma^*/2\right)+i\left(\omega_L-\omega_{m\mu'}\right)\right]\left[\left(\Gamma^*/2\right)+i\left(\omega_L+\omega_M-\omega_{m\mu}\right)\right]}\delta_{\mu'm} \quad (2.40)$$

$$+\frac{\left(\omega_{1\mu'm}/2\right)(b/2)}{\left[\left(\Gamma^*/2\right)+i\left(\omega_L-\omega_{m\mu'}\right)\right]\left[\left(\Gamma^*/2\right)+i\left(\omega_L+\omega_M-\omega_{m\mu}\right)\right]}\delta_{\mu\mu'}$$

In these expressions, the following definitions have been introduced:

$$\Gamma_{p\mu} = \frac{\left|\omega_{1\mu m}/2\right|^2 \Gamma^*}{\left(\Gamma^*/2\right)^2+\left(\omega_L-\omega_{mL}-\omega_{m\mu'}\right)^2}, \qquad \text{pumping rate from level } \mu \quad (2.41)$$

$$\Gamma_{p\mu'} = \frac{\left|\omega_{1\mu'm}/2\right|^2 \Gamma^*}{\left(\Gamma^*/2\right)^2+\left(\omega_L+\omega_M-\omega_{m\mu}\right)^2}, \qquad \text{pumping rate from level } \mu' \quad (2.42)$$

$$\Delta\omega_{l\mu} = \frac{\left|\omega_{1\mu m}/2\right|^2\left(\omega_L-\omega_M-\omega_{m\mu'}\right)}{\left(\Gamma^*/2\right)^2+\left(\omega_L-\omega_M-\omega_{m\mu'}\right)^2}, \qquad \text{light shift of level } \mu \quad (2.43)$$

$$\Delta\omega_{l\mu'} = \frac{\left|\omega_{1\mu'm}/2\right|^2\left(\omega_L+\omega_M-\omega_{m\mu}\right)}{\left(\Gamma^*/2\right)^2+\left(\omega_L+\omega_M-\omega_{m\mu}\right)^2}, \qquad \text{light shift of level } \mu' \quad (2.44)$$

We take note that:

$$\omega_L-\omega_M-\omega_{m\mu'} = \omega_L-\omega_{m\mu} \quad (2.45)$$

and

$$\omega_L+\omega_M-\omega_{m\mu} = \omega_L-\omega_{m\mu'} \quad (2.46)$$

The above expressions for the light shift can of course be derived in several other ways, for example, by means of the operator formalism developed by Happer and Mather (1967) or still by considering optical pumping as a relaxation process (Vanier 1969). In the present case, that is laser optical pumping, keeping the optical coherence allows the evaluation of the light shift and of the light transmitted by means of Equation 2.32, even though the final mechanism is intensity optical pumping.

With the values assumed for the Rabi frequencies, several approximations can be made. Since we have assumed ω_{1g} to be of the order of 10^3, $\omega_{1\mu m}$ of the order of 10^6, and Γ^* of the order of 10^9, the two last terms on the right-hand side of Equations 2.39 and 2.40 are negligible. We are thus left with:

$$\delta_{\mu m} = -i \frac{\omega_{1\mu m}/2}{\left[\left(\Gamma^*/2\right) + i\left(\omega_L - \omega_{m\mu}\right)\right]} \rho_{\mu\mu} \qquad (2.47)$$

This essentially implies that the coherence introduced in the transition from μ to m is driven mainly by the direct interaction of the laser with that pair of levels and not by feedback from the other transition μ' to m through the ground-state coherence $\delta_{\mu\mu'}$.

The value of $\rho_{\mu\mu'}$ is obtained from the set of Equations 2.24 through 2.30. Within the approximations just made, we finally obtain:

$$\rho_{\mu\mu} = \left(\frac{1}{2}\right)\frac{\gamma_1 + \Gamma_{p\mu'}}{\gamma_1'} + \frac{1}{\gamma_1'}\frac{S}{S+1}\frac{\left(1/4\right)\left(\Gamma_{p\mu} - \Gamma_{p\mu'}\right)}{1 + \left(\omega_M - \omega_{\mu'\mu}'\right)^2 / \left[\gamma_2'(S+1)\right]} \qquad (2.48)$$

where:

$$\gamma_1' = \gamma_1 + \frac{1}{2}\left(\Gamma_{p\mu} + \Gamma_{p\mu'}\right) \qquad (2.49)$$

$$\gamma_2' = \gamma_2 + \frac{1}{2}\left(\Gamma_{p\mu} + \Gamma_{p\mu'}\right) \qquad (2.50)$$

$$\omega_{\mu'\mu}' = -\left(\Delta\omega_{l\mu} - \Delta\omega_{l\mu'}\right) + \omega_{\mu'\mu} \qquad (2.51)$$

and S is the microwave saturation factor defined as:

$$S = \frac{\omega_{1g}^2}{\gamma_1'\gamma_2'} \qquad (2.52)$$

We note that Equation 2.48 is identical to the result obtained in *QPAFS*, Volume 1 (1989), for a calculation made in a phenomenological way in terms of pumping rates only. In the present case, a complete analysis in terms of the coherence introduced by the laser radiation provides exact expressions for the light shift and pumping rates and gives insight on the relative importance of various coherences introduced by the laser and the microwave radiation. The first term of Equation 2.48 gives the equilibrium population under the effect of optical pumping alone, assuming the laser tuned close to the transition from the lower ground-state level μ to the excited level m. The second term gives the resonance signal under the influence of the applied microwave radiation represented by the Rabi angular frequency ω_{1g}. An equation similar to that of Equation 2.48 can be obtained for the laser tuned close to the transition from the upper ground-state level μ' to the excited level m. Equation 2.48 is valid locally. Since the laser radiation is absorbed in the cell, the pumping light intensity decreases with the distance z travelled within the cell. Consequently, the pumping rates Γ_{pi} are functions of distance within the resonance cell. In the case that the laser is tuned exactly to the transition μ to m, $\Gamma_{p\mu'}$ is very small compared to $\Gamma_{p\mu}$ and can be neglected. $\Gamma_{p\mu}$ is then given by Equation 2.41, which becomes:

$$\Gamma_{p\mu} = \frac{|\omega_{1\mu m}|^2}{\Gamma^*} \tag{2.53}$$

where:

$\omega_{1\mu m}$ is a function of z

Equation 2.47 is replaced in Equation 2.32, and at optical resonance we have:

$$\frac{d\omega_{1\mu m}}{dz} = -\alpha \frac{\omega_{1\mu m}}{\Gamma^*} \rho_{\mu\mu} \tag{2.54}$$

with $\rho_{\mu\mu}$ given by Equation 2.48. On the other hand, the FWHM is obtained from Equation 2.48 and is given by:

$$\Delta \nu_{1/2} = \frac{1}{\pi} \gamma_2'(S+1)^{1/2} \tag{2.55}$$

This is the essential result that can be derived analytically for laser optical pumping in a cell. The final solution cannot be written analytically. We will solve the equations numerically in Section 3.4 where we will apply the present results to obtain signal amplitude, contrast, and light shifts as a function of various parameters in a practical application with cells of various dimensions used to implement a frequency standard optically pumped by means of laser radiation.

2.4 COHERENT POPULATION TRAPPING

The demand for small atomic frequency standards with moderate frequency stability has led to the search for new phenomena at the atomic level that could be used as an alternative to the classical approaches just described. Coherent Population Trapping (CPT), a quantum phenomenon observed in the 1970s (Alzetta et al. 1976), turns out to be appropriate for such a purpose. We have introduced it briefly in *QPAFS*, Volume 1 (1989). In early studies, the phenomenon was perceived as an inhibition to efficient optical pumping in optically pumped Cs beam frequency standards and attempts were made to avoid it by slightly detuning the laser radiation from exact resonance (Gray et al. 1978). It was soon realized, however, that the effect led to rather narrow resonance lines, particularly in alkali metal atoms and that the coherence introduced in the ground state had rather interesting characteristics similar to those encountered in the field of microwave atomic frequency standards (QPAFS 1989). In the 1960s, those characteristics were exploited to implement a passive atomic frequency standard (Cyr et al. 1993; Levi et al. 1997; Vanier et al. 1998) and even a maser (Godone et al. 2000).

The CPT phenomenon is observed by irradiating an alkali atom, for example, with two coherent laser radiation fields applied in a so-called Λ scheme. CPT makes possible the preparation of the atomic ensemble and resonant excitation at the same time and same region of space. In the case of alkali atoms, the coherent fields are applied to an ensemble of atoms in resonance with the transitions between the two hyperfine levels of the $S_{1/2}$ ground state and one of the P state hyperfine levels, forming the Λ scheme. Due to internal physical quantum properties, coherence is created in the ground state and interference appears in the quantum excitation process. The ensemble is placed in a non absorbing state called a *dark state*. The phenomenon was called CPT because the atoms appeared to be trapped in their respective levels since no actual transitions take place. The phenomenon, observed at a hyperfine frequency, was observed for the first time by Alzetta et al. (1976) in sodium by means of a multimode dye laser. A related phenomenon was also reported earlier at the Zeeman frequency of Cs and Rb vapours using an amplitude modulated spectral lamp (Bell and Bloom 1961). (Some of the material presented in this section is re-edited with appropriate changes from Vanier [2005] with the kind permission of Springer Science+Business Media.)

The phenomenon may be observed by means of several techniques. Since in CPT no atoms are excited to the P state at exact resonance, a narrow dark line in the fluorescence spectrum of the optically pumped ensemble is observed. Furthermore, since the atomic ensemble does not absorb energy it becomes transparent at resonance. Consequently, the phenomenon can either be observed on the fluorescence as a dark line or on the transmitted radiation as a bright line. On the other hand, the ensemble is placed in a superposition state of the two alkali atom hyperfine ground levels and coherence at the hyperfine frequency is created. This coherence, in turn, creates an oscillating magnetization that can be detected directly in a cavity as in a maser, but without population inversion.

Early proposals were made on the use of the phenomenon in several applications such as magnetometry (Scully and Fleischhauer 1992; Nagel et al. 1998), induced transparency (Kasapi et al. 1995; Harris 1997; Scully and Zubairy 1999), atom cooling (Aspect et al. 1989), precision spectroscopy (Wynands and Nagel 1999), and its properties were studied in connection to state selection improvement in optical pumping (Avila et al. 1987).

On the other hand, the resonance phenomenon reflects all the properties of the hyperfine resonance in the ground state of the alkali atoms and may be used to implement a passive atomic frequency standard in the same way as is done in the classical approach using double resonance microwave-optical pumping or still an active maser exploiting the coherence that is created at the hyperfine frequency. A first effort was reported by Thomas et al. on the use of the phenomenon towards implementing an atomic clock (Thomas et al. 1982). In those experiments, the phenomenon was used on a sodium beam to implement a Ramsey type of separated interaction regions entirely with light beams without state selector magnets and the classical Ramsey microwave cavity. Serious efforts were also started in the mid-1990s on the application of the CPT phenomenon towards the use of sealed-off cells to implement small-scale atomic frequency standards (Cyr et al. 1993; Levi et al. 1997; Vanier et al. 1998). The last effort has led to the industrial implementation of a small, completely autonomous, passive frequency standard using ^{87}Rb (Vanier et al. 2004, 2005) as well as a laboratory-type CPT maser (Godone et al. 1999; Levi et al. 2002; M. Delaney 2005, pers. comm.). The phenomenon was also studied in a pulse excitation mode in a cell, with the goal of implementing later a

Ramsey type of time separated pulse excitation in a laser-cooled ensemble (Zanon et al. 2004).

It is possible to interpret the phenomenon and most of the observations by means of a simple three-level model. In the case of atomic beams, the problem is simplified due to the free evolution of the atoms in the beam and the absence of relaxation. In the case of cells, with the presence of relaxation, the rate equations for the evolution of the energy levels population and of the coherence created by the laser radiation fields in the ensemble may be solved exactly in the case of low alkali metal atom density (Orriols 1979). The final results of that analysis, however, are rather complex and are not transparent to easy interpretation. In order to interpret more easily experimental data in connection to atomic frequency standards applications, a simpler analysis has been developed (Cyr et al. 1993; Vanier et al. 1998). In that analysis, a three-level model is also used, but the rate equations are solved approximately by means of a first-order approximate analysis relative to optical coherence. The expressions obtained for the hyperfine resonance line shape, width, and amplitude are transparent and easy to interpret. The results can explain most experimental data in the case of a dilute or optically thin absorbing medium. In the case of an optically thick ensemble ($T > 50°C$ in the case of Rb, $T > 40°C$ in the case of Cs), a more elaborate approach using concepts developed in the study of the electromagnetically induced transparency (EIT) phenomenon (Harris 1997) is required (Godone et al. 2002). Furthermore, in the case where relatively intense radiation fields are used, it is found that the three-level model is not adequate (Vanier et al. 2003a, 2003b, 2003c). The lower manifolds of an alkali metal atom contain many energy levels and the Λ system is no longer closed as in the three-level model generally used (Renzoni et al. 1999). For example, when circularly polarized radiation is used, as required to observe the so-called ground-state field-independent 0-0 transition, atoms are optically pumped to a level not involved in the CPT phenomenon. They are transferred to that level, and trapped in it. They no longer contribute to the CPT resonance phenomenon. This effect has consequences on the observed signal amplitude. This effect is also encountered in several other experimental studies related to atom cooling (Berkeland et al. 1998; Wallace et al. 1992; Sortais 2001; Sortais et al. 2001) and absorption of laser radiation (Vanier et al. 2003d).

Reviews have been made on the use of CPT in high resolution spectroscopy (Arimondo 1996; Wynands and Nagel 1999). The present section will concentrate on the basic physics of CPT in connection to applications to the field of atomic frequency standards (Vanier 2005).

2.4.1 PHYSICS OF THE CPT PHENOMENON

We will first establish the basic equations describing the CPT phenomenon in an alkali metal atom such as the ^{87}Rb isotope and later we will adapt those to various situations. To do so, we will assume the presence of two coherent radiation fields resonant with the transitions from the levels $F = 1$ and $F = 2$ of the ground state to one of the hyperfine levels of the excited P state. In practice it is best to choose level $P_{1/2}, F' = 2$ (D_1 radiation), since in that case the transition probability is the same for both transitions making the system symmetrical, avoiding direct optical pumping and simplifying the analysis (Levi et al. 1999). The analysis can be easily adapted

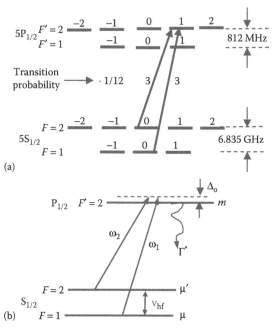

FIGURE 2.16   (a) Illustration of the lower energy levels manifold involved in the CPT excitation of the alkali metal atom ^{87}Rb using circularly polarized radiation σ^+. (b) Closed three-level model generally used to analyze the CPT phenomenon.

to other pairs of levels and to the case of other alkali metal atoms. The scheme is illustrated in Figure 2.16.

In the scheme shown, only those transitions starting from the levels $m_F = 0$ of the ground state are shown, although the spectral width of the lasers generally used would allow transitions from other Zeeman levels. In all applications concerned, however, a small magnetic field of the order of 10 µT is applied to the atomic ensemble and the other Raman resonances are well resolved. Consequently, a given Λ scheme may be selected to be resonant with only two of the ground-state levels. In the present case, the scheme starting at levels $m_F = 0$ is chosen since the energy of these levels is independent of the applied magnetic field in first order, a property desired in frequency standards applications. It is also noted that σ^+ polarization exciting the $\Delta m_F = +1$ transition is used. Left-hand circular polarization can also be used. Circular polarization is required by the fact that, due to states symmetry, transitions involving simultaneously $\Delta F = 0$, $\Delta m_F = 0$ are forbidden.

However, other Zeeman levels may play a role in the evolution of the system. Decay from the excited state takes place to all ground-state Zeeman sublevels due to spontaneous emission and due to atomic collisions in the case a buffer gas is used. Consequently, atoms may find a path to a Zeeman sublevel not involved in the Λ scheme. In such a case, the system is no longer closed and atoms may be trapped in a level such as $m_F = +2$ or $m_F = -2$, depending on the polarization used as in a classical optical pumping process (Kastler 1950). Those atoms are lost for the

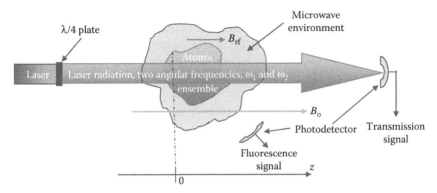

FIGURE 2.17   General configuration used in the development of the rate equations.

CPT phenomenon. This effect takes place in a buffer gas and becomes important at laser intensities corresponding to excitation levels comparable to or larger than the ground-state relaxation rates. Consequently, it is necessary to use a four-level system including a trap to explain some of the results obtained (Vanier et al. 2003b). The CPT phenomenon analysis is best done in the density matrix formalism. The general configuration shown in Figure 2.17 is used to establish the rate equations.

The atomic ensemble may be contained in a closed cell and imbedded in a buffer gas. In that case, the population of the Zeeman levels within the ground state and the coherence that may exist at the hyperfine frequency relax towards equilibrium at rates γ_1 and γ_2, respectively. If the atomic ensemble is an atomic beam, those rates may be assumed to be zero. The buffer gas also has important effects on the excited state, increasing the decay rate Γ^* of the P state by a factor that depends on its pressure. Consequently, the optical resonance line is homogeneously broadened. At a buffer gas pressure of about 20 Torr, the rate Γ^* is of the order of $3 \times 10^9 \mathrm{s}^{-1}$ as compared to a free space decay rate Γ (spontaneous emission) of the order of $3 \times 10^7 \mathrm{s}^{-1}$, as in a beam.

2.4.2   BASIC EQUATIONS

The analysis of the CPT phenomenon is done based on concepts similar to those developed in Section 2.3. However, two laser radiation fields are present. That laser radiation is assumed to propagate in the same direction as the applied magnetic field and the electric component E_n of the two laser radiation fields is assumed to be of the form:

$$E_n\left(\omega_n, z, t\right) = E_{on}\left(z\right)e_\lambda \cos\left(\omega_n t - k_n \cdot r\right) \qquad (n = 1, 2) \qquad (2.56)$$

where:
   E_{on} is the amplitude of the laser radiation field component n at position z
   ω_n is its angular frequency
   k_n is its wave vector
   e_λ is the polarization vector assumed to be the same for both radiation fields
   z is the distance travelled by the wave as measured from the entrance of the cell
       containing the atomic ensemble

These two radiation fields may be produced from a single laser through modulation of the laser driving current or by means of an electro-optic modulator, creating frequency modulation and sidebands in the laser spectrum. The two radiation fields may also be created by means of two phase-locked lasers with a frequency difference corresponding to the hyperfine frequency.

We introduce the optical Rabi angular frequency ω_{R1} and ω_{R2} as a measure of the laser field intensity and of the transition probability:

$$\omega_{R1} = \frac{E_1}{\hbar}\langle\mu|er\cdot e_\lambda|m\rangle \tag{2.57}$$

$$\omega_{R2} = \frac{E_2}{\hbar}\langle\mu'|er\cdot e_\lambda|m\rangle \tag{2.58}$$

where:
  e is the electronic charge
  $\hbar$ is Planck's constant over 2π
  $<i|e\ r|m>$ is the electric dipole moment of the transition from level i to level m
       $(i = \mu$ or $\mu')$

We assume that the ensemble is placed in an environment where a microwave field $B_{\mu w}(z,t)$ at angular frequency $\omega_{\mu w}$ may be present. This field may be excited by means of an external source or created by the atoms themselves through stimulated emission. It is assumed to be parallel to the dc magnetic field (z direction), a condition required for $\Delta m_F = 0$ transitions. In that condition, the microwave field is assumed to have the form:

$$B_{\mu w}(z,t) = B_z(z)\sin\left(\omega_{\mu m}t + \phi\right)z \tag{2.59}$$

where:
  ϕ is its phase

To simplify notation, the associated Rabi angular frequency describing the interaction of this field with the atomic ensemble is defined as:

$$b = b(z) = \frac{\mu_z B_z(z)}{\hbar} \tag{2.60}$$

where:
  μ_z is the atom magnetic moment

This definition makes $b = \omega_{1g}$ used in Section 2.3 concerned with a single laser optical pumping. We also define:

$$\omega_{12} = \omega_1 - \omega_2 \tag{2.61}$$

The rate equations for the population of the levels and for the coherence in the ground state are obtained by means of Liouville's equation as in the case of optical pumping with a single laser:

$$\frac{\partial \rho}{\partial t} = \frac{1}{i\hbar}[\mathcal{H}, \rho]$$

(2.62)

where:

ρ is a density matrix element

$\mathcal{H}$ is the interaction Hamiltonian

We assume a solution for the off diagonal matrix elements of the form:

$$\rho_{\mu\mu'}(z,t) = \delta_{\mu\mu'}(z,t)e^{i\left[(\omega_1 - \omega_2)t - (k_1 - k_2)z\right]}$$

(2.63)

$$\rho_{\mu m}(z,t) = \delta_{\mu m}(z,t)e^{i(\omega_1 t - k_1 z)}$$

(2.64)

$$\rho_{\mu' m}(z,t) = \delta_{\mu' m}(z,t)e^{i(\omega_2 t - k_2 z)}$$

(2.65)

and we expand Liouville's equation into its various terms. The resulting equations are:

$$\frac{d}{dt}\rho_{mm} = -\omega_{R1}\,\mathrm{Im}\,\delta_{\mu m} - \omega_{R2}\,\mathrm{Im}\,\delta_{\mu' m} - \Gamma^*\rho_{mm}$$

(2.66)

$$\frac{d}{dt}\rho_{\mu'\mu'} = -\mathrm{Im}\left(be^{-i\phi}\right)\delta_{\mu\mu'} + \omega_{R2}\,\mathrm{Im}\,\delta_{\mu' m} + \Gamma^*_{m\mu'}\rho_{mm} - \left(\frac{\gamma_1}{2}\right)\left(\rho_{\mu'\mu'} - \rho_{\mu\mu}\right)$$

(2.67)

$$\frac{d}{dt}\rho_{\mu\mu} = +\mathrm{Im}\left(be^{-i\phi}\right)\delta_{\mu\mu'} + \omega_{R1}\,\mathrm{Im}\,\delta_{\mu m} + \Gamma^*_{m\mu}\rho_{mm} - \left(\frac{\gamma_1}{2}\right)\left(\rho_{\mu\mu} - \rho_{\mu'\mu'}\right)$$

(2.68)

$$\frac{d}{dt}\delta_{\mu m} + \left[\frac{\Gamma^*}{2} + i\left(\omega_1 - \omega_{m\mu}\right)\right]\delta_{\mu m} = +i\left(\frac{\omega_{R1}}{2}\right)\left(\rho_{mm} - \rho_{\mu\mu}\right) + \frac{b}{2}e^{i\phi}\delta_{\mu' m} - i\left(\frac{\omega_{R2}}{2}\right)\delta_{\mu\mu'}$$

(2.69)

$$\frac{d}{dt}\delta_{\mu' m} + \left[\frac{\Gamma^*}{2} + i\left(\omega_2 - \omega_{m\mu'}\right)\right]\delta_{\mu' m} = +i\left(\frac{\omega_{R2}}{2}\right)\left(\rho_{mm} - \rho_{\mu'\mu'}\right)$$

$$+ \frac{b}{2}e^{i\phi}\delta_{\mu m} - i\left(\frac{\omega_{R1}}{2}\right)\delta_{\mu'\mu}$$

(2.70)

$$\frac{d}{dt}\delta_{\mu\mu'} + \left[\gamma_2 + i\left(\omega_{12} - \omega_{\mu'\mu}\right)\right]\delta_{\mu\mu'} = i\left(\frac{b}{2}\right)e^{i\phi}\left(\rho_{\mu'\mu'} - \rho_{\mu\mu}\right)$$

$$+ i\left(\frac{\omega_{R1}}{2}\right)\delta_{m\mu'} - i\left(\frac{\omega_{R2}}{2}\right)\delta_{\mu m}$$

(2.71)

with the condition:

$$\rho_{mm} + \rho_{\mu'\mu'} + \rho_{\mu\mu} = 1 \qquad (2.72)$$

The above equations apply to a group of atoms that have a specific velocity. Doppler broadening of the optical interaction is of the order of 500 MHz. In a cell containing a buffer gas, as mentioned above, collisions cause optical line broadening of several hundreds of MHz and in that case homogeneous broadening represented in the equations by the decay rate Γ^* is assumed. In such a case, it is also assumed that Dicke narrowing takes place at the microwave frequency corresponding to the hyperfine frequency of the ground state (Dicke 1953). When the phenomenon is observed in an atomic beam, the experimental setup is such that the laser radiation traverses the beam at right angle, avoiding first-order Doppler effect. Questions relative to the phase of the various fields are taken into account in the derivation when appropriate.

In Chapter 3, we will apply this set of equations to a few cases representing experimental approaches that have been studied towards the implementation of a frequency standard. The parameters in the above set of equations will be adjusted to the appropriate experimental situations encountered in those various approaches. The set of Equations 2.66 through 2.72 cannot be solved exactly in the general case. However, in many situations, there is no microwave field applied and b may be set equal to zero. Figure 2.17 then reduces to a laser radiation beam traversing an ensemble of atoms. In that case the solution of this set of equations can be done exactly (Orriols 1979; Zanon-Willette et al. 2011). However, such a solution is rather complex and is not transparent easy interpretation. In order to understand the result it is then best to evaluate the behaviour of the equation for the transmitted light for a given situation. It is then possible to get some insight into the result by making approximations. The physical parameters measured in practice are the fluorescence power and power transmitted as illustrated in Figure 2.17. The total fluorescence emitted is readily evaluated as:

$$P_{fl} = \hbar\omega_l \Gamma_{fl} N \rho_{mm} \qquad (2.73)$$

while the power absorbed in a slice of length dz is given as:

$$\Delta P_{abs}(z) = n\hbar\omega_l \Gamma^* \rho_{mm} dz \qquad (2.74)$$

where:
   N is the total number of atoms in interaction with the laser radiation
   n is the density
   Γ^* is the decay rate from the excited state caused by all types of relaxation, buffer
       gas collisions, and spontaneous emission
   Γ_{fl} is the decay from the excited state resulting in fluorescence

These parameters are functions of the density matrix elements representing the population of the excited state ρ_{mm}. It is thus the term to be calculated. This can be done approximately by means of a step technique in which first it is assumed that two laser radiation fields have the same intensities, $\omega_{R1} = \omega_{R2} = \omega_R$. In a first-order step, we then assume that $\delta_{m\mu'}$ and $\delta_{\mu m}$ are small and that ρ_{mm} is zero. The solution obtained is then replaced in the steady-state equation of the ground-state hyperfine coherence and after some algebra an approximate solution for ρ_{mm} is obtained in terms of the real part of the ground-state coherence $\delta_{\mu\mu'}$:

$$\rho_{mm} = \frac{\omega_R^2}{\Gamma^{*2}}\left(1 + 2\delta_{\mu\mu'}^r\right) \tag{2.75}$$

On the other hand, the real part of the ground-state coherence is given by:

$$\delta_{\mu\mu'}^r = -\frac{\omega_R^2}{2\Gamma^*}\frac{\left(\gamma_2 + \omega_R^2/\Gamma^*\right)}{\left(\gamma_2 + \omega_R^2/\Gamma^*\right)^2 + \Omega_\mu^2} \tag{2.76}$$

where Ω_μ stands for the detuning of the two laser frequency difference from the hyperfine frequency or

$$\Omega_\mu = \left(\omega_1 - \omega_2\right) - \omega_{\mu'\mu} \tag{2.77}$$

As is readily seen from these results, the fluorescence and the transmitted power are affected by a sharp resonance of width:

$$\Delta\nu_{1/2} = \frac{\left(\gamma_2 + \omega_R^2/\Gamma^*\right)}{\pi} \tag{2.78}$$

which is characteristic of a hyperfine resonance and is proportional to the laser radiation intensity through the optical Rabi frequency ω_R. On the other hand, in this closed three-level model, the ensemble becomes totally transparent for a light intensity such that ω_R^2/Γ^* becomes much larger than the relaxation rate γ_2. This simple homogeneous model gives some insight into the behaviour of the system and the importance of the size of the parameters. For a value Γ^* of $3 \times 10^9 s^{-1}$ and a γ_2 of $1000 s^{-1}$, typical values encountered in practical setups for a cell volume of a cubic centimetres and a buffer gas pressure of the order of 2 kPa (15 Torr), the Rabi angular frequency required to double the line width is $1.7 \times 10^6 s^{-1}$. The laser power required for providing such a Rabi frequency is of the order of $10 - 100$ µW depending on the laser radiation spectral width and the cell cross section. It is also readily seen that in equilibrium, using these values, the population ρ_{mm} of the excited state does not exceed 3×10^{-7}, a value much smaller than that of the ground state. In fluorescence, the phenomenon is observed as a dark line in the emitted radiation. The situation has been called a dark state: since no transitions take place, there is no fluorescence present at resonance. In transmission, we observe a bright line.

We have outlined a simple solution which makes explicit the properties of the CPT phenomenon. In Chapter 3, we will apply this set of equations to the cases where the ensemble of atoms has the form of an atomic beam or is contained in a cell with a buffer gas or a wall coating. In that case we will study the situation where the ensemble is placed in a cavity resonant at the alkali atom hyperfine frequency, leading to the concept of the CPT maser. We note that the analysis can be adapted easily to other alkali atoms through a proper adjustment of the frequencies, wavelength, and other decay and relaxation parameters.

2.5   LASER COOLING OF ATOMS

On the basis of the various analysis made in Chapter 1 and the conclusions drawn, it is rather obvious that a primary condition for improving the characteristics of a particular atomic frequency standard is to reduce or possibly cancel Doppler effect, which broadens the clock resonance line, reduces its quality factor (first-order Doppler), and causes a frequency shift (second-order Doppler). An improvement in line quality factor has a direct effect on frequency stability and accuracy and a reduction of the second-order Doppler shift in most cases improves frequency accuracy. Major efforts have been made towards implementing methods for increasing line quality factor, either by means of a buffer gas, storage in a container with proper coating to inhibit relaxation, or still use of the Ramsey separate interaction zones approach in the time or space domain, all these being techniques that reduce the line width of the observed atomic resonance line. We have seen that those various solutions have limitations. One proposed brute force approach in relation to that question was to reduce the global temperature of the ensemble of atoms by means of cryogenic techniques, reducing the random speed of atoms and consequently Doppler effect. Such an approach also has a direct effect on frequency stability through a reduction of thermal noise. A typical example using such an approach is the cryogenic H maser described earlier. However, cryogenic techniques are very sophisticated, require rather cumbersome equipment and are not easily adaptable to the kind of simplicity and dimensions wished for in the implementation of atomic frequency standards. Furthermore, such cryogenic techniques are not applicable to all atomic ensembles such as alkali atoms, for example. Also, they do not considerably improve line quality factor and, furthermore, introduce other unexpected frequency shifts through the various storage techniques used.

Soon after the development of lasers, it was proposed to use them to directly affect the motion of atoms in a very effective way (Letokhov 1968; Hansch and Schawlow 1975; Wineland and Dehmelt 1975). The atom–light interaction would then act directly on Doppler effect if it could be used to reduce the thermal agitation of atoms. This idea was based on the consideration that the direction and speed of atoms could be altered by an exchange of energy and momentum with photons in resonance with those atoms (Frisch 1933). In practice, one can say that an atom is slowed down by the radiation pressure created by a counter-propagating laser photon. It is readily calculated that about 10^5 collisions with counter-propagating photons can reduce to a negligible value an atom average velocity associated to its room temperature motion. The process does not look efficient at first sight, but due to the number of photons carried by a laser beam of moderate intensity (~mW) and the fast internal response of atoms, it is possible to alter considerably the velocity of atoms in very short periods (~ms or less). The technique could thus be used on a beam of atoms to reduce their speed at least in one direction. It could also be used in principle on an ensemble of atoms, using multiple laser beams oriented in orthogonal directions, to reduce their average speed and at the same time reduce the width of the velocity spectrum, thus reducing the temperature of the ensemble. The process has been given the name *laser cooling*.

The expression *laser cooling* that indirectly refers to the concept of temperature has to be used with caution. Temperature is a physical concept that is associated with thermal equilibrium, that is to say equilibrium of an ensemble of atoms with its interacting environment. In the case of so-called laser cooling, temperature is not normally associated with the laser radiation with which the atoms interact. Furthermore, after interaction with the laser radiation, the atomic ensemble may not be in thermal equilibrium itself, velocities having a spectrum very different from that of a Maxwell distribution. Another way of looking at the question is to observe the behaviour of an atom in a frame of reference moving with that atom. The observer sees that atom at rest. The same reasoning can be made relative to a beam of atom emerging from a source at temperature T and moving at an average speed v. An observer moving at that average speed sees atoms moving forward or receding from him, reflecting the spectrum of the random speeds of atoms within the beam. It is the width of that spectrum that is connected to the concept of temperature and to the temperature of the source or oven. Nevertheless, it has become convenient in common language to use the concept of temperature, even in cases where atomic speed is reduced in a single direction, making reference to the relation between kinetic energy of atoms E_k and thermal energy through the relation:

$$E_k = \left(\frac{1}{2}\right) M v^2 = \left(\frac{1}{2}\right) k_B T \tag{2.79}$$

where:

M is the mass of the atom whose speed v has been reduced by laser radiation

k_B is Boltzmann constant

T is then a temperature characterizing that speed

Sometimes, we will use that concept to characterize an atomic velocity that has been reached in a given experimental situation, but we will use it with caution, reserving normally the concept temperature to a situation where we can identify the width of the atomic velocity spectrum of an ensemble.

The technique of laser cooling has had a major impact in several fields of physics, for example, improving precision in spectroscopy and the observation of long predicted phenomena such as Bose–Einstein condensation. It has also altered in an irreversible way the development of atomic frequency standards, particularly those primary standards used to implement the definition of the second. It has opened new promising avenues of research, particularly in the domain of optical frequency standards. This is the subject that interests us. We will now outline the physics behind the concept and outline various methods that have been developed using laser radiation to alter the velocities of atoms, reduce their temperature by means of atom–radiation interaction and provide means of trapping and store ensembles of cooled atoms. There has been a tremendous amount of work done in that field over the last few decades. We will limit the outline to the aspects that are important to atomic frequency standards and are used in their implementation. For a more complete description, the reader is referred to some excellent books that have been written on the subject (Cohen-Tannoudji et al. 1988; Metcalf and van der Straten 1999; Letokhov 2007; Cohen-Tannoudji and Guéry-Odelin 2011). Excellent early reviews have also been made on the subject (see, e.g., Adams and Riis 1997; Phillips 1998; Srinivasan 1999). The following sections provide an outline of the main concepts, which will be necessary for understanding Chapters 3 and 4.

2.5.1   ATOM–RADIATION INTERACTION

In Sections 2.3 and 2.4, we have examined in some detail the interaction between electromagnetic radiation and atoms. In those cases we have looked essentially at the effect of that radiation on the internal state of the atoms such as energy level population and coherence introduced particularly in the ground state. This led us to the study of the so-called double resonance phenomenon, using microwave and optical radiation simultaneously, as in the case of optically pumped Rb frequency standards. Similarly we have introduced the CPT phenomenon in which two laser radiation fields are used simultaneously to force the atom into an internal state with rather interesting properties. For doing so we have used Bloch equations for analyzing energy levels populations and coherence in atomic systems character-ized by an excited state m and a ground state which consisted of two hyperfine lev-els μ and μ' separated by an energy in the microwave range. We could characterize the internal properties of the atoms rather completely and identify the parameters that were important in the implementation of atomic frequency standards. The external properties such as atomic velocities and ensemble temperature were dealt with in a phenomenological way through the introduction of relaxation parameters and Doppler effect. We did not have much control on them. We will now examine how we can alter those external properties by means of atom-photon interactions.

2.5.1.1   Effect of a Photon on Atom External
Properties: Semi-Classical Approach

In the present section we wish to examine the effect of optical radiation generated by a laser on the external properties of an atom, that is, to say its effect on the atom speed and kinetic energy.

2.5.1.1.1   Elementary Approach and Interpretation

We assume that the atom is characterized by two internal states, m and μ separated by an energy $\Delta E = \hbar\omega_{m\mu}$, where $\omega_{m\mu}$ is an angular frequency in the optical range, as shown in Figure 2.18a.

Level μ is the ground state and if the atom is excited to state m by the optical radiation, it normally decays to the ground state at the spontaneous rate Γ. The opti-cal radiation produced by a laser at an angular frequency ω_L close to $\omega_{m\mu}$ consists of photons with energy E and momentum p:

$$E = \hbar\omega_L \tag{2.80}$$

$$p = \hbar k_L \tag{2.81}$$

where:

k_L is the radiation wave vector $(2\pi/\lambda)$ $\mathbf{z}$ for a wave propagating in the z direction

We consider first an atom at rest. The atom-photon interaction is represented pictorially in Figure 2.18b. If the frequency ν_L of the laser radiation is close to the atom resonant

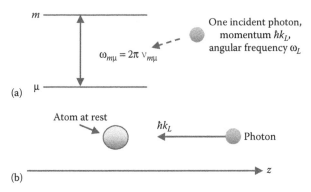

FIGURE 2.18   (a) Representation of an atom with energy levels m and μ in interaction with a photon of energy $\hbar\omega_L$. (b) Pictorial interpretation of a photon with momentum $\hbar k_L$ incident on an atom.

frequency $v_{m\mu}$, a photon may be absorbed. The photon disappears and the atom is excited to level *m*. The atom receives a kick and recoils in the direction of propagation of the radiation. In agreement with the principle of conservation of momentum, the photon momentum $p = h/\lambda$ is transferred to the atom, which acquires a momentum in the negative *z* direction equal to that of the photon that we call Δp_a. The atom finds itself in the excited state m and recoils at velocity v_{rec} satisfying the relation:

$$\Delta p_a = \hbar k_L = M v_{rec} \tag{2.82}$$

where:
   M is the mass of the atom

The recoil velocity, upon interaction with the photon, is thus:

$$v_{rec} = \frac{\hbar k_L}{M} \tag{2.83}$$

In addition to its internal energy gain $\hbar_L$, the atom gains an external energy δE_{rec} per absorption cycle given by:

$$\delta E_{rec} = \frac{1}{2} M v_{rec}^2 = \frac{1}{2} \frac{\hbar^2 k_L^2}{M} \tag{2.84}$$

which can also be written as:

$$\delta E_{rec} = \frac{1}{2} \frac{\hbar^2 \omega_L^2}{Mc^2} \tag{2.85}$$

where:
   c is the speed of light

The atom is in the excited state, in interaction with the vacuum radiation field reservoir. Its lifetime in the excited state m is limited by that interaction and it decays by spontaneous emission of a photon. The photon is emitted in a given direction but has equal probability of being emitted in the opposite direction. On the other hand, in practice, the atom being exposed to a laser coherent radiation beam, we assume that it makes a large average number $\langle N \rangle$ of absorption–emission cycles per second at the rate $\partial \langle N \rangle / \partial t$. Consequently, upon averaging over several cycles, no net average recoil is perceived by the atom in the spontaneous emission process since the direction of emission of individual photons is random with equal probability in opposite directions. Aside from a small residual effect from random walk since emission takes place in discrete steps, the atom does not gain momentum on average from the spontaneous emission process. On the other hand, the maximum rate of absorption–emission cycles is limited by the actual rate of return of the atom to the ground state. This is controlled by the vacuum radiation field reservoir and this rate is Γ, the spontaneous emission rate. Consequently, since the change of velocity is v_{rec} at each cycle, the rate of change of the atom velocity is

$$\frac{d}{dt} v = \frac{\partial \langle N \rangle}{\partial t} v_{rec} = \Gamma \frac{\hbar k_L}{M} \tag{2.86}$$

This change in velocity can be interpreted classically as caused by a force

$$F = \frac{d}{dt} Mv = M \frac{d}{dt} v \tag{2.87}$$

which using Equation 2.86 becomes:

$$F = \hbar k_L \Gamma \tag{2.88}$$

In this analysis we have essentially assumed that the laser intensity is low and the atoms spend most of their time in the ground state. We will see later, in the quantum mechanical approach, that in the case of a high rate of transition, saturation takes place and the number of atoms in interaction is half the value assumed in the present approach. They spend half their time in the excited state and half in the ground state. The maximum force is then half the value just calculated because the rate of transitions is reduced by half.

The interaction of the atom with a continuous radiation field thus appears classically as if a continuous force was applied to the atom accelerating it in the direction of the laser beam. It can be interpreted as a radiation pressure pushing on the atom in the direction of propagation of the laser radiation.

2.5.1.1.2 Size of the Effect

We may question the size of the effect for its practical use in providing a measurable speed to an atom at rest or for a more interesting use, that of changing the velocity of an atom already in motion. As mentioned earlier, this is actually our purpose: changing the velocity of atoms or the spread of those velocities and

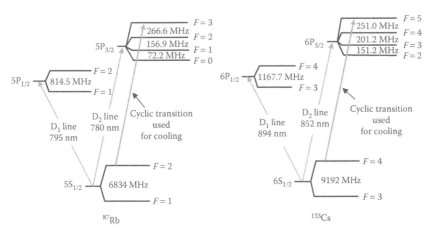

FIGURE 2.19 Energy level manifold of S ground state and first excited P states of ^{87}Rb and ^{133}Cs. (Data from Steck, D., *Cesium D Line Data, Rubidium 87 D Line Data*, 2003, 2008. http://steck.us/alkalidata.)

thus altering their temperature. For the moment we will limit ourselves to atoms commonly used in atomic frequency standards, that is Rb and Cs. We will examine later the case of other atoms and ions when required. We will use selected data contained in Tables 1.1.2 and 4.2.1 of Volume 1 of *QPAFS* (1989) and from a report by Steck (2003, 2008). The lower energy levels of Rb and Cs are shown in Figure 2.19. To facilitate an evaluation of the order of magnitude of the parameters involved, we reproduce some of the basic data in Table 2.1. We then calculate from the previous analysis various parameters of importance in the interaction process. These results are given in Table 2.2.

It is important to evaluate the rate or the time constant at which the change in velocity introduced by the interaction takes place. We have to fix a kind of reference

TABLE 2.1
Selected Atomic Properties of ^{133}Cs and ^{87}Rb Useful In the Calculation of Some of the Parameters of Importance in the Absorption–Emission Cycle Discussed in the Text

	M (kg) 10^{-25}	λ (m) (air) 10^{-9}	v (Hz) 10^{12}	ω (rad/s) 10^{15}	k (m^{-1}) 10^{6}	τ (s) 10^{-9}	Γ (s^{-1}) 10^{7}	$\Delta v_{1/2}$ (Hz) 10^{6}
^{87}Rb	1.44	(D$_2$)780.0(P$_{3/2}$)	384	2.41	8.06	26.2	3.82	6.08
		(D$_1$)794.8(P$_{1/2}$)	377	2.37	7.91	27.7	3.61	5.76
^{133}Cs	2.21	(D$_2$) 852.1(P$_{3/2}$)	351.7	2.21	7.37	30.5	3.28	5.26
		(D$_1$) 894.4(P$_{1/2}$)	335.1	2.10	7.02	34.9	2.87	4.58

Source: Steck, D., *Cesium D Line Data, Rubidium 87 D Line Data*, 2003, 2008. http://steck.us/alkalidata.

TABLE 2.2

Values of Various Recoil Effects Caused by the Interaction of Laser Radiation at the Resonance Frequency of the D_2 Transition with Cs and Rb Atoms at Rest

	δv_{rec} (per cycle) (mm/s)	δE_{rec} (per cycle) (J) 10^{-30}	$\delta E_{rec}/k_B$ (K) 10^{-6}	τ_{ext} (for $\Delta v_{rec} = \Gamma/k_L$) (s) 10^{-6}
Formula	$\hbar k_L/M$	$(1/2)(\hbar^2 k_L^2/M)$	$(1/2)(\hbar^2 k_L^2/M\, k_B)$	$\hbar/2\delta E_{rec}$
^{87}Rb	5.9	2.51	0.181	21
^{133}Cs	3.5	1.36	0.099	38.8

Note: The value of the time required, τ_{ext}, for reaching a speed that causes a Doppler shift equivalent to a full natural line width is also given.

for this evaluation. For example, we may set arbitrarily as an *external* time constant, τ_{ext}, the time it takes for an atom at rest to acquire a speed that, through Doppler effect, changes its D_2 frequency by about one natural line width following an accumulation of several absorption-emission cycles. At such a speed, the atom is off resonance with the radiation and the probability of absorbing a photon is reduced considerably (factor ~5). This frequency change is given by:

$$\Delta\omega = k\Delta v_{rec} \approx \Gamma \qquad (2.89)$$

where:

Δv_{rec} is the total speed acquired during the time τ_{ext}

That cumulative speed is given from Equations 2.87 and 2.88 as:

$$\Delta v_{rec} = \frac{F}{M}\tau_{ext} = \frac{\Gamma\hbar k_L}{M}\tau_{ext} \qquad (2.90)$$

and from the condition 2.89 and Equation 2.84, we have:

$$\tau_{ext} \approx \frac{1}{2}\frac{\hbar}{\delta E_{rec}} \qquad (2.91)$$

Several remarks can be made regarding the values of the various parameters calculated and reported in the last table. First, we have limited our evaluation of the effect of the interaction to D_2 radiation since it is a spectral line that originates from a cycling transition. This transition is closed and is more efficient: in principle atoms are trapped in that transition and no optical pumping to another level of the ground state takes place. We have also compared the external energy acquired by the atom to that of thermal energy $k_B T$. The table shows that the energy involved in one absorption-emission cycle is extremely small of the order of 10^{-30} J. This is to be compared to thermal energy which at room temperature is of the order of (1/40) eV, that is 4×10^{-21} J. The process thus requires a large number of excitation cycles to have a visible effect. Fortunately, the possible rate of transitions that can be excited is very large. On the other hand, the time constant τ_{ext} calculated for a notable increase in speed is of the order of 30 μs while the internal response τ of the atom is in the range of 30 ns. Thus, internal parameters follow very rapidly external perturbations and we can say that at all times the atom's internal properties are in a stationary state while external properties are evolving. This property will allow us to simplify a quantum analysis that we will make below to obtain a more complete picture of the process.

We recall that we have assumed that the atom interacting with the radiation field was at rest and the effect of the incident photon and momentum transfer was to accelerate the atom in the direction of the laser radiation. In a sense, we can say loosely that the atom is *heated* by the radiation since its speed increases in the laboratory frame. On the other hand, we note that the elementary analysis can be applied as well to an atom moving at speed v in positive z direction, which is in opposite direction to the laser radiation. In that case, the atom sees the laser frequency shifted upwards by Doppler effect. To be resonant with the atom frequency, the laser must then be tuned to an angular frequency lower than $\omega_{m\mu}$ by the amount $k_L v$. Then, the above elementary analysis can be used to calculate the force applied on the atom and its change in speed and energy. In that case we can say loosely that the atom is cooled by the incident radiation since its speed is decreased by the interaction. In Appendix 2.A, we outline a concept that we have used in *QPAFS*, Volume 2 (1989), and which is based on relativistic equations that are based on considerations regarding the energy or frequency of photons absorbed and emitted by spontaneous emission. The results are the same as those obtained above in the elementary approach based entirely on momentum exchange.

2.5.1.2   Quantum Mechanical Approach

The above elementary approach, although rather instructive regarding the connection between external and internal atomic parameters, does not provide the full picture of the interaction between the atom and the radiation field. It is based entirely on an exchange of momentum and considers in a phenomenological way the absorption of the photon. It does not address, for example, the details of the absorption–emission process and the difference existing in the interaction with a travelling wave and a standing wave. In fact, in its interaction with the atom, the radiation induces an oscillating dipole moment and the interaction is sensitive to the relative phase of the field and the induced dipole moment. When properly handled in a quantum mechanical approach, it is found that new effects added to cooling are introduced, such as trapping the atoms in specific circumstances and altering the energy level structure of the atom that leads to new cooling mechanisms. These effects are important in the implementation of atomic frequency standards particularly in the optical range. We will now introduce a quantum mechanical approach to the atom–radiation interaction and make explicit its effect on the behaviour of the external properties of the atom. We will follow the main lines of the analysis made in the excellent texts mentioned in Section 2.2 (Cohen-Tannoudji et al. 1988; Metcalf and van der Straten 1999; Cohen-Tannoudji and Guérin-Odelin 2011).

2.5.1.2.1   *Atom at Rest*

2.5.1.2.1.1   The Force as a Function of Internal Variables   As in Figure 2.18, we assume that the atom is at rest and is exposed to a laser radiation field propagating in the negative z direction. We need an expression connecting the classical force exerted by the field on the atoms, as introduced in the previous elementary approach, to the internal properties of the atom altered by the radiation field. Classically, the force is given by the expression:

$$F = Ma = \frac{d}{dt} Mv = \frac{d}{dt} p \qquad (2.92)$$

This force is interpreted as the expectation value of the operator F_{op} that can be calculated by means of Heisenberg equation (*QPAFS* 1989, Vol. 1). We have thus:

$$F = \langle F_{op} \rangle = \frac{d}{dt} \langle p \rangle = \frac{i}{\hbar} [\mathcal{H}, p] \qquad (2.93)$$

where:
   p is the quantum mechanical momentum operator
   $[\mathcal{H}, p]$ is the commutator of the Hamiltonian representing the atom–radiation interaction and the momentum operator

In the context of the assumed geometry (Figure 2.18), we take the z-axis as the axis of propagation of the radiation and we replace p by its operator $i\hbar \partial/\partial z$. Equation 2.93 may then be written as (Schiff 1968, Section 24):

$$F = -\left\langle \frac{\partial}{\partial z} \mathcal{H} \right\rangle \qquad (2.94)$$

We thus need to evaluate $\mathcal{H}$, the interaction Hamiltonian. We are dealing with transitions between an S state and a P state, which involves an interaction between the electric dipole $d = -er$ and the electric component of the radiation field $E(R, t)$ present at the site of the atom. The vacuum radiation field is assumed to have zero average value and its gradient ($\partial/\partial z$) is zero. Consequently, it is taken care of in a phenomenological way by means of the introduction of spontaneous emission phenomenon in the master equations describing the evolution of the atomic state and is assumed to have no effect on the mean radiation force. The interaction Hamiltonian is thus (*QPAFS* 1989, Volume 1):

$$\mathcal{H}(R_o, t) = -er \cdot E(R, t) = -d \cdot E(R, t) \qquad (2.95)$$

where:
   $E(R, t)$ is the applied field

Equation 2.94 becomes:

$$F = \left\langle \left(\frac{\partial}{\partial z} d \cdot E(R, t) \right) \right\rangle \qquad (2.96)$$

The evaluation of the expectation value of the interaction $d \cdot E(R, t)$ can be done by means of a density matrix approach, the expectation value of an operator Q being $\langle Q \rangle = Tr \rho Q$.

Our atomic system is a two-level isolated system and its density matrix, as driven by the laser radiation, is:

$$\rho = \begin{pmatrix} \rho_{mn} & \rho_{\mu m} \\ \rho_{m\mu} & \rho_{\mu\mu} \end{pmatrix} \tag{2.97}$$

We define the optical Rabi frequency $\omega_{Rm\mu}$ as in Equation 2.21, of Section 2.3, but for our two-level system:

$$\omega_{Rm\mu} = \left(\frac{E_L}{\hbar} \right) \langle m | er \cdot e_\lambda | \mu \rangle = \left(\frac{E_L}{\hbar} \right) d_{m\mu} \tag{2.98}$$

The force becomes:

$$F = \hbar Tr \left(\rho \frac{\partial}{\partial z} \omega_{Rm\mu} \right) \tag{2.99}$$

which may be made explicit as:

$$F = \hbar \left\{ \rho_{m\mu}^* \frac{\partial}{\partial z} \omega_{Rm\mu} + \rho_{m\mu} \frac{\partial}{\partial z} \omega_{Rm\mu}^* \right\} \tag{2.100}$$

Using the definition of the Rabi frequency (Equation 2.98), we finally obtain the force as a function of internal variables and $E(z)$:

$$F = -\left\{ \rho_{m\mu}^* d_{\mu m} \nabla E(z) + \rho_{m\mu} d_{\mu m}^* \nabla E(z) \right\} \tag{2.101}$$

where we have assumed that the field propagates in the z direction as in Figure 2.18 and have represented the derivative $\partial/\partial z$ by the gradient symbol ∇.

2.5.1.2.1.2   Density Matrix Elements: Optical Bloch Equations   In order to elaborate further, we need to determine the density matrix, ρ, that represents the state of the atom under the influence of the radiation field. We can make as usual the approximation that the atom is very small and that due to its large mass it is well-localized. We thus make the long wavelength approximation. We also make the rotating wave approximation in which the counter-rotating wave in the radiation field incident on the atoms is neglected as we have done in our earlier analysis of optical pumping with a laser. We can thus use the set of optical Bloch equations derived in Section 2.3 assuming this time a single level in the ground state as shown in Figure 2.18. We thus have:

$$\frac{d}{dt} \rho_{mm} = -\omega_{Rm\mu} \, \text{Im} \, \delta_{\mu m} - \Gamma \rho_{mm} \tag{2.102}$$

$$\frac{d}{dt} \rho_{\mu\mu} = \omega_{Rm\mu} \text{Im} \delta_{\mu m} + \Gamma \rho_{mm} \tag{2.103}$$

$$\frac{d}{dt} \delta_{\mu m} + \left[\frac{\Gamma}{2} + i \left(\omega_L - \omega_{m\mu} \right) \right] \delta_{\mu m} = i \left(\frac{\omega_{R\mu m}}{2} \right) \left(\rho_{mm} - \rho_{\mu\mu} \right) \tag{2.104}$$

We have assumed that the off diagonal elements of the density matrix ρ are driven by the laser radiation field and have the form:

$$\rho_{\mu n}(z,t) = \delta_{\mu n}(z,t)e^{-i\omega_L t}$$

$$\rho_{m\mu}(z,t) = \delta_{\mu n}^*(z,t)e^{i\omega_L t}$$

(2.105)

with the condition for the diagonal elements (populations):

$$\rho_{mm} + \rho_{\mu\mu} = 1 \qquad (2.106)$$

We assume the atom to be at rest at $z = 0$ and assume as mentioned earlier that the internal response is much faster than the evolution of the external properties. We thus solve the equations for a stationary state with:

$$\frac{d\rho_{mm}}{dt} = \frac{d\rho_{\mu\mu}}{dt} = \frac{d\delta_{\mu n}}{dt} = 0 \qquad (2.107)$$

To simplify notation, we define:

$$\delta_{\mu m} = \delta^r + i\delta^i \qquad (2.108)$$

$$\Delta = \rho_{mm} - \rho_{\mu\mu} \qquad (2.109)$$

We also define the saturation factor as:

$$S = \frac{\omega_{Rm\mu}^2/2}{(\Gamma/2)^2 + (\omega_L - \omega_{m\mu})^2} \qquad (2.110)$$

and obtain:

$$\delta^r = \frac{\omega_{Rm\mu}}{2} \frac{(\omega_L - \omega_{m\mu})}{(\Gamma/2)^2 + (\omega_L - \omega_{m\mu})^2} \frac{1}{(1+S)} \qquad (2.111)$$

$$\delta^i = -\frac{\omega_{Rm\mu}}{4} \frac{\Gamma}{(\Gamma/2)^2 + (\omega_L - \omega_{m\mu})^2} \frac{1}{(1+S)} \qquad (2.112)$$

$$\rho_{mm} = \frac{S/2}{1+S}, \qquad \Delta = \frac{1}{1+S} \qquad (2.113)$$

We may now replace in Equation 2.101 the assumed solutions 2.105 for $\rho_{\mu m}$. We obtain:

$$F = -\nabla E(z,t)d_{\mu m}\left\{2\cos(\omega_L t)\delta^r - 2\sin(\omega_L t)\delta^i\right\} \qquad (2.114)$$

At this point we need to assume a form for the radiation electric field component. We take the coherent radiation field as:

$$E(z,t) = E_o(z)\cos[\omega_L t + \phi(z)] \qquad (2.115)$$

where:

$\phi(z)$ is the phase of the radiation field to which the atom is exposed

We may set the origin of the z-axis arbitrarily and set the phase $\phi(0) = 0$ at that origin. We replace $E(z,t)$ in Equation 2.114 to obtain:

$$F = -d_{\mu m}\left\{2\delta'\nabla E_o(z)\cos^2 \omega_L t + 2E_o\delta'\nabla\phi\sin^2 \omega_L t\right\} \tag{2.116}$$

This force can be averaged over one period of the radiation field to give:

$$F = -d_{\mu m}\left\{\delta'\nabla E_o(z) + E_o\delta'\nabla\phi\right\} \tag{2.117}$$

We may analyze two cases that help in understanding the meaning of the two terms in this last expression for the force.

2.5.1.2.1.3 Case of a Travelling Wave This is the simplest case illustrated in Figure 2.18 and encountered experimentally. This is the case, for example, of the first experiments in which a beam of atoms at a given temperature, such as sodium emitted by an oven and collimated to form a narrow beam, is slowed down by a counter-propagating laser radiation beam (Balykin et al. 1979; Phillips and Metcalf 1982; Phillips and Prodan 1983). We will examine later in some detail the conditions for implementing such an experiment. For the moment it is only necessary to examine its general characteristics. The travelling wave may be written as:

$$E(z,t) = E_o \cos(\omega_L t + k_L z) \tag{2.118}$$

This wave has no amplitude gradient (E_o is constant in space). Thus $\nabla E_o = 0$. However, the phase is $k_L z$ and its gradient $\nabla\phi = k_L$. We assume that the atom is situated at $z = 0$. There is thus a force present and we will see below that this force is associated with a dissipation of energy. We call it F_{diss}. We obtain:

$$F_{\text{diss}} = -d_{\mu m}E_o k_L \delta^i \tag{2.119}$$

Using the value found for δ^i we obtain:

$$F_{\text{diss}} = \hbar k_L \frac{\omega_{R\mu m}^2}{2} \frac{\Gamma/2}{(\Gamma/2)^2 + (\omega_L - \omega_{m\mu})^2} \frac{1}{(1+S)} \tag{2.120}$$

which can also be written using the definition of S introduced in Equation 2.110:

$$F_{\text{diss}} = \hbar k_L \Gamma \frac{\omega_{R\mu m}^2}{4} \frac{1}{(\Gamma/2)^2 + (\omega_L - \omega_{m\mu})^2 + (\omega_{R\mu m}^2/2)} \tag{2.121}$$

In the process, a photon is absorbed by the atom and the energy of the photon is dissipated into spontaneous emission and energy acquired by the atom originally at rest. In a classical approach, we can calculate the work done on the atom as $dW = \boldsymbol{F}\cdot\boldsymbol{dr}$

where dr is the displacement of the electron under the influence of the force F. Assuming the field applied at the site of the electron to be:

$$E_{op} = E_o \cos \omega_L t \tag{2.122}$$

the force is then $(e\, E_{ap})$. We identify the term $(e\, dr)$ as an elementary dipole and we calculate the expectation value of the rate of energy absorption dW/dt as:

$$\left\langle \frac{dW}{dt} \right\rangle = E_o \cdot \left\langle \dot{d} \right\rangle \cos \omega_L t \tag{2.123}$$

where:

$\dot{d}$ is the time derivative of d

The expectation value of the dipole moment is calculated as in the case of the force through the relation $<d> = Tr(\rho\, d)$ and we obtain from the equilibrium density matrix calculated above:

$$\left\langle d \right\rangle = 2d_{m\mu} \left(\delta^r \cos \omega_L t - \delta^i \sin \omega_L t \right) \tag{2.124}$$

Finally, replacing d in Equation 2.123 we obtain:

$$\left\langle \frac{dW}{dt} \right\rangle = \hbar \omega_{Rm\mu} \omega_L \delta^i \tag{2.125}$$

It is thus clear that the force expressed in Equation 2.119, proportional to the imaginary part δ^i of $\rho_{m\mu}$, is the result of the absorption of a photon as made explicit in Equation 2.125 for the rate of energy absorbed. It is also clearly seen, by means of Equation 2.124, that it is the component of the dipole moment that is out of phase with the field that causes the absorption.

2.5.1.2.1.4 Case of a Standing Wave This is the situation that can be created, for example, when laser radiation is reflected back on itself by a mirror. In that case, the radiation field may be written:

$$E(z,t) = E_o \cos\left(k_L z\right) \cos\left(\omega_L t\right) \tag{2.126}$$

There is thus an amplitude gradient and no phase gradient. The radiation field pattern is fixed in space. Equation 2.117 is reduced to the first term that we call F_{react}:

$$F_{react} = -d_{\mu m} \left\{ \delta^r \nabla E_o(z) \right\} \tag{2.127}$$

where the subscript, react, indicates that we are in effect in presence of a reactive force that we will make more explicit below. Using the definitions introduced earlier and the solution of the Bloch equations for δ^r, this becomes:

$$F_{react} = -\hbar \frac{\nabla \omega_{R\mu m}^2}{4} \frac{\left(\omega_L - \omega_{m\mu}\right)}{\left(\Gamma/2\right)^2 + \left(\omega_L - \omega_{m\mu}\right)^2 (1+S)} \tag{2.128}$$

which, using the definition for S, can also be written as:

$$F_{react} = -\hbar \frac{\nabla \omega_{R\mu n}^2}{4} \frac{\left(\omega_L - \omega_{m\mu}\right)}{\left(\Gamma/2\right)^2 + \left(\omega_L - \omega_{m\mu}\right)^2 + \left(\omega_{Rm\mu}^2/2\right)} \tag{2.129}$$

2.5.1.2.1.5 Properties of F_{diss} and F_{react} We may examine each type of force by setting actual experimental conditions. In the case of the dissipative force, if we set the laser frequency at resonance, $\omega_L = \omega_{m\mu}$, and make the laser radiation intensity weak, $(\omega_{R\mu m} \ll \Gamma)$, then we have:

$$F_{diss} \cong \hbar k_L \frac{\omega_{R\mu n}^2}{\Gamma} \qquad \text{(low laser radiation intensity)} \tag{2.130}$$

The force is proportional to $\omega_{Rm\mu}^2$ thus to the laser radiation intensity. At high intensity, $\omega_{R\mu m} \gg \Gamma$ and Equation 2.121 shows that the force is proportional to the decay rate of the excited state, the atom not being able to absorb photons at a rate larger than that at which it decays to the ground state, thus spending half its time in either of the excited state or the ground state.

$$F_{diss} = \hbar k_L \frac{\Gamma}{2} \qquad \left(\text{high laser radiation intensity}\right) \tag{2.131}$$

In the present case of a travelling wave, the force F is called *dissipative*. The atom absorbs a photon from the laser radiation field and receives a *kick* through the exchange of momentum as we calculated earlier. Since it was assumed at rest to start with, it is thus accelerated to a given velocity under the force just calculated. If the atom travels at speed v in an atomic beam, its speed is reduced and, thus, in principle, is *cooled* in the laboratory frame.

The case of the reactive force F_{react} is rather different. For the laser tuned to the transition $(\omega_L = \omega_{m\mu})$, there is no force present. On the other hand, the force changes sign with the sign of the detuning. Equation 2.129 is represented graphically by a dispersive curve: for $\omega_L < \omega_{m\mu}$ the atom is pushed by the gradient to a field of higher intensity. In the opposite case, $\omega_L > \omega_{m\mu}$, the atom is pushed to a field of lower intensity. For a field of large intensity we have:

$$F_{react} = -\hbar \frac{\nabla \omega_{R\mu m}}{\omega_{R\mu m}} \left(\omega_L - \omega_{m\mu}\right) \tag{2.132}$$

If we set $\omega_{R\mu m} \sim \omega_L - \omega_{m\mu}$, we then have $F_{react} = -\hbar \nabla \omega_{R\mu m}$ and the force increases with the applied field. This is different from the dissipative force which saturates at high values of the field due to the fact that the rate of photon absorption cannot be larger than the rate at which the atoms fall back in the ground state where they are available for absorption. In the case of high intensity, the atom may be stimulated in emission. In that case, the process of spontaneous emission does not play the major role. In the case of the dissipative force, the atom emits a photon in phase with the incident wave and there is no net transfer of momentum. In the case of the reactive force, F_{react}, present in a standing wave, the atom may be excited by the wave propagating in the positive direction and stimulated into emission by the wave of the same frequency propagating in the negative direction. There is no exchange of energy, but the atom receives a kick from the absorption and the emission processes. The term $\nabla \omega_{R\mu n}$ for the case $\omega_{R\mu n}$ created by a standing wave of the form $E_o \cos k_L z$ may have a maximum for a given field amplitude. This maximum exists for $\omega_{R\mu n} \sim \omega_L - \omega_{m\mu}$, and F_{react} becomes:

$$F_{react} \sim \hbar k_L (\omega_{R\mu n})_o \tag{2.133}$$

where:

$(\omega_{R\mu n})_o$ stands for the Rabi frequency $(1/\hbar)d_{\mu n}E_o$

We see readily that the force is proportional to an exchange of momentum $\hbar k_L$ at the rate $(\omega_{R\mu n})_o$, which is different from the dissipative force, for which the exchange of momentum takes place at maximum at the rate Γ.

It should finally be noted that the force can be represented as the negative gradient of a potential U $(F_{react} = -\nabla U)$ having the form (Cohen-Tannoudji and Guéry-Odelin 2011):

$$U = \frac{\hbar(\omega_L - \omega_{m\mu})}{2} \ln \left[1 + \frac{\omega_{R\mu n}^2/2}{(\omega_L - \omega_{m\mu})^2 + (\Gamma/2)^2} \right] \tag{2.134}$$

In view of the form of this potential and the description just made, the reactive force provides the possibility of creating a situation appropriate for trapping atoms. A standing wave is thus such a situation. Some conditions of course must be fulfilled and this will be examined later when we will describe in more detail the type of traps that are appropriate for atomic frequency standard implementation.

It should be mentioned that such trapping is not created by the dissipative force. We will see later that it is possible to cool an ensemble of atoms by means of an arrangement consisting of several laser radiation beams travelling at right angle. The arrangement is called a *molasses* and the cooling is created primarily by the process of exchange of momentum that we described in the calculation of the dissipative force. Consequently, although the atoms are cooled, they tend to diffuse away randomly in such an absorption–emission process and are not trapped as such.

2.5.1.2.2 *Atom in Motion: Doppler Cooling*

Up until now, except for an occasional mention of the effect of laser radiation on atoms in motion, we have analyzed essentially the case of the interaction of laser radiation with an atom at rest. Since it is the purpose of reducing Doppler effect that we have evoked for introducing the subject of cooling, we wish to extend the analysis to the case where the atoms are in motion. We may consider several configurations. For example, we may have an ensemble of atoms in a cell of arbitrary shape at a temperature T. The velocity spectrum is characterized by a Maxwell–Boltzmann distribution around an average speed v_a. It is that speed that we wish to reduce as well as the width of the velocity distribution or velocity spectrum, which determines the temperature of the atomic ensemble. On the other hand, we may have a collimated

beam of atoms originating from a source at a given temperature. The construction of
the passive Cs standard that is used for implementing the primary standard of time
is based on the use of a beam of Cs atoms. In that case as in the case of beams of
other alkali atoms, for example, Rb, the source may be a small heated glass or metallic capsule with a small orifice having the form of a cylinder to act as a collimator.
The speed of the atoms in the beam reflects the temperature of the oven and the
velocity spectrum is characterized by an altered Maxwell–Boltzmann distribution.
In the case of the Cs beam clock, the width of the resonance line reflects the length
of the transit time that the atoms spent in the interaction region. Consequently it
would be desirable to reduce the speed of the atoms in such a beam and at the same
time reduce the line width. Successful experiments in alteration of atomic speeds
by means of laser radiation were done on an atomic beam, in occurrence sodium, in
the early 1980s (Phillips and Metcalf 1982). Let us examine how we can adapt the
analysis we have done above, for atoms at rest, to the case of atoms in motion.

The atom moving in the laboratory frame is at rest in its own reference frame
attached to its centre of mass. Assume it is moving towards the laser and thus opposite to the propagation of the laser radiation beam as in Figure 2.20.

In the elementary approach done at the beginning of Section 2.5.1.1.1, we
have calculated that the exchange of momentum between the radiation and the
atom causes recoil of the atom, which gains a speed v_{rec} whose value is given by
Equation 2.83. The same reasoning applies to the atom in motion except that now
that recoil velocity is added vectorially to the velocity of the atom. However, a
condition is imposed on the laser frequency. If the laser is tuned at the resonance
frequency of the atom at rest, it is no longer resonant with the atom in motion at
speed v. If the atom moves towards the laser or in other words moves against the
direction of propagation of the wave, it sees a wave shifted up in frequency by
Doppler effect by the amount:

$$\Delta \omega_D = k_L v \qquad (2.135)$$

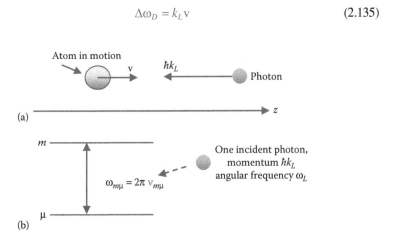

(a)

(b)

FIGURE 2.20 (a) Pictorial interpretation of a photon with momentum $\hbar k_L$ incident on an
atom moving at velocity v against the photon or laser radiation propagation. (b) Representation
of an atom with energy levels E_m (excited state) and E_μ (ground state) in interaction with a
photon of energy $\hbar \omega_L$.

Consequently, in order to be resonant with the frequency of the transition E_μ to E_m, that is $\omega_{m\mu}$, the laser must be tuned to a frequency given by:

$$\omega_L = \omega_{m\mu} - k_L v \tag{2.136}$$

that is a lower frequency or in other words tuned towards the red relative to $\omega_{m\mu}$. In that situation the laser tuning is such as to be resonant with the atom and the transition probability is at maximum. The atom absorbs the photon and the recoil velocity adds vectorially to the original atom velocity. The atom motion is thus slowed down and a similar reasoning regarding energy and spontaneous emission as we did in the case of the atom at rest may be done. We would like, however, to examine the details of the effect of this interaction and apply it later to an ensemble of atoms in two- (2D) and three-dimensional (3D) spaces and exposed to counter-propagating waves. To do this, we re-examine the quantum mechanical approach we have used previously to derive expressions for the two types of forces that result from the atom–radiation interaction.

Since that atom sees a frequency different from the one it sees when it is at rest, we can simply express this fact by adding the Doppler shift (Equation 2.135) to the frequency ω_L of the laser. The dissipative force is then given by Equation 2.121, which we reproduce here with laser frequency now displaced by Doppler effect:

$$F_{\text{diss}}(v) = -\hbar k_L \Gamma \frac{\omega_{R\mu n}^2}{4} \frac{1}{\left(\Gamma/2\right)^2 + \left(\omega_L + k_L v - \omega_{m\mu}\right)^2 + \left(\omega_{R\mu n}^2/2\right)} \tag{2.137}$$

We have thus a force that is dependent on v. As mentioned above, it is maximum when the laser frequency ω_L is tuned towards the red relative to $\omega_{m\mu}$ to compensate for Doppler effect, that is when:

$$\omega_L + k_L v = \omega_{m\mu} \tag{2.138}$$

The resonance line shape is Lorentzian. The force is represented in Figure 2.21 as a function of the laser frequency for an atom travelling against the wave direction at a velocity v_1.

The equation representing the force is relatively complex. Atoms with different velocities respond in a similar way to the laser radiation except that the peak is displaced

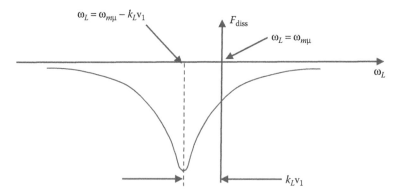

FIGURE 2.21 Schematic representation of the dissipative force as a function of the laser frequency for an atom travelling against the wave at a velocity v_1 as in Figure 2.20.

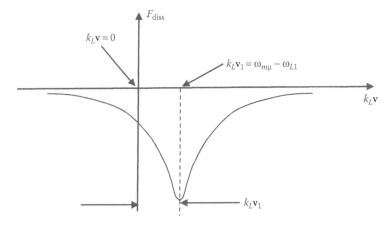

FIGURE 2.22 Schematic representation of the dissipative force as function of the atom velocity for a laser detuned to the red relative to resonance frequency of an atom at rest.

relative to the resonance frequency of an atom at rest. One can also plot the force as a function of v and a similar curve is obtained. This is represented in Figure 2.22. In the case represented in the figures, the laser is tuned at $\omega_L = \omega_{m\mu} - k_L v_1$. We see that atoms at $v = 0$ are on the wings of the resonance curve but still experience a force.

It is readily observed that since the width of the resonance is controlled by Γ ($\Delta\omega_{1/2} = \Gamma$), a limited number of atoms within a velocity range are in interaction with the laser radiation. From Table 2.1, the natural line width at half the height is about 5 MHz for Cs giving $\Delta\omega_{1/2}$ equal to $30 \times 10^6 \text{s}^{-1}$. Detuning the laser to the red by 2Γ would thus displace the resonance peak to a velocity given by $k_L v = 60 \times 10^6 \text{s}^{-1}$ and, with $k_L \sim 7 \times 10^7 \text{m}^{-1}$, atoms with velocities of the order of 1 m/s would experience maximum force. Atoms in a cell at a given temperature have a spectrum of velocities given by a Maxwell distribution. If exposed to such a radiation field, those atoms within a range 0.5 m/s to 1.5 m/s centred on that value would thus be submitted to a force at least 50% of the maximum given by Equation 2.137.

On the other hand, an atom with negative velocity or travelling in the same direction as the travelling wave will receive an impulse in the same direction as it is travelling and thus will be accelerated. However, due to the added Doppler shift and the recoil velocity acquired, it will be submitted to a smaller force. One way of understanding better the situation is to expand Equation 2.137 in power series around $v = 0$. After some elementary algebra, one finds in first order of v:

$$
F_{diss} = - \left[\frac{\hbar k_L \Gamma}{2} \frac{\omega_{R\mu n}^2/2}{\left(\Gamma/2\right)^2 + \left(\omega_L - \omega_{m\mu}\right)^2 + \left(\omega_{R\mu n}^2/2\right)} \right]
$$

$$
+ \left\{ \hbar k_L^2 \Gamma \frac{\omega_{R\mu n}^2}{2} \frac{\left(\omega_L - \omega_{m\mu}\right)}{\left[\left(\Gamma/2\right)^2 + \left(\omega_L - \omega_{m\mu}\right)^2 + \left(\omega_{R\mu n}^2/2\right)\right]^2} \right\} v
$$

(2.139)

It is standard practice to write this expression as:

$$F_{\text{diss}} = F_0 - \alpha v \tag{2.140}$$

with

$$F_0 = -\frac{\hbar k_L \Gamma}{2} \frac{\omega_{R\mu n}^2/2}{\left(\Gamma/2\right)^2 + \left(\omega_L - \omega_{m\mu}\right)^2 + \left(\omega_{R\mu n}^2/2\right)} \tag{2.141}$$

and

$$\alpha = -\hbar k_L^2 \Gamma \frac{\omega_{R\mu n}^2}{2} \frac{\left(\omega_L - \omega_{m\mu}\right)}{\left[\left(\Gamma/2\right)^2 + v\left(\omega_L - \omega_{m\mu}\right)^2 + \left(\omega_{R\mu n}^2/2\right)\right]^2} \tag{2.142}$$

We note that these expressions can also be written in terms of the saturation parameter S as:

$$F_0 = -\frac{\hbar k_L \Gamma}{2} \frac{S}{(1+S)} \tag{2.143}$$

$$\alpha = -\hbar k_L^2 \Gamma \frac{\left(\omega_L - \omega_{m\mu}\right)}{\left[\left(\Gamma/2\right)^2 + \left(\omega_L - \omega_{m\mu}\right)^2\right]} \frac{S}{(1+S)^2} \tag{2.144}$$

F_0 is seen to be a force independent of the velocity of the atom and acts on the atom in the direction of the propagation of the wave while the second term, αv, appears as a force proportional to the speed of the atom as in a friction environment, opposing the atom motion.

If a reduction in speed of an atom is desired, the force F_0 must be compensated for to start with. There are various ways of accomplishing that (Cohen-Tannoudji and Guéry-Odelin 2011). We have seen that it is possible to trap ions, for example, by means of a Paul trap, and consequently it is possible to oppose that force by the restoring force of the trap. F_0 is very small as we have shown earlier since it acts by an exchange of momentum through the recoil effect. On the other hand, it is readily realized that if another radiation field of the same frequency and same intensity is created travelling in the opposite direction as shown in Figure 2.23a, a force independent of velocity opposite to the original one is then created (Hansch and Schawlow 1975). The process is totally symmetrical and the two forces F_0, independent of velocity introduced by each wave, cancel each other.

On the other hand, in a given situation where an atom travels at speed v_1 in the $+z$ direction and the laser is tuned to the red relative to $\omega_{m\mu}$ by the amount $-k_L v_1$, the transition probability (the force) is maximum for the wave travelling from the right, with wave vector $-k_L$. In that case, the laser radiation to the right with positive wave vector k_L is off resonance and the transition probability (the force) is smaller.

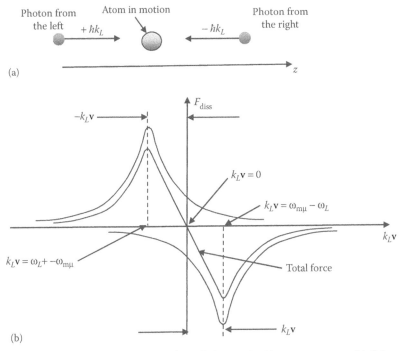

(a)

(b)

FIGURE 2.23 (a) Pictorial representation of photons incident on an atom. (b) Schematic representation of the force created on an atom as a function of its velocity for an interaction created by the presence of two identical, but symmetrical, laser radiation fields travelling in opposite directions.

The difference between the two implies more absorption–emission cycles by the wave travelling from the right creating a net force proportional to the velocity of the atom and opposing its motion. The same argument applies to the symmetrical situation, which is to say for the atom travelling to the left and the laser radiation travelling to the right in the figure. Since the speed reduction mechanism relies on an imbalance caused by Doppler effect, it is generally called *Doppler cooling*. As calculated from Equation 2.141, the force is maximum for $\omega_L - \omega_{m\mu} = -\Gamma/2$. In that case, for weak laser radiation intensity, that is $S \ll 1$, the friction coefficient is maximum and is given by:

$$\alpha = 2\hbar k_L^2 S \qquad\qquad (2.145)$$

The factor 2 originates from the presence of the two travelling waves forming a standing wave and assumed to act independently in the case of weak laser field intensity.

From the analysis just made, it would appear that there is no cooling limit in the process and that an atom would be totally stopped in its motion. However, as is realized, the processes of absorption and emission are discrete and random. Consequently, these processes lead to fluctuations in spontaneous emission and absorption of photons. For example, spontaneous emission may fluctuate in direction and average momentum acquired by the atom may not be zero. This effect can lead to random walk with a residual velocity (Dalibard 1986). This effect may be associated to heating and opposes cooling. When the two processes are equal, one reaches the limit of cooling. As we will see below, the minimum energy that can be reached is given by:

$$E_{min} = \frac{1}{2}\hbar\Gamma \qquad (2.146)$$

This appears to be a limit in laser cooling. However, we will see later that nature has provided means for attaining lower energies and thus by extension, lower temperatures.

2.5.1.2.3 Molasses

The arrangement shown above is for one dimension, the z direction of Figure 2.23a. The wave travelling to the left with wave vector $-k_L$ may originate directly from the laser and the wave travelling to the right with wave vector $+k_L$ may be created by reflecting the same wave on a mirror. In a cell containing an alkali atom vapour and exposed to such one-dimensional interaction, cooling would take place in one direction and some heating would take place through the random spontaneous emission process in all directions. It is readily visualized, however, that a similar situation can be created for the two orthogonal axes x and y. Consequently, a set up in which three lasers emitting in directions oriented at right angles, creating six radiation beams by reflections, can be used to create a 3D environment to reduce the speed of atoms in a gas in all directions over a given volume. This reduction takes place over a certain range of velocities close to zero and in such an arrangement those atoms are in a sense *captured* by the interaction process. At every absorption–emission cycle their speed is reduced, as if submitted to a friction mechanism in a viscous substance. The arrangement has been called a *molasses*, in reference to the high pseudo-viscosity of the arrangement (Chu et al. 1985). The size of the small ball formed by the molasses may be of the order of mm to cm depending on the size of the overlapping volume of the six radiation beams. A schematic of a typical arrangement is shown in Figure 2.24.

As shown in the figure, the properties of the molasses can be studied by means of a narrow laser probe beam. If the cooling lasers are turned off, the molasses falls in the earth's gravitational field and its properties such as size, density, number of atoms, and temperature can be determined as a function of time. From those studies one concludes that with cooling lasers detuned to the red by a few Γ, about 10^9 atoms can be captured in the molasses. On the other hand, the calculation we did previously for a single beam concerning the capture velocity still applies. Depending on the detuning of the lasers, the technique can capture velocities of the order of several metres per second. Furthermore, the same reasoning as the one used in one dimension relative to randomness of the processes involved can be applied to the three-dimensional case. Since in three dimensions the process of cooling relies on the discrete absorption–emission cycles that are random, it is submitted to statistics that result in random walk and atoms

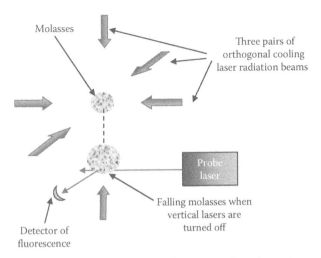

Molasses

Three pairs of
orthogonal cooling
laser radiation beams

Probe
laser

Falling molasses when
vertical lasers are
turned off

Detector of
fluorescence

FIGURE 2.24   Schematic arrangement used for the creation of a molasses by means of three pairs of radiation beams oriented at right angle to each other. Upon turning off of the lasers the ball falls in the earth gravitational field and its properties are analyzed by means of the probe beam.

diffusion in space. The minimum energy (or equilibrium temperature) is given by (Phillips 1992):

$$E_{\min} = \frac{\hbar \Gamma}{4} \left[\frac{2 \left(\omega_L - \omega_{m\mu} \right)}{\Gamma} + \frac{\Gamma}{2 \left(\omega_L - \omega_{m\mu} \right)} \right] \tag{2.147}$$

For $\omega_L - \omega_{m\mu} = -\Gamma/2$, we obtain the same result as that of Equation 2.146. An approximate value for the minimum temperature that can be reached in principle may be obtained by equating this $E_{\min}$ to $k_B T_{\min}$. For Cs, one obtains $T \sim 230\,\mu K$. We will see below that in practice temperatures much lower than that value are obtained.

The small ball created by the molasses possesses extremely interesting properties relative to its use in atomic frequency standards. In particular, it can be used to implement a Cs frequency standard based on Zacharias' original idea of creating a so-called fountain of atoms in the earth's field (see Forman 1985). By changing slightly the frequency of the lasers, the ball can be given a small vertical velocity. When the lasers are turned off, the ball continues its vertical motion up to a height given by classical mechanics laws and falls back under gravity. Its dispersion or increase in size depends on the temperature reached in the cooling process. On that question, a most interesting phenomenon was discovered when the temperature of the molasses was determined experimentally: the measured temperature reached by the atomic ensemble was two orders of magnitude lower than that calculated above on the basis of equilibrium between the cooling effect of the lasers and the heating caused by the randomness of the absorption–spontaneous emission cycles as made explicit in Equation 2.147. A most convincing experimental result obtained on that matter is shown in Figure 2.25 (Phillips 1992).

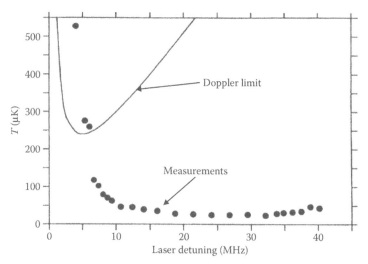

FIGURE 2.25 Variation of temperature of a sodium atoms molasses as a function of the laser red-detuning from the sodium atom rest resonance. The upper solid line shows the Doppler limit as calculated from Equation 2.147. The points are obtained experimentally. (Courtesy of W.D. Phillips 2014, pers. comm.)

At the time of observation, those results appeared to be problematic in the sense that they did not agree with the predictions based on the basic physics connected to Doppler cooling (Lett et al. 1988). The phenomenon required some serious thinking on the behaviour of atoms in the presence of intense laser fields. It was known for some time that such fields cause displacements of the atom energy levels, that is, light shifts which we calculated in Section 2.3. The analysis shows that the atom in its motion in space is forced to spend extra energy when this motion is correlated to the space modulated pumping rate of the absorption–emission cycle and the energy valleys and hills created by the light shifts. It has been given the name Sisyphus cooling (Dalibard and Cohen-Tannoudji 1989). The effect will be described semi-quantitatively in Section 2.5.3.

2.5.2 EFFECT OF FLUCTUATIONS IN LASER COOLING AND ITS LIMIT

Absorption and spontaneous emission are random discrete processes. Consequently, during a time Δt these processes are affected by fluctuations. Spontaneous emission takes place in random directions and the process leads to a random walk of momentum of minimum step, $\hbar k_L$.

This leads in momentum space to a diffusion of the atom momentum characterized by a dispersion which increases with the time of observation Δt. The number of steps is equal to the number of absorption cycles dN/dt per unit time. The mean quadratic dispersion of momentum $\langle \Delta p^2 \rangle$ can thus be written as the product of three terms: the square of the minimum momentum value, the number of absorption cycles per unit time, and the time of observation:

$$\langle \Delta p^2 \rangle = \hbar^2 k_L^2 \left(\frac{dN}{dt} \right) \Delta t \qquad (2.148)$$

The number of transitions per second is limited by the rate of absorption which cannot be larger than the rate of spontaneous emission. It was evaluated previously to have a maximum value of:

$$\frac{dN}{dt} = \Gamma \rho_{mm}$$ (2.149)

where:

ρ_{mm} is the stationary fractional population of the excited state that we have calculated earlier and is given by:

$$\rho_{mm} = \frac{1}{2}\frac{S}{S+1}$$ (2.150)

It is a standard practice to write the momentum dispersion given by Equation 2.148 as:

$$\left\langle \Delta p^2 \right\rangle = 2D_{\text{spont}}\Delta t$$ (2.151)

where:

D_{spont} is the so-called spontaneous emission diffusion coefficient

From the above reasoning, D_{spont} can thus be written as:

$$D_{\text{spont}} = \frac{1}{4}\hbar^2 k_L^2 \Gamma \frac{S}{S+1}$$ (2.152)

A similar analysis can be done in connection to the absorption process, fluctuations being present in the number of photons absorbed. The calculated dispersion is similar to the one just calculated and the dispersion coefficient is approximately the same as the one calculated for spontaneous emission (Cohen-Tannoudji and Guéry-Odelin 2011).

These considerations on dispersion of momentum lead to a limit in the cooling process, which can be calculated as follows. The damping of the atomic velocity, or cooling, can be calculated from the damping force acting on the atoms through the friction coefficient α. We have $F = -\alpha v$, which through simple algebra becomes:

$$\frac{dp}{dt} = -\frac{\alpha}{M}p$$ (2.153)

where:

M is the mass of the atom

This can be written in terms of fluctuations of p as:

$$\frac{d\Delta p^2}{dt} = -2\frac{\alpha}{M}\Delta p^2$$ (2.154)

The friction coefficient was evaluated earlier and is given by Equation 2.144. If we assume a laser frequency detuning equal to $(1/2)\Gamma$, which makes the cooling rate maximum, and weak laser intensity such as to make $S \ll 1$, then α becomes:

$$\alpha = 2\hbar k_L^2 S \qquad (2.155)$$

The rate of change of the momentum steps Δp in the cooling process can thus be written as:

$$\left.\frac{d\Delta p^2}{dt}\right)_{cool} = -4\frac{\hbar k_L^2 S}{M}\Delta p^2 \qquad (2.156)$$

On the other hand, the diffusion of the momentum due to random emission and absorption is interpreted as a heating process. From Equation 2.151, we have:

$$\left.\frac{d\Delta p^2}{dt}\right)_{heat} = 2D \qquad (2.157)$$

A stationary state is reached when heating rates and cooling rates are equal. This leads to the situation that momentum fluctuations $(\Delta p)^2$ becomes stationary. Considering fluctuations in both absorption and emission it reaches a value equal to $(1/2)\hbar\Gamma M$. In terms of residual energy this can be interpreted as one degree of freedom residual thermal energy $(1/2)\,k_B T_{eq}$ leading to:

$$k_B T_{eq} = \frac{1}{2}\hbar\Gamma \qquad (2.158)$$

This result can also be interpreted in terms of residual equilibrium velocity:

$$\Delta v_{min} = \left(\frac{\hbar\Gamma}{2M}\right)^{1/2} \qquad (2.159)$$

This analysis thus leads to a limit in the temperature that can be reached through so-called Doppler cooling. In the case of Cs, for example, this limit is 130 μK.

2.5.3 COOLING BELOW DOPPLER LIMIT: SISYPHUS COOLING

We have reported earlier that in practice a much lower temperature than the one just calculated was obtained in practice. Actually the temperature reached experimentally was two orders of magnitude lower than that just calculated. The result appeared of course to be most interesting since it proved the existence of phenomena not known up to that time, phenomena that opened the door to techniques of reaching much lower temperatures than those expected. The task, however, was to explain the origin of such phenomena in order to control them as well as possible and possibly reach still lower temperatures.

2.5.3.1 Physics of Sisyphus Cooling

The origin of the phenomenon causing such an efficient cooling was established as a coupling of the external parameters of the atom, such as its speed, to internal interactions causing a light shift created in the atom energy levels by the laser radiation field. We will outline this explanation in a semi-quantitative way and evaluate its limit. This outline is based on texts published by Dalibard, Cohen-Tannoudji, and Phillips (Dalibard and Cohen-Tannoudji 1989; Cohen-Tannoudji and Phillips 1990; Cohen-Tannoudji and Guéry-Odelin 2011).

Let us re-consider the configuration consisting of two plane waves counter-propagating along the z-axis that we have used up until now. Let us assume at this time that those waves have orthogonal linear polarizations and with the same frequency and the same intensity. The phase of the two waves varies linearly with z and interference results. This interference causes not a standing wave of the kind observed in a cavity, but rather a standing pattern of polarizations distributed in space as illustrated in Figure 2.26a. The ellipticity of the wave varies linearly with z passing from circular to linear polarization in a stationary pattern.

Up until now, our analysis relative to Doppler cooling has been done with an atom having two energy levels and the laser radiation was assumed to be nearly resonant with the frequency corresponding to a transition between these two levels. In practice, as with alkali atoms, this is not the case. Atoms, such as Rb or Cs, for example, have an unpaired electron with spin 1/2 in interaction with a nucleus having a nuclear spin I. This creates manifolds of Zeeman levels in both the ground and excited states. Those levels are usually degenerate, but that degeneracy can be lifted by various effects such as interaction with and electromagnetic radiation field, a static electric field or static magnetic field. It turns out that the presence of those energy levels manifolds is essential in establishing a comprehensive explanation of the sub-Doppler cooling observed. However, in order to simplify the analysis, we will use a simpler system in which the atom has an angular momentum $J_\mu = 1/2$ in the ground state and $J_m = 3/2$ in the excited state. We assume that the separation between these two states corresponds to radiation in the optical range. This will allow us to examine, without ambiguity, the behaviour of the internal state of such a simple atom when exposed to radiation of the kind illustrated in Figure 2.26.

The various transitions that can take place between the levels of the ground state μ and those of the excited state m are illustrated in Figure 2.26b. The two Zeeman sublevels $m_\mu = +1/2$ and $m_\mu = -1/2$ undergo different light shifts, shifts that depend on the laser polarization, because of changes in interaction set by Clebsch–Gordon coefficients. Consequently, the degeneracy is lifted. Figure 2.26d makes explicit the spatial modulation of the splitting between the two sublevels with a period $\lambda/2$, reflecting the periodicity of the polarization. Thus, an ensemble of atoms distributed uniformly along the z-axis experiences an interaction that causes energy shifts that depend on their location in space as if submitted to a spatially modulated potential, with *hills* and *valleys*, corresponding to a modulation of the light shift of the ground-state levels. On the other hand, we first realize that atoms submitted to a field such as that shown in Figure 2.26a can be optically pumped from one level to the other of the ground state. The transitions causing that optical pumping are shown in Figure 2.26b and the probability of those transitions is calculated in a straightforward way (*QPAFS* 1989). The intensity of those transitions is dictated by symmetry and the associated Clebsch–Gordon coefficients. The optical pumping is characterized by a pumping rate Γ_p given by the expression derived earlier, Equation 2.41, with the Rabi frequency weighted by a Clebsch–Gordon coefficient. The resulting weight is given in Figure 2.26 for the transitions associated with the

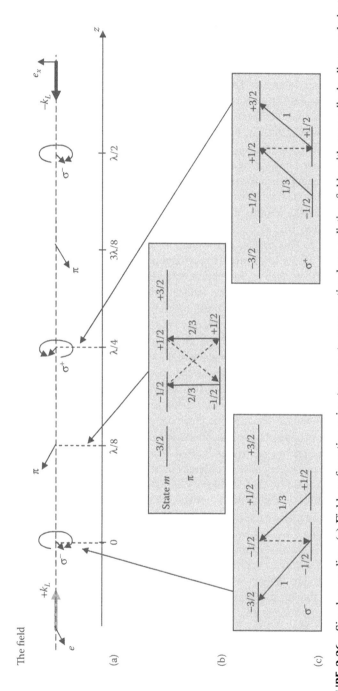

FIGURE 2.26 Sisyphus cooling. (a) Field configuration using two counter-propagating laser radiation fields with perpendicular linear polarization. (b) States of an atom with two states $J = 0$ and $J = 1$ exposed to that field. The transition probabilities in the various sectors are identified. (c) Field configuration as in (a). *(Continued)*

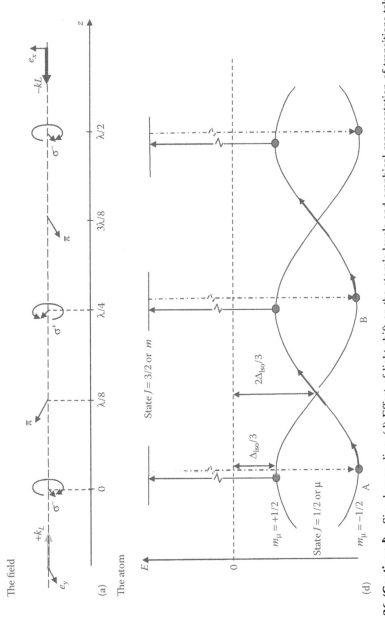

FIGURE 2.26 (Continued) Sisyphus cooling. (d) Effect of light shift on the atom's levels and graphical representation of transitions taking place in Sisyphus cooling.

illustrated polarization. Consequently, the pumping rate is modulated. A simple analysis of the situation leads to the conclusion that in the region of circular polarization an atom is pumped at a rate that is different depending in which level it is in. For example, in the region where the radiation is polarized σ^- the atoms are pumped to the level $m_\mu = -1/2$. This has the net effect of pumping the atoms in the lower energy level with the consequence that in stationary state the population itself is modulated along the z-axis. The polarization modulation thus leads to modulated pseudo-potential where atoms with sufficiently low energy (low speed) could be trapped. We can associate with Γ_p a pumping time $\tau_p = 1/\Gamma_p$. This time is much longer, by several orders of magnitude, than the lifetime of the atom in the excited state $J = 3/2$, which is $1/\Gamma$, Γ being a spontaneous decay rate of the order of tens of nanoseconds. During the optical pumping time, in the millisecond range, the atom has the time to travel a certain distance. Let us assume that such an atom, at position A in Figure 2.26d and in state $m_\mu = -1/2$ travels at speed v_p. Let us assume that this speed is such that during the time t_p it is sufficient for the atom to climb up the hill and travel a distance $\lambda/4$, to region B as illustrated. In such a situation, due to change in polarization, the atom in level $m_\mu = -1/2$ is pumped to level $m_\mu = +1/2$. In the process, the atom climbs a potential hill. Its kinetic energy is transformed into potential energy. It has slowed down. It decays from level $m_\mu = -1/2$ to level $m_\mu = +1/2$ in a very short time $1/\Gamma$ and the photon emitted has an energy larger than the original pumping photon by the quantity $2/3\ \Delta_{ls}$. Through the absorption–emission cycle, the atom has lost kinetic energy. It may then subsequently climb a new hill, slowing it down still further. Like Sisyphus in the Greek mythology, who was always rolling a stone up the hill, the atom is running up potential hills more frequently than down since it has overall greater probability of being in the lower ground-state energy level as explained above.

2.5.3.2   Capture Velocity

The speed v_p used above in the explanation may effectively be considered as a capture velocity of the cooling process since above that velocity the atom would not stay long enough in the appropriate polarization to be pumped to the other ground-state level. During the pumping time τ_p the atom travels a distance $\lambda/4$ and we can write:

$$v_{capt}\,\tau_p = v_{capt}\left(\frac{1}{\Gamma_p}\right) \approx \frac{\lambda}{4} \tag{2.160}$$

The definition of k_L in terms of λ leads to a capture velocity:

$$v_{capt} \approx \frac{\pi\Gamma_p}{2k_L} \tag{2.161}$$

This is to be compared to the capture velocity effective in Doppler cooling, $\Gamma/2k_L$, which gives:

$$\frac{v_{capt}(\text{Sis.})}{v_{capt}(\text{Dop.})} \approx \frac{\pi\Gamma_p}{\Gamma} \tag{2.162}$$

This ratio is extremely small and it shows that Sisyphus cooling is effective only on atoms of low velocities or having already been cooled by another process such as Doppler cooling.

2.5.3.3 Friction Coefficient

On the other hand we may attempt to define a friction coefficient α_{Sis} as we have done in the case of Doppler cooling. We thus write:

$$F = \alpha_{Sis} v \qquad (2.163)$$

The energy W dissipated by the atom in climbing the potential hill is $\hbar(2/3)\Delta_{ls}$. At speed v_p optimum for cooling it does that in time $\tau_p = 1/\Gamma_p$. Thus:

$$\frac{dW}{dt} = -\frac{\hbar(2/3)\Delta_{ls}}{\tau_p} = (2/3)\hbar\Delta_{ls}\Gamma_p \qquad (2.164)$$

On the other hand, the work done by the atom is equal to force × distance travelled. We may thus write:

$$\frac{dW}{dt} = F\frac{d}{dt}(\text{distance travelled}) = Fv \qquad (2.165)$$

Using Equation 2.163, we obtain:

$$\frac{dW}{dt} = -\alpha_{Sis} v^2 \qquad (2.166)$$

Identifying v as v_{capt} and using Equation 2.164 we finally get:

$$\alpha_{Sis} \approx -\hbar k_L^2 \frac{\Delta_{ls}}{\Gamma_p}\left(\frac{8}{3\pi^2}\right) \qquad (2.167)$$

We identify the light shift by means of Equation 2.43 and the pumping rate by means of Equation 2.41 and obtain the following order of magnitude value for the friction coefficient for Sisyphus cooling:

$$\alpha_{Sis} \sim -\hbar k_L^2 \frac{\omega_L - \omega_{m\mu}}{\Gamma} \qquad (2.168)$$

We see readily that for a laser detuning much larger than Γ, α_{Sis} may be much larger than the Doppler cooling friction coefficient, which is given by $\alpha_{Dop} \sim -\hbar k_L S$, where S is the saturation factor assumed much smaller than 1. What must be realized, however, is that the range of velocities over which Sisyphus effect is effective or in other words the capture velocity is very small compared to that of Doppler cooling. Sisyphus cooling thus acts on very small velocities and, in practice, one relies on Doppler cooling to feed the Sisyphus cooling step.

2.5.3.4 Cooling Limit Temperature

The limit of Sisyphus cooling is readily estimated by means of Figure 2.26 by realizing that at each absorption–emission cycle the kinetic energy of the atom is reduced by the work done in climbing the pseudo-potential hill. The depth of the well is of the order of Δ_{ls}. Once the atom has reached a kinetic energy of the order of $\hbar\Delta_{ls}$, it is essentially trapped in the well and can no longer be optically pumped. We can thus say that at that point we have:

$$k_B T_{Sis} \sim \hbar\Delta_{ls} \tag{2.169}$$

Using again the value calculated for Δ_{ls} and assuming a large detuning relative to the line width we obtain:

$$k_B T_{Sis} \sim \hbar \frac{\left(\omega_{1\mu n}\right)^2}{\left(\omega_L - \omega_{m\mu}\right)} \tag{2.170}$$

One readily observes that a decrease in Rabi frequency $\omega_{1\mu n}$, that is to say laser intensity, could lead to lower temperature. This can be understood by the fact that the light shift is smaller for a lower radiation intensity. Laser detuning has the same effect. Thus, in principle one could reach very low temperatures by simply reducing the laser intensity or by increasing the detuning. The conclusion of this analysis is that although Sisyphus cooling acts on a small number of atoms, that is to say those having already a low velocity, it can reduce that velocity even more rather efficiently and cool a sample of atoms to very low temperatures.

It should be mentioned that polarizations different from those of the radiation beams illustrated in Figure 2.26a can be used in order to obtain a similar result. For example, the two counter-propagating laser radiation fields could be polarized circularly in a σ^+–σ^- configuration. In such a case, a linear polarization that rotates in space is created. This is different from the configuration of Figure 2.26a where the polarization alternates from σ^+ to linear to σ^-. However, although details of the interactions that take place are quite different from the case just analyzed, the final results obtained are similar and in practice the two schemes of polarization can be used (see Dalibard and Cohen-Tannoudji 1989 and references therein).

2.5.3.5 Recoil Limit

Sisyphus cooling, however, is limited by the recoil effect. It should be realized that in Sisyphus cooling as in Doppler cooling, fluorescence cycles never cease. Since the random recoil $\hbar k$ communicated to the atom by the spontaneously emitted photons is random and cannot be controlled, it seems impossible to reduce the atomic momentum spread Δp below a value corresponding to the photon momentum $\hbar k$. The condition $\Delta p = \hbar k$ defines the single photon recoil limit, the corresponding recoil temperature being set as $\left[\left(k_B T_R\right)/2\right] = E_R$, E_R being the recoil energy of a spontaneously emitted photon. If the cooling process reaches that temperature, heating takes place through that recoil energy and Sisyphus cooling is no longer efficient (see Appendix 2.A). The lowest temperatures that can be achieved with such a scheme is thus of the order of a few E_R/k_B. This is of the order of a few microkelvins for heavy atoms such as Rb or Cs. This analysis is confirmed by a full quantum theory and is in good agreement with experimental results.

2.5.3.6 Sub-Recoil Cooling

There are ways of reducing the temperature of an ensemble of atoms still further. This is a subject known as sub-recoil cooling. What one realizes is that the recoil energy results from spontaneous emission. If one could make spontaneous emission velocity dependent, one could in principle manipulate the recoil effect. Consequently, if the atom could be put in a state where fluorescence vanishes at $v = 0$, one would create a favourable situation where cooling could be done to a temperature below recoil since no fluorescence would be present. Such a situation can be created by means of coherence population trapping in which two laser radiation fields are used to create a non absorbing state (see Section 2.4). In a specific condition where the laser beams are circularly polarized and propagate in opposite directions, atoms with speed $v = 0$ being in a dark state are not absorbing radiation while those atoms with a given small velocity can interact with the radiation. These atoms are excited and can randomly fall upon cooling into the dark state condition close to zero velocity. This approach has been used by Aspect (Aspect et al. 1989). A similar approach consists in using stimulated Raman transitions to favour trapping in velocity space atoms with near zero velocity (Kasevich and Chu 1992). These techniques rely essentially on rare events that are functions of random walk processes. Another approach for reducing temperature still lower consists in using the so-called evaporative technique (Hess 1986). In that technique, the depth of the potential well used to trap an ensemble of atoms is gradually reduced and atoms with energy larger than that of the depth of the potential well are left to escape leaving behind those atoms with lower energy. The ensemble then consists of atoms with lower energy or lower temperature. These techniques are used in experiments connected to fundamental research on Bose–Einstein condensation, for example. They will not be developed further here since they fall outside the direct interest of this text.

2.5.4 MAGNETO-OPTICAL TRAP

In a molasses, the atoms appear to be trapped in a region of space whose volume is about the size of the crossing volume of the six laser beams, that is a cm or so. However, this trapping is rather loose in the sense that there is no spatial restoring force as such. The force applied to the atoms acts essentially in the velocity domain through a friction mechanism while in a real trap the force should act in real space forcing the atoms to a single region of space or ideally to a single point. It is possible to compensate for that situation by applying to the ensemble of atoms, in conjunction with the cooling lasers, a magnetic field whose distribution in space is such as to introduce a new force that pushes the atoms towards a given limited region of space or a point.

The arrangement is called a magneto-optical trap (MOT) and shown in Figure 2.27 (Raab et al. 1987). The field produced by the two coils is represented in Figure 2.28a. The two horizontal coils with current in opposite directions produce a quadrupolar field. The field lines are axially symmetric and the field is zero in the centre of the

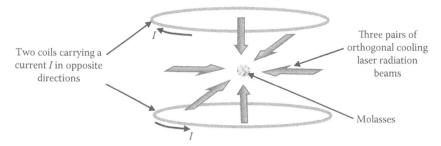

FIGURE 2.27 Configuration of a system that exposes a molasses to a so-called quadrupolar magnetic field that introduces a new force that pushes the atom towards the central region where the magnetic field is zero. The arrangement is called a magneto-optical trap.

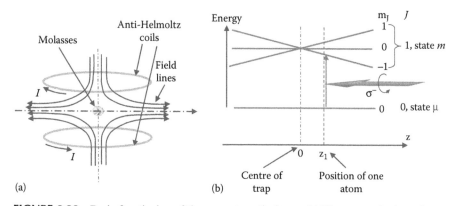

FIGURE 2.28 Basic functioning of the magneto-optical trap. (a) The current in the coils are opposed such as to produce a quadrupolar magnetic field. (b) Two-dimensional representation of the energy levels of an atom with a ground state $J = 0$ and excited state $J = 1$ submitted to such a field around the central part.

structure. The operation of the system is best understood by examining what happens in one-dimension as shown in Figure 2.28b. In order to simplify the analysis, we will assume an atom with two states with angular momentum $J = 0$ for state μ and $J = 1$ for state m. There is no hyperfine splitting and the m state splits into three levels whose energy increases linearly with increasing magnetic field intensity. An atom positioned at z_1 is exposed to a magnetic field that splits the upper state $J = 1$ into three Zeeman levels. The laser radiation from the right, polarized σ^-, and tuned to the red relative to the resonance frequency of the atom at the centre of the trap, where the magnetic field is zero, excites the transition from level $J = 0$, $m = 0$ to $J = 1$, $m = -1$. This interaction causes an exchange of momentum. This is reflected by the creation of a force as we have calculated earlier. In the calculation it was assumed that the atom in motion was exposed to radiation of a frequency displaced by Doppler effect. In the present case it is the atom frequency itself that is displaced by the magnetic field. We can thus use the same equation for the force (Equation 2.137).

and adapt it to the present situation where the atom resonance frequency is displaced relative to its value in zero magnetic field. The magnetic component of the interaction Hamiltonian is (*QPAFS* 1989, Volume 1):

$$\mathcal{H} = g_J \mu_B S_z B \tag{2.171}$$

creating the level structure shown in Figure 2.28b. In the present situation, for the transition considered in the figure ($m_J = -1$ to $m_J = 0$), the resonance frequency of the atom is altered by Zeeman effect from its resonance in zero field $\omega_{m\mu}$ by the quantity $-2\mu_B B/\hbar$, in which g_J has been made equal to 2. In such a situation, the force exerted by the incident photon on the atom may be written as:

$$F_{diss}(v) = -\hbar k_L \Gamma \frac{\omega_{R\mu n}^2}{4} \frac{1}{\left(\Gamma/2\right)^2 + \left\{\omega_L + k_L v - \left[\omega_{m\mu} - 2\mu_B B(z)/\hbar\right]\right\}^2 + \left(\omega_{R\mu m}^2/2\right)} \tag{2.172}$$

On the other hand the magnetic field at the site of the atom is assumed to vary linearly as:

$$B(z) = Az \tag{2.173}$$

where:
   A is the gradient and expressed in T/m

We thus have a force that depends on position, a situation required for the creation of a trap. We may assume that the change in frequency is small close to the centre of the trap and as we did for the case of the term that depends on velocity, we may expand the equation in a power series and keep the linear term only. We then obtain a force that depends on velocity and on position within the trap:

$$F = F_0 - \alpha v_z - \kappa z \tag{2.174}$$

where α has the same definition as previously (Equation 2.144) and κ is defined as:

$$\kappa = 2k_L \mu_B AS \frac{-\left(\omega_L - \omega_{m\mu}\right)\Gamma}{\left(\omega_L - \omega_{m\mu}\right)^2 + \left(\Gamma/2\right)^2} \tag{2.175}$$

The force αv_z creates a friction as mentioned earlier and the force $-\kappa z$ pushes the atoms towards $z = 0$ where that component of the force is zero. The wave in its travel to the left can interact with an atom that would be situated on the negative side of the z-axis. However, its frequency would be off resonance due to its polarization that inhibits its interaction with the pair of levels that has the proper resonance frequency. The same reasoning applies to a radiation field coming from the left with equal intensity and same frequency but with the polarization σ^+. The force is symmetrical and the atom is pushed towards the centre of the trap. On the other hand, in the case when radiation is applied from the left and the right simultaneously, the force F_0 cancels as in our previous analysis. We thus have a situation in which atoms on each side of the centre of the structure of Figure 2.28a are pushed towards the centre.

We can re-create the three-dimensional arrangement used for creating the molasses and obtain the same type of force proportional to distance from zero on the other axes x and y. The atoms are now within a molasses with their speed reduced by friction and at the same time within a real trap with its centre situated at the zero of the magnetic quadrupole field. The technique was described for an atom with a single ground level. In practice, however, atoms used in atomic frequency standards have a nuclear spin and a ground-state total angular momentum larger than 0. That ground state splits into several Zeeman levels. Consequently, the cooling of atoms even using a cycling transition may lead to optical pumping (due to various causes such as imperfect circular polarization). Atoms may accumulate in a ground level that does not take part in the laser cooling, creating a situation similar to that observed in the case of CPT analyzed earlier. Such a trapping state is sometimes called a *dark state*, a notation that should not be confused with the CPT phenomenon itself also called a dark state. In that case, it is necessary to re-pump the atoms out of that level by an additional laser beam. This is generally implemented without much difficulty. The MOT turns out to be one of the most efficient laser-cooled atom trap. At the application of the quadrupole magnetic field, the volume of the molasses is seen experimentally to reduce in size by an order of magnitude. It has become the main tool for research in atomic physics and implementation of new systems in the field of atomic frequency standards.

As the reader certainly realized, the use of radiation emitted by a laser and reflected on a mirror leads to the creation of standing waves in the environment in which the radiation propagates. The standing wave created is stable in space and has a constant phase. We have analyzed that situation previously and have arrived at the conclusion that the force exerted on the atoms was due, not to a gradient of phase since it is constant, but to a gradient of field amplitude.

The force is given by Equation 2.129. It increases with the intensity of the field present to which the Rabi frequency $\omega_{Rm\mu}$ is proportional, is function of the laser detuning $(\omega_L - \omega_{m\mu})$ and its behaviour looks like a dispersion curve. It was called a *reactive force*. It acts effectively like a trap for the atoms. In the present case, however, we assumed that the laser radiation field was weak.

Atoms are not trapped then as such in the Doppler cooling that we described and that led to the creation of a cooled ensemble that we called a molasses. It is found that in such a weak field situation, interference terms appear in the interaction of the atomic dipoles with the two waves travelling in opposite directions. These terms are spatially modulated over a half wavelength and their spatial average is zero. In weak fields the atoms in their motion have sufficient energy to traverse several of those modulation cycles and average those terms. The interaction from the two counter-propagating waves contributing to the friction coefficient α can thus be considered independently as we did and can be summed up (Cohen-Tannoudji and Guéry-Odelin 2011).

2.5.5 Other Experimental Techniques in Laser Cooling and Trapping

We have described above the so-called technique of laser Doppler cooling. The technique is applied to an ensemble of atoms in a vapour state. Historically, the first experiments on laser manipulation of atomic speeds and cooling were done on beams of atoms such as sodium. With atomic beams, a complication arises in the deceleration process because of the changing Doppler shift itself with deceleration. As the atom velocity changes, the laser frequency seen by the atom in its rest frame also changes and the resonance condition is no longer fulfilled, the laser frequency becoming off resonance with the atom.

A compensation for that changing Doppler shift during the atom deceleration can be done by modifying either the laser frequency ω_L or the atomic resonance frequency $\omega_{m\mu}$. Along this line of thought, several solutions have been proposed and demonstrated. So-called Zeeman-slowing and laser chirping have been the most commonly used methods. In the first case, frequency tuning is achieved with a spatially varying Zeeman frequency shift induced by an inhomogeneous static magnetic field. Experiments using such an approach were done on a sodium beam (Phillips and Metcalf 1982; Prodan et al. 1982). The atoms could even be stopped producing a small volume of cold atoms (Prodan et al. 1985). In the second case, the laser frequency is swept (chirped) so as to follow the changing frequency shift due to Doppler effect during deceleration (Letokhov et al 1976; Balykin et al. 1979; Phillips and Prodan 1983).

These cooling techniques, although interesting on their own, turned out to be useful in the field of atomic frequency standards. As we mentioned, molasses have limitations in capturing a large number of atoms, since the cooling lasers interact through detuning of a few Γ with a limited number of atoms at rather low velocities. The number of atoms captured is a function of the laser detuning and is limited by the natural line width of the atoms in question. Consequently, if the atomic ensemble is pre-cooled by some technique, more atoms are captured by the molasses. In practice, this can be done by feeding the atomic ensemble, from which the molasses is extracted, by means of an atomic beam consisting of atoms whose speed has been reduced by transfer of momentum from laser radiation. The two techniques just mentioned can thus in principle be used for that purpose. On the other hand, we have seen that the molasses is rather efficient in cooling an ensemble of atoms in a small volume particularly if a quadrupole magnetic field is applied to the atomic ensemble. An interesting development took place when it was realized that the MOT could be altered in such a way that a beam of slow atoms could be extracted from it. The arrangement is called a 2D MOT (Monroe et al. 1990; Riis et al. 1990; Dieckmann et al. 1998). It could then be used for feeding an actual molasses increasing considerably the number of atoms captured. We will introduce at this point these techniques by which low velocity atomic beams can be created.

2.5.5.1 Laser Atom-Slowing Using a Frequency Swept Laser System: Chirp Laser Slowing

In the chirping technique, a laser beam counter-propagating to the atomic beam, tuned slightly to the red relative to the rest frequency of the atoms, is used to reduce the speed of a group of atoms within a range of velocities. During slowing, the atoms are driven out of resonance, their Doppler shift changing with their changing speed. To compensate this effect, a modulation having the form of a ramp is

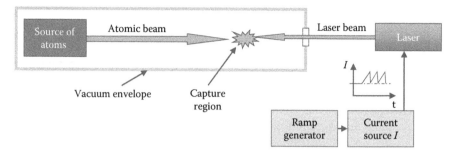

FIGURE 2.29 Schematic diagram of the slowing of an atomic beam by means of laser chirping. The central region is called loosely the *capture region* although the atoms are not trapped as such. The atoms in that region, whose speed has been considerably reduced, or have even been stopped in their motion in the z direction, can be used for other experiments. Although not made explicit, the laser frequency is generally stabilized by means of a technique outlined earlier.

applied to the laser diode driving current. This has the consequence of sweeping the laser frequency, maintaining resonance with the atoms at every instant. A schematic diagram of a laser chirping system is shown in Figure 2.29. In the particular setup shown, atoms of Cs, for example, originate from a reservoir heated at ~90°C. The pressure difference between the oven and the loading chamber produces an effusive jet which is collimated generally by an array of capillaries. In a typical case, the thermal beam travels over a distance of the order of 50 cm during which the speed of the atoms is reduced by the laser radiation. The atoms enter a region called capture region where they can be used for other experiments. The laser current is driven by a source modulated by a ramp generator. The laser radiation is thus chirped and, when the chirping rate is appropriate, the radiation propagating against the beam direction is tuned to compensate for the changing speed or Doppler shift seen by the atoms. The radiation interacts with the atoms over the full length and reduces their longitudinal velocity through the momentum exchange mechanism described above.

In a simple calculation, the time required to stop an atom with initial velocity v_0 is calculated as $\Delta t = v_0/a$, and the required length of travel, ΔL, is given by:

$$\Delta L = \frac{1}{2}a(\Delta t)^2 = \frac{v_0^2}{2a} \qquad (2.176)$$

where:
   a is the average deceleration rate of the atom

As just explained, in order for the laser to compensate for the changing speed of the atom, it must be continuously retuned. The Doppler shift as seen by the atom is given by:

$$\omega_D = k_L v \qquad (2.177)$$

where:
   v is the atom speed

The change in speed by Δv upon deceleration and the change in laser angular frequency must be equal:

$$\Delta\omega_D = k_L \Delta v \qquad (2.178)$$

The laser frequency must then be changed at the same rate as the change in speed of the atoms. We thus set:

$$\frac{d\omega_L}{dt} = \frac{\Delta\omega_D}{\Delta t} = k_L \frac{dv}{dt} \qquad (2.179)$$

where:
dv/dt is the rate of change of velocity

Assuming that the laser radiation intensity is very large such as to make the Rabi frequency $\omega_{R\mu m}$ much larger than $\Gamma/2$, the force applied to the atom is given by Equation 2.121 leading to a maximum rate of change of velocity given by

$$\left(\frac{dv}{dt}\right)_{max} = a_{max} = \frac{F}{M} = \frac{\hbar k_L}{M}\frac{\Gamma}{2} = v_{rec}\frac{\Gamma}{2} \qquad (2.180)$$

We thus obtain a required sweep frequency equal to

$$v_{sweep} = \left(\frac{dv_L}{dt}\right)_{max} = \frac{1}{2\pi}\frac{d\omega_L}{dt} = \frac{k_L v_{rec}}{2\pi}\frac{\Gamma}{2} \qquad (2.181)$$

We need to apply this sweep for a time τ_{rep} that depends on the original velocity v_0 of the atoms in the beam velocity spectrum selected by the choice of the laser initial frequency. This, of course, is also a function of the temperature of the source. That time is calculated as:

$$\tau_{rep} = \frac{v_0}{v_{rec}(\Gamma/2)} \qquad (2.182)$$

The sweep can thus be repeated at the rate $1/\tau_{rep}$ in order to accumulate a larger number of atoms in the region called *capture region* in Figure 2.29. From the analysis just made, we note that, selecting a velocity v_0 of 300 m/s, the length required for reducing that speed to zero is 0.76 m. On the other hand, we require for Cs a frequency sweep v_{sweep} = 68.35 MHz/ms; for Rb we need a frequency sweep v_{sweep} = 145 MHz/ms.

The disadvantage of the frequency chirping techniques lies in its pulsed nature. As a result, a smaller number of slow atoms per second is generally produced as compared to other techniques to be described below.

2.5.5.2 Laser Atom-Slowing Using Zeeman Effect: Zeeman Slower

The Zeeman slower technique uses a spatially varying magnetic field to shift the energy levels of an atom. The atomic transition frequency during the complete atom's forward trajectory is thus changed automatically in order to compensate for the Doppler shift reduction caused by the deceleration.

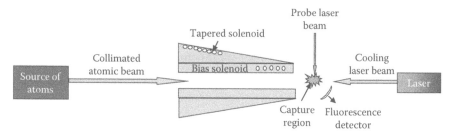

FIGURE 2.30 Block diagram illustrating the principle of laser Zeeman slowing.

A Zeeman slower thus consists of an atomic beam tube inside which a magnetic field is applied to shift the energy levels of the atoms moving along the axis. With the appropriate field profile, atoms moving through the tube can be decelerated efficiently by a counter-propagating laser beam of constant frequency. A schematic diagram of a Zeeman slower is shown in Figure 2.30. In order to illustrate the technique, the variation of the $S_{1/2}$ and $P_{3/2}$ states of ^{85}Rb with the applied field is sketched in Figure 2.31. For the two transitions shown, the resonance frequency varies linearly with the applied magnetic induction B according to:

$$\Delta\omega_z = \pm\frac{\mu_B}{\hbar}B \tag{2.183}$$

When both the Doppler shift and the magnetic field are taken into account, the frequency of the atom and the frequency of the laser seen by the atom are shifted. We define:

$$\Omega_o = \omega_L - \omega_{m\mu} \tag{2.184}$$

and

$$\Omega = \Omega_o + k_L v - \Delta\omega_x \tag{2.185}$$

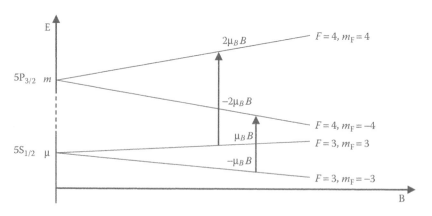

FIGURE 2.31 Energy level diagram of ^{85}Rb used for illustrating the laser Zeeman slowing technique.

or

$$\Omega = \Omega_{0} + k_{L}v \mp \frac{\mu_{B}}{\hbar}B \qquad (2.186)$$

where we have introduced the Doppler shift $k_{L}v$ seen by the atom moving against the laser radiation beam. With an appropriate field profile, atoms moving through the tube can be decelerated efficiently by the counter-propagating laser beam maintained at a constant frequency. The resonance condition ($\Omega = 0$) for Zeeman-tuned slowing can then be written as:

$$\Omega_{0} + k_{L}v \mp \frac{\mu_{B}}{\hbar}B = 0 \qquad (2.187)$$

To maintain the atom in resonance with the counter-propagating laser radiation field, the magnetic field must be shaped accordingly in order to satisfy the equation:

$$B(z) = \frac{\hbar}{\mu_{B}}\left(\Omega_{0} + k_{L}v\right) \qquad (2.188)$$

where v is altered by the laser radiation according to the analysis we have made earlier. For a constant deceleration a, the velocity changes as v^2 and the field must be shaped accordingly. Using Equation 2.183, $B(z)$ must be shaped as:

$$B(z) = \frac{\hbar}{\mu_{B}}\left(\Omega_{0} + k_{L}\sqrt{v_{0}^{2} - 2az}\right) \qquad (2.189)$$

In order to avoid optical pumping and increase efficiency, v circularly polarized light is used to drive a transition between the extreme hyperfine energy levels, that is to say a cycling transition. In σ^{+} slowing, the atoms decouple from the field when the magnetic field gradient becomes larger than the allowable value determined by the maximum deceleration. In σ^{-} slowing, the atoms decouple from the field at the maximum field location since the resonance retreats to higher velocities after that. These important differences are manifested in real magnetic fields (as opposed to ideal fields) and make it difficult if not impossible to extract slow atoms from a σ^{+} slower. The coils are somewhat voluminous, require a large current to shift the energy levels and thus sometimes, depending on construction, require water cooling.

It should be mentioned that the type of Zeeman atom slower just described is relatively complex to build and furthermore cannot be altered easily when built. The setup is generally not demountable. In order to avoid this complication and create a system easily adaptable to various atoms and situations, a Zeeman slower configuration based on the use of an array of permanent magnetic dipoles (MD) has been proposed (Ovchinnikov 2007). The system has the extra desirable advantage of not requiring electrical power and water cooling. In addition, the whole system can be assembled and disassembled at will, a property that is desirable regarding questions related to vacuum and bakeout. The magnetic field may be oriented transversely or longitudinally in relation to the direction of the atomic beam and laser radiation beam. A typical experimental goal of such Zeeman slowers was to slow Sr atoms, for example, from an initial velocity of 420 m/s to a final velocity of 25 m/s over a length of 25 cm. Such a transverse slower using permanent MD has been successfully implemented for use in a Sr optical clock (Ovchinnikov 2008).

In the transverse Zeeman slower, laser radiation with linear polarization is used, which can be decomposed into σ^+ and σ^- radiation for cooling by means of the appropriate cycling transition. Due to the fact that only one polarization is used, the transverse slower demands twice as much laser power, when compared to standard longitudinal slowers. On the other hand, the transverse Zeeman slower, in which the magnetic field changes its sign (spin-flip type), is working only for atoms without a large splitting of their ground state (e.g., Sr or Yb). For atoms like Rb and Cs, where there is significant splitting of the ground state, linear polarization prevents excitation to the proper cooling state, which results in the loss of the decelerated atoms in the region where the magnetic field becomes zero. A transverse Zeeman slower has recently been perfected through the use of a MD distribution adapted from a so-called Halbach configuration (Cheiney et al. 2011). In the Halbach configuration, the direction of each of the MD within an array is such as to create a field in a preferred direction in a region of space and cancel the field in another region (Halbach 1980). In the setup used by Cheiney et al., 8 MD, 6 × 6 mm cross section and 14.8 mm length, with specific Halbach orientation are distributed in an annulus around the beam (Cheiney et al. 2011). Such a section is repeated eight times along the beam and forms a slower about 1 m long. The diameter of the annulus varies from about 50 mm to about 30 mm along the beam axis. A rather uniform transverse field is created, which varies along the length by $\Delta B = 388$ G. In order to avoid low field level crossing in the P state of Rb, for example, at about 120 G, a bias field of 200 G is applied on the full length. The authors report a very smooth varying field all along the slower and homogeneity of the order of 1 G across the atomic beam. The slower operates on the transition from $S_{1/2}$ to $P_{3/2}$ (D_2 line, 780 nm) tuned to -800 MHz below the atom rest resonance. Because of the transverse configuration, π polarization is used of which the σ^- component is used to excite the cycling transition $F = 2$, $m_F = -2$ to $F = 3$, $m_F = -3$. The presence of the σ^+ polarization causes other transitions and some atoms decay to the level of the ground state that is not used in the cooling process. For this reason, a repumping laser is used to remove atoms from that so-called dark state. The slower was used to feed a MOT. The captured velocity of the slower was evaluated to be 450 m/s and the measured output flux was up to 5×10^{10} atoms/s at a speed of 30 m/s. A MOT could be loaded with more than 10^{10} atoms in one second.

The design of a longitudinal Zeeman slower, in which the magnetic field changes sign along the beam length has been found to be very efficient when properly designed (Slowe et al. 2005). In that particular design the field varied from 200 to -100 G and used laser repumping from the ground $F = 2$ state to the excited P state $F = 2$. The capture velocity was of the order of 320 m/s and the exit velocity was 40 m/s. In their case the exit velocity could be altered by varying the current in the last section of the winding, which could be driven independently of the main solenoid. Their special design including a source and initial transverse cooling led to a flux over 3×10^{12} atoms/s with a transverse temperature of 3 mK.

On the other hand a longitudinal Zeeman slower using MD arrays, which does not have the disadvantage of the transverse slower, such as limitation to Sr and Yb

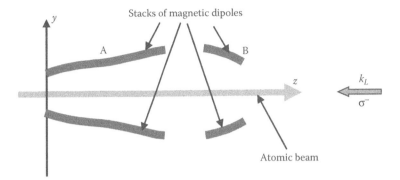

FIGURE 2.32   Cross section (y0z plane) of the MD horizontal Zeeman slower proposed for slowing Sr atoms. The longitudinal magnetic field goes through a zero between the two pairs of stacked MDs. A similar symmetrical arrangement is mounted in the x0z planes. In the case of Sr, stacks A are made of 16 MD and stacks B are made of 6 MD. A simpler version of this set up was tested in an actual Sr clock setup. (Data from Hill, I.R. et al., A simple, configurable, permanent magnet Zeeman slower for Sr. In *Proceedings of the European Forum on Time and Frequency* 545, 2012.)

atoms and the requirement of a larger laser power, has been proposed (Figure 2.32) (Ovchinnikov 2012). The slower is typically made of two groups of four arrays of MD distributed symmetrically around the atomic beam axis with a space between the two groups. For example, in the Sr slower, each array of the slower entrance group consists of 16 MD (8 mm long and 20 mm diameter). The second group of arrays made of similar MD, with polarity opposed to the first group, is situated at the other end of the slower or exit and is spaced from the first group by 58 mm. It is made of 6 MD. A schematic diagram of the arrangement is shown as a cross section in Figure 2.33. The distance of the individual MD from the beam axis is given as a table and the resulting field is calculated in the reference (Ovchinnikov 2012). The atoms go through a zero field (so-called spin-flip) between the two groups. The field changes from −300 G to +300 G and is rather homogeneous. The system is designed for a capture velocity of 410 m/s and an exit velocity of 25 m/s with a cooling frequency tuned to the red, 476 MHz below the atom resonance at rest. In the case of Rb, the system is designed in a similar way but with the arrays consisting of 19 and 6 MD for the input and output groups respectively, separated by 125 mm. The MDs have a diameter of 30 mm and 20 mm length. The calculated field changes from −118 to 116 G avoiding the crossing of the P state levels. The design is made for a capture velocity of 275 m/s, an output velocity of 25 m/s and a cooling laser frequency tuned to the red by 196 MHz. These slowers appear to be rather promising regarding simplicity compared to the solenoid approach and should give a beam flux of a similar intensity.

2.5.5.3   2D Magneto-Optical Trap

A device providing a slow beam of atoms can be implemented directly from the magneto-optical trap described previously. For example, the top vertical laser beam in Figure 2.27 can be replaced by a laser beam with a so-called narrow *dark column* in its centre. The atoms are accelerated out of the trap through the dark column by the detuned counter-propagating laser beam, which acts as a pusher. With a 0.6 mm diameter hole, a flux of the order of 5×10^9 atoms/s at a speed of 14 m/s and a spectral width of 2.7 m/s could be realized (Lu et al. 1996).

Another approach used to form a beam of slow moving atoms is the so-called two-dimensional magneto-optical cooling. Since it traps the atoms by means of a transverse magnetic field gradient in two of the dimensions of the MOT described earlier, the device has been given the name 2D-MOT. The technique was first used to cool and compress atomic beams transversally (Riis et al. 1990). It was soon realized that two-dimensional cooling could be used to produce a beam of cold atoms out of a vapour cell (Monroe et al. 1990; Dieckmann et al. 1998; Camposeo et al. 2001). Several types of 2D-MOT have been implemented and can be classified grossly as follows.

2.5.5.3.1 Basic 2D-MOT

That type uses two orthogonal pairs of laser radiation beams that transversally cool atoms in a vapour cell. Two pairs of elongated coils provide a quadrupole magnetic field for the MOT with zero field along the z-axis on which the atoms are confined. The atomic beam emerges from a collimator centred on the z-axis. A flux of up to 6×10^{10} atoms/s has been achieved and geometrical filtering can limit the mean longitudinal velocities to below 30 m/s.

2.5.5.3.2 2D-MOT⁺

In that type of 2D-MOT, an additional laser radiation beam pair along the z-axis is added to cool the atoms longitudinally. The setup is given the name MOT⁺. The mean velocity along the z-axis may reach a value below 10 m/s, which means that more atoms can be captured in a 3D-MOT when fed from such a source of atoms. Typically, the flux has been found to be of the order of 10^{10} atoms/s.

2.5.5.3.3 2D-MOT with Pusher

A thin red-detuned *pushing* laser beam along the z-axis may be added to push the atoms of a 2D-MOT towards a chosen spot, hole, or collimator, from which they emerged and can be used. The arrangement is particularly efficient for those atoms with negative or small longitudinal velocity components. This pusher technique can increase the atomic flux by a factor of 2 - 5.

The principle of operation of a 2D-MOT is in general similar to that of the MOT described earlier, that is to say Doppler cooling of an atomic vapour, but in two (x and y) rather than three dimensions, and the use of magnetic field gradients along the x- and y-axes. Two orthogonal pairs of counter-propagating laser beams having opposite circular polarization and with a frequency below the cooling transition (red-detuned) are directed into a vapour cell. In combination with the two-dimensional magnetic quadrupole field, the laser radiation creates a radial restoring force towards the region of zero magnetic field and thus encloses a finite volume. The magnetic field splits the excited state into its Zeeman sub-states. In such a situation, as in the MOT described earlier, the transitions to the lower Zeeman sub-states (which differ on each side of the MOT) are closer to resonance with each inward beam, pushing atoms effectively initially slow enough towards the centre and trapping them along the z-axis. A schematic of such a 2D MOT is shown in Figure 2.33. The component of velocity of the atoms in the longitudinal direction is not altered. The cooling takes place in the radial direction. Hence, upon interaction with the laser radiation, the atoms travel on a skewed trajectory into the centre of the 2D MOT. It is assumed that the pressure inside the cell is low and that the mean free path of the atoms is larger than the cell dimension. In that case, atoms are not removed from the cooling region through collisions. In such a situation, atoms with a high velocity component in the z direction need to be nearly on the axis to escape through the output tube if it is long. Consequently, only atoms with very low speeds on the z-axis escape through the tube and a beam of slow atoms is formed.

The beam has no privileged direction along the z-axis. However, it is possible to make it unidirectional and improve the output flux by means of a so-called weak *pushing* laser beam oriented along the $+z$-axis, that is in the same direction as that of the desired atom beam. In a typical case, an atom flux at a speed of 25 m/s with a narrow speed spectrum (of the order of 7.5 m/s) is observed. A flux of 6×10^{10} atoms/s was obtained with laser power of 160 mW per cooling beam for a MOT length of 90 mm (Schoser et al. 2002). A 2D-MOT constructed at the SYRTE laboratory in Paris is shown in Figure 2.34.

The trap is made of titanium, which is non-ferromagnetic and equipped with five windows. The windows are treated with an antireflection coating at the wavelength

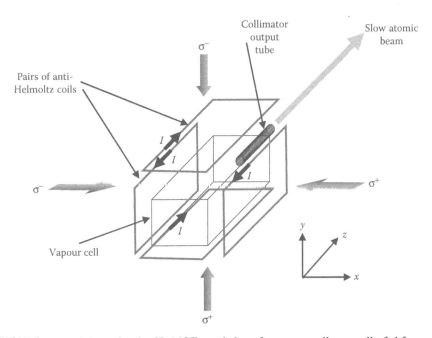

FIGURE 2.33   Schematic of a 2D MOT consisting of a vapour cell, normally fed from a beam of atoms, two pairs of anti-Helmholtz coils, and two pairs of counter-propagating laser radiation beams with circular polarization. The collimator output tube forms a beam of slow atoms. (Data from Schoser, J. et al., *Phys. Rev. A* 66, 023410, 2002.)

FIGURE 2.34   Photograph of the 2D MOT$^+$ constructed at the SYRTE laboratory of Paris.

of the laser used. The pressure inside the trap is of the order of 10^{-9} mbar. The system uses a laser beam pusher of a few microwatts. The velocity of the atoms at the exit of that particular trap was measured to be ~9.5–10 m/s and the atomic flux was evaluated to be ~4.2×10^9 atoms/s (Chapelet 2008).

2.5.5.4   Isotropic Cooling

Isotropic cooling is a technique that was suggested as a possible approach for slowing down atoms in a beam. It was experimented first in two dimensions on a Na beam (Ketterle et al. 1992) and on ^{85}Rb (Batelaan et al. 1994). It was realized experimentally in three dimensions on Cs atoms (Aucouturier 1997; Guillemot et al. 1997). We will describe this last case where isotropic cooling is accomplished in a cell.

There are some advantages of cooling in isotropic light. In the conventional cooling configuration, atoms undergo a radiative friction force only in the intersection region of the collimated laser beams travelling in three orthogonal directions. In the isotropic cooling scheme, the whole cell is irradiated by light coming from all directions or in other words by isotropic radiation. As a consequence, all atoms contained in the cell undergo a radiative friction force. The main advantages of the isotropic cooling relative to the classical geometry are thus: large number of cooled atoms and considerable simplification of the optical setup.

An *isotropic* laser field can be realized, for example, by storing energy of a laser field in a cavity with a highly reflecting internal surface, relying on the multiple reflections (or scattering) of the laser light inside the cavity. In the same manner as in the case of the so-called *integral sphere* used in metrology laboratories, the reflectivity or diffusivity chosen for the appropriate optical wavelength (e.g., 852 nm in the case of Cs) is used to build an isotropic laser field, recreating grossly the conditions for a 3D optical molasses as shown in Figure 2.35.

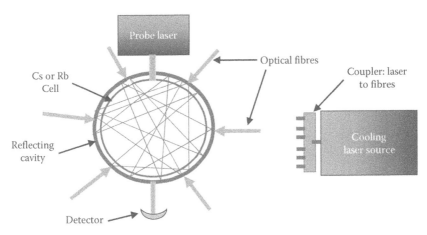

FIGURE 2.35 Setup for cooling an ensemble of atoms in a cell by means of isotropic laser light. Laser radiation is reflected by the highly polished surface of the enclosure and all atoms are submitted to the laser radiation. The case shown is realized by means of 6 optical fibres feeding the cavity with laser radiation. In a first experimentation, 14 optical fibres were used to feed the cavity. (Data from Guillemot, C. et al., 3D cooling of cesium atoms with isotropic light. In *Proceedings of the European Forum on Time and Frequency* 156, 1997.)

It is important to note that in practice exact field isotropy is difficult to reach. It can also be added that the geometry in which the cavity is fed with six fibres is somewhat close to the standard configuration of an optical molasses. The advantages of isotropic cooling relative to the classical geometry reside essentially in a simplification of the optical system and the large number of atoms trapped.

In the following elementary analysis, we consider the case of an atom that possesses two energy levels m and μ and moving at velocity v_0. It interacts with a laser beam with an angle θ with respect to v. The atom in its motion sees the laser light with an altered frequency displaced by Doppler effect by the quantity $k \cdot v$. The interaction may be illustrated as in Figure 2.36.

We have derived earlier the force exerted on the atom by a photon incident in a direction in opposite direction to the atom velocity v (Equation 2.137). In that case, in order to have resonance between the laser radiation and the atom, the frequency of the laser must be detuned to the red by the quantity $k_L v$. Similarly, in the present case, referring to Figure 2.36, and the definition of the angle θ, the frequency of the

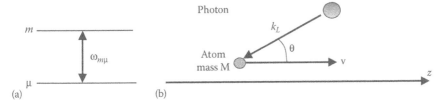

FIGURE 2.36 Illustration of the concept of isotropic cooling discussed in the text. (a) The level structure of the atom in interaction with the incident photon. (b) Photon incident on the atom at angle θ measured from a direction when the propagation is opposite to the atom motion.

laser must be detuned to the red by the quantity $k_L v \cos\theta$ to obtain resonance in order to take into account the component of velocity along the direction of propagation of the radiation. The force may then be written as:

$$F_{diss}(v) = -\hbar k_L \Gamma \frac{\omega_{R\mu m}^2}{4} \frac{1}{\left(\Gamma/2\right)^2 + \left(\omega_L + k_L v \cos\theta - \omega_{m\mu}\right)^2 + \left(\omega_{R\mu m}^2/2\right)} \qquad (2.190)$$

Exact resonance, for an atom speed v_a and laser tuning ω_{L1}, is obtained for an incident photon direction θ_1 satisfying the relation:

$$\omega_{L1} + k_L v_a \cos\theta_1 - \omega_{m\mu} = 0 \qquad (2.191)$$

or

$$\cos\theta_1 = \frac{\omega_{m\mu} - \omega_{L1}}{k_L v_a} = \frac{\Omega_{L1}}{k_L v_a} \qquad (2.192)$$

This is a rather interesting condition for resonance. Since radiation comes from all directions or all angles θ, we may question the exact effect of the presence of such radiation on the final state of the atom. What is concluded directly from that equation is that all photons coming at the angle θ_1 are at exact resonance with the atom since they have all the same frequency detuning. Thus, they form a cone of angle θ_1 in space as in Figure 2.37a. After a photon-emission cycle, the atom slows down and v_a is smaller. From Equation 2.192 it is observed that the atom is now in exact resonance with the radiation coming at a smaller angle θ since v_a is smaller and the radiation frequency remains the same. Consequently, the cone has a smaller angle θ. The process takes place until the angle becomes 0 and the cone closes on itself as shown in Figure 2.37b.

We realize that this process takes place for all directions of the atomic motion and, in principle, all directions of velocities are reduced. The process takes place on a sphere and is thus isotropic. From Equation 2.192, the closing of the cone takes place when v_a reaches a limiting value:

$$v_{a\,lim} = \frac{\Omega_{L1}}{k_L} \qquad (2.193)$$

If the laser detuning is such as to make $\Omega_{L1} = \Gamma$, a detuning by one natural line width, we have for Cs, with radiation tuned to the D_2 transition (852 nm, $k_L = 7.37 \times 10^6$),

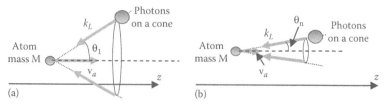

(a)   (b)

FIGURE 2.37 Illustration of the cone formed by the atom-photon interaction for exact resonance for an atom at velocity v_a and angle of incidence of the photon. (a) Initial situation before the absorption–emission cycle. (b) Closing of the cone upon cooling.

$v_{lim} = 4.4$ m/s. Cooling takes place efficiently until that speed is reached since radiation at the proper angle is always present in such an isotropic situation. The system is auto regulating, radiation coming from all angles. Below the limiting speed, light is no longer exactly resonant with the Doppler shifted laser frequency and cooling takes place less efficiently. The cooling takes place as in a molasses with collimated light. In practice the number of atoms cooled is of the order of a few $\times 10^9$.

The whole process of course is a type of Doppler cooling and has the same limit as in the case of the molasses. Because of the discreteness of the process, there is random walk and diffusion in the spontaneous emission that takes place. The limit is of the order of 150 μK. In practice, however, the temperature obtained is more than an order of magnitude lower. Other slowing mechanisms take place such as Sisyphus cooling that was mentioned earlier.

Such atom cooling in a cell appears to open possibilities regarding the implementation of a small frequency standard, since it addresses the exact goal of reducing Doppler effect (Guillot et al. 1999). This avenue has been studied and will be covered in Chapter 3 (Pottie 2004; Esnault et al. 2008; Esnault 2009).

2.5.5.5   Optical Lattice Approach

We have insisted on the difficulties created by Doppler effect in using atoms in motion for implementing in practice a frequency standard. Classically, Doppler effect shifts the resonance frequency of an atom in motion at speed v by the quantity kv, which, in H, for example, is 12 kHz at the ground-state hyperfine frequency of 1.4 GHz at room temperature. This corresponds to a fractional shift of parts in 10^{-5} and is of course very large. Furthermore, the speed is spread over a Maxwell distribution, function of temperature, and it is even unthinkable to resolve broadened spectral lines to the extent required to reach 10^{-17}–10^{-18} accuracy and 10^{-16} stability desired at one second. We have just shown that laser cooling is a means to considerably reduce the effect. However, several other approaches have been developed, which were described in *QPAFS* (1989) and reviewed in Chapter 1, to simply get rid of Doppler effect. In the microwave classical frequency standards described earlier, Dicke approach is used in several cases. For example, in H masers, a storage container that restricts the motion of atoms to a field of the same phase is used. In some other cases, a buffer gas is used in which the motion of the atoms is limited by diffusion. Absorption and emission then take place in a field of the same phase and consequently Doppler effect is absent. There are other approaches such as pulse techniques either in space or in time taking advantage of the preservation of phase by an atom in motion. This is the technique used in the Cs clock, in which the atoms form a beam that interacts with the microwave radiation for a short time at different regions in space. The atoms, being free, keep their phase in their motion and interferences take place which leads to fringes and a narrow resonance line. All these techniques have been perfected to the point where conceptual improvements can hardly be imagined. We have seen that attempts at improving the H maser technology by reducing its temperature to the cryogenic region hardly improved its performance because of the presence of other effects inherent to the actual storage concept used, which appeared at low temperatures. We have just studied atom laser cooling in this chapter. We will see in Chapter 3 how that technique, reducing atomic velocity and temperature considerably, has led to orders of magnitude improvement of the classical Cs beam clock through the use of the atomic fountain concept. Similarly, laser cooling in small electromagnetic traps has led to improvements in microwave frequency standards using ions such as Yb^+ and Sr^+.

It is realized, however, that in all those implementations the atomic resonance phenomenon used and observed takes place at microwave frequencies. A simple reasoning leads to the conclusion that if a higher frequency were used, the fractional frequency stability and accuracy would be increased at the condition that the perturbations that cause the instabilities and inaccuracies remain at the same level than those observed at microwaves. Consequently, if atomic transitions at frequencies of the order of 500 THz in the optical range could be used, a substantial gain would result. Several types of lasers emit radiation at those frequencies. However, their emission line width is very large due to Doppler effect just emphasized and their frequency is essentially determined by the FP cavity needed for their operation. Solid-state semiconductor lasers are also rather unstable, their frequency being determined by the driving current and the temperature, which alter the dimensions of the cavity which is formed normally from the substrate itself. In practice all those lasers must be stabilized somehow by locking their frequency to external cavities and narrow atomic resonance references. We have described earlier the stabilization of lasers on such external cavities. The techniques described improve their spectral purity. In *QPAFS*, Volume 2 (1989), and in the present chapter, we have also outlined techniques for locking such lasers to atomic resonance lines, either by means of broad linear absorption (fully Doppler broadened) or by means of saturated absorption (narrowed to their natural line width) in cells containing atomic vapours with quantum resonance transitions corresponding to the laser frequency. We have reviewed these techniques in the present chapter. However, even in the case of saturated absorption, those techniques have limitations, reflecting the properties of the atomic resonance cells, affected by various perturbations. The main problem in using atomic transitions at optical wavelengths is that it is not possible to construct a classical mechanical trap that would be small enough to fulfil Dicke criteria to inhibit Doppler effect that broadens the line. It appears that the only approach is to reduce their speed. This is what we have described above through the concept of laser cooling leading to the MOT. In a MOT, with an atomic cloud of the order of several millimetres in diameter, atoms may have a residual temperature of the order of several tens of microkelvins leading, for example, to a residual velocity of a few centimetres per second for an atom with a mass of a few hundred atomic mass units. This corresponds to a Doppler shift of the order of $\Delta v = k\, v_d/2\pi \sim 100$ kHz at a frequency in the range of 400–500 THz or a fractional frequency shift of the order of 10^{-10}. This is very large and cannot be controlled or evaluated at the level of frequency stabilities and accuracies of the magnitude of those obtained in microwave standards. Consequently another method has to be imagined to get rid of Doppler effect at optical frequencies.

Thus, atoms must be essentially at rest or, if moving, not experience a change of phase of the radiation they are exposed to. Consequently a travelling wave is out of the question and the dimension of the storage box, such as a cavity, must have dimensions less than 1/2 wavelength. Mechanical boxes of the order of 1/2 wavelength at optical frequencies are not practical.

We have seen that it is possible to influence the motion of atoms by means of radiation fields. Forces on atoms created by static magnetic fields are very small and their use to store atoms would cause important frequency shifts because of the field amplitude required. In MOT, upon application of a magnetic field, the force is relatively large because it originates from the incidence of photons on the atom causing a transfer of momentum, at a rate which may be as large as the decay rate of the excited state. The effect takes place through absorption–emission cycles with appropriate polarization rather than by means of a direct effect of the interaction of the applied magnetic field to whatever magnetic moment may exist in the atom in question. However, we have noted that the interaction of the electric field of that radiation with the atom is much stronger than the interaction with the magnetic field. The electric field induces an electric dipole in the atom, interacts with this dipole and, if its frequency is the same as the difference in energy from the ground state to an excited state, it may excite transitions if they are allowed by symmetry of the wave function of the states in question. We have analyzed that phenomenon in detail in *QPAFS* (1989). However, we have found another effect that appears when the frequency applied is not exactly equal to the resonance frequency of the atom: the interaction causes a shift of the energy levels in question, a light shift which was used in the Sisyphus cooling process described above. This shift in the energy level is similar to the shift of a static electric field on the energy levels of an atom. The shift is proportional to the square of the electric field and for this reason the alternating electric field of the incident wave has a net effect on the position of the levels. We have calculated the effect of this phenomenon on the hyperfine levels of an atom and found that for the microwave hyperfine transition that we were interested in for implementing a frequency standard, its effect was actually undesirable since it shifted the hyperfine resonance frequency of the transition.

However, the effect can be used for the purpose of trapping if an arrangement in which this shift in the energy levels is made to vary in space to produce a potential. If that potential shows periodic minimum values, small wells or traps can be created. This is readily feasible by means of a standing wave. Such a wave has nodes of zero field and crests of high field whose positions in space do not vary. Consequently their phase is constant. The interaction of an atom with that field is then periodic in space and small potential wells are created, where atoms could be trapped if their kinetic energy is smaller than the well depth.

We have calculated such effect in the present chapter using a laser as a source of the radiation creating a standing wave. We wish to extend that analysis to the present purpose of trapping. The situation is made explicit in Figure 2.38 in which an atom with two energy levels is exposed to a standing wave radiation field.

We adapt our notation to that generally used in that field and express the displacement of an energy level in second order in the electric field by the expression:

$$\delta E_i = \alpha_i \left(\omega_L \right) \left(\frac{E_L}{2} \right)^2 \qquad (2.194)$$

where:
α_i is the dynamic polarizability of state i
E_L is the electric field amplitude of the laser radiation at the site of the atom

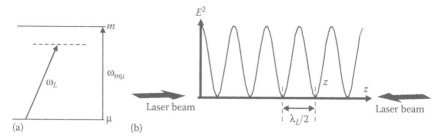

FIGURE 2.38 (a) Schematic representation of the two energy levels used in the present analysis showing the laser frequency ω_L largely detuned from the atom resonance frequency $\omega_{m\mu}$. (b) Representation of the standing wave (electric field only).

The polarizability of the atom is proportional to the oscillator strength or the electric dipole matrix element $d_{m\mu}$. The displacement can also be expressed in terms of the Rabi frequency $\omega_{R\mu m}$ defined earlier in Equation 2.57.

$$\omega_{Rm\mu} = \frac{E_L}{\hbar}\left\langle m\left|er\cdot e_\lambda\right|\mu\right\rangle = \frac{E_L}{\hbar}d_{m\mu} \tag{2.195}$$

On the other hand, the standing wave represented in Figure 2.39 may be written as:

$$E_L(z) = 2E_{Lo}\exp-\left[\frac{r}{w(z)}\right]^2\cos(k_L z) \tag{2.196}$$

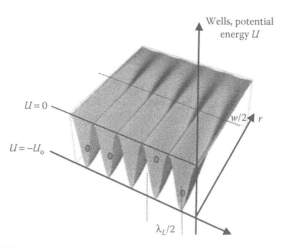

FIGURE 2.39 Illustration of a 1D laser lattice. The graph is essentially a representation of Equation 2.197. Only one half of the breadth of the lattice is shown in order to make well shapes explicit. w is the laser beam waist. The vibration states in the wells are not shown. Although only one atom is shown in a given well it is possible that a well contains several atoms. For low electric fields (low laser power) the wells depth is shallow and the atoms may tunnel from one well to the next.

where:

$w(z)$ is the beam waist at position z

The atom frequency shift in the presence of such a spatial field distribution may be interpreted as a potential:

$$U(r,z) = U_o \exp \left\{ -2 \left[\frac{r}{w(z)} \right]^2 \cos^2 (k_L z) \right\} \qquad (2.197)$$

where U_o is given by:

$$U_o = -\alpha(\omega_L) E_L^2 \qquad (2.198)$$

For the purpose of illustration, for a far detuned laser frequency, the polarizability can be written as:

$$\alpha_\mu(\omega_L) = \frac{\langle \mu | er \cdot e_\lambda | m \rangle}{\omega_{m\mu} - \omega_L} \qquad (2.199)$$

leading to the potential

$$U_\mu = \hbar \frac{(1/4) \omega_{R\mu m}^2}{\left(\omega_L - \omega_{m\mu} \right)} \qquad (2.200)$$

and a similar expression for U_m. The calculation we did earlier regarding the force exerted on an atom by a laser radiation beam led to a dissipative and a reactive force. As was made explicit, the dissipative force is used in atomic beam cooling and molasses. On the other hand, we noted that the reactive force obtained could be derived from a potential given by Equation 2.134. Making the same kind of approximation as we did above, that potential for large detuning:

$$\omega_L - \omega_{mv} \gg \Gamma \qquad (2.201)$$

leads to the same potential as the one obtained above.

We take note that the radiation used for the trapping may be detuned to the red or to the violet relative to the resonance frequency. The potential well changes sign from one case to the other and the force exerted on the atoms being the gradient of the potential also changes sign. The atoms are thus forced towards either a minimum or a maximum of the field depending on the sign of the polarizability.

It is thus concluded that, with a standing wave, a periodic structure can be created containing potential wells, which, in principle, can trap atoms. For moderate laser power, the depth of these potential wells is somewhat shallow, less than 100 μK in temperature unit. Consequently, in order to trap an atom in such a well, its kinetic energy must be reduced to an equivalent temperature or in other words it must be cooled by one of the processes described earlier.

The periodic structure is called an *optical lattice* and can be implemented in one, two, or three dimensions by means of counter-propagating orthogonal laser radiation beams. A 1D optical lattice is illustrated in Figure 2.39. Atoms are trapped in such a structure in the Lamb–Dicke regime that we described earlier. They are trapped in quantum vibration levels identified by means of quantum numbers n and oscillate harmonically in such wells. That oscillation may be observed as a phase modulation on a narrow laser probe beam that is resonant with one transition and it creates sidebands on the fluorescence signal detected, an effect predicted in Dicke atom storing technique. Depending on the depth of the well controlled by laser power and their energy, atoms may tunnel from one microtrap to the next. The structure is sometimes compared to a solid-state crystal lattice. However, considerable differences exist (Cohen-Tannoudji and Guéry-Odelin 2011). Firstly, the optical lattice is very flexible since changing wavelength changes cell spacing. Thus, the lattice dimension can be altered easily, which cannot be done in a crystal. Secondly, the depth of the wells is very shallow compared to that of a crystal since it is not created by atomic interaction but by laser radiation. Finally, the fact that the wells are rather more distant in an optical lattice (100–1,000 nm) than in a crystal (~0.1 nm) leads to very little interaction between the individual atoms in an optical lattice.

In the use of this approach to create an optical frequency standard, laser radiation is used to interrogate the atoms by excitation of a specific very narrow transition between a ground state and a metastable state as illustrated in Figure 2.40. It is thus necessary that the pair of levels involved not be displaced by the trapping radiation. On the other hand, in the above discussion, the presence of the wells was interpreted as a spatial variation of the light shift created by intense trapping radiation at a long wavelength in a standing wave. These two requirements seem to be contradictory. The problem was addressed by a careful study of the light shift itself as a function of frequency or the dependence of the polarizability on frequency, which must take into account all the levels connected by the off resonance radiation causing the trapping (Katori 2001). It was found theoretically that wavelengths

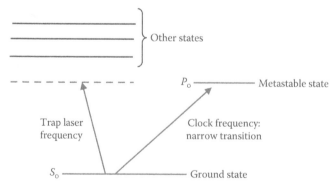

FIGURE 2.40 Illustration of the various frequencies and energy levels of a typical atom used in the implementation of an optical frequency standard based on laser cooling and optical trapping.

existed for which the two levels involved in the clock interrogation transition were shifted by exactly the same amount by the trapping radiation. These were called *magic wavelengths* (Ovsiannikov et al. 2007; Derevianko and Katori 2010; Mejri 2012). The prediction and calculation were verified experimentally leading to the possibility that an accuracy determination in the parts in 10^{17} range could be realized in a clock implementation.

Optical standards based on that property, using Yb, Sr, and Hg were developed.

We will study those standards in Chapter 4, outline their construction, address the magic wavelength question in more details, and review their state of the art characteristics.

APPENDIX 2.A: LASER COOLING—ENERGY CONSIDERATIONS

In *QPAFS*, Volume 2 (1989), we have evaluated the change of energy of an atom absorbing a photon by means of relativistic energy considerations (Wineland and Itano 1979). Let us examine that approach briefly and show that it leads to similar conclusions as those obtained in the present chapter. The atom is excited from the ground state g to the excited state m by the laser radiation of propagation vector $\mathbf{k}_L$. It is assumed for the moment that the atom travels at speed $\mathbf{v}_\mu$ in the ground state and $\mathbf{v}_m$ in the excited state. The angular frequency of absorption (ω_{abs}) and emission (ω_{emit}) of a photon by the atom including Doppler effect as well as recoil effect can be written (*QPAFS* 1989, Volume 1) as:

$$\omega_{abs} = \omega_{m\mu} + \mathbf{k}_L \cdot \mathbf{v}_\mu - \frac{1}{2}\frac{\omega_0 v^2}{c^2} + \frac{\hbar k_L^2}{2M} \tag{2.A.1}$$

$$\omega_{emit} = \omega_{m\mu} + \mathbf{k}_{emit} \cdot \mathbf{v}_m - \frac{1}{2}\frac{\omega_0 v^2}{c^2} - \frac{\hbar k_L^2}{2M} \tag{2.A.2}$$

where the second term on the right-hand side of these equations is the first-order Doppler effect and the third term is the second-order Doppler effect (time dilation). The last term is the recoil effect affecting the atom dynamics upon emission or absorption of a photon. Upon absorption, momentum is conserved. Since the photon disappears its momentum, $\hbar k_L$, is transferred to the atom. The atom recoils and gains speed $\hbar k_L/M$ in the direction of propagation of the incident photon. On the other hand, the emission direction is random and upon averaging over several absorption-emission cycles, the term $k_{emit} \cdot \mathbf{v}_m$ in Equation 2.A.2 averages to zero. Consequently, the average energy change over several cycles experienced by the atom is the negative of the energy change by the photon, which is $\Delta E(\text{photon}) = \hbar(\omega_{emit} - \omega_{abs})$, or neglecting time dilation effect:

$$\langle \Delta E(\text{atom}) \rangle = \hbar k_L \cdot \mathbf{v}_\mu + \frac{k^2 \hbar^2}{M} \tag{2.A.3}$$

For k_L and $\mathbf{v}_\mu$ in opposite directions, there is a net reduction in energy $\hbar^2 k_L^2/M$ of the atom and of its velocity caused by the recoil. This recoil energy is very small and it takes many steps to have a visible affect. There is net cooling until that recoil energy becomes larger than the residual Doppler effect. On the other hand, the emission process is random and the atom experiences a random walk. The limiting energy of this random walk is at least the recoil energy and it appears that the atom velocity cannot be reduced to a value lower than that caused by the recoil effect. Consequently, these simple energy considerations lead to the same general conclusions as those reached by the more elaborate calculations made in the main text.

3 Microwave Frequency Standards Using New Physics

In this chapter, we will describe the impact that the recent developments in atomic physics, which were described in Chapter 2, have had on the development of microwave atomic frequency standards. Some of those developments were based on earlier theoretical knowledge but started only in the 1980s because of the unavailability of appropriate technology. Developments in the field of lasers, particularly those lasers using solid-state technology, have provided new tools for addressing some of the limitations encountered in the approaches used up to that time in the development of classical atomic frequency standards. In particular, three new different approaches were studied for implementing Cs beam frequency standards. Firstly, an old idea of using laser optical pumping for state preparation was resurrected, thanks to advances in solid-state laser diodes operating at room temperature (Picqué 1974; Arditi and Picqué 1980). That approach resulted in the laboratory development of a very stable and accurate clock. Another approach was studied in the early 1980s using coherent population trapping (CPT) for both states preparation and microwave excitation (Hemmer et al. 1983). Unfortunately, that approach remained a laboratory study. Finally, a real revolution took place through the use of lasers to cool atomic ensembles in the form of molasses in the microkelvin range, resulting in a Cs standard with narrow Ramsey fringes with line width in the Hz range (Kasevich et al. 1989; Clairon et al. 1996).

The advent of those same laser diodes has also made possible the realization of more efficient optical pumping in passive Rb standards using the double resonance technique (Levi 1995; Mileti 1995) and has opened the door to the use of CPT for implementing small atomic frequency standards not requiring a microwave cavity as in the double resonance approach (Cyr et al. 1993; Levi et al. 1997; Vanier et al. 1998).

Laser technology has also made possible the cooling of ions in traps, opening the door to the implementation of frequency standards with good frequency stability in the microwave range, an approach that had been proposed previously for operation at room temperature (Dehmelt 1967).

We will now describe these various developments based on the theoretical advances described in Chapter 2.

3.1   Cs BEAM FREQUENCY STANDARD

3.1.1   Optically Pumped Cs Beam Frequency Standard

3.1.1.1   General Description

Early work had already been done on optical pumping of a Rb beam in the 1970s by means of spectral lamps (Arditi and Cerez 1972). That approach was not efficient due to the properties of the radiation emitted by such lamps, particularly their spectral width. With the advent of solid-state laser diodes emitting narrow spectral lines, it has become possible to excite specific narrow optical transitions in alkali metal atoms, connecting a single hyperfine level of the ground state to either of the excited $P_{1/2}$ or $P_{3/2}$ states. In such a process, it is possible to populate a given ground-state hyperfine level at the expense of the others.

A block diagram of a typical implementation of such a standard is given in Figure 3.1. The lower energy levels $S_{1/2}$ and $P_{3/2}$ of the Cs atom are illustrated in Figure 3.2.

Several optical pumping schemes are possible. In one of them, atoms are excited from level $F = 4$ of the $S_{1/2}$ ground state to level $F' = 3$ of the $P_{3/2}$ excited state. The lifetime of atoms in the excited state is of the order of 30 ns and they fall back to either hyperfine levels of the ground state with nearly equal probabilities. This process thus tends to increase the population of level $F = 3$ at the expense of level $F = 4$. It can thus be used to pump atoms in that energy level. Similarly, at the exit of the Ramsey interaction region, the same technique can be used to analyze the composition of the beam in the various levels. In that case, a particular transition is again

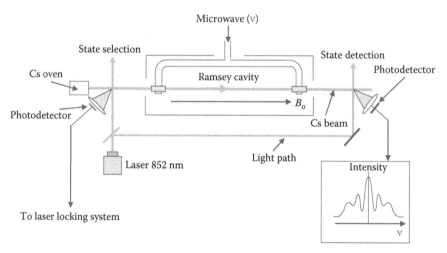

FIGURE 3.1   Simplified conceptual diagram of the Cs beam frequency standard using optical pumping for state selection and detection. In the case shown only one laser is used for both selection and detection. The inset illustrates the amplitude of the fluorescence observed at the detection end when the microwave frequency is swept slowly across the resonance line. (Data from Vanier, J., Atomic frequency standards: basic physics and impact on metrology. In *Proceedings of the International School of Physics Enrico Fermi Course CXLVI*, J. Quinn, S. Leschiutta, and P. Tavella, Eds., IOS Press, 2001. With kind permission of Societa Italiana di Fisica.)

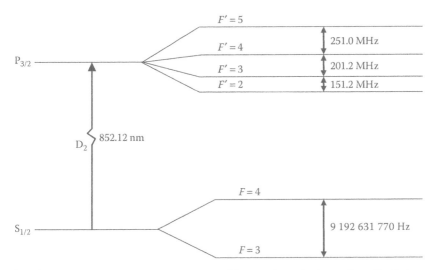

FIGURE 3.2 Illustration of the lower states of interest in optical pumping of a Cs beam. (Data from Tanner, C.E. and Wieman, C., *Phys. Rev. A*, 38, 1616, 1988.)

excited, for example, from $F = 4$ to $F' = 5$, and the fluorescence emitted by the atoms upon decay to the ground state is used as a measure of the population in level $F = 4$. It is a measure of the number of atoms that have made a transition in the Ramsey interaction region. Several other schemes such as two laser pumping for accumulating all atoms in the same state are possible.

We will not give the details of the calculation involved in the optical pumping state selection. The reader is referred to *QPAFS*, Volume 2 (1989), for details on the possible schemes of pumping and their efficiency in populating one level at the expense of the others. It is sufficient to say that one advantage of the approach lies in the fact that the system is highly symmetrical and avoids magnetic inhomogeneities that can be present when magnetic states selectors are used. The inherent symmetry of the optical pumping state selection results in the absence of asymmetry in the amplitude of the field-dependent transitions and reduces considerably the effect of Rabi pulling mentioned previously. The Rabi pedestal frequency shift is essentially eliminated. On the other hand, in the case of magnetic selection, for example, with Stern–Gerlach-type dipole magnets, an offset geometry is required resulting in a beam velocity distribution different from the normal Maxwell distribution. As mentioned in Chapter 1, the distribution must be deduced from experimental data to evaluate certain frequency shifts. In the case of optical pumping, the velocity distribution is known and the averaging required in the calculation of the second-order Doppler frequency shift, for example, can be done analytically. Furthermore, in the case of optical pumping, it is possible, by means of several laser diodes for pumping and detection, and proper optical geometry, to transfer almost all the atoms to one of the m_F levels of the ground state (Avila et al. 1987). A somewhat better signal-to-noise ratio (S/N) is then possible at the detection, leading to improved frequency stability (Makdissi et al. 1997). Its limitation by the frequency noise of the laser used for state selection has also been evaluated (Dimarcq et al. 1993).

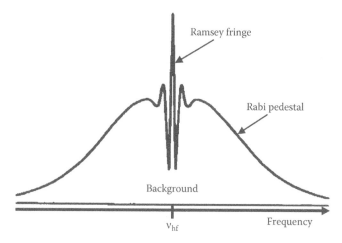

FIGURE 3.3 Ramsey pattern obtained in a short optically pumped caesium beam tube. The line width of the central fringe is ~500 Hz. (Reproduced with permission from Cérez, P. et al., *IEEE Trans. Instrum. Meas.*, 40, 137, 1991. Copyright 1991 IEEE.)

The device can be implemented as a laboratory primary standard (de Clercq et al. 1989) or a small field application device (Cérez et al. 1991). For the purpose of comparison with the kind of fringes observed with magnetic state selection illustrated earlier, Figure 3.3 shows the resonance pattern obtained in a short optically pumped experimental setup (Cérez et al. 1991). Only two fringes on each side of the central fringe are clearly visible since in that case the velocity distribution is broad.

We will now outline the practical characteristics of a device based on that optical pumping approach in reference to those standards using magnetic state selection described earlier.

3.1.1.2 Frequency Shifts and Accuracy

As mentioned above, the structure of the optically pumped Cs beam standard is symmetrical and in principle its frequency should not be affected by as many spurious effects as in the case of magnetic state selection due to the symmetry introduced in the optical pumping process. However, several of those shifts are still present and a new one, the light shift, is introduced. We will examine those and review some results obtained recently relative to the question of phase shift evaluation.

3.1.1.2.1 Light Shift

In early development stages of the standards, it was realized that in the case of the optical pumping approach, light from fluorescence emitted in the pumping region as well as light scattered by the various internal surfaces could cause, through diffusion to the microwave cavity, a frequency shift known as a *light shift* as described earlier (Barrat and Cohen-Tannoudji 1961). The effect of these light shifts on the frequency of the central Ramsey fringe have been calculated in some detail in *QPAFS*, Volume 2 (1989). We will recall the actual calculation made and expand the analysis to the case where square wave frequency modulation is used to detect the central Ramsey fringe and lock an external oscillator to that fringe.

The light shift originates from the interaction of the light used for optical pumping and detection with the atoms traversing the Ramsey cavity, including both the interaction regions of the two arms and the drift section. This light originates from two sources. Firstly, it may arrive at the Ramsey cavity by diffusion from the surrounding surfaces which may reflect laser radiation. The exact characteristics of that scattered light are not well-known. There is not much to do about it except to attempt to reduce reflection and diffusion as much as possible. Calculation shows that to avoid fractional frequency shifts larger than 10^{-14} the reflected light diffusing to the cavity should be less than 10^{-6} W/m^2.

On the other hand, fluorescence originating from the pumping and detecting regions may penetrate inside the interaction regions and even diffuse through the arms into the drift region. Such fluorescence causes light shifts that may be important and need to be evaluated carefully. An approximate evaluation based on the calculation done earlier (Equations 2.43 and 2.44) and using general results developed in *QPAFS* (1989) gives the light shift as:

$$\Delta\omega_\mu = \sum_\mu \frac{e^2 \left|\langle \mu | \mathbf{r} \cdot \mathbf{e}_\lambda | m \rangle \right|^2}{2\varepsilon_o c \hbar^2} \times \int_0^\infty I f(\omega^e) \left(\int_0^\infty \frac{\omega_L - \omega_{m\mu} - \mathbf{k} \cdot \mathbf{v}}{\left(\omega_L - \omega_{m\mu} - \mathbf{k} \cdot \mathbf{v}\right)^2 + \left[(1/2)\Gamma\right]^2} p(v) dv \right) d\omega^e \quad (3.1)$$

This equation gives the shift of level μ of the ground state, expressed in frequency units $\Delta\omega_\mu$, caused by radiation of intensity $I f(\omega^e)$ that originates from the pumping and detection regions as fluorescence. Since two levels, μ and μ', of the ground state are involved in the frequency standard clock transition, a similar expression applies to the other level and the actual light shift is the difference between the shifts of these two levels, $(\Delta\omega_\mu - \Delta\omega_{\mu'})$. The spectrum of the fluorescence radiation, $f(\omega^e)$, is a function of the velocity distribution of the atoms in their excited state. In the case of optical pumping in a beam, the velocity distribution reflects the velocity of the atoms in the atoms source and may be approximated by a Maxwellian shape weighted by 1/v or 1/v^2, depending on the type of transition excited and the intensity of the laser radiation. This specification reflects the time the atoms in the beam stay in the pumping and detecting laser light beam. The term within large brackets on the right-hand side of Equation 3.1 is a measure of the intensity of the light–atom interaction, through the lifetime (1/Γ) of the excited state and is a measure of the spread of that interaction through the velocity distribution of the absorbing atoms $p(v)$. The fluorescence light interaction thus consists of a double integral over the spectrum of the moving fluorescence atoms and over the moving absorbing atoms in the interaction regions of the Ramsey cavity. The intensity of this fluorescence light is proportional to the number of fluorescence photons emitted per second. This is a function of the pumping and detecting laser intensity and, in the case of cycling transitions, of the saturation of the transition excited in the process, $S/(1 + S)$, where S is the saturation parameter. This can be verified experimentally and the question of saturation that may affect linearity between the lasers, intensity and the light shift observed can be taken care of directly. We recall that the velocity v of the atoms may be expressed as l/τ where l is the interaction length and τ is the resulting time of interaction due to speed v. We may finally write the light shift as:

$$\Delta_L = I_L \times g(\tau, \omega) \tag{3.2}$$

where $g(\tau, \omega)$ includes the double integrals of Equation 3.1 taking care of both the spectrum of the fluorescence light and the speed distribution of the absorbing atoms. On the other hand, I_L includes all the constants involved in the light–atom interaction given above.

The effect of this light shift Δ_L can be finally evaluated by integrating it into the Ramsey fringe shape as was done for the second-order Doppler effect calculated in Chapter 1. Doing a similar calculation as was done in that case, we obtain:

$$f_L(\omega_m, b) = \frac{1}{2\pi} \frac{\int_0^\infty \Delta_L(\tau) T f(\tau) \sin \omega_m \tau \sin^2 b\tau d\tau}{\int_0^\infty f(\tau) T \sin \omega_m \tau \sin^2 b\tau d\tau} \tag{3.3}$$

Using Equation 3.2, this can be written as:

$$f_L(\omega_m, b) = I_L F_L(\omega_m, b) \tag{3.4}$$

where $F_L(\omega_m, b)$ contains all the terms included in the numerator and denominator of Equation 3.3 and the expression for Δ_L given by Equation 3.2.

3.1.1.2.2 Phase Shift

We note that a similar expression exists for the frequency shift introduced by the phase shift ϕ that may exist between the arms of the Ramsey cavity (Equation 1.19). We have:

$$\omega_\phi - \omega_o = -\frac{\phi}{L} \frac{\int_0^\infty f(\tau) \sin^2 b\tau \sin \omega_m T d\tau}{\int_0^\infty (1/v) f(\tau) \sin^2 b\tau \sin \omega_m T d\tau} \tag{3.5}$$

which can also be written as:

$$f_\phi(\omega_m, b) = \phi F_\phi(\omega_m, b) \tag{3.6}$$

where $F_\phi(\omega_m, b)$ contains similarly all the functions in the numerator and denominator of Equation 3.5.

3.1.1.2.3 Doppler Shift

We recall that the second-order Doppler shift is still present and is given by Equation 1.18, repeated here for convenience:

$$\frac{\omega_D' - \omega_o}{\omega_o} = \frac{\int_0^\infty v f(\tau) \sin^2 b\tau \sin \omega_m T d\tau}{2c^2 \int_0^\infty (1/v) f(\tau) \sin^2 b\tau \sin \omega_m T d\tau} \tag{3.7}$$

and must also be evaluated carefully before any conclusion can be drawn on the frequency shifts caused by cavity phase shift and light shift.

3.1.1.3 Experimental Determination of Those Shifts

The experimental problem that is raised by the presence of these shifts is their independent evaluation. It is readily seen that they are functions of the same velocity distribution, $f(\tau)$, and of the same parameters such as the modulation depth ω_m and the Rabi frequency b. In principle, the individual shifts have to be evaluated independently. It is also realized that the evaluation of a given shift, say the phase shift, for a given choice of parameters will be invalid for another choice since the other shifts, Doppler and light shift, will also be altered. This interdependence has been a major difficulty in the correct evaluation of those shifts.

One approach consists usually in the careful evaluation first of the Doppler shift. This can be done quite exactly by first determining the interaction time distribution (velocity distribution) by means of techniques described earlier. An approach using the cosine transform of the Ramsey fringe pattern usually provides sufficient accuracy. With this knowledge, the Doppler shift can then be calculated for various values of b and ω_m. This frequency shift is of the order of 5×10^{-14} and is normally evaluated with an accuracy of 1×10^{-15}.

This Doppler shift being identified and evaluated, it is then possible to evaluate the two others, phase shift and light shift. The phase shift is generally evaluated by means of beam reversal. The difference between the two frequencies as measured for example for an east–west (EW) direction and for a west–east (WE) direction is then used as a means, by simple averaging, for determining the actual correction to be applied to the servo-locked frequency. However, as mentioned above, this technique relies on the exact retrace of the beam in the two directions. Although the proposed ring cavity considerably reduces the phase gradient within the cavity, there remains a small effect which limits the accuracy below the expected value.

One approach for addressing that question has been the use of a so-called *parameter approach*. For example, the frequency shift as measured experimentally can be plotted directly against the function $F_\phi(\omega_m,b)$ evaluated for various values of ω_m and b. The value of this function is calculated by numerical analysis. As made explicit by Equation 3.6, the slope of the straight line obtained is the phase shift ϕ. It is noted that care must be taken in the measurements that corrections are applied for the frequency shift caused by Doppler effect varying with modulation depth ω_m and Rabi frequency b.

On the other hand, it is readily observed that the light shift, through Equation 3.4 appears to act very much like a phase shift. In fact, a plot of $F_L(\omega_m,b)$ against $F_\phi(\omega_m,b)$ calculated for various values of ω_m and b, gives very closely a straight line and it can be concluded that the light shift acts effectively like a phase shift (Makdissi et al. 2000). It is concluded that the light shift, if important, may cause a rather important error and uncertainty in the evaluation of the phase shift between the arms of the Ramsey cavity. In a particular device studied, the light shift could be assimilated to a phase shift of 8 µrad, a value that leads to a frequency error of 2.6×10^{-14} in evaluating the frequency shift caused by the phase shift (Makdissi et al. 2000).

It may be mentioned that the technique just described is rather laborious, requiring days and weeks for evaluation of the many parameters involved by means of in-line computers. Furthermore, the approach applies particularly well when the velocity distribution is rather broad, which is the case when state selection is done by means of optical pumping. This appears to be the main reason why it was not applied to the case of state selection by means of dipole magnets, particularly in the case of the hexapole–quadrupole tandem magnets, which lead to a rather narrow velocity distribution (A. Bauch 2012, pers. comm.).

On the other hand, some other laboratories have used successfully a more straightforward approach at evaluating those various shifts. It was found that in their particular design, the fluorescent light shift as calculated was negligible. The cavity phase shift could thus be determined by beam reversal, the retrace of the beam being entirely satisfactory through the use of ring cavities (Shirley et al. 2001). Other laboratories have also come to the conclusion that in their design, the light shift was very small and did not affect the overall accuracy to a large extent (see, e.g., Hagimoto et al. 1999; Jun et al. 2001; Hasegawa et al. 2004).

3.1.1.4 Frequency Stability

The short-term frequency stability of the optically pumped standard depends on the same parameters as in the case of the standard using state selection and is given by Equation 1.20. Consequently, it is a function of the line Q and the S/N. The line Q is a function of the distance between the arms of the Ramsey cavity. There is thus no improvement to be expected on that parameter over the magnetic selector approach. However, an improvement may be expected from the S/N. A larger signal is expected in optical pumping since a greater inversion of population can be established in the optically pumped standard relative to the magnetic state selection. Furthermore, more atoms contribute to the signal since the technique makes use of the whole velocity spectrum and the atomic beam has generally a larger cross section. Finally, optical detection is rather efficient and detection may be limited primarily by detector noise. In practice, in a particularly well-designed system, a frequency stability of $3.5 \times 10^{-13} \, \tau^{-1/2}$ has been realized (Makdissi and de Clercq 2001).

The long-term frequency stability of the optically pumped Cs beam frequency standards depends on the stability of the various frequency shifts and offsets enumerated above. Consequently, the frequency of a unit is dependent to a certain extent on its environment. Temperature, humidity, atmospheric pressure, and magnetic field, depending on construction, play a role to various degrees in determining long-term frequency stability. Temperature fluctuations appear to have the most important effect. In general, best results are obtained in a temperature controlled environment. Fortunately most of those frequency biases are small and consequently their fluctuation with time is small.

Other fluctuations of unknown origins generally limit the frequency stability in the long term. When the averaging time τ is increased, the frequency stability as given by Equation 1.20 improves and reaches a plateau called the *flicker floor*. The level of this flicker floor is a function of unknown parameters. Better quality in construction and design lowers this flicker floor to nearly undetectable levels.

Results of selected implementations using optical pumping state selection are reported in Table 3.1 and compared to the results obtained with some of the best performing laboratory Cs beam standards using magnetic state selectors. As is readily

TABLE 3.1

Comparison of Optical Pumping to Magnetic State Selection in Implementing a Cs Beam Primary Frequency Standard

	PTB Cs1 (Germany)	PTB Cs2 (Germany)	PTB Cs3 (Germany)	SYRTE JPO (France)	NIST NIST-7 (the United States)	NIICT CRL-01 (Japan)	NRLM NRLM-4 (Japan) (NMIJ)
Distance between Ramsey cavities (m)	0.8	0.8	0.77 (vertical construction)	1.03	1.53	1.53	0.96
Microwave-magnetic field direction/beam	$=$	$=$	$=$	$\perp$	$=$	$=$	$\perp$
State selector-analyzers	Hexapole + quadrupole	Hexapole + quadrupole	hexapole	Optical pumping: selection: $F = 4\text{-}F' = 4$ detection: $F = 4\text{-}F' = 5$	Optical pumping: selection: $F = 4\text{-}F' = 3$ detection: $F = 4\text{-}F' = 5$	Optical pumping: selection: $F = 4\text{-}F' = 3$ detection: $F = 4\text{-}F' = 5$	Optical pumping: selection: $F = 4\text{-}F' = 4$ detection: $F = 4\text{-}F' = 5$
Mean atom velocity (m/s)	93	93	72	215	230	250	100
Line width (Hz)	59	60	44	100	77	62	100
$\sigma_y(\tau)$ $(\tau^{-1/2})$	5×10^{-12}	4×10^{-12}	9×10^{-12}	3.5×10^{-13}	1×10^{-12}	3×10^{-12}	8×10^{-13}
Accuracy	7×10^{-15}	12×10^{-15}	1.4×10^{-14}	6.4×10^{-15}	5×10^{-15}	6.8×10^{-15}	6.7×10^{-15}
Reference	Bauch et al. 1998, 2003	Bauch et al. 2003	Bauch et al. 1996	Makdissi and de Clercq 2001	Shirley et al. 2001	Hasegawa et al. 2004	Hagimoto et al. 2008

concluded from that table, a better accuracy and stability is obtained with optically pumped Cs beam frequency standards than with the best Cs beam frequency standards using state selector magnets in operation at the time of comparison. However, as will be seen below, laser cooling has led to the realization of the so-called atomic fountain with an order of magnitude accuracy better than that characterizing all these standards.

3.1.1.5   Field Application

The development of a Cs standard using state selection by means of optical pumping led to the implementation of laboratory primary standards with great stability and accuracy comparable to that of standards using magnetic dipole state selection (see Table 1.1). On the other hand, use of the approach has also been studied towards the implementation of small standard for field application, particularly in space for satellite navigation system. A compact breadboard unit has been developed based on knowledge acquired at SYRTE, Paris, described above in the development of a large scale primary standard. The size of the unit proposed is in the range of the units presently planned for the navigation system (passive H maser and passive Rb standard). Results confirmed the feasibility of the project: a line Q of 1.2×10^7 was obtained and a frequency stability of $2.3 \times 10^{-12} \, \tau^{-1/2}$ was observed (Ruffieux et al. 2009).

3.1.2   CPT APPROACH IN A BEAM

3.1.2.1   General Description

In the classical approach of implementing a Cs beam frequency standard, the state selection and microwave interaction are effectuated in separate regions of space. As illustrated in Figures 1.2 and 3.1, this approach requires either a dipole magnet or optical pumping done in one region and microwave interaction in another. This last operation is done at the first arm A of the Ramsey cavity. In fact, at the first arm A of the cavity, the operation essentially places the atoms in a coherent superposition of the states $F = 3$, $m_F = 0$ and $F = 4$, $m_F = 0$. This requires that Cs atoms be first placed in a single state, say $F = 3$, $m_F = 0$. When dipole magnets are used, this is accomplished simply by means of deflecting atoms that are in that state and use appropriate collimation. In the case of optical state selection, atoms are optically pumped in that state. Those atoms are then excited for a given interaction time $\tau = l/v$ by microwave radiation at the hyperfine frequency in the first arm of the cavity. Here l is the length of the either arm of the Ramsey cavity and v the speed of the atoms. If the power of the microwave radiation is adjusted such as to produce a $\pi/2$ pulse during time τ, the atoms find themselves in a pure superposition state at the exit of the first arm of the Ramsey cavity. They are in a coherent state. They drift to the second arm B where they are submitted again to microwave that acts again as a $\pi/2$ pulse. If everything is well-adjusted, a full transition corresponding to a π pulse takes place. This is detected by means of a selector dipole magnet followed by a counter of atoms or by means of optical detection (fluorescence).

These operations done in succession are somewhat complex and require construction and assembly of components such as magnets and a special microwave double-arm cavity. These components, as we have seen, are somewhat delicate and require high precision in their dimensions particularly with regard to phase shifts between the two cavity arms. One is tempted to ask if it were not possible to simplify the design of such a device by regrouping operations. Optical detection provides some simplification through the use of atom counting by means of optical excitation and fluorescence at the exit of the second arm of the Ramsey cavity, rather than state selection and atom ionization and counting. However, the gain is rather modest.

We have described in Chapter 2 a phenomenon called CPT that does exactly the two operations that we just described. This is done by applying, to an ensemble of atoms, two optical coherent radiation fields whose frequency corresponds to the transition from the atom ground S state to the excited P state. If the difference in frequency of the two laser radiation fields is that of the hyperfine transition of the Cs atom, coherence is created in the ground state at the hyperfine frequency. There is no need for state selection. If the field is applied for an appropriate time $\tau = l/v$, where l is now the width of the laser beam, the atoms are found in a coherent superposition state as if they had been prepared first in a single state and then submitted to a $\pi/2$ pulse. This is exactly the operation we are effectuating in the classical approaches described previously.

3.1.2.2 Analysis

The first reports on the use of CPT for implementing a frequency standard describe a laboratory system in which a beam of alkali atoms is excited by means of CPT at two zones separated by a distance L, simulating the Ramsey two-zone approach with a microwave cavity as in the classical Cs beam device (Thomas et al. 1982; Hemmer et al. 1983, 1985, 1986; Shahriar and Hemmer 1990; Shahriar et al. 1997). The arrangement is shown in Figure 3.4.

In the arrangement shown, two co-linear laser radiation fields at frequencies ω_1 and ω_2 are used. The light beam incorporating these two fields is made to cross the thermal alkali-metal atomic beam at right angle. The interaction lasting for a time τ, which is a function of the size of the beam waist of the combined laser beams

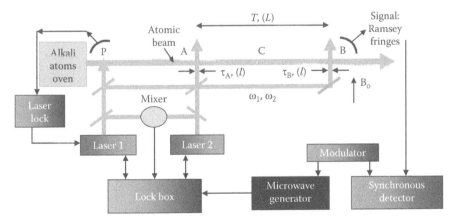

FIGURE 3.4 Cs beam atomic frequency standard implemented using coherent population trapping excitation. (With kind permission from Springer Science+Business Media: *Appl. Phys. B*, Atomic clocks based on coherent population trapping: a review. 81, 2005, 421, Vanier, J.)

and the speed of the atoms, places the atoms in a superposition state through the CPT phenomenon. The atoms emerge from the zone of interaction in a coherent state as if they had been put in a single state and then submitted to a microwave $\pi/2$ pulse at the hyperfine frequency. At their exit from the excitation zone, the atoms evolve freely in region C in that superposition state. Zone A essentially combines the state selection of the first dipole magnet and the first branch of the Ramsey cavity used in the classical Cs beam standard approach. The main advantage that results over the classical technique described earlier is the absence of a microwave interaction structure and of a selector magnet. No state preparation is required and no microwave excitation is required. The whole system gains in simplicity over the classical approach and over the optical pumping approach described above.

We have provided a somewhat general analysis of CPT in Chapter 2. The set of Equations 2.63 through 2.72 describing the interaction of the optical radiation fields with the atomic system applies, except that in the present circumstances there is no microwave field applied, and there is no buffer gas present. Consequently, the Rabi frequency b is set equal to 0. The relaxation rates in the ground state, γ_1 and γ_2, are also both zero since in principle, except for weak spin-exchange interactions, no relaxation takes place in the ground state of atoms in a beam. Atoms are free, no buffer gas is present and the decay rate from the excited state takes place by spontaneous emission at the rate Γ. In order to obtain easily tractable expressions, we assume that the optical Rabi frequencies ω_{R1} and ω_{R2} are equal. We set them equal to ω_R.

In the analysis, we use an operator formalism somewhat different from that used by Hemmer et al. (Hemmer et al. 1989). First, we apply the adiabatic approximation to the evolution of the excited state. In that case it is assumed that the excited state response is so rapid that it follows the evolution of the ground state and is essentially in stationary equilibrium at all times. The set of Equations 2.66 through 2.71 is then reduced to three equations containing only ground-state matrix elements and atomic parameters. We thus obtain for the evolution of the density matrix elements in a given zone.

$$\dot{\delta}^r_{\mu\mu'} = -\frac{\Gamma\xi}{1+3\xi}\delta^r_{\mu\mu'} - \frac{1}{2}\frac{\Gamma\xi}{1+3\xi} + \Omega_\mu\delta^i_{\mu\mu'} \tag{3.8}$$

$$\dot{\delta}^i_{\mu\mu'} = -\Gamma\xi\delta^i_{\mu\mu'} + \frac{1}{2}\Gamma\xi\delta_o(\rho_{\mu\mu} - \rho_{\mu'\mu'}) - \Omega_\mu\delta^r_{\mu\mu'} \tag{3.9}$$

$$\frac{d}{dt}(\rho_{\mu\mu} - \rho_{\mu'\mu'}) = -\Gamma\xi(\rho_{\mu\mu} - \rho_{\mu'\mu'}) - 2\Gamma\xi\delta_o\delta^i_{\mu\mu'} \tag{3.10}$$

The fluorescence power in zone B is given by:

$$P_{fl} = \int \hbar\omega N\Gamma\rho_{mm}dt \qquad \text{(integral over zone B)} \tag{3.11}$$

with the population of the excited state given by:

$$\rho_{mm} = \frac{\xi}{1+3\xi}(1 + 2\delta^r_{\mu\mu'}) \tag{3.12}$$

We have introduced several terms whose definitions are:

$$\delta_o = \frac{2\Delta_o}{\Gamma} \tag{3.13}$$

$$\xi = \frac{\omega_R^2/\Gamma^2}{1+\delta_o^2} \tag{3.14}$$

$$\Omega_\mu = \omega_{12} - \omega_{\mu'\mu} \qquad \omega_{12} = \omega_1 - \omega_2 \tag{3.15}$$

Reference is made to Figure 2.16 for the identification of the various frequencies involved. The set of Equations 3.8 through 3.10 can be written in matrix form as:

$$\frac{d}{dt}\begin{pmatrix} \delta^r \\ \delta^i \\ \rho_{\mu\mu} - \rho_{\mu'\mu'} \end{pmatrix} = \begin{bmatrix} \dfrac{-\xi\Gamma}{(1+3\xi)} & \Omega_\mu & 0 \\ \Omega_\mu & -\Gamma\xi & \left(\dfrac{1}{2}\right)\Gamma\xi\delta_o \\ 0 & -\Gamma\xi\delta_o & -\Gamma\xi \end{bmatrix} \times \begin{pmatrix} \delta^r \\ \delta^i \\ \rho_{\mu\mu} - \rho_{\mu'\mu'} \end{pmatrix} \\ + \begin{pmatrix} \dfrac{-\xi\Gamma}{(1+3\xi)} \\ 0 \\ 0 \end{pmatrix} \tag{3.16}$$

where we represent the state of the ensemble by a vector u:

$$u = \{\delta^r, \delta^i, \rho_{\mu\mu} - \rho_{\mu'\mu'}\} \tag{3.17}$$

and the CPT interaction in a zone as the application of an operator represented by the above matrix to the atom state vector. The characteristics of the atoms in the beam, such as levels population difference and coherence created by the interaction can be studied further down the beam path by means of a probe laser beam and detection of the resulting fluorescence as illustrated in Figure 3.4. The case of several successive zones of CPT application can thus be treated by a repeated application of the matrix operator. We note that the atomic beam is thermal and the velocities of the atoms within the beam are spread over an altered Maxwell–Boltzmann distribution (QPAFS 1989):

$$f(\tau) = \frac{2}{\tau_o}\left(\frac{\tau_o}{\tau}\right)^5 e^{-(\tau_o/\tau)^2} \tag{3.18}$$

which in the evaluation of a fluorescence signal leads to a requirement of averaging over the times of interaction.

The effect of CPT in the two zones is obtained by solving Equation 3.16 and may be represented by a pseudo-rotating operator $M_A(\omega_R, \Omega_\mu, \delta_o, t)$ affecting the state vector given by Equation 3.17. The state vector in zone B is obtained directly by transformation of the original state vector u_o by means of the relation:

$$u_B = M_B(\omega_R, \Omega_\mu, \delta_o, \tau_B) M_C(0, \Omega_\mu, 0, T) M_A(\omega_R, \Omega_\mu, \delta_o, \tau_A) u_o \qquad (3.19)$$

where:

M_i are the rotating matrices corresponding to each zone

τ_A, T, and τ_B are the time spent by the atoms in each zone, respectively

In zone A and B, M is obtained from Equation 3.16 by means of Laplace transform and has the form:

$$M(\omega_R, \Omega_\mu, \delta_o, t) = \begin{pmatrix} \exp-\dfrac{t\Gamma\xi}{1+3\xi} & 0 & 0 \\[2ex] 0 & (\exp-t\Gamma\xi)\cos(t\Gamma\xi\delta_o) & \dfrac{1}{2}(\exp-t\Gamma\xi)\sin(t\Gamma\xi\delta_o) \\[2ex] 0 & 2(\exp-t\Gamma\xi)\sin(t\Gamma\xi\delta_o) & (\exp-t\Gamma\xi)\cos(t\Gamma\xi\delta_o) \end{pmatrix} \qquad (3.20)$$

In the operation, however, a term appears that is not coupled to the components of u_o. It is not included in $M(\omega_R, \Omega_\mu, \delta_o, t)$ and is thus introduced as a vector u_s added as a source term:

$$u_s = \left\{ \frac{1}{2}\left[-1 + \exp\left(-\frac{\Gamma\xi t}{1+3\xi} \right) \right], 0, 0 \right\} \qquad (3.21)$$

This vector must be added to the state vector resulting from the transformation of u_o by means of $M(\omega_R, \Omega_\mu, \delta_o, t)$. At the exit of zone A, the state vector is thus:

$$u_A = M_A \cdot u_o + u_s \qquad (3.22)$$

In zone C where the atoms evolve freely, the operator has the form:

$$M(0, \Omega_\mu, 0, t) = \begin{pmatrix} \cos t\Omega_\mu & \sin t\Omega_\mu & 0 \\ -\sin t\Omega_\mu & \cos t\Omega_\mu & 0 \\ 0 & 0 & 1 \end{pmatrix} \qquad (3.23)$$

In zones A and B, we have assumed that the intensity of the laser radiation is very large and that the CPT resonance line is broadened to the extent that the small detuning Ω_μ is negligible. We also recall that we have assumed that the system evolves in the rotating frame attached to the difference in frequency between the two laser radiation fields. In that frame, the ground-state off-diagonal matrix elements present in zone C appear to rotate at Ω_μ. We call ϕ_A the phase of the coherence at the exit of the first zone, defined through the relation:

$$\delta_{\mu\mu'} = |\delta_{\mu\mu'}| e^{i\phi_A} \tag{3.24}$$

After performing the operation of Equation 3.22, we obtain $\delta_{\mu\mu'}$ and the phase is given by:

$$\tan(\phi_A) = \frac{\alpha^i}{\alpha^r} = \frac{\exp(-\Gamma\xi\tau_A)(\rho_{\mu'\mu'} - \rho_{\mu\mu})_o \sin(\Gamma\xi\delta_o\tau_A)}{1 - \exp\left[-(\Gamma\xi\tau_A/1 + 3\xi)\right]} \tag{3.25}$$

On the other hand, the fluorescence integrated over zone B is:

$$P_{fl}(\text{zone } B) = \left\{ 1 - \exp\left[-\frac{\Gamma\xi\tau_B}{(1+3\xi)} \right] \right\}$$

$$\times N\hbar\omega \left(1 + \left\{ -1 + \exp\left[\frac{-\Gamma\xi\tau_A}{(1+3\xi)} \right] \right\} |\sec(\phi_A)| \cos(\phi_A - T\Omega_\mu) \right) \tag{3.26}$$

In those equations, N is the total number of atoms in interaction with the laser beam in zone B and $(\rho_{\mu'\mu'} - \rho_{\mu\mu})_o$ is the population difference of states μ' and μ at the entrance of zone A. It is noted that the expression given by Hemmer et al. (1989) for the phase shift is different from ours by a factor of 2 in the argument of the sin term. The appropriate expression obtained from Equation 3.26, using Hemmer's et al.'s notation is:

$$\tan(\phi_A) = \frac{-\exp(-\Omega^2 S\tau_A)(\rho_{\mu'\mu'} - \rho_{\mu\mu})_o \sin(2\Omega^2 D\tau_A)}{\exp(-f\Omega^2 S\tau_A) - 1} \tag{3.27}$$

where the various terms are connected to those used in the present article by the relations:

$$\Gamma\xi = \Omega^2 S \qquad\qquad \frac{\Gamma\xi}{(1+3\xi)} = f\Omega^2 S \qquad\qquad \Gamma\xi\delta_o = 2\Omega^2 D$$

An analysis by Shahriar et al. (1997) made later is compatible with this conclusion.

The phase shift is caused by the ac Stark effect taking place in zone A (light shift) and causes a frequency shift as in the classical Ramsey separated zone approach, although the physical origin of this phase shift is quite different. The times τ_A, τ_B, and T are functions of the atomic velocities. Consequently, the actual shape of the Ramsey fringe is obtained through integration over all interaction times in the beam, whose spread is given by an altered Maxwell–Boltzmann distribution given by Equation 3.18. This expression is normalized to unity upon integration over τ. For a ratio (L/l) equal to 100, numerical integration with $\Gamma = 3 \times 10^7 \, s^{-1}$ gives a fringe shape as shown in Figure 3.5.

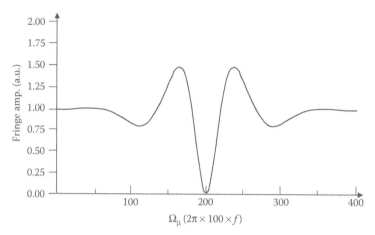

FIGURE 3.5   Ramsey fringe calculated for the CPT two zones approach in an alkali metal atomic beam. The abscissa zero is chosen in computer calculation as to make exact resonance at 200.

The effect of the phase shift on the central fringe is to displace its centre by the amount:

$$\Delta\omega\left(\phi_A\right) = \left(\frac{\phi_A}{T}\right) \tag{3.28}$$

This phase shift is referenced to the phase imbedded in the laser radiation and supported in the second zone. Its behaviour as a function of detuning δ_o is shown in Figure 3.6 for several values of ω_R and a value of $(\rho_{\mu'\mu'} - \rho_{\mu\mu})$ equal to 0.2. It has been averaged over atomic velocities using the same distribution as that used in the case of the fringe calculation, with $\Gamma = 3 \times 10^7\,\mathrm{s}^{-1}$ and a ratio of $L/l = 100$.

As is clearly seen in this graph, the dependence of the phase on tuning is much reduced around the value $\delta_o = 50$. The reduction varies with Rabi frequency or laser intensity. This behaviour appears essentially as an optical saturation effect, although the excited state m never gets much populated even for $\omega_R = 0.25\,\Gamma$.

With the intent of making more transparent the physics involved in this two-zone CPT setup, we have limited the above analysis to the case of equal intensities for the laser radiation fields at ω_1 and ω_2. The problem of unequal intensities is more complex. However, it is shown that the general behaviour regarding phase shift is not much dependent on this asymmetry (Hemmer et al. 1989).

3.1.2.3   Experimental Results

Laboratory setups using such an approach have been implemented with alkali atoms Na and Cs (Thomas et al. 1982; Hemmer et al. 1993). In the case of Na, a dye laser was used as the source of the two radiation fields, while for Cs, diode lasers were used. The results confirm the observation of Ramsey fringes in the fluorescence of the second zone with a shape in agreement with that shown above.

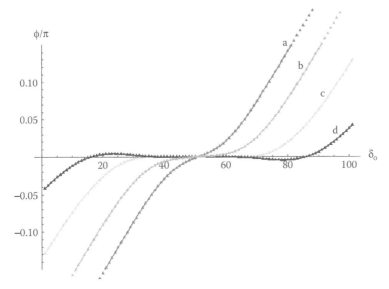

FIGURE 3.6 Phase shift affecting the Ramsey fringe in the case of a two zone CPT approach in an atomic beam (theory). The phase shift is averaged over velocities in zone A as for the Ramsey fringes. The parameters used for this graph are, $\Gamma = 3 \times 10^7$ s^{-1}, $\tau_A = 10^{-5}$ s. (a): $\omega_R = 0.1\ \Gamma$, (b): $\omega_R = 0.15\ \Gamma$, (c): $\omega_R = 0.2\ \Gamma$, (d): $\omega_R = 0.25\ \Gamma$. The optical resonance condition corresponds to $\delta_o = 50$ on the axis chosen.

3.1.2.3.1 Frequency Stability

In Na, with a distance $L = 15$ cm between the interaction zones, a fringe line width at half maximum of 2.2 kHz was obtained experimentally. On the other hand, an *S/N* of 4500 for 1 s averaging time was also measured. The expected frequency stability for such a setup in the limit of shot noise was calculated to be 5×10^{-10} $\tau^{-1/2}$. The measured frequency stability confirmed this evaluation (Thomas et al. 1982; Hemmer et al. 1986). In the case of Cs, a signal-to-noise ratio of 1800 was observed for an averaging time of 1 s. The authors projected that the measured fringe width (1 KHz) and this *S/N* translated into a frequency stability of 6×10^{-11} $\tau^{-1/2}$ (Hemmer et al. 1993).

3.1.2.3.2 Frequency Shifts

3.1.2.3.2.1 The ac Stark Phase Shift The phase shift was measured for different values of the optical Rabi frequency and various values of the ground-state population difference at the entrance of first zone (Hemmer et al. 1989). Their results are in general agreement with the model developed and summarized above, the phase shift introduced by the first interaction zone following the trend shown in Figure 3.6. It is found that for certain adjustments of the parameters, such as large pumping rates, the slope of the phase shift as a function of detuning δ_o changes sign. The authors explain this behaviour by means of differences between the decay rates from the excited state to various levels of the ground state. It should be mentioned, however, that the actual variation of ϕ_A with δ_o is a very sensitive function of the Rabi frequency w_R and shows a complicated behaviour regarding the sign of the slope at large values of the Rabi frequency as shown in Figure 3.6. The authors have also evaluated that the minimum slope observed in their setup (Na beam) lead to a frequency shift of the order of 2×10^{-11} for a laser detuning of 0.01 Γ, corresponding to a detuning of the order of 300 kHz. Recent theoretical work has considered the multiplicity of levels entering into interaction with the laser radiation (Kim and Cho 2000).

3.1.2.3.2.2 Other Frequency Shifts There are several other frequency shifts present in the approach proposed. In particular, a frequency shift may be present if a phase shift is introduced physically through a difference in path lengths between the two radiation beams before entering the interaction regions. Such a phase shift was commented upon (Mungall 1983). It has also been observed experimentally (Hemmer et al. 1983). Other important frequency shifts may be caused by misalignment of ω_1 and ω_2 radiation beams, phase shifts due to laser polarization, atomic beam versus laser beam misalignment, and overlapping of Ramsey fringes. The importance of these shifts has been evaluated and although they are not negligible, it is believed that they would not cause a major impediment in the practical realization of an operational frequency standard with useful properties (Hemmer et al. 1986).

3.1.3 Classical Cs Beam Standard Using Beam Cooling

We have raised the importance of an appropriate determination of the atomic velocity spectrum in the atomic beam several times for the evaluation of various frequency biases. The spread of this spectrum has also some effect on the width of the Ramsey fringes and reduces the number of observed fringes. Furthermore, selection of lower velocities reduces the importance of shifts such as the second-order Doppler effect. The matter was addressed in developments made at Physikalisch Technische Bundesanstalt (PTB), Germany, by using combinations of multipole magnets that resulted in a Cs beam with a mean atomic velocity of 72 m/s and a full width at half maximum of speed distribution of 12 m/s (Table 3.1) (see Bauch et al. 1996). This beam property improves the general physical characteristics of that particular Cs frequency standard.

A question comes to mind immediately. Is it not possible to still do better simply by selecting slower atoms in the beam? The main problem in this approach is that the spectrum of velocities contain less atoms at lower speed for a given oven temperature. Just raising oven temperature to increase beam flux has a limit due to background raising pressure and the actual scattering in the beams of slow atoms by the faster ones. However, we have seen in Chapter 2 that it is possible to reduce the speed of an atomic beam using laser radiation pressure. Actually the atoms can even be stopped in their motion (Phillips and Metcalf 1982; Ertmer et al. 1985).

That approach was studied experimentally for slowing down a Cs atomic beam and implementing directly an atomic clock (Lee et al. 2001, 2004). The technique was actually used in connection to the approach where the state selector magnets were replaced by laser optical pumping, an approach that was examined above.

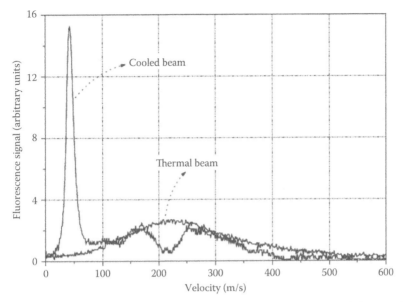

FIGURE 3.7 Velocity distribution obtained by means of cooling of a beam of Cs atoms. The cooled beam has a very narrow distribution of 0.9 m/s around a mean speed of 30 m/s. The beam was used in a classical standard with a 21 cm long Ramsey cavity providing a central Ramsey fringe 62 Hz wide. (Courtesy of S.H. Lee; Data from Lee, H.S. et al., *J. Korean Phys. Soc.*, 45, 256, 2004; Lee, H.S. et al., *IEEE Trans. Instrum. Meas.*, 50, 531, 2001.)

Figure 3.7 illustrates the results obtained for the velocity distribution when such cooling is accomplished. The mean velocity of the atoms in the beam was 30 m/s with a spread of 0.9 m/s. This is a rather slow atomic beam with a compressed velocity spectrum. For all practical purposes the beam looks monokinetic. To make it more explicit, Figure 3.7 includes also the velocity spectrum of the same beam at room temperature without cooling. It is obvious that such a slow beam cannot be used with a long Ramsey cavity, since atoms fall in the earth's gravitational field as mentioned earlier. In a 1 m long Ramsey cavity, the atoms would fall by 5 mm before they reach the second arm of the cavity. In the particular case cited, the Ramsey cavity used was reduced to a length of 21 cm and the width of the observed central fringe was 62 Hz. This is actually four times narrower than the width of the line observed for a room temperature thermal beam and a 37-cm-long cavity. It is about the same width as that obtained in clocks with a 1.5 m-long cavity using a beam emerging from an oven at 100°C. There is thus some gain in the approach compared to the conventional approach using dipole state selector magnets and a thermal beam. However, when this is compared to the results obtained with a thermal beam and selection of slow atoms by means of a combined hexapole-quadrupole state selector, the gain is marginal (see Chapter 1, Table 1.3). Furthermore, the complexity and size of the cooling section of the beam makes the system rather large and in view of other developments that will be described below it appears that the approach has not raised further interest.

3.2   ATOMIC FOUNTAIN APPROACH

3.2.1   In Search of a Solution

Following the line of thought of the previous paragraphs as well as earlier discussions on the question of the influence of atomic beam velocity spectrum on the accuracy of Cs beam standards, it appears that a totally new approach is required to fully take advantage of the properties of cooled atoms. We have seen that the atoms can be used in a cell and cooled by means of isotropic cooling. In principle, such an approach should open the door to various solutions to the problems introduced by atomic motion. We will look at that technique later in the case of atomic frequency standards implementation in a cell. For the moment we wonder if laser-cooled atoms, such as those in a molasses, could not be used to implement an atomic beam frequency standard. In satellites in orbit around the earth, where gravity is reduced to the microlevel, a beam of slow atoms, as described in the previous paragraphs, could be used since it would not be deflected by external forces. We will see later that such a frequency standard called PHARAO (PHARAO stands for *Projet d'Horloge A Refroidissement d'Atomes en Orbite* or *Project of a clock in orbit using cold atoms*) has been constructed specifically for the purpose of being part of an ensemble of clocks in orbit around the earth. At the surface of the earth, it appears that the only solution is to use the beam vertically. We have seen in Chapter 1 that such an approach was proposed at PTB with relatively slow atoms (Bauch et al. 1996). However, it was found that the gain relative to the horizontal approach was not substantial because the beam speed was still relatively large (70 m/s) and because of other constraints in the particular setup used that made this gain marginal relative to the standard horizontal approach. However, with laser-cooled atomic beams to the level of m/s, the gain should be large. Such a vertical approach was suggested in the 1950s by Zacharias (see Forman 1985). The system proposed was a vertical apparatus, in which Cs atoms in a thermal beam, flying upwards from an oven, would be slowed by gravity. The slower atoms would attain a certain height and fall back down into a detector alongside the oven. Unfortunately, the system, called *Fallotron*, failed. Slow atoms in the Maxwell-Boltzmann distribution within the beam were not present, being scattered out of the beam by fast atoms in the vicinity of the oven.

However, in view of the analysis made earlier and the experimental results reported in the 1980s on laser cooling, Zacharias' concept appeared to be feasible. The main idea is to laser-cool to a very low temperature ($\sim\mu K$) an ensemble of atoms in a molasses as described in Chapter 2. The ensemble of atoms cooled in such a molasses forms a small ball ($\sim$cm in diameter) that can be sent upwards by appropriate adjustment of the frequency of the cooling laser beams, a technique to be described below. The atoms within the ball, having a very small velocity ($\sim$1 cm/s), do not diffuse much in space during their travel upwards and downwards. The ball expands of course but only to a few centimetres and the density at detection is still large. The launching speed being set at the level of $\sim$m/s, the ball being decelerated by gravity reaches a height of the order of one metre or so and falls back under gravity. If a microwave cavity is placed on the path of the ball, the atoms pass twice through that cavity, once on their way up and again on their way down. If the cavity is fed with microwave at the hyperfine frequency, atoms therefore experience two coherent microwave pulses. The time interval between the two pulses can be of the order of 1 s, that is to say about two orders of magnitude longer than with usual thermal beam atomic frequency standards. Those successive events are identical to those experienced by atoms in a standard room temperature atomic beam frequency standard but in a pulse mode, rather than in a continuous mode. The ball cooling and launching can be repeated at a rate compatible with the time of travel up and down of the ball, and with appropriate averaging a totally integrated frequency standard similar to that using a thermal beam can be implemented. The system is known as an *atomic fountain*. The first atomic fountains have been realized with Na atoms at Stanford University, USA (Kasevich et al. 1989), and with Cs atoms in Paris within a collaboration between Laboratoire Primaire du Temps et des Fréquences (LPTF) and Laboratoire Kastler-Brossel (LKB) (Clairon et al. 1991). The first Cs fountain capable of acting as a primary standard, Cs-FO1, was constructed in 1995 at LPTF (Clairon et al. 1995). Following those early successes, similar fountains were assembled in several laboratories in various countries such as: Brazil, Canada, China, Germany, India, Japan, Russia, United Kingdom, and United States.

3.2.2 GENERAL DESCRIPTION OF THE Cs FOUNTAIN

A schematic diagram of a Cs fountain clock is shown in Figure 3.8. We will first give a general description of the various physical parts or components forming the fountain. Excellent reviews have been written on the subject (see, e.g., Wynands and Weyers 2005; Bize et al. 2009). In Section 3.2.3, we will describe in more details their exact functions with emphasis on the temporal sequence of those functions.

> *Zone A*: Formation and launching of a small cloud of cold atoms looking like a small ball a few millimetres in diameter. This zone is the region where an ensemble of atoms that exists as a vapour is exposed to laser radiation coming from six orthogonal directions. They are cooled as a molasses and in some designs are trapped in a magneto-optical-trap (MOT). They are also launched upwards at a slow velocity.
>
> *Zone B*: Preparation of the atoms. In this zone, the atoms are prepared in a pure state while they travel upwards in the vertical direction. The region consists of a cavity exposing the atoms in the cloud to a microwave radiation field that forces transitions at the hyperfine frequency, acting as state selector. It also contains a laser radiation beam, called a *pusher*, completing the state selection.
>
> *Zone C*: Interrogation of the atoms. This is the region containing the microwave interrogation cavity for exciting the clock transition. The resonant cavity (mode TE_{011}) is made of oxygen-free high conductivity copper (OFHC). The loaded Q is of the order of 10,000.
>
> *Zone D*: Free motion zone. The atomic cloud is moving up freely while exposed to the earth's gravitational field. The atoms reach a height h and fall back traversing the interrogating cavity again, thus simulating the function of the double arm Ramsey cavity in the classical approach.

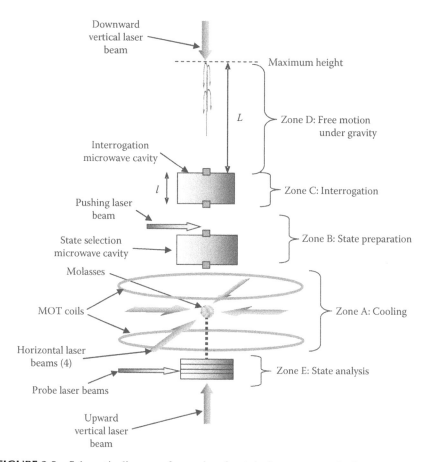

FIGURE 3.8   Schematic diagram of a caesium fountain frequency standard.

Zone E: Detection zone—This is the region where the atoms are detected by means of fluorescence created by a probe laser beam and analyzed in order to determine their state. In some design, that region may be situated above the cooling region.

The space surrounding the cavity and the region of free travel of the atoms is shielded from environmental magnetic field fluctuations by means of several layers of magnetic shields (cylinders and caps, three to five layers). With an internal solenoid and some compensation coils, the vertical magnetic induction is around 10^{-7} T, homogeneous to better than a few percent over the entire volume of the area of microwave interaction.

Atomic fountains require a very good vacuum to prevent loss of cold atoms by collision with residual thermal atoms (Rb, Cs, H, N, and rare gases). The rate of thermal atoms-cold atoms collisions is about one per second at a residual pressure of 10^{-6} Pa. The vacuum level obtained is some 10^{-8} Pa after several months of pumping which ensures very little loss of cold atoms. The pumping is generally done by ion pumps and getter pumps.

The system is operated in a temporal sequence, small clouds of atoms being sent up individually at a rate of the order of one per second or so. We now describe the details of the functions of the various zones and the time sequence of the various operations.

3.2.3 FUNCTIONING OF THE CS FOUNTAIN

3.2.3.1 Formation of the Cooled Atomic Cloud: Zone A

Zone A is the region where a cold cloud of atoms is created in a MOT as described in Chapter 2. That region is exposed to three pairs of orthogonal laser beams. In the configuration shown in Figure 3.9, which is that used in the first operational fountain FO1 implemented at SYRTE, Paris, France, the first two pairs propagate horizontally and one pair vertically. That configuration is somewhat easy to implement. However, the vertical beams traverse the interrogating microwave cavity and their cross section or beam waist is limited by the size of the holes made in that cavity. In turn, the size of those holes is limited by restrictions dictated by cavity performance particularly relative to Q and microwave radiation leakage. Furthermore, special care must be taken in the manipulating of the time sequence of the lasers to avoid the presence of radiation in the cavity while the atoms interact with the microwave field since a light shift would be introduced in the frequency of the clock transition. Another arrangement, called the *(1,1,1) configuration* was used in later

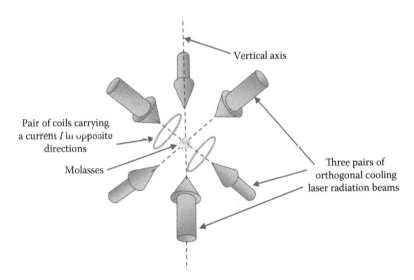

FIGURE 3.9 Configuration of the laser beams used to avoid the presence of a radiation beam along the trajectory of the cooled atomic cloud on the vertical axis. The laser beams are oriented perpendicular to the faces of an imaginary cube with one diagonal of the cube oriented in a direction along the vertical axis of the fountain.

developments. It consists in orienting the pairs of laser beams perpendicular to the faces of a cube whose orientation is such as to have one diagonal oriented along the desired vertical atomic cloud direction. In such a case there is no direct laser radiation traversing the interrogating microwave cavity or propagating along the vertical axis of the moving cold atomic cloud. This is illustrated in Figure 3.9. Two coils carrying current in opposite directions are part of that zone and complete the MOT. Those coils, forming an anti-Helmholtz configuration, produce a magnetic field gradient of ~10 G/cm in the centre of the capture area. The counter propagating lasers used to form the MOT are polarized in a σ^- to σ^+ configuration. The radiation fields consist of counterpropagating laser beams, which can be created by means of reflectors or still by means of beam splitters including coupling to optical fibres. These fields create the molasses cooling arrangement described in Chapter 2. The waist of the laser beams at the site of formation of the cooled cloud is of the order of 10–20 mm. The cloud is formed out of the Cs vapour pressure already existing in region A. The lasers are detuned from resonance by $-3\,\Gamma$ (to the red) and the capture velocity limit is about 30 m/s. In a typical setup, a MOT captures between 10^8 and 10^9 atoms in 1 s. The result is a small cloud with a total number of atoms of the order of 10^8 to 10^9 atoms having speeds of the order of $\Delta v \cong 2v_{rec}$ or ~7 mm/s. In a MOT, the cloud has a Gaussian velocity distribution and may have a diameter of the order of a few millimetres. It is possible to create a cloud with a higher density of atoms by feeding the space where the MOT is created by means of a beam of atoms already slowed down in a 2D MOT as described in Chapter 2.

The cold cloud of atoms is launched by means of the moving molasses technique. In the setup shown in Figure 3.8, this is done by detuning the vertical laser beams relative to each other. If the upward-directed beam is detuned by $+\delta v$ and the downward-directed one by $-\delta v$ from the set cooling frequency v_c, the resulting interference pattern moves vertically at a velocity $c\delta v/v_c$. We may visualize the effect in the reference frame of the atomic cloud itself, forming the molasses, which is essentially locked to the interference pattern and moves with it. When this detuning is done, the magnetic field gradient forming the MOT is turned off and the molasses as a whole acquires speed in the vertical direction, thus its name moving molasses. The atomic cloud, under the form of a small ball, thus acquires speed vertically in a short time (~1–2 ms) as it follows the interference pattern and finds itself launched vertically at a speed given by:

$$v(\text{launch}) = \frac{c\delta v}{v_c} \qquad (3.29)$$

In a typical case, the launching is done at a speed of the order of 4 m/s by appropriate detuning. The same technique for launching the atoms can be applied in the case of the arrangement shown in Figure 3.9 where the laser beams are oriented in a (1,1,1) arrangement. The detuning δv is then applied to all pairs of laser beams and from geometrical considerations on the orientation of the laser beams relative to the vertical axis, it is shown that the ball is launched at a speed equal to:

$$v(\text{launch}) = \frac{\sqrt{3}c\delta v}{v_c} \qquad (3.30)$$

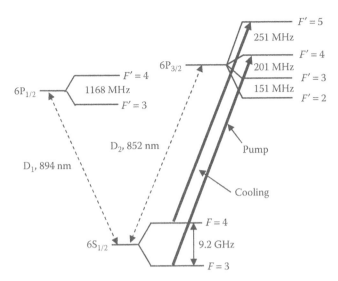

FIGURE 3.10 Lower manifold of energy levels of Cs^{133} used in the analysis of the cooling and interrogation in the Cs fountain.

The cooling process itself may be examined by means of Figure 3.10 representing the lower manifolds of levels of interest of ^{133}Cs. The actual transition used for cooling atoms in the MOT is the cycling transition $\left|S_{1/2}, F = 4\right\rangle \leftrightarrow \left|P_{3/2}, F' = 5\right\rangle$ as illustrated. In principle, atoms are trapped in an absorption-fluorescence cycle involving these two states since transitions from state $\left|P_{3/2}, F' = 5\right\rangle$ to other states than $\left|S_{1/2}, F = 4\right\rangle$ are forbidden. The cooling process should thus be very efficient. However, laser radiation is applied for a relatively long time (0.1–0.5 s) and off resonance excitation may take place to state $\left|P_{3/2}, F' = 4\right\rangle$. In that case, decay may then take place to state $\left|S_{1/2}, F = 3\right\rangle$ by means of spontaneous emission. When that occurs the atoms involved are lost for cooling. It is possible to reintegrate those atoms into the cooling cycle by means of radiation at the frequency of the transition $\left|S_{1/2}, F = 3\right\rangle \leftrightarrow \left|P_{3/2}, F' = 4\right\rangle$ and optically pump the atoms out of state $\left|S_{1/2}, F = 3\right\rangle$ as shown in the figure. Those atoms fall back by spontaneous emission to both levels $F = 4$ and $F = 3$ of the $S_{1/2}$ ground state but since they are pumped out of state $\left|S_{1/2}, F = 3\right\rangle$ they finally end up in level $F = 4$ and are thus cooled again. This optical pumping is done in practice by superposing to the cooling laser beams one or two pumping beams at the appropriate frequency corresponding to the $\left|S_{1/2}, F = 3\right\rangle \leftrightarrow \left|P_{3/2}, F' = 4\right\rangle$ transition.

A typical time sequence of the various operations is illustrated in Figure 3.11. For the purpose of illustration of the functioning, we will assume that the fountain is operated with a pair of laser beams propagating vertically, the two other pairs propagating in a horizontal plane. The particular sequence illustrated should be considered as one of several possible sequences with different periods of cooling and laser detuning and intensity. In practice, several other sequences, lapse times and tunings may be chosen. In the sequence shown, the laser beams, vertical and horizontal, are detuned to the red from the unperturbed transition frequency by 3 natural line widths (-3Γ) during the early period of cooling. This is called the *capture period*, identified as (1) in

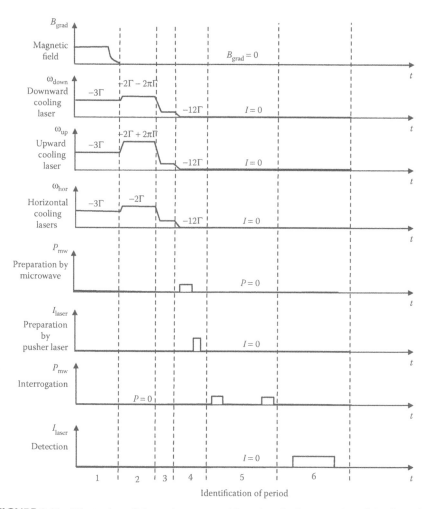

FIGURE 3.11 Illustration of the various steps taking place in the operation of the fountain. The diagram does not respect the actual length of the various periods. The various periods may be identified as follows: 1—A molasses is created in the MOT. 2—The atomic cloud created is sent up vertically by detuning of the vertical laser beams. 3—The cloud is cooled to a sub-Doppler limit temperature. 4—The atoms are prepared by means of state selection. 5—The atoms pass through the interrogating cavity on their way up and their way down. 6—Finally the state of the atoms in the falling cloud is analyzed.

Figure 3.11. It may last for a time between 0.1 and 1 s. The temperature of the atoms is reduced by both Doppler and Sisyphus cooling and may reach a few microkelvins. Near the end of that period, the magnetic field, creating a gradient for the operation of the MOT, is turned off. All the laser beams are then detuned by -2Γ, while the vertical upward- and downward-propagating laser beams are detuned by a small amount $+\delta v$ and $-\delta v$, respectively. This is shown as period 2 in Figure 3.11. During those early periods, repumping radiation between states $\left|S_{1/2}, F = 3\right\rangle$ and $\left|P_{3/2}, F' = 4\right\rangle$ is added to one or several of the horizontal beams. As explained previously, due to the detuning of the vertical lasers, the atoms find themselves in a reference frame moving at the same velocity as the moving standing wave mode of the two interfering radiation fields. During the launching phase the horizontal and vertical beams tend to warm up the atoms. To re-cool the atoms in their moving frame, the lasers frequency detuning is increased linearly to $\sim-12\ \Gamma$ and the intensity of the radiation beams is slowly reduced down a ramp. This is shown as period 3 in Figure 3.11. In the whole process, sub-Doppler cooling takes place through Sisyphus cooling as described in Chapter 2. Finally, the beams are blocked by mechanical shutters with a fall-time of about 1 ms. The final temperature of the cloud being sent up vertically is a few microkelvins corresponding to an atomic velocity of the order of 1 cm/s or so. This is small compared to the vertical speed of the cloud at the exit of the MOT, which is about 4 m/s. At that velocity, the cloud of atoms reaches a height of about 1 m before falling back under gravity.

It should be mentioned that the cooling of atoms in an MOT is not absolutely required. The system also functions with a cloud created by means of a molasses alone. In such a case, a laser radiation polarization lin $\perp$ lin configuration is used. The size of the cloud is larger to start with since it is not confined in space by the magnetic trap. It is created in the intersection volume of the cooling lasers and is thus of the order of 1 cm. The density of atoms in the cloud is then smaller than that existing in the case of a cloud created by means of an MOT. These properties have advantages in practice regarding frequency shifts, a question to be addressed later. Furthermore, the magnetic field needed to create the MOT, being relatively large, may cause other problems since it may be felt at the site of the detection cavity.

3.2.3.2 Preparation of the Atoms: Zone B

At the exit of the molasses, the atoms find themselves distributed equally among all Zeeman sublevels of the $F = 4$ level of the ground state. This ensemble of atoms could be used as such. However, atoms in field dependent sublevels do not contribute to the clock signal and may cause undesirable added noise in the detection. The cooling region is thus followed by a state preparation zone called B in Figure 3.8 and period 4 in Figure 3.11. This region is exposed to a weak magnetic field of the order a few mG that removes the degeneracy of the ground-state Zeeman sublevels. On their way up, the atoms first pass through a microwave cavity tuned at the hyperfine frequency $v_{hf} = 9.2$ GHz. The cavity is excited at that frequency by an external generator at a level such that during the time of the cloud passage, the radiation is perceived by the atoms as a π pulse for the transition to $\left|F = 3, m_F = 0\right\rangle$. Consequently atoms originally in state $\left|F = 4, m_F = 0\right\rangle$ exit that cavity in state $\left|F = 3, m_F = 0\right\rangle$ with other atoms distributed in all Zeeman sublevels $\left|F = 4, m_F \neq 0\right\rangle$. The cloud is then submitted to a radiation beam travelling at right angle to its trajectory and tuned at the frequency corresponding to the transition at the $\left|S_{1/2}, F = 4\right\rangle \leftrightarrow \left|P_{3/2}, F' = 5\right\rangle$. The atoms left in levels $\left|F = 4, m_F \neq 0\right\rangle$ are pushed out of the beam by transfer of momentum. In the case that cooling is done by vertical beams, this action can be done by means of those beams tuned properly. At the exit of that preparation zone, essentially all atoms in the cloud are in state $\left|F = 3, m_F = 0\right\rangle$.

3.2.3.3   Interrogation Region: Zone C

Essentially all Cs atoms are thus in state $|F = 3, m_F = 0\rangle$ and are launched up at a velocity around 4 to 5 m/s. In the frame of reference moving at that speed, their temperature is of the order of 1 μK. On their way up, they pass through the interrogation cavity, which is tuned to the hyperfine frequency in the TE_{011} mode. This is zone C in Figure 3.8. Energy is fed to the cavity at a frequency of 9.2 GHz, close to the hyperfine frequency, and the atoms are put in a superposition of the two ground states $|F = 3, m_F = 0\rangle$ and $|F = 4, m_F = 0\rangle$. If the microwave power and the launching speed are set appropriately, the atoms find themselves at the exit of the cavity in a state as if they had been submitted to a $\pi/2$ pulse. The cloud pursues its flight up the system, is slowed down under gravity and reaches a height L of the order of 1/2 m where it reverses its path. This is represented as zone D in Figure 3.8. At the height h the atoms fall back as in a fountain and pass again in the same cavity at the same speed as in their way up. This action corresponds to period 5 of the sequence illustrated in Figure 3.11. The atoms thus achieve a Ramsey interaction as in the case of the classical horizontal beam approach. However, the interaction takes place in the same cavity. If a longitudinal phase gradient exists in the cavity due, for example, to losses in the copper walls, its effect on the frequency of the standard is cancelled due to the reversal of the speed of the atoms in traversing the cavity. Thus, in principle, there should be no so-called end-to-end frequency shift caused by a cavity phase gradients. This statement is valid at the condition that the atoms retrace the same path in the cavity on their way up and down in their free flight.

3.2.3.4   Free Motion: Zone D

In this region the atoms are left free to move under gravity. They reach a maximum height and fall back.

3.2.3.5   Detection Region: Zone E

After the interrogation zone, the atoms fall in the detection area. The atoms exit from the cavity in a coherent superposition of the two states $|F = 3, m_F = 0\rangle$ and $|F = 4, m_F = 0\rangle$ the nature of which depends on the frequency tuning of the radiation applied to the cavity. The number of atoms in each of the two states may be determined in the following way (Chapelet 2008). The atoms cloud is excited by multiple laser beams in the form of thin layers travelling at right angle to the falling atomic cloud. The first beam, in the form of a standing wave excites the transition $|F = 4\rangle \leftrightarrow |F' = 5\rangle$ and the atoms traversing that beam emit fluorescence as a pulse of light reflecting the shape of the atomic cloud. This pulse is called the *time-of-flight detection pulse*. The number of atoms $N_{F=4}$ in level $F = 4$ is calculated from the area of this time-of-flight signal. Following that operation, the falling atomic cloud passes through a region where it is submitted to a laser travelling wave at the same frequency and again as a layer. The atoms in level $F = 4$ are pushed out of the falling cloud. The atoms are then submitted to a standing wave radiation field tuned at the frequency of the transition $|F = 3\rangle$ to $|F = 4\rangle$ and are pumped into level $|F = 4\rangle$. The cloud then falls in a region where they are submitted to the same radiation field as at their entrance in the detection region. The fluorescence pulse that results represents the number of atoms that have been pumped in that level and thus determines the number of atoms $N_{F=3}$ in level $F = 3$ in the cloud at the entrance of the detection region. The probability of transitions caused by the interrogating cavity is then obtained as:

$$P = \frac{N_{F=4}}{N_{F=4} + N_{F=3}} \tag{3.31}$$

It is then possible to analyze this probability as a function of various parameters, in particular, the frequency applied to the interrogating microwave cavity. The Ramsey fringes of the pattern obtained are very narrow, reflecting the time the atoms have spent in free zone D or the time sequence of the microwave pulses they have experienced on their way up and down to and from height h, which may be of the order of 0.5 m.

Figure 3.12 shows the Ramsey fringes observed with fountain FO2 designed and assembled at SYRTE, Paris, France. The insert of the figure makes explicit the experimental results appearing as points representing the transition probability deduced from fluorescence levels at the detector for a given frequency applied to the interrogating cavity. An excellent signal-to-noise ratio of approximately 5000/point is obtained.

3.2.4 PHYSICAL CONSTRUCTION OF THE Cs FOUNTAIN

In the following paragraphs we will describe a typical implementation of a fountain. We will use as example the Cs fountain developed at SYRTE, Paris, France.

3.2.4.1 Vacuum Chamber

A fountain is essentially composed of two parts: the cold atom manipulation zone and the interrogation zone. They are enclosed in a high-quality vacuum chamber.

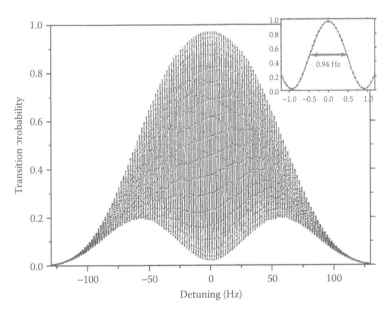

FIGURE 3.12 Ramsey fringes observed in SYRTE's FO2 fountain. The insert shows the central fringe, which has a width of 0.94 Hz corresponding to a period of free flight above the cavity of about 500 ms. (Courtesy of SYRTE Paris, France.)

The interrogation region is enclosed in a tube that provides a high temperature stability and homogeneity. In the particular fountain constructed at SYRTE, Paris, France, its diameter is 150 mm and its length is 700 mm. The top of this cylindrical tube is closed by a glass window with a diameter of 40 mm.

The atom manipulation zone is made of stainless steel. It consists of two parts: capture, Zone A of Figure 3.8, and detection Zone E of the same figure. The two zones are separated by 15 cm. The first part contains the glass windows for passing the cooling laser beams and observing the cold atom cloud. The second one contains the glass windows for the detection system. All glass windows have an antireflection coating and are soldered on the vacuum flanges. A vacuum tube connects the capture zone with the Cs source.

3.2.4.2 Microwave Cavity

One of the most critical parts of a fountain clock is the microwave interrogation cavity made of OFHC copper through which the atoms pass twice, once on their way up and again on their way down, providing a Ramsey type of interaction. That microwave cavity is cylindrical and operates in the TE_{011} mode. It has axial symmetry and a relatively high Q. That fundamental TE_{011} mode allows one to open relatively large holes in the end caps of the cavity, with only a slight perturbation of the electromagnetic field within. The fundamental mode of a TE_{011} structure has a small transverse phase gradient and very small magnetic field transverse components that could excite undesirable

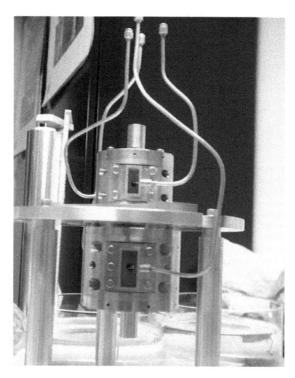

FIGURE 3.13 The dual Rb and Cs resonant cavities. (Courtesy of SYRTE, Paris, France.)

$\Delta m = \pm 1$ transitions. Figure 3.13 is an illustration of a double cavity fabricated at SYRTE, which is used in a fountain operating with both atomic species Rb and Cs.

3.2.4.3 Magnetic Field

A solenoid made of copper wire wound around an aluminium tube is used to produce the required static magnetic field. Three cylindrical magnetic shields made of μ-metal (thickness 2 mm) are placed around the solenoid. Each of them is closed by two end caps. In some designs, several compensation coils are placed on the end-caps of the magnetic shield to ensure the continuity and the smoothness of the magnetic field changes in those transition regions in order to avoid rapid field variations that might induce Majorana transitions. The coils are also used to improve the C-field homogeneity.

3.2.4.4 Temperature Control

A typical working temperature of the fountain may be 29°C. In the first implementation, a heating wire made of *ARCAP* (nonmagnetic and high resistivity) was wound in double spiral coils on an aluminium tube placed between the first and second magnetic shields. The temperature was regulated and maintained within ± 0.5°C. As the metal cavity resonance frequency has a temperature sensitivity of 150 kHz/°C, it is possible to use it as a sensor element, and a measurement of its resonance frequency allows the control of temperature in the interrogation region with a resolution of 10^{-3} °C. However, in a room well-controlled in temperature (within 0.5°), it is possible to operate the fountain without temperature control with satisfactory results. The approach has an added advantage of reducing temperature gradients within the structure (A. Clairon 2014, pers. comm.)

3.2.4.5 Capture and Selection Zone

In atomic fountains, gravity is used to accomplish a Ramsey interrogation type of interaction, the atoms passing twice through the same cavity. In this way, the frequency shift due to a difference between the phases of the microwave field in the two cavities used in the standard room temperature beam approach is greatly reduced. As the parasitic light from the detection and the cooling region can penetrate the interaction region and produce a frequency shift of the atomic resonance frequency as in the optically pumped beam standard, the laser beams used for manipulation of atoms are blocked by mechanical shutters during the flight of the atoms.

3.2.4.6 Detection Zone

The detection zone is situated approximately 15 cm below the capture centre. Three laser beams and two low-noise photodiodes are used in the detection process. Two ion pumps of 25 l/s each are used to produce a vacuum of 10^{-7} Pa.

3.2.4.7 Supporting Systems

3.2.4.7.1 Microwave

The microwave signal feeding the cavity for probing the atomic transition is either synthesized from a low-noise BVA quartz-crystal oscillator or from a cryogenic-sapphire oscillator (CSO, α Al$_2$O$_3$) (Vian et al. 2005), the latter exhibiting an extraordinary high frequency stability of $5.4 \times 10^{-16} \tau^{-1/2}$ for τ between 1 to 4 s and $2 - 4 \times 10^{-16}$ up to 800 s, before a slow drift sets in (Bize et al. 2004). In a typical operation, the voltage-controlled oscillator (VCO), or the CSO, is phase-locked weakly to an H maser which serves as a frequency reference for the fountain clock. The 9.2 GHz microwave signal for the interrogation is generated by means of proper multiplication and mixing of the VCO or the CSO frequency with the frequency of a synthesizer, usually of commercial origin.

3.2.4.7.2 Laser Beams

The radiation beams required for cooling, pushing, and detecting originate from solid-state diode lasers. These lasers are generally stabilized by means of techniques described earlier in Chapter 2, such as saturated absorption. The various laser frequencies and detuning required in the operation of the fountain are obtained by means of accousto-optic-modulators (AOM). The various components are mounted on an optical table and the lasers radiation is fed to optical fibres by means of couplers. The optical fibres are connected to the body of the fountains by means of special couplers and collimators that form beams with a waist of the order of 1 to 2 cm (Chapelet 2008).

3.2.4.8 Advantages and Disadvantages of a Pulsed Fountain

At this stage, we will mention some of the main advantages and disadvantages of the fountain in pulse operation. We will come back in more details later on some of them.

3.2.4.8.1 Frequency Stability

Advantages of the pulsed fountain regarding short-term frequency stability reside mainly in the good S/N obtained and the narrowness of the central Ramsey fringe due to the long time of interrogation as shown in Figure 3.12. Consequently the line Q is very large, of the order of 10^{10}. The number of atoms detected is of the order of 10^7 and the S/N may be of the order of 5000. The resulting short-term stability may be as high as $10^{-14} \tau^{-1/2}$, which is an improvement of the order of 20 to 30 over the room temperature optically pumped standard realized at SYRTE as reported in Table 3.1. On the other hand, a main disadvantage of the pulsed operation resides in the sensitivity of the short-term stability to phase noise of the microwave signal acting on the transition probability during only part of the cycle (Dick 1987).

3.2.4.8.2 Frequency Accuracy

In relation to frequency accuracy, the fountain has several advantages due to the narrowness of the resonance line observed and the temperature of the atoms in the cloud. The frequency of the fountain is relatively insensitive to cavity pulling and the second-order Doppler shift is reduced to 10^{-17} with an uncertainty of 10^{-18}. In pulsed operation, there is no light during the interrogation phase and the light shift can be completely eliminated. As the atoms pass twice the same microwave cavity with opposite velocity, the *end-to-end phase shift* effect no longer exists in a fountain clock. The only remaining effect is caused by the spatial variation of the phase of the oscillatory field in the cavity. Another advantage of the atomic fountain resides in the fact that it is possible to map the magnetic field as a function of height by launching the cooled cloud at different initial velocities. With this knowledge, it is possible to determine the second-order Zeeman frequency shift of the clock transition to an uncertainty better than 10^{-16}. This shift is of the order of 10^{-13}. It should also be mentioned that it is possible to vary the cycle time and the interaction time making possible various tests to improve frequency accuracy. One disadvantage of the low temperature (1 μK) involved in the atomic cloud of the fountain resides in the fact that the collision frequency shift becomes important. Calculation by Tiesinga et al. (1992) and his co-workers estimate this shift at $10^{-22}/(\text{atom/cm}^3)$ for Cs. For the transportable fountain clock at SYRTE, the shift is $<3.4 \times 10^{-15}$, and its uncertainty is $<5.8 \times 10^{-16}$ (Ghezali et al. 1996; Abgrall 2003).

It should finally be mentioned that the fountain is a rather complex device and that its operation requires an optical table for lasers operation and stabilization and that its construction is rather large since the path of the atoms in their way up and down is of the order of 1 m.

3.2.5 FREQUENCY STABILITY OF THE Cs FOUNTAIN

As made evident in Chapter 1, frequency stability and accuracy are the two most important properties of primary atomic frequency standards. We will examine the question of accuracy below with all known phenomena and perturbations that can affect it. In this section, we wish to examine the limit of frequency stability of such a fountain. The question was examined in detail in Volume 2 of *QPAFS* (1989) in connection to room temperature Cs beam standards. In particular, the question of transferring to a reference oscillator, such as a quartz oscillator, the inherent frequency stability of a resonance signal was examined in detail and the reader is referred to that book on that matter. In the case of the fountain the problem is different since we are now in the presence of a pulsed system, the resonance signal being obtained from a small ball of atoms in free motion with a repetition period of the order of one second. This subject was addressed in several articles (Audoin et al. 1998; Greenhall 1998; Santarelli et al. 1998, 1999) and we will outline the main characteristics of the resulting frequency stability when an external oscillator is locked to the fountain resonance line.

Frequency stability is limited by the presence of random fluctuations in the detection of the atomic resonance line and on the time behaviour of various biases that affect the frequency of the resonance. We will examine later the origin of those biases and their stability with time. For the moment we wish to concentrate on the random fluctuations on the number of photons arriving at the detector, or noise, as seen by the detection system. In the case of the fountain, frequency locking of a reference oscillator generating the interrogating radiation is done in the following way. In the first part of the locking cycle, a small cloud or ball of cooled atoms is prepared, sent up the instrument, passed twice in the microwave cavity on its way up and down, and is analyzed relative to atoms distribution in the two states $F = 3$ and $F = 4$ in the detector region. A continuous microwave interrogation signal is fed to the interrogation cavity. The frequency of that interrogation signal, ω_{osc}, may be displaced from the atomic resonance frequency ω_M (maximum of the central fringe) by a small value $\Delta\omega$. Furthermore, its frequency is square wave modulated at a low rate with an amplitude ω_m. This modulation is set equal to about half the line width of the central Ramsey fringe. With the interrogation frequency set on the low frequency side of the fringe, at $(\omega_{osc} + \Delta\omega - \omega_m)$, the number of atoms in level $F = 3$ and $F = 4$ is measured at the detector as described previously. The normalized number of atoms having made a transition is then determined by means of Equation 3.31. It is recalled that the number of atoms is determined by means of fluorescence photons counting. In the second part of the locking cycle, a second ball is prepared, sent up in free flight and the same process is effectuated with the microwave frequency set to the high frequency side of the central Ramsey fringe, that is $\omega_{osc} + \Delta\omega + \omega_m$. Again the normalized number of atoms having made a transition is measured at the detector. The difference in the two results indicates the agreement of the chosen ω_{osc} with the maximum of the Ramsey fringe. The process is repeated and the results can be averaged over several cycles. The signal obtained is processed digitally and an error signal is constructed as a function of the interrogation frequency, ω_{osc}. This error signal is then used to lock the frequency ω_{osc} of the generator to the maximum of the central Ramsey fringe making the detuning $\Delta\omega$ virtually zero. This is all done digitally by means of a processor that determines the sequence of all settings and operations.

Unfortunately, those measurements are affected by noise in the detection of atoms. It is important to identify properly the origin of that noise, which degrades the signal observed and affects the frequency stability as measured. Various contributions have been identified and expressions for those contributions have been presented by various authors. We provide in Appendix 3.A, for some of those contributions, an analysis describing important steps in the derivation of those expressions. The main noise components contributing to the instability of the fountain are discussed in the following sections.

3.2.5.1 Photon Shot Noise

In detecting a large number of fluorescence photons for identifying the resonance line, random fluctuations appear as shot noise in the number counted. This type of noise was studied in detail in *QPAFS* (1989) for various devices, in particular optically pumped Rb and Hg$^+$ standards. In fact, the frequency locking technique described in the case of Hg$^+$ trap is very similar to that of the fountain approach aside from the fact that the resonance line in the case of the ion trap was obtained through a continuous Rabi interrogation type rather than double pulse Ramsey interrogation technique. Shot noise is proportional to the number of photons $N_{at}n_d$ counted, where N_{at} is the number of atoms in the ball and n_d is the number of fluorescence photons n_{ph} emitted per atom multiplied by the efficiency of collection of the photons ε_c at the detector and the efficiency ε_d of the detector itself.

We define:

$$n_d = n_{ph}\varepsilon_c\varepsilon_d \tag{3.32}$$

The analysis done in Appendix 3.A gives the following expression for the frequency stability of the fountain as degraded by shot noise:

$$\sigma_{ySN}(\tau) = \frac{1}{\pi}\frac{1}{\sqrt{N_{at}n_d}}\frac{1}{Q_l}\sqrt{\frac{T_c}{\tau}} \tag{3.33}$$

3.2.5.2 Quantum Projection Noise

The frequency of the microwave interrogation signal is set off the maximum of the central fringe by about one half the linewidth. In that situation, the atoms find themselves in a superposition state of the two ground-state hyperfine levels. The probability of an atom being in either state is the same and the measurement process sends the atom in either state. In one interpretation, this process is said to collapse the wave function representing the state of the atom, or in another interpretation it may be said that the measurement projects the wavefunction on either of the state vectors forming the superposition state. In quantum mechanics, this process is undetermined and mathematically is represented by a probability. This leads to randomness in the emission process and noise in the detection. It is often called *projection noise*. It is proportional to the number of particles N_{at} in the ball. The contribution of this type of noise as calculated in Appendix 3.A is given by:

$$\sigma_{ySN}(\tau) = \frac{1}{2\pi}\frac{1}{\sqrt{N_{at}}}\frac{1}{Q_l}\sqrt{\frac{T_c}{\tau}} \qquad (3.34)$$

These results are similar to those originally derived by Wineland (Wineland and Itano 1981) in the context of an ion trap. In their case, the cloud of atoms was submitted to a Ramsey type of microwave interrogation similar to the one used in the case of the fountain. However, that cloud of ions was static and only the preparation was done in a discontinuous way.

It is noted that in the case that the number of photons n_{ph} emitted by each atoms in the ball is very large, n_d becomes large and the fountain frequency stability is controlled mainly by quantum projection noise.

3.2.5.3 Electronic Noise

Noise may also be internally generated in the detection components as well as in the amplifying electronics. Generally this noise is made negligible in a well-designed system by means of a proper choice of components.

3.2.5.4 Reference Oscillator Noise: Dicke Effect

Finally since the fountain operates in a pulse mode, the reference oscillator fluctuations may not be corrected or filtered by the servo system which in such a case has variable loop gain. As mentioned by Dick (Dick 1987), this effect has a very different character from those effects due to local oscillator fluctuations treated by Vanier et al. which are due to finite loop gain (Vanier et al. 1979). In the present case, in a pulse mode, high-frequency phase fluctuations of the interrogation oscillator may be down converted in the servo system and may appear as white frequency noise. This type of noise is important in the case of the fountain. When a quartz crystal oscillator is used as the reference oscillator in the synthesizer chain with an important $1/f$ noise contribution, the frequency stability of the fountain may be limited by this effect. It has been found that the use of a CSO as reference oscillator with low short-term noise improves the measured frequency stability considerably.

Characterization of the fountain is often done from a measurement of the *S/N* of the fringe observed. It is often desired to conclude on the expected frequency stability of such a fountain. An expression can be written for the frequency stability in terms of that S/N as obtained in the measurements:

$$\sigma_y(\tau) = \frac{2}{\pi} \frac{1}{Q_{at}} \left(\frac{S}{N}\right)^{-1} \sqrt{\frac{T_c}{\tau}} \qquad (3.35)$$

In practice, when measured at the maximum of the fringe (frequency ω_M), the S/N may be of the order of 1000. However, at the frequency $\omega \pm \omega_m$, at half the height of the fringe, the slope of the line shape acts as a discriminator of phase noise of the interrogating signal and the S/N is degraded to something like ~600. The calculated frequency stability is $\sigma_y(\tau) = 1.4 \times 10^{-13} \tau^{-1/2}$ in agreement with experimental data (Szymaniec et al. 2005). The expression leads to an obvious way of increasing frequency stability. The line Q can be increased by increasing the launching velocity, which increases the launching height. However, there are limits to that avenue. There is dispersion of the cloud when launched at a higher height reducing the number of atoms detected and the difficulty of generating clouds with high density. This last limit has been resolved to some extent by feeding with slow atoms, using a 2D MOT beam, or other similar means, the space where the MOT is formed. On the other hand, the effect of noise generated by the detection system is resolved by means of design and choice of low noise electronics.

The effect of the reference oscillator noise, unfortunately, is difficult to control. Most groups use very high quality synthesizers and quartz oscillators for generating the interrogating signal. However, real improvements can be achieved by using a CSO (Luiten et al. 1994), or still high quality sources such as optically stabilized microwave sources (Millo et al. 2009; Weyers et al. 2009) with low phase noise, although adding some complexity to the approach. Using the CSO approach, a fractional frequency instability of $1.6 \times 10^{-14} \tau^{-1/2}$ was realized at SYRTE, Paris, France, using that approach combined with a specially designed synthesizer characterized by low phase noise (Bize et al. 2004). The stability was limited by quantum projection noise (Santarelli et al. 1999).

Another strategy for reducing the influence of the Dicke effect is to accelerate the loading and preparation of the cold atomic cloud making possible the reduction of the fraction of the fountain time cycle where no atoms are in or above the microwave cavity. Loading times of the MOT (or molasses) can be greatly shortened when the atoms are collected from a slow atomic beam. For the same loading time, a molasses loaded from a slow beam captures more than 10 times as many atoms as a molasses loaded from the background vapour. Low speed atoms can be fed in the molasses region by means of a chirp-slowed atomic beam (Vian et al. 2005) or by means of a 2D MOT (Chapelet 2008) as mentioned above.

Finally, the ideal approach would be to operate the fountain in a continuous mode (Berthoud et al. 1998). This approach has been studied but is rather difficult to implement. It will be described below and recent results will be outlined.

3.2.6 RUBIDIUM AND DUAL SPECIES FOUNTAIN CLOCK

In implementing frequency standards, it is important to compare the behaviour of different atomic species under the same conditions or to use one as a reference for the other in order to identify possible fundamental biases or simply to verify the advantages of one species relative to the other. For example, comparison of H used in a maser and Cs used in a beam device showed the superiority of Cs regarding accuracy while H in a maser was better regarding frequency stability. In the case of the fountain, Rb is a natural other candidate because the atomic physics regarding that atom is very similar to that of Cs and because laser cooling and manipulation of Rb are just as conveniently possible as with Cs. Furthermore, in practice, the set-up and the operation of a Rb fountain clock are basically the same as those of a Cs fountain clock (Bize 2001).

One of the most obvious things to do is a determination of a Rb fountain output frequency with respect to that of a Cs fountain and the relative frequency stability of the two devices. Those measurements would first establish the quality of each standard as an actual time standard as well as their relative quality as a primary standard. Furthermore, various experiments could be done to determine some fundamental properties of the ratio of their hyperfine frequencies with time addressing the question of stability of fundamental constants.

The lower manifolds of levels of interest of the [87]Rb atom are shown in Figure 3.14, which can be compared to those of Cs shown in Figure 3.11. As can be seen, this level structure is slightly different from that of the Cs atom but it is readily observed that what has been described above for the Cs fountain, construction and functioning, can be transposed directly to the case of the implementation of a Rb fountain. A fountain was realized at SYRTE, Paris, incorporating all the components required to operate both a Cs and an Rb fountain in the same unit. A diagram of such a double fountain is given in Figure 3.15. It can operate either alternatively with [133]Cs and [87]Rb or with both atoms simultaneously. The environment, magnetic field and temperature, being the same a comparison of the two standards is relatively easy to address, the results having high credibility. The precise and repeated measurements of ν_{Cs}/ν_{Rb} have led to the adoption of the hyperfine frequency $\nu_{Rb} = 6{,}834{,}682{,}610.904312$ Hz for [87]Rb in

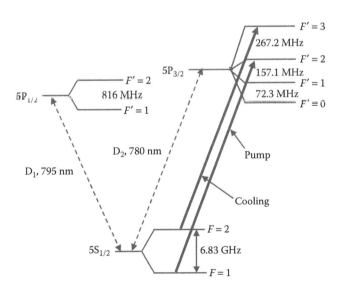

FIGURE 3.14 Lower energy manifolds of the [87]Rb atom.

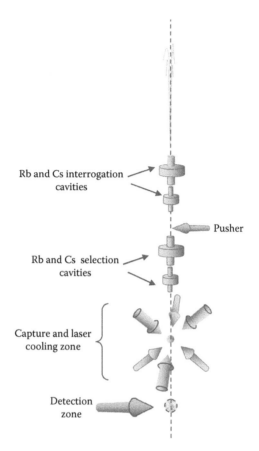

Rb and Cs interrogation cavities

Pusher

Rb and Cs selection cavities

Capture and laser cooling zone

Detection zone

FIGURE 3.15 Conceptual diagram of the double, Cs and Rb, fountain implemented at SYRTE, Paris, France. The detection zone is similar to that implemented for single species fountain. The cooling laser radiation beams are created by means expanding collimators situated at the entrance of the vacuum envelope and fed by means of optical fibres. The whole arrangement regarding vacuum, magnetic field and shielding is similar to that of the single fountain described earlier.

term of the SI definition of the second and as a secondary representation of the second with a relative uncertainty of 1.3×10^{-15} (CCTF 2004; Guena et al. 2014).

Figure 3.16 shows the actual dual Rb-Cs fountain designed and constructed at SYRTE, Paris, France. As in the single species Cs fountain, laser beams required for cooling the two species cross at right angle in the centre of the preparation zone, where the cold atomic cloud of each species is produced. The laser radiation fields are fed to the system by means of optical fibres. Laser frequencies and intensities required by the time sequence illustrated in Figure 3.11 are controlled on two optical tables by means of AOM and mechanical shutters. Microwave interrogation for each species takes place in a double cavity as the one shown in Figure 3.13.

FIGURE 3.16   Photograph of the dual, Rb-Cs, fountain. (Courtesy of SYRTE, Paris, France.)

The two atomic species of Rb and Cs are captured in the same region by dual optical molasses operating in a lin ⊥ lin configuration. For simultaneous operation, dichroic collimators are used (Chapelet 2008). Two separated optical benches are used to generate laser beams at 780 and 852 nm respectively for Rb and Cs. The optical molasses are created at the same region of space. Both species are launched at the same instant with a slightly different velocity, which is chosen to avoid collisions between the two atomic clouds during the entire ballistic flight.

3.2.7 FREQUENCY SHIFTS AND BIASES PRESENT IN THE FOUNTAIN

We have examined in detail in Chapter 1 the sources of frequency shifts in the classical room temperature Cs beam standard. Most of these shifts are still present in the fountain but some are reduced considerably. However, since the frequency of the fountain is very stable, some of the shifts that were not visible in room temperature Cs beam standards are now easily detected. Furthermore, since the accuracy of the fountain is much greater than that of those standards, some of the smaller frequency shifts have to be taken into account in the evaluation of its possible accuracy. We will now examine in detail the physics of those frequency shifts, determine

their value and discuss their effect on the overall performance of the fountain as a
primary standard.

3.2.7.1 Second-Order Zeeman Shift

In the fountain, since the interaction times in the cavities are long, the width of
the Rabi resonances is narrow and the system can operate with a smaller Zeeman
splitting of the ground-state levels than in conventional classical Cs standards. The
magnetic field applied is then set usually in the range of 10^{-7} T (~1 mG). The shift
Δv_{oo} of the frequency v_c of the clock transition $|F = 4, m_F = 0\rangle \rightarrow |F = 3, m_F = 0\rangle$ in
the presence of a small field is given by:

$$\Delta v_{00} = v_c - v_{HFS} = \frac{v_{HFS}}{2} x^2 \tag{3.36}$$

where x stands for:

$$x = \frac{(g_I + g_J)\mu_B}{E_{HFS}} B_o \tag{3.37}$$

B_o is the magnetic induction in Tesla and v_{HFS} is the hyperfine splitting in zero field.
When the constants are introduced in the equation, the shift is actually given by
Equation 1.21 of Chapter 1. A field of 10^{-7} T gives a fractional shift of about 5×10^{-14}.
This shift is very small compared to the shift that results for the field applied in the
classical approach, which as shown in Table 1.1 of Chapter 1 is 10^4 times larger.
Nevertheless, in order to reach frequency accuracy better than 10^{-16}, this magnetic
field has to be determined accurately.

The value of the magnetic field in the region of the free trajectory above the cavity can be determined by several means. This may be done by exciting a hyperfine
transition that is field dependent, such as $|F = 4, m_F = 1\rangle \rightarrow |F = 3, m_F = 1\rangle$. Ramsey
fringes are observed above a Rabi pedestal and the position of the central fringe of
this field dependent transition can be used to determine the average field seen by the
atoms. Neglecting second order terms, the frequency shift of the $\Delta m_F = 0$ transition
for the transition given above is given by:

$$\frac{\Delta v_{11}(B_o)}{v_{HFS}} = \frac{1}{4} x \tag{3.38}$$

Thus, if the field is homogeneous, a measurement of the frequency shift of the field-
dependent transition provides a value of the magnetic field through the path followed
by the cloud. In practice, the shift of the field independent line v_{oo} is evaluated by
combining Equations 3.36 and 3.38 to obtain:

$$\frac{\Delta v_{00}(B_o)}{v_{HFS}} = 8 \left[\frac{\Delta v_{11}(B_o)}{v_{HFS}} \right]^2 \tag{3.39}$$

However, the field may not be homogeneous. The shift of the field dependent
resonance is linear with the field and the position of the resultant fringe reflects
the time average of that field or $\langle \Delta v_{11} \rangle$. From Equation 3.36, the shift of the field

independent transition Δv_{oo} should be the time average of the square of the field and Equation 3.39 is not quite appropriate. This is the whole question about the actual difference between $\langle B_o^2 \rangle$ and $\langle B_o \rangle^2$, which causes an uncertainty:

$$\delta(\Delta v_{oo}) = 427 \times 10^8 \left(\langle B_o^2 \rangle - \langle B_o \rangle \right)^2 \tag{3.40}$$

This problem can be resolved by making a map of the field in the trajectory followed by the atomic cloud. It is thus important to know the actual value of the field at every point in the cloud and to do the appropriate averaging. In the classical thermal beam approach, the atoms travel horizontally at constant velocity. The average over time can be replaced by an average over space. However, in the case of the fountain, the atoms are launched vertically and the speed of the atomic cloud varies from a launch velocity of the order of 4 m/s to a value equal to zero at the top of the trajectory. The effect of the magnetic field on the clock transition is thus written as:

$$\Delta v_B = 427.45 \times 10^8 \langle B_z^2(t) \rangle_{traj} \tag{3.41}$$

Consequently, the time average takes place over length sections of various sizes due to the change in speed of the atoms. A map of the field can be done using the field dependent transition by adjusting the launch velocity of the cloud and thus varying the height that is reached by the cloud. For example, a given atomic cloud, n, may be launched to a given height and $\langle \Delta v_{11} \rangle_n$ can be measured for that height or launch n. A subsequent cloud, $n + 1$, may be launched at a larger speed to a height larger by a small amount and the frequency $\langle \Delta v_{11} \rangle_{n+1}$ measured again for that launch $(n + 1)$. The difference in frequency provides an indication of the variation of the field along the path and, interpreting the time averaging as a function of the changing velocity with height, a map of the field deduced from the frequencies measured may be made.

   Another method that could be used is the so-called time domain Zeeman spectroscopy. In that case the launching velocity is kept constant and the power is adjusted in the interrogating cavity to excite the atoms as if they were submitted to a π pulse. In the approach described previously in the standard operation of the fountain, the atoms are first prepared in state $F = 3$, $m_F = 0$ and enter the interrogating cavity in that state. A π pulse sends them into state $F = 4$, $m_F = 0$. An rf field at the Zeeman frequency corresponding to transitions $\Delta m_F = \pm 1$ is applied for a short time, say 10 ms, over the whole region of the free flight of the cloud. In the process, atoms are excited to levels $F = 4, m_F \neq 0$ at a given time corresponding to the height they have reached at the time of application of the rf pulse. In traversing the microwave cavity on their way down, the atoms in the cloud are submitted again to a microwave field of the same intensity and frequency as in their way up and are thus submitted to a π pulse again. The atoms that have not been excited by the Zeeman field are still in state $F = 4, m_F = 0$ and are excited back to level $F = 3, m_F = 0$. The measurement can be completed at the detector by exciting atoms in the cloud at the cycling transition frequency $F = 4$ to $F' = 5$ and measuring the fluorescence. The intensity of the fluorescence is an indication of the number of atoms that have been excited to $m_F \neq 0$ levels. The measurement can be done as a function of the delay between the cloud arrival and the time the Zeeman pulse was applied. A measurement of the applied Zeeman frequency for which the effect on the fluorescence is a maximum gives a direct indication of the value of the magnetic field at the height of the atomic cloud when it was excited. Consequently a map of the magnetic field in the region above the interrogating cavity can be done as a function of time delay providing an absolute value of the magnetic field at the number of points desired.

Another somewhat related technique consists in operating the fountain normally and excite Zeeman transitions within state $F = 4$ between levels $m_F = 0$ and $m_F = \pm 1$ like is done, for example, in the determination of the field in the H maser described earlier. An rf pulse, say of duration of 100 ms, exciting such transitions is applied at the time the atoms reach their maximum height (Meekhof et al. 2001; Levi et al. 2004, 2009). When the frequency of the applied rf pulse is resonant with those transitions, the 0-0 Ramsey resonance is affected, the coherence of the atoms in their transit time above the interrogation cavity being destroyed. The effect on the amplitude of the central field independent Ramsey fringe is then measured as a function of the rf frequency applied. This is done for various launching velocities and the time of application of the rf pulse is correlated to the launching velocity in such a way as to be applied at the atoms apogee in their free flight. A detailed map of the magnetic field in the region above the interrogating cavity is thus obtained and Equation 3.40 can be used to obtain the magnetic field shift of the clock transition. The method is rather efficient but care should be taken on the interpretation of the data since the atomic cloud has a spread in space and an integration of the field over space and time duration of the pulse is present.

It is recalled that the transverse shielding factor of cylindrical magnetic shields is generally larger than their longitudinal factor. Consequently, a vertical construction, which orients the shield cylinders nearly parallel to the earth's field, is less efficient than a horizontal construction used in the case the thermal beam approach described in Chapter 1. However, it is found that field inhomogeneities in a well-designed fountain do not cause major problems. This is accomplished by using sufficient shielding layers (>3), end caps and possibly correction coils near those caps to compensate for end effects of the solenoid. In practice, although inhomogeneities and magnetic field are present, they are found to be small. For example, in a particular system constructed at the National Physical Laboratory (NPL), UK, it is found that for a field of say 176 nT, the variance of the field may be as low as 0.35 nT and the resulting bias introduced by the field inhomogeneity is less than 10^{-18} (Szymaniec et al. 2005). Similar results were obtained by other groups. Fluctuations of the magnetic field with time (which may be caused by thermoelectric currents in the region of the cavity) may be more important causing shifts of the order of 5×10^{-17}, but this is still small.

3.2.7.2   Black Body Radiation Shift

We have shown earlier that electric fields have the properties of altering the internal energy levels structure of atoms, either lifting degeneracy or shifting those levels. It is the well-known Stark effect. Both dc and ac electric fields have an effect. In the fountain, atoms are also submitted to environmental radiation called Black Body radiation (BBR). It causes an ac Stark effect and shifts the ground-state hyperfine levels of alkali atoms. It is one of the largest biases and a significant source of uncertainty in either Rb or Cs atomic fountains. We now outline our present knowledge on the subject and its effect on the accuracy of the fountain.

3.2.7.2.1   Static Stark Effect

The calculation of such an effect on the hyperfine levels of H and alkali atoms was summarized in Volumes 1 and 2 of *QPAFS* (1989). An applied static electric field E_o shifts those energy levels. The effect may be interpreted as an interaction that creates an admixture of the excited states, in particular P states, into the ground S state. This causes a different shift for each of the hyperfine levels creating a shift of the hyperfine frequency by the amount:

$$\left.\frac{\Delta\nu_{hf}}{\nu_a}\right|_{st} = k_{st}E^2 \tag{3.42}$$

where:
k_{st} is the static Stark effect coefficient

That coefficient is a function of the polarizability of the atoms. It is generally calculated in a perturbation approach intermixing upper states with the hyperfine ground state. The accuracy of the theoretical determination of the effect of a static electric field on the hyperfine frequency depends on the extent of that summation over excited states. For example, Anderson (1961) limits the choice to those excited states that yield a proper value when used in the calculation of the polarization α_e. Such a calculation yielded a fractional hyperfine frequency shift for Cs equal to $-3.5 \times 10^{-20} E^2$ while a fractional shift of $-2.44 \times 10^{-20} E^2$ was measured (Mowat 1972). The coefficient given above is often expressed in absolute frequency units. In the case of Cs, Mowat's measurement gives $\Delta\nu_{hf} = -2.25 \times 10^{-10} E^2$ (Hz). A somewhat more elaborate analysis using a different approach consisting of perturbed wave functions and differential equations (Lee et al. 1975) gave a value of k_{st} equal to 2.23×10^{-10}, while still more elaborate perturbation calculations extending to higher continuum states gave k_{st} equal to $2.26\,(2) \times 10^{-10}$ (Angstmann et al. 2006), and $2.271\,(8) \times 10^{-10}$ (Beloy et al. 2006). Calculations made by Micalizio et al. (2004) giving a smaller value for k_{st} were recognized as having a limited accuracy since they were not extending to continuum states.

3.2.7.2.2   ac Stark Effect and BBR

An important consideration is that an oscillating electric field, whose frequency is low relative to the frequency of the transitions from the ground state to the excited states of the atom, causes a similar effect. It is then assumed that this ac Stark shift is proportional to the average value of the square of the oscillating electric field, $\langle E^2(t) \rangle$. Consequently, BBR being electromagnetic radiation in thermal equilibrium with a Black Body at temperature T and being always present in the region of the fountain where the atoms evolve is expected to cause such an effect. That radiation has a spectrum of energy per unit volume in interval v and $v + \delta v$ given by:

$$\rho(v)dv = \frac{8\pi h v^3}{c^3\left[\exp(hv/kT)-1\right]}dv \tag{3.43}$$

where:
  h is Planck's constant
  k is Boltzmann constant

The radiation satisfies the criterion that most of the energy is concentrated at low frequencies. In fact, at room temperature, the energy spectrum is maximum around 30 THz (10 μm), much lower than the frequency corresponding to transitions from levels that cause an admixture of the excited states into the ground state, which in Cs are above 350 THz (850 nm). In such a case, one can assume that the BBR frequency shift of the hyperfine frequency is given by the mean square value of the BBR electric field. Equation 3.42 is then written as:

$$\left.\frac{\Delta v_{hf}}{v_{hf}}\right|_{BBR} = k_{BBR}\left\langle E^2(t)\right\rangle \tag{3.44}$$

where $\left\langle E^2(t)\right\rangle$ is the mean square value of the BBR electric field and k_{BBR} is the Stark coefficient that is calculated as in the case of a static electric field. The problem is thus reduced to an evaluation of k_{BBR}. A first approximate evaluation can be done by calculating directly the mean square value of the BBR electric field and using k_{BBR} as the measured dc value. The total energy per unit volume is obtained from Equation 3.43 and is equal to the energy in the electric and magnetic fields of that radiation. Using the property that the stored magnetic and electric energies are equal and integrating Equation 3.43 over all frequencies, we obtain the root mean square value of the electric field as (*QPAFS* 1989, Vol. 2):

$$\left\langle E^2(t)\right\rangle^{1/2} = 831.9\left(\frac{T}{300}\right)^2 \tag{3.45}$$

It should be mentioned that in the analysis it is found that the contribution of the magnetic component of the BBR radiation does not cause a shift that is important in the context of the accuracy reached with room temperature beam frequency standards and even fountains. The shift calculated is of the order of a few in 10^{-17}.

We could use the coefficient calculated by Anderson for k_{st} to calculate k_{BBR}. However, a better approximation consists in using the coefficient measured experimentally by Mowat as reported above for the shift of the hyperfine frequency caused by a dc electric field. The result is:

$$\left.\frac{\Delta v_{hf}}{v_{hf}}\right|_{BBR} = -1.69\times10^{-14}\left(\frac{T}{300}\right)^4 \tag{3.46}$$

This is a rather large shift and for a desired accuracy of 10^{-16}, the question of accuracy of our knowledge of coefficient k_{BBR} must be addressed. Fortunately, more exact calculations than the empirical one just outlined using an experimental result have been done. In particular, it was found that the frequency distribution of the BBR introduced a small correction factor ε to the shift (Itano et al. 1982). The shift can actually be written as:

$$\left.\frac{\Delta v_{hf}}{v_{hf}}\right|_{BBR} = \beta \left(\frac{T}{300}\right)^4 \left[1 + \varepsilon \left(\frac{T}{300}\right)^2\right] \tag{3.47}$$

where

$$\beta = \frac{k_{st}}{v_{hf}}(831.9)^2 \tag{3.48}$$

As mentioned above, accurate theoretical calculations have been done for several ions and alkali atoms. We reproduce the results obtained by Angstmann et al. (2006) and Beloy et al. (2006) in Table 3.2 for Rb and Cs.

As is readily seen, the factor ε introduces a small correction a little larger than 1% or one part in 10^{-16}, which is at the limit of the accuracy desired. However, the main term β causes a frequency shift to the order of parts in 10^{-14} and is rather important. This shift has to be known to better than 1% in order for the standard to reach an accuracy goal of 10^{-16}. Furthermore the shift depends on temperature to the fourth power and an error in T of 1 K represents a change in frequency of 2.26×10^{-16}. It should also be realized that temperature gradients as well as exposure to outside radiation traversing holes in the fountain structure could also play a role in the final accuracy reached.

In principle, the above analysis could be used to determine the bias caused by BBR simply by measuring accurately the temperature around the region of the atomic cloud free flight. The result would be used in the determination of the accuracy of a fountain in implementing the definition of the second in terms of the defined hyperfine frequency of Cs, 9,192,631,770 Hz now established as a definition at 0 K (BIPM. 2006). Thus, in principle, for a standard operating at a room temperature of 300 K, a correction of ~1.73 (1) $\times 10^{-14}$ would need to be applied to the measured

---

TABLE 3.2

Values of the Parameters Characterizing the Hyperfine Frequency Shift Caused by BBR for Rb and Cs[a]

Atom	$k_{st} \times v_{hf}$ (10^{-10})	β (10^{-14})	ε	Reference
^{87}Rb	−1.24 (1)	−1.26 (1)	0.011	Angstmann et al. (2006)
^{133}Cs	−2.26 (2)	−1.70 (2)	0.013	Angstmann et al. (2006)
^{133}Cs	−2.271 (8)	−1.710 (8)		Beloy et al. (2006)

[a] The numbers in parentheses are the claimed accuracy in the calculation.

frequency. The contribution of BBR to the inaccuracy of the standard would be ~1 × 10^{-16}. However, that approach relies essentially on theoretical calculations and in order to be able to use the above analysis to determine the actual accuracy of a particular fountain, it is important to verify somehow if the above result is in agreement with experimental data.

A high accuracy measurement of the coefficient k_{st} was made early in LNE-SYRTE FO1 Cs fountain. These measurements where made by applying electric fields ranging from 50 to 150 kV/m with metallic plates in the region of free flight of the atoms (Simon et al. 1998). Those fields are significantly higher than those corresponding to the BBR field at 300 K, which is 831.9 V/m rms as shown above. Deviations from a purely quadratic dependence on the electric field were also investigated. The test did not show any such deviations. Other measurements (Zhang 2004; Rosenbusch et al. 2007) at field strengths ranging from 1.5 to 25 kV/m, confirmed those early measurements. The experimental value for the static Stark coefficient was determined finally as: $k_{st} = -2.282(4) \times 10^{-10}$, in agreement with the theoretical value reported in Table 3.2. It is worth mentioning also a measurement of k_{st} by means of CPT resonances in closed cells containing a buffer gas (Godone et al. 2005). The experimental result was $k_{st} = 2.06(6) \times 10^{-10}$, a value significantly different from the value reported above on Cs beam standards.

Those measurements were made directly by applying a dc field to the ensemble of atoms. The form of Equation 3.47, however, leads naturally to a verification of the theory by varying the temperature of the environment in which the ensemble of atoms is embedded. Neglecting for the moment the effect of ε (T^6 dependence) the slope of the variation of the hyperfine frequency against temperature at the 4th power gives directly the value of β. One such experiment was done by means of a classical Cs beam frequency standard operating at room temperature (Bauch and Schröder 1997). The unit was equipped with cylinders enveloping the Cs beam and their temperature was varied. The result obtained is: $\beta = 1.66 (20) \times 10^{-14}$. The stability of the standard operating at room temperature did not allow a better accuracy but, nevertheless, the result is in agreement with the theoretical result of Table 3.2. Another determination was also done with a fountain using a similar approach and varying the temperature of the environment surrounding the atoms in their free trajectory by means of a graphite tube surrounding the region of free flight of the atoms (Rosenbusch et al. 2007). The range of measurements extended from room temperature to 440 K. The result fitted well a T^4 dependence and k_0 was found equal to 2.23 (9) × 10^{-10} or a fractional 1.68 (7) × 10^{-14}, a value in agreement with the theoretical value within experimental error. However, in a similar experiment, Levi et al. (2004) found $\beta = -1.43 (9) \times 10^{-14}$, a value significantly smaller, which raised questions.

In view of those experimental results, it appears that the BBR shift is a large frequency shift, but its value, as will be confirmed below appears to be known with an uncertainty comparable to other shifts present in the fountain. However, Equation 3.47 suggests that if a fountain could be operated at very low temperatures (cryogenic), this shift could effectively be made very small and be evaluated with a negligible uncertainty. In such a case, due to the 4th power dependence of the shift on temperature, this shift in a fountain with the Ramsey cavity and free flight region operated at 77 K would be reduced by a factor of 230, relative to a room temperature unit. This would make the uncertainty negligible on the determination of the shift. Such a fountain was developed at the National Institute of Standards and Technology (NIST), USA, and the Istituto Nazionale di Ricerca Metrologica (INRIM), Italy (Levi et al. 2009; Heavner et al. 2011). In those fountains, the source and detection regions are maintained at room temperature. The magnetically shielded interrogation region is enclosed in a liquid nitrogen Dewar vessel and operates around 80 K. The microwave cavities are tuned to be resonant at the cryogenic operating temperature. The total BBR shift at that temperature can be evaluated by means of Equation 3.46 and is found to be of the order 10^{-16} with uncertainty in the 10^{-18} range. All other biases outlined in this section are present in such a fountain and can be evaluated in the same standard approach as in a room temperature unit.

Three fountains were used in precise measurements of their relative frequencies: NIST-F1 was operated at 317.35(10) K, NIST-F2 at 81.0 K at NIST, USA, and IT-CSF2 at 89.4(10) K in INRIM Italy (Jefferts et al. 2014). In NIST, an H maser was used in the comparison of the two fountains. The comparison of the two cryogenic units, at NIST, USA, and INRIM, Italy, was done by means of two-way time transfer. It should be mentioned that extreme care was taken in the temperature determination of the various regions of the interrogation section. In particular, the effect of penetration of external radiation through windows was evaluated as well as emissivity of various regions. Each unit was the object of precise determination of accuracy and were reported in Levi et al. (2014) and Heavner et al. (2014). Some of the results are reproduced in Table 3.3. Measurements extended over several years. It appears that the question of the BBR shift was resolved to a level of accuracy that answered the question raised above. The value of β was evaluated as $\beta = -1.719$ (16) $\times$ 10^{-14} in agreement with the calculated value of Beloy et al. (2006) and the measurement of Rosenbusch et al. (2007).

3.2.7.3 Collision Shift

In the fountain, a small cloud containing 10^7 to 10^9 laser-cooled atoms is used. The diameter of the cloud may be a few millimetres in the case of a MOT or 1 cm in the case of a molasses. The temperature inside that cloud is of the order of several μK, which results in an average speed of the order of a few centimetres per second. The original density within that cloud is thus 10^7 to 10^8 atoms/cm^3. That cloud may be visualized as an ensemble of atoms in the same sense as the ensemble of atoms contained in a cell studied earlier in the case of a passive standard such as the optically pumped Rb standard. In such an approach, the ensemble of alkali atoms, Cs and Rb, are seen as a vapour in a vacuum environment, moving at random and colliding with each other at a rate that is a function of their relative speed (temperature), their density and the cross section appropriate for the phenomenon considered. If the vacuum is not good enough, residual gases are present and the alkali atoms may collide with those parasitic atoms with the consequence that relaxation and frequency shifts may result as in the case when the alkali vapour is in a buffer gas. Usually the residual pressure in the system is so low that this effect does not contribute to the inaccuracy of the clock within the accuracy presently reached and the effect of residual gases is neglected. However, although the density of atoms within the cooled cloud is low, the collision rate, due to the high accuracy reached in the fountain, may be large enough to create visible effects either on the coherence created by the Ramsey interrogation scheme or on the hyperfine frequency itself. It may be mentioned that the analysis is complicated by the fact that the atom density changes continuously during the path of the atoms, due to expansion of the atomic cloud in its free flight. In fact, the density in the cloud decreases by more than a factor 10 during its flight between its successive passage in the cavity and consequently the interaction is much more important in the first part of the flight.

This is a similar situation as the one we have studied in detail in *QPAFS*, Volume 1 (1989), which we have applied in the analysis of the functioning of various frequency standards such as the H maser, the Rb passive standard based on the double resonance technique or CPT approach, as well as the Rb maser. The phenomenon was introduced in the context of so-called spin-exchange interaction and we have studied the phenomenon both in the classical and in the quantum mechanical approach. The general idea is as follows. At high temperature, two alkali atoms approaching each other may find themselves in either a repelling or an attracting state depending on the symmetry property of the combined wavefunction including orbital and spin coordinates. They form a triplet or a singlet state, which are usually represented by Lennard–Jones potentials or so-called 6–12 potentials. The collision leads to the creation of a phase shift in the wave function, whose size depends on the form and amplitude of the potential energy of these two states. The phenomenon is interpreted as spin-exchange since, when a collision leads to a phase shift of π, the wave function is altered in such a way as if the unpaired electron of each atom had been exchanged. In the analysis made in that context, the atomic ensemble was assumed to be at room temperature corresponding to an energy kT of about 4×10^{-21} J. In the present case, however, the temperature of the atoms in the laser-cooled cloud is in the μK range and the energy is of the order of 10^{-29} J. The separation of the hyperfine levels in the ground state corresponds to an energy of 6×10^{-24} J. This consideration leads to the conclusion that the energy of the colliding atoms is not sufficient to cause large phase shifts that would cause transitions between the hyperfine levels. In that case, a collision introduces only a small phase shift in the wavefunction of the atoms in the hyperfine ground-state energy levels. Calculation then shows that those phase shifts, being random, introduce a broadening of the hyperfine resonance and a small average frequency shift without affecting the populations. Identifying the two hyperfine states as a and b, the effect on the density matrix may be written as (Cohen-Tannoudji et al. 1988) (see also Equations 1.42 and 1.43 of Chapter 1):

$$\frac{d}{dt}\rho_{\alpha\alpha} = \frac{d}{dt}\rho_{\beta\beta} = 0 \qquad (3.49)$$

for the populations and for the coherence:

$$\frac{d}{dt}\rho_{\alpha\beta} = nv_r(\sigma + i\lambda)\rho_{\alpha\beta} \qquad (3.50)$$

We recall that n is the atomic density and v_r is the relative velocity of the atoms. Broadening is caused by the first term of Equation 3.50 and is proportional to the cross section σ. This term is essentially a coherence relaxation term. The second term is imaginary and introduces a frequency shift. It is proportional to the cross section λ. Early measurements on a Cs fountain showed that the frequency shifts were larger than expected (Gibble and Chu 1993; Ghezali et al. 1996). In view of their importance in connection to primary standard applications, interest was raised on exact values of those cross sections and there was a fair amount of work done on their actual evaluation (Kokkelmans et al. 1997). In particular, the interest was raised in connection to the evaluation of the effect for different isotopes since the size of the effect was rather sensitive to the interaction potential determining the so-called scattering length (see Appendix 3.B). For example, the frequency shift $\Delta v/v$ for Cs was predicted to be of the order of 20×10^{-13} in standard operation while for ^{87}Rb its value was predicted to be 15 times smaller. The choice of Cs as the basis of the definition of the second was then questioned and work was initiated on the construction of a fountain using ^{87}Rb in SYRTE, Paris, as described earlier (Sortais 2001; Bize et al. 2009).

Traditional methods used to evaluate that frequency shift consist, as indicated by Equation 3.50, of measuring the hyperfine frequency as a function of the cloud density n, which can be altered experimentally in some particular way and then extrapolated to zero density. The density is usually measured by means of the fluorescence emitted in the detection zone. The problem with that technique, which has been used by many laboratories, is that an alteration of the density of the fountain cloud is usually accompanied by a change of the velocity of the atoms in the cloud and its distribution. This inevitably leads to an inaccurate extrapolation to zero density, especially in the case that the collision shift depends critically on the collision energy. Various experiments have shown that in practice the experiment is not clean and that it is difficult with that method, to determine the collision shift to better than 5%–10%. Since, as mentioned above, the shift expected in Cs for example is of the order of 17×10^{-13}, the resulting accuracy would be of the order of 10^{-13} or slightly better. This is far from the 10^{-16} accuracy hoped for.

SYRTE developed a new technique allowing the preparation of a pair of two successive samples or balls with relative densities very close to 2 (Pereira et al. 2003). The technique could then be used with pairs of samples of different densities, for verifying the reproducibility of the results. This accurate density ratio of 2 was achieved in the following way. First the atoms in the cloud were set in a superposition of the two states $F = 3, m_F = 0$ and $F = 4, m_F = 0$ in the selection cavity shown in Figure 3.8. This was achieved by means of an appropriate microwave intensity and the use of a pushing laser radiation beam tuned to the transition connected to the $F = 4, m_F = 0$. We have shown earlier that it was possible to select only atoms in state $F = 3, m_F = 0$ by this scheme. However, in the simple technique described, the number of atoms is sensitive to the amplitude of the field in the selection cavity. Since this field is a function of position within the cavity, the state superposition obtained by simple passage within the cavity is not perfect. It is possible, however, to obtain a quasi perfect superposition of states by means of adiabatic fast passage using a pulse (called Blackman pulse) which has a special temporal shape and a frequency sweep that is turned off exactly at half resonance (Leo et al. 2000; Marion 2005). The advantage of this method is that it allows the preparation of successive atomic samples at high and low densities in a strict ratio equal to 2. The value of this ratio is largely insensitive to the amplitude of the microwave field present in the cavity. The only critical parameter in that method of preparation is that the microwave frequency sweep must be interrupted exactly at half resonance. The uncertainty obtained on the evaluation of the collision displacement is then between some 10^{-16} and 10^{-17} depending on the original atomic density used.

Other laboratories, NPL and PTB have proposed operation of the fountain in the conditions where the displacement of frequency vanishes (Szymaniec et al. 2007). This method uses the fact that the displacement of frequency for Cs depends on the collision energy in the atomic cloud and the population of the two hyperfine levels. This method is very sensitive to parameters of the fountain (geometry of the cloud of cold atoms, launching height, ratio of population ...) and a priori is used with a MOT as an initial source of cold atoms.

In order to decrease the collision shift without reducing the number of detected atoms, two fountain configurations have been proposed. The first is called the *juggling fountain* (Legere and Gibble 1998). Like in a human juggler exercise with balls, several atomic clouds are launched in succession with a time separation smaller than the flight time of an individual cloud. Each cloud is less dense than in the standard fountain and, consequently, the internal atoms collision rate is reduced as desired. The multi-ball scheme requires a precise control of the launch times, velocities and densities of the individual balls (Fertig et al. 2001). Another scheme has been proposed by Levi et al. (Levi et al. 2001). In this configuration, the balls are launched in such a way as to never meet in the free flight zone above the cavity where collisions would lead to a frequency shift. However, they all come together in the detection zone—where cold collisions no longer matter—to produce a strong signal. A technical difficulty is that a fast mechanical shutter must be constructed inside the vacuum system in order to protect the balls that have already been launched from the stray light produced while cooling and launching the next balls (Jefferts et al. 2003).

3.2.7.4   Cavity Phase Shift

Another phenomenon that affects the accuracy of the fountain is the cavity phase shift. Since the atoms pass in the same cavity in opposite directions, in principle the frequency shift introduced by the cavity end-to-end phase shift is absent. However, a distributed cavity phase shift (DCP) or a gradient of phase is present in the cavity. Depending on their trajectories, the atoms may not experience on their way down the same phase as in their way up: this corresponds, as we have seen earlier in room temperature Cs beam standards, to a residual first-order Doppler shift, which arises mainly from the inhomogeneities of phase in the interrogation cavity. It is realized that this effect makes the fountain sensitive to tilt and power. Although DCP shifts have been considered for many decades, the effect has remained a major obstacle to the improvement of the accuracy of the room temperature Cs beam standard. We have addressed in some detail the question in Chapter 1. In fact, agreement between measurements and theory has been marginal (De Marchi et al. 1988). On the other hand, progress on this subject in the case of the fountain has also been marginal because the phase of the field in the cavity cannot be mapped accurately. Furthermore, the calculations are difficult since the holes in the cavities, required for the passage of the atomic cloud, produce deformation of the field lines. It should be noted that, although the effect is visible in fountains because of their high accuracy, it is rather small and its measurement, in the presence of other perturbations, is rather delicate. On the other hand, an accurate calculation of the actual phase distribution in the cavity and its corresponding frequency shift is a challenging numerical problem. Nevertheless, recent calculations and experiments have been done in order to elucidate the problem as well as possible (Li and Gibble 2004, 2010, 2011; Guéna et al. 2011; Weyers et al. 2012).

Fountains clocks, in general, use cylindrical TE_{011}. Those are centimetre-size cavities with a microwave feed at the cavity mid-plane and holes in the end caps to let the atomic ball pass through as shown in Figure 3.17.

In the calculation, the field within the cavity is essentially treated as a sum of two components, one large standing wave and a small travelling component that is created by the cavity losses. That last component introduces localized phase shifts that are seen by the atoms in their path. In the analysis, the small field component is expanded in a series of which three terms are kept: one term represents power fed and being lost in the end caps of the cavity causing longitudinal phase gradients; a second term produces a transverse gradient made explicit in Figure 3.17. This term is very sensitive to a tilt of the cavity relative to the direction of the atom travel and is the term most studied. It is especially important in the initial passage of the atomic ball in its travel up, since the ball is small, well-localized and samples a small portion of the cavity field. The last component considered is that connected to the phase variation when energy is fed from two opposite directions as shown in the figure when the dotted feed is introduced opposite to the initial feed.

A rather complete investigation of the question was done at SYRTE in order to elucidate the experimental aspect of the problem (Bize et al. 2009; Guéna et al. 2011). For example frequency measurements were done with one feed as a function of the tilt angle between the atomic ball path and the cavity. Other measurements

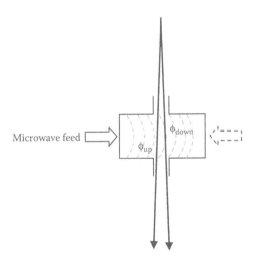

Microwave feed ϕ_{down} ϕ_{up}

FIGURE 3.17 Illustration of the way a cavity phase shift is introduced in the fountain due to the fact that the atoms do not replicate the same path on their way up and down and a gradient of phase exist within the cavity depending on the direction of the microwave feed. The figure shows the approach in which a second feed (dotted) can be introduced from an opposite direction.

were done with two feeds and relative power fed to these feeds. It is found that the shift can be evaluated with an accuracy somewhat better than 1×10^{-16}. It remains thus an important component of the total accuracy budget of the clock.

3.2.7.5 Cavity Pulling

The microwave cavity used for implementing a fountain must be tuned exactly to the atomic resonance frequency otherwise a frequency shift is observed in the detection of the clock transition. The apparent frequency of the atomic transition depends on the frequency offset between that of the cavity and the atomic resonance frequency. We have outlined this effect in Chapters 1 and 2, called generally *cavity pulling* for the case of various frequency standards as the room temperature Cs beam and the H maser. The question was addressed in detail in *QPAFS*, Volume 2 (1989). There are various ways by which a cavity detuning can affect the frequency of the detected resonance. A bias is introduced by the simple fact that if the cavity is detuned from the hyperfine frequency, the central Ramsey fringe appears distorted when observed by a scan of the applied frequency. This is due to the fact that the amplitude of the microwave field, as seen by the atoms, changes with the frequency applied, since for a detuned cavity the hyperfine resonance is situated on one side of the cavity mode resonance with a steep slope. The maximum of the Ramsey fringe is thus seen as displaced relative to the position it would have if the interrogating field amplitude were constant with frequency. On the other hand, if the detection of the Ramsey fringe is done by means of square wave frequency modulation of the applied microwave with a depth $\delta\omega$, the amplitude of the microwave at the two frequencies $\omega \pm \delta\omega$ will be different and the centre of the resonance as detected with a synchronous detector will appear displaced. These two effects were mentioned earlier in the case of the room temperature Cs beam standard. In the case of the fountain, these effects are very small because of the narrowness of the central Ramsey fringe relative to the width of the cavity resonance. Proper tuning of the cavity makes them negligible in comparison to the shift caused by the coupling of the magnetization to the cavity mode analyzed above.

In practice, cavity pulling exists when the magnetization created by the atomic ensemble becomes important and perturbs substantially the interrogation field introduced in the cavity. In *QPAFS* (1989), two cases were differentiated: when the detection of the resonance is done either by means of the microwave radiation and when it is detected by means of the population of the levels. The effect is sensitive to the atomic gain of the device considered, which is a function of the magnetization created in the cavity by the atomic ensemble at the hyperfine frequency. For example, in the case of the H maser, the hyperfine resonance is detected by means of microwave radiation. The magnetization is large, self-sustained, and the cavity pulling is simply proportional to the cavity detuning from hyperfine resonance multiplied by the ratio of the cavity Q to the atomic line Q, which is written as:

$$\Delta\nu_{\text{maser}} = \frac{Q_c}{Q_{at}} \Delta\nu_{\text{cav}} \tag{3.51}$$

This ratio Q_c/Q_{ai} is of the order of 10^{-5} and the sensitivity of the maser frequency at 1.4 GHz is of the order of 7×10^{-12}/kHz of cavity detuning. This is very large, and considerable efforts have been spent in attempts to stabilize the cavity frequency either passively or actively through external feedback. In the case of the passive standard like the passive room temperature Cs beam standard, the relative population of the levels is detected as an indication of resonance between the applied field and the hyperfine resonance. In that case the gain of the device is weak and an approximate calculation shows that the cavity pulling is proportional to the square of the ratio Q_c/Q_{at}. In view of the fact that in that case the cavity Q is a few hundreds and the line Q is of the order of 10^7, cavity pulling is of the order of 10^{-14} for a cavity detuning of 1 MHz, a shift that is negligible since the accuracy of those standards is at best a few parts in 10^{-14} and the cavity can be tuned to the hyperfine frequency much closer than 1 MHz.

In the case of the fountain, because of its inherent high accuracy reaching the 10^{-16}, it is important to evaluate more accurately the effect of that cavity pulling. We will thus review the subject, outline the physics involved and quantify the effect on the accuracy of the standard.

In the room temperature Cs clock described in Chapter 1, the atoms pass through the Ramsey cavity horizontally at constant speed and continuously. In the fountain a small ball of atoms is sent vertically at initial speed v, passes through the microwave cavity, reaches a given height, L, falls back and passes again through the same microwave cavity. Its speed varies continuously along the trajectory. This type of trajectory has properties that alter the response of the system to perturbations along the path of the atoms. In such a case, it is instructive to introduce several definitions and parameters that make explicit the physics involved and simplify calculation. We assume as in Figure 3.8 that the cavity has a length l and that the atomic ball has a very small size compared to that of the cavity. It takes a time τ to traverse that cavity and the ball travels a time T along the path above the cavity. In the cavity, the atoms are exposed to a microwave induction $B(\omega) = \mu_o H(\omega)$. On the other hand, the cavity dimensions are such as to support a TE_{011} mode, at ω_c, tuned close to the Cs or Rb hyperfine frequency, ω_{at}, depending on the element chosen. The microwave magnetic field in the cavity orientation is along z and is assumed to vary as:

$$H_{\ldots}\left(r(t)\right) = \sin \pi \frac{t}{\tau} = f(t) \qquad (3.52)$$

representing the mode shape. The field is thus a maximum at the centre of the cavity. The atomic ball is assumed small, traverses the cavity on its symmetry axis at speed v and samples that field on its trajectory. The function $f(t)$ introduced in Equation 3.52 is equal to H_{cz} as is shown above and is essentially a representation of the variation of the field within the cavity. We define the Rabi frequency as we did in Chapter 1 at the centre of the cavity as:

$$\omega_{R0} = \frac{\mu_B}{\hbar} B_o(\omega) \qquad (3.53)$$

The detuning of the field from atomic resonance, ω_{at}, is defined as $\delta = \omega - \omega_{at}$, which is generally very small compared to the cavity resonance width since, as said previously, the atomic line width is of the order of 1 Hz and the cavity Q is about 10,000 with a full width of ~1 MHz. It is convenient to define an effective time of duration of the interaction by means of:

$$\tau_{\text{eff}} = \int_0^\tau f(t)dt \tag{3.54}$$

which is a measure of the effect of the field distribution given by Equation 3.52 and thus of the Rabi frequency, ω_R. It is also an instrument determining the weight of the microwave field shape on the probability of a transition or, in a more visible picture, the angle of rotation in space of the magnetization, by means of the relation:

$$P(\tau) = -\int_0^\tau \Omega_o f(t)dt = -\Omega_o \tau_{\text{eff}} \tag{3.55}$$

We may then have $P(\tau) = -\pi/2$, an effective $\pi/2$ pulse in one passage in the cavity if the launch speed of the atomic ball and field amplitude are adjusted as to make $\tau = (\pi/2)^2(1/\omega_{R0})$.

The path of the atomic ball was outlined above and it is convenient to introduce a sensitivity function, $g(t)$, that describes that path. It is defined as:

$$g(t) = \begin{cases} (-1)^p \sin(-P(t)) & \text{for } 0 < t < \tau \\ +1 & \text{for } \tau < t < T + \tau \\ \cos(-P(t)) & \text{for } T + \tau < t < T + 2\tau \end{cases} \tag{3.56}$$

where:

p is an integer, 0, 1, 2 ... characterizing the pulse seen by the atoms

For $p = 0$, the ball is submitted to a $\pi/2$ pulse, the magnetization being tilted by $\pi/2$. $g(t)$ is essentially an impulse response to the field and is 1 along most of the path except in the cavity. It is then possible to define an effective interrogation time, T_{eff}, of the ball by means of the relation:

$$T_{\text{eff}} = \int_0^{T+2\tau} g(t)dt \tag{3.57}$$

It is then readily shown that a frequency shift (δv_{at}) present in a given region along the path is weighed by that function and affects the clock frequency as (Lemonde et al. 1998):

$$\Delta v_{\text{clock}} = \frac{1}{T_{\text{eff}}} \int_0^{T+2\tau} \delta v_{at}(t)g(t)dt \tag{3.58}$$

Similarly, if a shift $\Delta\phi(t)$ affects the phase between the microwave radiation and the atoms magnetization in a given region along the path of the ball, a clock frequency shift appears as:

$$\Delta v_{clock} = \frac{1}{T_{eff}} \int_0^{T+2\tau} \frac{d\Delta\phi(t)}{dt} g(t)dt \tag{3.59}$$

The coupling between the magnetization created by the atomic ensemble within the cavity and the magnetic field within that cavity is given by Equation 1.60 of Chapter 1. It is the equation that is used to evaluate the effect of the cavity tuning on the resonance frequency observed. Using expressions for the field and magnetization in complex form, that equation can be transformed to (QPAFS 1989):

$$\left(\omega_c^2 - \omega^2 - i\omega\frac{\omega_c}{Q_c}\right)H(\omega) = -\frac{\omega_c^2}{Q_c}H_e(\omega) + \frac{\omega^2}{V_{mode}}\int_{V_a} m(r,\omega)H_c(r,\omega)dv \tag{3.60}$$

where:
$H(\omega)$ is the field within the cavity
H_c is the field cavity mode
H_e is the field at frequency ω coupled into the cavity from an external source
$m(r,\omega)$ is the magnetization of the atomic cloud traversing the cavity

The first term on the right hand side thus represents the coupling of the external field to the cavity by means of a wave guide or a loop. The last term on the right hand side represents the coupling of the magnetization to the field created in the cavity and in the case of the fountain is not negligible because of the cavity high quality factor. The parameter, V_{mode}, is the so-called mode volume and is the integral over the volume of the cavity of the mode shape given by Equation 3.52. It is equal to

$$V_{mode} = \int_{Vol} \left\| H_c(r) \right\|^2 \tag{3.61}$$

For a TE_{011} cavity, it is equal to:

$$V_{mode} = \pi LR^2 \left[1 + \left(\frac{\pi R}{kL}\right)^2 \right] \frac{1}{2} J_o^2(k) \tag{3.62}$$

where:
$J_o(k)$ is the Bessel function and is equal to -0.403

For Rb with cavity resonant at 6.8 GHz of appropriate dimensions, V_{mode} is equal to 14.52 cm³.

In the case when the magnetization $m(r,\omega)$ is absent, the equation describes the response of the field $H(\omega)$ in the cavity to an external input $H_e(\omega)$. Due to the complex form of the left-hand member, the field within the cavity responds with given amplitude and phase relative to the input. This response is a function of the cavity Q and has a large frequency width. On the other hand, when the magnetization is present, there is another input and the field within the cavity is altered both in amplitude and phase. Since the resonance is very narrow, the variation in amplitude is small and has a negligible effect on transition probability. Consequently the presence of the magnetization causes essentially a phase shift between the atoms magnetization and the field. After its passage inside the cavity, the ball travels freely for a time T and enters the cavity a second time where it is submitted to the same field and in principle the same interaction takes place with its consequence on the phase. However, due to the type of frequency modulation used, in which the Ramsey fringe is probed at half maximum on each side of the resonance maximum, a cancellation of the phase shift takes place if the equivalent pulse seen by the atoms is $\pi/2$ (Sortais 2001). Furthermore, since the ball dimensions increases during its travel due to internal residual radial speed of the atoms, the density at the second cavity is much smaller, by a factor 5 to 10, and the second passage phase shift for an equivalent pulse different from $\pi/2$ is very small compared to the phase shift introduced by the first passage. It is thus sufficient to calculate only the phase shift introduced by the first passage of the atoms in the cavity.

The effect can thus be calculated from Equation 3.60 by evaluating the magnetization from a knowledge of the quantum mechanical properties of the atoms and the number of atoms in the ball. This can be done either in an approach using the fictitious spin approach (Bize et al. 2001; Sortais 2001) or still the density matrix approach used previously in the case of the H maser, for example. The result of the calculation provides an expression for the phase shift $\Delta\phi(t)$. The result is then introduced in Equation 3.59 and the frequency shift due to cavity mistuning is calculated as:

$$\Delta v_{clock} = K\frac{\tau_{eff}}{T_{eff}}\left[\frac{2\left(\omega_{at}^2/Q\right)\left(\omega_{at}^2-\omega_c^2\right)}{\left(\omega_c^2-\omega_{at}^2\right)^2+\left(\omega_{at}\omega_c/Q\right)^2}\right]\times\frac{\pi}{4\tau_{eff}}\int_0^\tau f^2(t)\sin\left[2\theta(t)\right]dt \quad (3.63)$$

where

$$\theta(t)=n\frac{\pi}{2}\frac{1}{\tau_{eff}}\int_0^t f(t')dt' \quad (3.64)$$

with

$$p=\frac{n-1}{2} \quad (3.65)$$

and

$$K=\frac{\mu_0\mu_B^2 N_{cav}Q}{2\pi^2\hbar V_{mode}} \quad (3.66)$$

The frequency shift thus appears to depend on cavity tuning as a dispersion form. When the cavity is tuned close to the hyperfine frequency, the slope of the fountain frequency dependence on cavity tuning is approximated as (Bize 2001):

$$\frac{d\Delta\nu}{d\nu_c}\bigg|_{\nu_c = \nu_{at}} = -8\tau_{\text{eff}} \times \frac{\mu_o\mu_B^2 N_{\text{cav}}}{2\pi^2\hbar V_{\text{mode}}} \times \frac{Q_c^2}{Q_{at}} \tag{3.67}$$

For the TE_{011} mode in the case of the Rb cavity, for $N = 10^8$ and $Q = 10,000$, K has a value of 3.578×10^{-3}. The slope is then 5.8×10^{-17}/kHz of cavity detuning. This calculation thus shows the relative importance of cavity tuning in the final evaluation of the accuracy of a given fountain. For example, in the case of Rb, a temperature change of ± 90mK may cause a shift of cavity frequency of the order of ± 10kHz with a clock frequency change of 5.8×10^{-16}.

The calculation just outlined makes explicit the offset that is generated directly from the coupling of the magnetization to the cavity field, which introduces essentially a phase shift of the field created in the cavity. It should be mentioned that in measurements, cavity pulling, and collision shift are interlaced and that in practice they are measured simultaneously.

3.2.7.6 Microwave Spectral Purity

If the applied microwave interrogation signal contains spectral components, these components may cause a frequency shift of the hyperfine frequency. This was studied in detail in *QPAFS*, Volume 2 (1989). The effect may be minimized by using an extremely pure microwave interrogation signal. It does not appear that, with the technology now available, this effect causes problems at the level of accuracy desired and presently achieved.

3.2.7.7 Microwave Leakage

A frequency shift may also result when any stray resonant microwave field is present outside the microwave cavity. The stray microwave radiation may be due to cavity leakage and microwave radiation reaching inside the fountain structure from the generating synthesizer via feeding circuitry or optical feedthroughs. This field is seen by the atoms in their free path outside the cavity. Being at the same frequency as that of the interrogation microwave in the cavity, strong interaction with the atoms takes place. It generally causes a frequency shift of the observed resonance. The exact clock bias is difficult to evaluate. Being unintended, such a residual field is by nature uncontrolled, with a potentially complex structure with both a large standing-wave as well as a large travelling-wave component. Such an extraneous field is difficult to exclude at the required level by means of purely electric means and leads to a frequency shift which generally has a complicated variation with the fountain parameters, especially the microwave power (Jefferts et al. 2005; Weyers and Wynands 2006). Furthermore, the frequency shift is generally difficult to disentangle from other systematic errors such as the distributed cavity phase shift. Attempts at evaluating the effect have consisted of changing the power fed to the interrogating microwave cavity. For example, the power can be altered in such a way that the atoms at the exit of the cavity are seen as having experienced a $(3/2)\pi$ pulse rather that a $\pi/2$ pulse. In principle, the atoms exit the cavity in a similar superposition state in both cases. However, the power is raised by a factor 9 and the effect, if it originates from cavity leakage, should be increased accordingly. Without leakage the measured frequency should, in principle, remain the same. Unfortunately, the distributed cavity phase shift may also introduce a power-dependent shift. Furthermore, the law by which the frequency is altered by the leakage is not well-known and the extrapolation to zero leakage may lead to errors because of the lack of knowledge of the actual physical behaviour of such leakage. For these reasons, a technique was developed in which the microwave feeding the cavity is turned on only when the cloud of atoms is traversing the cavity. To ensure total absence of microwave power when the cloud is travelling in its free path, the microwave is generated from a synthesized 8992 MHz by adding, by means of appropriate mixing, radiation at a frequency of 200 MHz. This radiation is turned on and off in correlation with the cloud motion and microwave is expected to be absent while the cloud is out of the cavity. The only remaining uncertainty is the generation of phase transients caused by switching. It was possible to test the technique with such a switched system and compare clock frequency with the case when the microwave was applied in a continuous mode (Santarelli et al. 2009). It was concluded that the microwave leakage effect was reduced to a level less than 10^{-16} (Bize et al. 2009).

3.2.7.8 Relativistic Effects

The second order Doppler effect, caused by relativistic time dilation is also present in the fountain. It is given by Equation 1.10 of Chapter 1, repeated here for convenience:

$$\frac{\Delta v_{D2}}{v_{hf}} = -\frac{v^2}{2c^2} \qquad (3.68)$$

The advantage of the fountain over traditional beam standards is that the atoms in the cloud of the fountain standard are at a temperature of the order of one microkelvin and thus move very slowly ~cm/s. On the other hand, the maximum speed of atoms during the interaction is about 3.5 m/s and varies continuously during the period of free flight. In principle, the effect needs to be integrated over the full trajectory of the cloud. Fortunately, the resulting second order Doppler shift is of the order of 10^{-17} or less and thus completely negligible in relation to other effects present in the clock.

On the other hand, the gravitational effect mentioned in Chapter 1 becomes rather important, in view of the accuracy reached by the fountain clock. It is given by Equation 1.17, also repeated here for convenience:

$$\Delta v_{gr} = \left(\frac{gh}{c^2}\right) v_{hf} \qquad (3.69)$$

where:
   g is the earth's constant of gravitation (9.8 m/s^2)
   h is the altitude of the clock above the geoid

The fractional frequency changes by about 10^{-16} m^{-1} of elevation above the surface of the earth. It is called the *red shift*. In the fountain, at first sight, the altitude of the clock is somewhat ambiguous since the positions of the atoms in interaction with the Ramsey cavity is changing during their free flight. The shift must thus be integrated over the path of the atoms between their passages inside the cavity. The evaluation of the shift is sufficiently precise as to not affect the accuracy of the clock.

However, the gravitational shift must be taken into account in comparing individual clocks at distant locations since they may be located at different altitudes relative to the geoid. Since altitude is generally known to better than one metre, the accuracy in evaluating that shift is better than 10^{-16}.

3.2.7.9 Other Shifts

3.2.7.9.1 Light Shift

Stray light from the laser used to cool and detect the atoms may enter the structure and interact with the atoms in the cloud either in their free flight or inside the cavity while interaction with the microwave radiation takes place. When the atoms are in interaction with the microwave field or still in free flight above the cavity, that stray light may cause a frequency shift, a so-called light shift, as in the case of the optically pumped Cs standard described earlier. This problem is generally avoided by turning off all radiation when the atoms are in free flight or in interaction with the microwave in the cavity. This is done normally by means of mechanical shutters in order to avoid perturbation of the lasers in their continuous operation while locked by means of servo systems to either an optical cavity or an absorption line. In general, when such precautions are taken, light shifts do not affect the clock frequency within the accuracy presently achieved.

3.2.7.9.2 Recoil Effect

Atoms interacting with an electromagnetic standing wave inside the interrogation cavity are subjected to multiple photon processes: absorbing photons from one travelling wave component of the field and emitting them into another. From our analysis done in Chapter 2, absorption of a photon leads to a momentum exchange and a corresponding change in kinetic energy. Thus, in principle, interaction of the atoms with the standing-wave field in the microwave cavity on the upward passage slightly modifies the atomic motion. This effect was initially called *microwave recoil* (Kol'chenko et al. 1969). It is also called *microwave lensing* (Weyers 2012). It leads to a shift of the clock frequency given by:

$$\frac{\delta\omega}{\omega_{at}} \cong \frac{\hbar k}{2mc} \tag{3.70}$$

For Cs, the fractional shift due to the photon at the hyperfine frequency is $\sim 1.5 \times 10^{-16}$. A numerical simulation for Cs fountains using a MOT has been carried out by Wolf et al. (2001). In normal operating conditions, this effect was evaluated to be of the order of 0.5×10^{-16} for an interrogation corresponding to a $\pi/2$ pulse.

3.2.7.9.3   Ramsey, Majorana and Rabi Transitions

Depending on the actual structure and shape of the magnetic field, various transitions may take place in the atomic cloud of the fountain. We have introduced the subject in Chapter 1. Those transitions may have different origins and in view of the accuracy reached with the fountain they need to be addressed carefully. We recall their nature:

- *Ramsey transitions:* If the static magnetic field is not parallel to the microwave field in the cavity, transition like $\Delta F = \pm 1$, $\Delta m_F = \pm 1$ can be induced. This situation may be created, for example, when the atoms do not travel exactly on the symmetry axis of the microwave cavity. The field lines are curved at the end plate of the cavity and an angle exists between the dc field and those field lines. This situation may cause transitions of the $\Delta m = \pm 1$ type mixing states and causing a frequency shift of the transition $F = 4$, $m_F = 0$ to $F = 3$, $m_F = 0$.
- *Majorana transitions:* When the magnetic field changes rapidly or changes sign, the atoms in their free path see a varying field. This may cause transitions of the atoms such as $\Delta F = 0$, $\Delta m_F = \pm 1$ of the states $F = 3$ or $F = 4$ again causing a mixing of state with a resulting frequency shift.
- *Rabi transitions:* We have shown in Chapter 1 that when the dc magnetic field is weak and the Rabi pedestal is large due to the relatively high speed of the atoms, the field dependent $\Delta F = 1$, $\Delta m_F = 0$ Rabi pedestal may be excited. Its wings, exhibiting a slope with frequency, may overlap with the desired field independent pedestal distorting the central Ramsey fringe.

It is generally concluded that these three biases have negligible effect on the accuracy of the fountain.

3.2.7.9.4   Background Gas Collisions

The background pressure of the vacuum system is below 10^{-9} Torr. The shift due to collisions with the residual thermal atoms (helium, hydrogen, etc.) is less than 10^{-16} as evaluated from published data (*QPAFS* 1989). Furthermore, collisions with background gas atoms are strong and the cold atoms that have suffered a collision are most probably ejected from the cloud before reaching the detector and thus do not contribute to the signal.

3.2.7.10   Conclusion on Frequency Shifts and Accuracy

Comparison of the characteristics of selected fountains are given in Table 3.3. It should be mentioned that the table should not be used to compare quality of actual fountains since evaluation of frequency shifts and uncertainties may be done differently from labs to labs. Table 3.3 essentially provides a summary of the results of measurements done with the goal of identifying the most important shifts and the accuracy with which they can be determined. It is readily seen from the table that the evaluation of these frequency shifts is estimated to lead to a maximum uncertainty of a few $\times 10^{-16}$.

TABLE 3.3

Accuracy Budget of a Few Selected Fountains[a]

Physical effect	Syrte FO1 (10^{-16})	Syrte FO1 (10^{-16}) (Guena et al. 2012)	PTB CsF2 (10^{-16}) (Weyers et al. 2012)	NPL CsF2 (10^{-16}) (Li et al. 2011; Szymaniec et al. 2014)	ITC-F2* (10^{-16}) (Levi et al. 2014)	NIST-F2* (10^{-16}) (Heavner et al. 2014)
	Typical correction	Uncertainty	Uncertainty	Uncertainty	Uncertainty	Uncertainty
Second-order Zeeman	−1274.5	0.4	0.59	0.8	0.8	0.2
Black Body radiation	172.6	0.6	0.76	1.1	0.12	0.05
Collisions + cavity pulling	70.5	1.4	3.0	0.4	0.3	<0.1
Distributed cavity phase shift	−1	2.7	1.33	1.1	0.2	<0.1
Rabi and Ramsey pulling	<1.0	<0.1	0.01	0.1		<0.1
Second-order Doppler shift	<0.1	<0.1		0.1		
Background gas collisions	<0.3	<0.3	0.5	0.3	0.5	<0.1
Spectral purity and leakage	<1.0	<1	1.0	0.6	1.5	0.5
Microwave lensing	−0.7	<0.1	0.42	0.3		0.8
Gravity + relativistic Doppler effect	−68.7	1.0	0.06	0.5	0.1	0.3
Total		3.5	4.1	3.3	2.3	1.1

[a] Both NIST-F2 and ITC-F2 operate with cryogenic cavity and flight region.

3.2.8 An Alternative Cold Caesium Frequency Standard: The Continuous Fountain

In order to avoid some of the disadvantages of the pulsed Cs or Rb fountains, a similar device operating in a continuous mode was proposed (Berthoud et al. 1998). Such a continuous fountain, using Cs as the atomic species, was developed by the Federal Office of Metrology (METAS), Switzerland, and Observatoire de Neuchâtel

Neuchâtel/Université de Neuchâtel, Switzerland (Dudle et al. 2000, 2001) taking into account that it can offer two main advantages:

- The density of Cs atoms in the beam would be about 50 times lower than that required in the cloud of atoms of a pulsed fountain. This would mean that the displacement of the frequency by collisions between cold atoms could be made much smaller than in the pulsed fountain. This effect would no longer limit the accuracy of the standard and would relax by a factor of 50 the compromise needed to be made between high frequency stability, which requires a high atomic flux, and high accuracy inasmuch as it is degraded at high atomic flux (Castagna et al. 2006).
- The interrogation of atoms would occur in a continuous manner as in a thermal jet. In this way, the degradation of the stability of a pulsed interrogation should disappear or at least be negligible relative to other sources of instability and the requirements relative to phase noise of the local oscillator would be less severe (Joyet et al. 2001). A quartz oscillator could then be used as conventional local oscillator rather than the CSO.

A continuous fountain poses, however, a number of technical and experimental challenges. First of all, since preparation and detection are continuous, the zones dedicated to these operations need to be spatially separated in order to maintain the correct sequence of preparation and interrogation/detection operation. The easiest way to do that is to have atoms fly along a parabolic path and to separate the detection and the cooling zone by an opaque light wall in order to prevent unwanted scattering of light between preparation and detection zones. A conceptual design of an arrangement showing that approach is shown in Figure 3.18.

The design of a continuous fountain requires two main physical components that differentiate it from a pulsed fountain. These are a light trap that avoids cooling and detection optical radiation from reaching the interrogation region and a special cavity through which the beam passes going up and going down in its parabolic path. The position of these components is shown in Figure 3.18. A 3D diagram of the device is shown in Figure 3.19.

3.2.8.1 Light Trap

In continuous operation, without a special mask in the path of the beam, atoms entering the microwave interaction region would be exposed to the fluorescence emitted from the continuous cooling region. The same perturbation would be present in connection to radiation originating from the detecting region. The atoms would experience a light-shift proportional to the fluorescence light emitted by the atoms in that region. Since the goal of this type of fountain is to reach a level of accuracy better than 10^{-15}, it is important to avoid such shifts which are evaluated to be of the order of 10^{-12}. It would be difficult to measure or extrapolate such a shift to the level of accuracy desired. In order to prevent that perturbation, a light trap, transparent to the atoms but opaque to the radiation has been designed. It is shown in Figure 3.20. A separating wall is also placed between the source of atoms and the detector to avoid interaction between the two regions.

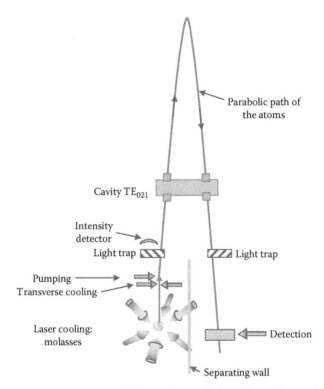

Cavity TE_{021}

Intensity detector

Light trap

Light trap

Pumping

Transverse cooling

Laser cooling: molasses

Detection

Separating wall

Parabolic path of the atoms

FIGURE 3.18 Conceptual diagram of the continuous Cs fountain making explicit the spatial separation of the cooling and detecting zones.

In the design chosen, the light trap is composed of 18 blades oriented at 45° (Joyet 2003). The trap is driven by an ultra-high vacuum (UHV) electrostatic motor (Füzesi et al. 2007) to prevent any undesirable stray magnetic fields. The rotation speed is adjusted such that the horizontal velocity of the blades matches the longitudinal velocity of the atoms in the beam. When this condition is satisfied, the system is transparent to the atomic beam and the atoms pass with a minimal attenuation. On the contrary, the photons travelling at speed c always see a blade in their path and are either absorbed by the blades or deviated into an absorber after two reflections on the blades. This light trap is important in the path between the source of cold atoms and the cavity, since the cooling region where the atomic beam is created and manipulated requires several lasers with relatively high intensity. However, one should not neglect the detection region where laser radiation, although of lower intensity, is also present. Ultimately, to avoid any light shift, a light trap should also be placed between the detection region and the microwave interrogating cavity.

3.2.8.2 Interrogation Zone, Microwave Cavity

A continuous beam of atoms requires two separate interrogation zones to effectuate the coherent stimulation required for a Ramsey type of interaction. An appropriate cavity with low phase variation appears to be of the ring type as described in Chapter 1. Such a cavity operating in the TE_{105} mode was used in the construction of the fountain

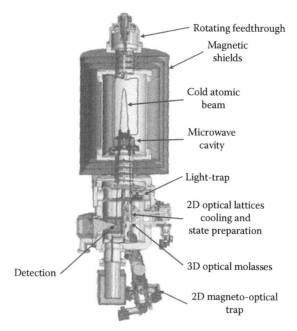

Rotating feedthrough

Magnetic
shields

Cold atomic
beam

Microwave
cavity

Light-trap

2D optical lattices
cooling and
state preparation

3D optical molasses

Detection

2D magneto-optical
trap

FIGURE 3.19 Detailed 3D view of the continuous Cs fountain developed at METAS. (Courtesy of P. Thomann and METAS.)

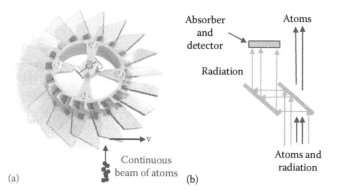

Absorber
and
detector

Atoms

Radiation

V

Continuous
beam of atoms

Atoms and
radiation

(a)   (b)

FIGURE 3.20 Light trap (a) the atomic beam pass through the trap when the blades horizontal velocity matches the atomic beam vertical velocity; (b) laser light is absorbed after reflexions on the blades. (Courtesy of P. Thomann and METAS.)

developed by METAS and University of Neuchâtel group (Devenoges et al. 2013). With a narrow parabolic flight, the two interrogation zones, which can be 1 m apart in the classical U-shaped Ramsey cavity, are separated by a few centimetres only. The system is thus designed with the atoms passing through the ring cavity on opposite sides of its axis but at the same radius r_p where the microwave magnetic field component is vertical. The field components are thus parallel to the applied dc magnetic field, satisfying the requirement for ground state $\Delta m_F = 0$ transitions. The mode chosen,

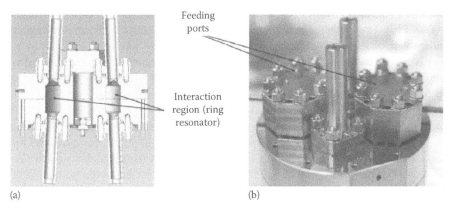

(a)   (b)

FIGURE 3.21   (a): Longitudinal cross section through the TE_{105} microwave cavity and (b): Top view of the constructed cavity. (Courtesy of METAS; Data from Devenoges, L. et al., Design and realization of a low phase gradient microwave cavity for a continuous atomic fountain clock. In *Proceedings of the Joint European Forum on Time and Frequency/IEEE International Frequency Control Symposium* 235, 2013.)

however, shows a π phase difference between the interaction regions, which may be an advantage since the central Ramsey fringe is minimum at its centre. In that case, atom detection may be characterized by lower shot noise (Audoin et al. 1994; Vanier and Audoin 2005). Such a cavity is shown in Figure 3.21.

The cavity is mounted on a support that can be rotated vertically by 180°. This allows exchange of the two interaction zones, and evaluation of the end-to-end phase shift between the two parts of the cavity can be done. It may be mentioned that such an operation is slightly different from the one done in the room temperature beam approach described in Chapter 1 for evaluating the cavity end-to-end phase shift. In that case, the beam itself is reversed relative to the cavity and not the cavity, which keeps its orientation. The openings in the cavity for the atomic fountain beam fix the geometry of the parabolic trajectory.

3.2.8.3 Preliminary Results

Preliminary results have been obtained on such a fountain in both modes, without light traps and with a light trap in the path of the fountain beam between the source and the cavity. In the system used, the atoms are first fed in the system from a continuous slow beam produced by a 2D-MOT and are then cooled in the molasses to a temperature of the order of 70 µK. This temperature corresponds to a rms atomic speed of ~7 cm/s. The beam of slow atoms is created by means of the moving molasses as described earlier by detuning the 45° cooling lasers by about 3 MHz. The beam is sent vertically. The resulting vertical speed is about 4 m/s, which in the design chosen allows a free flight time $T = ~0.5$ s between the passages through the two sides of the cavity. The total time of flight between launch and detection is of the order of 800 ms. It is readily realized that during that time, due to residual transverse speeds of atoms at 70 µK, corresponding to several centimetres per second, the beam would spread to several centimetres during its free flight. For this reason, a further stage of cooling is introduced in the path of the beam by means of transverse laser radiation fields (Di Domenico et al. 2004, 2010). This radiation is also used to deflect slightly (1.2°) the beam direction in order to create the required parabolic flight. The equivalent transverse temperature is ~3 μK corresponding to a speed of the order of a few millimetres per second. Finally, the atoms are optically pumped in their path by means of a laser tuned to the $F = 4$, $F' = 4$ transition in order to remove residual $F = 4$ atoms in the beam.

With the light trap in place, transfer of atoms with an attenuation less than 15% and blocked light by a factor of 10^4 was observed (Di Domenico et al. 2011a). In preliminary measurements, a frequency stability of $6 \times 10^{-14} \tau^{-1/2}$ was measured for an averaging time $10 < \tau < 4000$ s (Devenoges 2012). Frequency shifts particular to such a fountain were evaluated. The light shift was first measured without light trap, by means of changing lasers intensity in the cooling region and measuring the fluorescence level. It was found to be of the order of -1.6×10^{-12} at the nominal intensity of operation of the cooling lasers. Since the light trap attenuates the light transmitted in the path of the atomic beam by a factor of 10^4, it is thus expected that the light shift created by radiation originating in the region of the source should affect the frequency by no more than 1.6×10^{-16}. On the other hand, the light shift that could originate from fluorescence in the detector region was evaluated by means of the effect on the fountain frequency of a variation of the detector laser intensity by a factor 2 (1 mW and 0.5 mW). Elimination of the residual uncertainty associated to this experimental measurement and evaluated to be of the order of 10^{-15} requires a second light trap between the interrogation region and the detection region.

The interaction zones of the TE_{105} cavity are situated symmetrically relative to the feeding ports. A phase shift between the two zones may exist due to geometry. This phase shift may then be evaluated by rotating the cavity itself by 180°. In such an approach, the mounting was such as to reproduce the absolute position of the cavity to 0.01 mm. The frequency shift caused by this end-to-end phase shift was evaluated to be 1.7×10^{-15}. However, spatial phase variation within the cavity can also cause a frequency shift. This effect was mentioned above in the case of the pulsed fountain. In the TE_{105} cavity, this phase shift was calculated by means of 3D finite-elements simulations. It was shown that the phase variation across the interaction regions follows a saddle pattern relative to orthogonal directions, with a peak to peak variation of 30 μrad, about the same size as reported in the case of the ring cavity used in room temperature beam standards (see Chapter 1). For a free flight time of the atoms of the order of 0.5 s, the fractional frequency shift associated with this DPS should then be less than 10^{-15}.

The second-order Zeeman shift of the hyperfine frequency could not be evaluated only by changing the launch velocity, sending atom clouds at various heights, as in the pulse approach since the launching speed can be varied over a restricted range only. The static magnetic field map was evaluated by means of two complementary methods (Di Domenico et al. 2011b): firstly by means of a Fourier analysis of the field dependent Ramsey patterns recorded for launch velocities ranging from 3.7 to 4.2 m/s; secondly by means of a time resolved Zeeman spectroscopy. In that last technique, the atoms during first transit through the interrogating cavity are submitted to a π microwave pulse and transferred to level $F = 4$, $m_F = 0$ by proper adjustment of the rf power. During their free flight parabola, a short rf pulse (~12 ms) at the Zeeman frequency is applied to exciting transitions at $\Delta m = (\pm)1$ over the whole ensemble of atoms in the free flight region. Atoms are thus transferred to $m = (\pm)1$ levels during their flight, depending on the field at their location and the rf frequency applied. During their passage downwards through the interrogation cavity the atoms are then submitted to another π pulse transferring atoms to level $F = 3$, $m_F = 0$. Finally, the fluorescence signal is observed at the detector measuring the total population of the $F = 4$ level. The measurement of the fluorescence intensity is done as a function of time after the Zeeman rf pulse and as a function of its frequency. It gives full information on the transition probability for $\Delta m = (\pm)1$. Time analysis, rf frequency and knowledge of the beam trajectory provide a complete measurement of the magnetic field distribution in the space travelled by the atoms. The accuracy of determination of the second order Zeeman shift through time-averaging of the magnetic field squared was determined to be of the order of 0.2×10^{-15}.

These shifts, light shift, cavity phase shift, and magnetic field shift, are the main ones characteristic of the continuous fountain, which raised questions relative to the possible accuracy that would result with such an approach. The other shifts are similar to those encountered in a pulsed Cs fountain, but the collisional shift is much smaller. It is thus concluded at this stage of development that the evaluation of those important shifts and their control will lead to a final accuracy of the order of 10^{-15}. This evaluation, coupled to other advantages mentioned above relative to reduced collision shift, reduced Dicke effect making possible the use of conventional local oscillator and relatively good short-term frequency stability, appears to open the door to the implementation of a very interesting approach to realize a primary standard fountain clock operating in continuous mode, according its original concept.

3.2.9 COLD ATOM PHARAO CS SPACE CLOCK

French space agency CNES and SYRTE developed a cold atom space clock named PHARAO (Projet d'Horloge Atomique par Refroidissement d'Atomes en Orbite) for the space mission ACES (Atomic Clocks Ensemble in Space) managed by the European Space Agency (ESA). The project consists of the assembly and operation of highly stable and accurate atomic clocks in the microgravity environment of the International Space Station. A view of the PHARAO clock is shown in Figure 3.22.

This PHARAO clock takes advantage of two factors:

1. A very low temperature obtained by laser cooling techniques providing a slow atomic beam,
2. A micro-gravity environment provided by a satellite in Earth orbit.

The PHARAO clock uses cold Cs atoms that are launched at very low speed and made to traverse a Ramsey cavity made with two arms as in the standard room temperature approach (Laurent et al. 2006).

As mentioned several times earlier, in an atomic clock, the duration of the coherent interaction between the atoms and the electromagnetic field is a fundamental limit to the resolution of the frequency measurement. For practical reasons, this duration cannot exceed 1 s for Cs and Rb fountains. Due to the absence of gravity

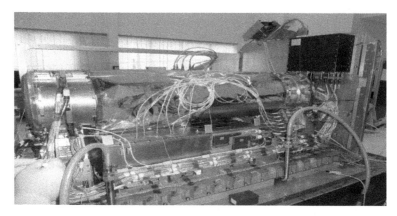

FIGURE 3.22   PHARAO space Cs frequency standard, final assemblage showing the complexity of the system implemented. (Courtesy of P. Laurent, SYRTE, Paris, France.)

in space, the speed of the atomic beam in the PHARAO device can be lowered to a value of the order of 5 cm/s or less by adjusting the moving molasses launch parameters. For a Ramsey cavity with 50 cm between the arms, the interaction time becomes 10 s giving a central fringe with a width of 0.1 Hz. This has to be compared to the case of a thermal beam clock of the same dimension with a line width larger than 500 Hz. As of 2015, the PHARAO clock has been constructed and behaves according to specifications.

3.3   ISOTROPIC COOLING APPROACH

In Chapter 2, we have shown that it is possible to cool an ensemble of atoms contained in a cell by means of isotropic laser radiation. One can imagine that a small cloud of low temperature atoms can be formed within a cell with essentially all the properties of a molasses as used in the fountain. However, in the present case, the cooled ensemble is easily created by means of isotropic radiation fed into a cell by means of optical fibres rather than six overlapping laser beams. The atomic ensemble can be used directly in the cell to implement a frequency standard as in the case of classical Rb frequency standard. Such a standard would have all the properties of the classical Rb standard, but improved by a lowering of Doppler effect and lowering of the collision rate between atoms since their speed would be considerably decreased. Furthermore, the device would be much smaller than the fountains described earlier. The main difficulty, however, is that intense laser radiation is used to cool the ensemble and consequently it is expected that the hyperfine frequency would be altered by that radiation if continuous interrogation at the microwave frequency is done in the same region of space as the cooling.

3.3.1   EXTERNAL CAVITY APPROACH: CHARLI

There are several ways to bypass the difficulty just mentioned. In a first approach, the system can be operated by letting the atomic cloud fall under gravity traversing a microwave cavity as shown in Figure 3.23.

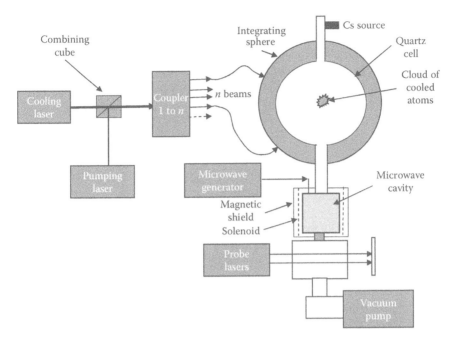

FIGURE 3.23 Implementation of a frequency standard using isotropic cooling and an external cavity (CHARLI). (Data from Guillemot, C. et al. A simple configuration of clock using cold atoms. In *Proceedings of the European Forum on Time and Frequency* 55, 1998.)

The idea essentially consists of using the system as in a reverse fountain. Instead of launching the cloud up in the vertical direction, it is left falling under gravity and traverses a microwave cavity. Unfortunately in reaching the cavity the cloud has acquired speed and does not stay long in the cavity microwave field. Interrogation is thus limited to a short period. Nevertheless, such an arrangement was implemented. It was given the pseudonym CHARLI (for Configuration d'Horloge à Atomes Refroidis en Lumière Isotropique) (Guillemot et al. 1998). In the design chosen, the bulb containing the Cs vapour is made of quartz while the integrating sphere is made of Spectralon, a high reflectance material. The cooling radiation originating from a distributed Bragg reflector (DBR) laser (150 mW) is injected inside the integrating sphere by means of several optical fibres connected to the sphere at symmetrical locations in order to obtain cooling radiation as isotropic as possible. Cooling is done through the cycling transition $|F = 4\rangle \rightarrow |F' = 5\rangle$. As in the case of the molasses, repumping is also done by means of the transition $|F = 3\rangle \rightarrow |F' = 4\rangle$. The period of Doppler cooling is of the order of a few hundred microseconds followed by a sub-Doppler cooling period of the order of a few microseconds. State preparation by means of optical pumping by means of the transition $|F = 3\rangle \rightarrow |F' = 4\rangle$ is of the order of 10 μs. The temperature of the cloud was evaluated to be about 40 μK. While the atomic cloud falling under gravity traverses the cavity, the atoms are exposed to the interrogating microwave field and Rabi transitions are excited.

Atoms having made a transition are detected by means of a double laser configuration, including a pushing action similar to the one described in the case of the fountain. This allows a normalization of the number of atoms which may vary from one cycle to another. The amplitude of the field seen by the atoms traversing the cavity is sinusoidal. In that case the width of the Rabi resonance is not constant and the line width is given by (*QPAFS* 1989):

$$W \approx \frac{5}{\tau_c} \qquad\qquad (3.71)$$

where:

τ_c is the time the atoms spend in the cavity

The expression is valid for the case when the atoms are submitted to a field intensity such that the atoms exit the cavity as if submitted to a π pulse. In the configuration chosen above, the cavity was 5 cm long and the measured width was 35 Hz in general agreement with the predicted width. This result and several measurements on time of flight parameters provided a basis for further experimentation on the proposal of using isotropic cooling for implementing a practical frequency standard with characteristics reflecting the properties of a cloud of atoms having an internal temperature of the order of 50 μK.

3.3.2 APPROACH INTEGRATING REFLECTING SPHERE AND MICROWAVE CAVITY: HORACE

Another approach which was then used consisted of having a microwave cavity surrounding the cell where isotropic cooling takes place and in interrogating the atoms at the early stage of their free fall (Esnault et al. 2010, 2011). In such a case, a Ramsey interrogation scheme in the time domain can be used with a possibility of obtaining a line width narrower than in the case where the atoms traverse a cavity. The system implemented at SYRTE, Paris, is shown in Figure 3.24. The arrangement has been given the pseudonym HORACE (for Horloge à Refroidissement d'Atomes en Cellule). In that case, a spherical cavity was used with walls polished at a level such as to provide a reflectivity of 96%. The atoms were confined to a quartz cell within that cavity. The cavity Q was 5000. The cooling scheme and state preparation was similar to that used in the previous approach (CHARLI). The cycle period was 80 ms. The temperature observed was of the order of 35 μK. Although a much lower temperature could be obtained by further sub-Doppler cooling, this was not needed because the expansion of the cloud within the cavity did not limit the performance of the clock. The Ramsey interrogation scheme consisted of two 5 ms pulses separated by a free evolution time of 25 ms. It should be mentioned that gain is obtained in the scheme simply due to the fact that at the end of the cycle the atoms in the cell are already cooled and, having fallen a relatively short distance, they are used again in the subsequent cycle. It is found that after 5 to 10 cycles a steady state recapture regime is reached. The resonance signal is detected by means of a vertical probe beam superimposed on one cooling beam and retro reflected as shown in the figure. The number of atoms in the proper state, involved in the absorption, is of the order of 1.5×10^6 and results in an

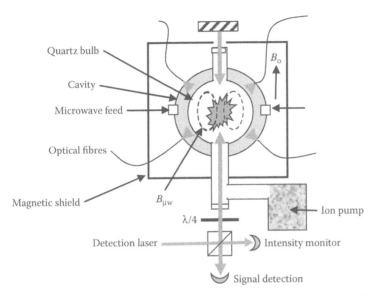

FIGURE 3.24 Configuration used in the implementation of a frequency standard based on isotropic cooling and the use of a time domain Ramsey interrogation scheme (HORACE). (Data from Esnault, F.X. et al., *Phys.Rev. A*, 82, 033436-1, 2010.)

absorption of 2.5% while absorption by background vapour is about 20%. The laser intensity is monitored and the result is used in reducing laser intensity fluctuations in the detection of the useful signal. A typical result showing the field independent Ramsey fringes is shown in Figure 3.25. The measured line width of the central fringe is 18 Hz in agreement with calculation and the contrast is 90%. The arrangement was used as the heart of an atomic frequency standard using as reference a CSO for the microwave interrogation radiation as in the fountains described earlier. The measured frequency stability was $2.2 \times 10^{-13} \tau^{-1/2}$. The authors claimed that a simulation showed that a quartz crystal oscillator with frequency fluctuations spectral density given by $S_y(f) = 10^{-25} f^{-1} + 3.3 \times 10^{-28} f + 3.3 \times 10^{-31} f^2$ would result in a minor degradation with an small increase in frequency instability to $2.4 \times 10^{-13} \tau^{-1/2}$.

3.3.3 DIFFERENT **HORACE** APPROACH

It is possible to implement a frequency standard similar to the one just described, but using the molasses technique described earlier on which the functioning of the fountain is based. The idea is simply to use a small cavity with windows such as to allow penetration of the six beams of cooling laser radiation. The cloud of cooled atoms forms an ensemble of atoms similar to that obtained in the case of isotropic cooling. One disadvantage may be that the system requires large holes in the cavity to allow penetration of the beams with waists of the order of one cm. Another disadvantage may be that the creation of a molasses requires accurate orientation of laser beams while isotropic cooling is done with light injected by means of optical fibres. Nevertheless, a small frequency standard was implemented using a molasses in a

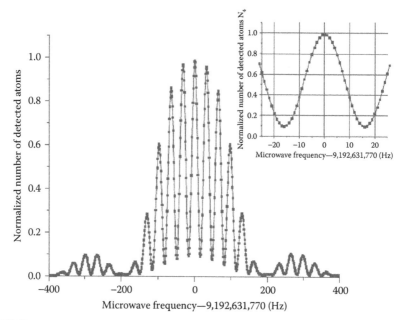

FIGURE 3.25 Ramsey fringe observed in the HORACE clock concept described in the text. (Reprinted with permission from Esnault, F.X. et al., *Phys.Rev. A*, 82, 033436-1, 2010. Copyright 2010 by the American Physical Society.)

cavity and a time domain Ramsey double pulse approach (Müller 2010; Müller et al. 2011). A central Ramsey fringe with a width of 47 Hz was obtained using a pulse separation of 8 ms. The system was operated as a clock and its frequency stability was evaluated as $\sigma_y(\tau) = 5\times10^{-13}\tau^{-1/2}$ for integration times larger than 100 s.

3.4 ROOM TEMPERATURE RB STANDARD APPROACH USING LASER OPTICAL PUMPING

In Chapter 1, we have reviewed the progress made during recent years in the improvement of optically pumped passive Rb frequency standards using the classical approach in which a spectral lamp is used as the source of radiation. We have shown that the frequency stability of such a standard was limited to the order of 10^{-11} at 1 s integration time due to shot noise, limiting the observable *S/N*. We have shown in Chapter 2 that the advent of solid-state semiconductor lasers opened a new door in the field of optical pumping, improving efficiency, the energy being concentrated around desired specific transitions. This is an obvious advantage over a spectral lamp for which energy is spread over a relatively large spectrum of frequencies, which did not contribute to the useful signal.

We have developed a laser optical pumping theory in relatively general terms, using a three-level atom. However, the choice of an atom for the implementation of a frequency standard using laser optical pumping is rather limited. This is due

TABLE 3.4

Alkali Metal Isotopes and Required Wavelengths in Vacuum for Optical Pumping

Atom	Hyperfine Frequency	D_1 Radiation	D_2 Radiation
^{87}Rb	6.835 GHz	794.978 nm	780.241 nm
^{85}Rb	3.035 GHz	794.979 nm	780.241 nm
^{133}Cs	9.192 GHz	894.592 nm	852.347 nm

to various environmental considerations such as temperature of operation, power consumption and the obvious condition that a solid-state laser with the appropriate wavelength be readily available at reasonable cost and size for the atom considered. This reduces essentially the choice to the isotopes ^{87}Rb, ^{85}Rb, and ^{133}Cs. The required wavelengths are tabulated in Table 3.4 (*QPAFS* 1989). The availability of lasers at these wavelengths and their characteristics were discussed in Chapter 2. We will assume that such lasers are available. In practice, ^{87}Rb has remained the element of choice due to considerations such as operation at a reasonable temperature of 60 to 70°C, and lower nuclear spin ($I = 3/2$) compared to Cs ($I = 7/2$) and ^{85}Rb ($I = 5/2$), a property that reduces the number of energy levels in the ground state. This property is advantageous since more atoms find themselves in the proper hyperfine state $m_F = 0$, reducing frequency perturbations which depend on density.

3.4.1 CONTRAST, LINE WIDTH, AND LIGHT SHIFT

A typical implementation of a ^{87}Rb cell frequency standard using laser optical pumping is shown in Figure 3.26. As in the case of the spectral lamp approach, the microwave generator used to feed the cavity and excite the hyperfine transition of ^{87}Rb is locked by means of a frequency control loop to the frequency of the resonance line as observed at the photodetector.

In that approach, the microwave generator frequency is modulated at a low frequency of the order of 100 Hz to a depth of a fraction of the hyperfine resonance line width. The signal at the photodetector, processed by means of synchronous detection, approximates the derivative of the resonance line and provides a frequency discrimination pattern. This signal can be used to lock in frequency of the microwave generator to the center of the resonance line. In practice, the microwave generator consists of a quartz crystal oscillator synthesized numerically to the desired frequency v_{hf}. As is readily visible in Figure 3.26, the system is thus very similar to that using the spectral lamp approach. However, the use of a laser requires some additional control. As mentioned in Chapter 2, the frequency of those solid-state diode lasers is dependent on several parameters such as temperature and driving current. For this reason, they must be controlled in order to maintain their frequency tuned to the selected optical transition. In the arrangement shown in the figure, this is done directly by means of linear absorption in the same cell as that used for detection of

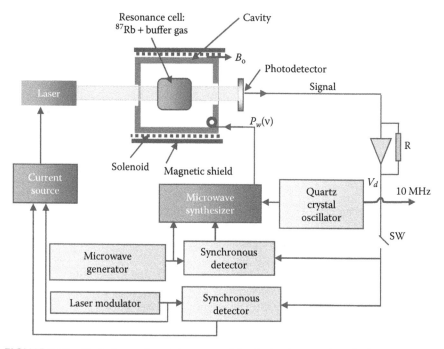

FIGURE 3.26 Typical experimental setup used for the implementation of a frequency standard with a sealed cell and laser optical pumping. The system can be used as a spectrometer (switch SW opened), or a frequency standard (switch SW closed).

the microwave hyperfine resonance. A different modulation frequency is used for the microwave locking and laser stabilization. One goal in the use of a laser for optical pumping is the improvement of S/N, which should improve short-term frequency stability of the system. We have provided in Chapter 1 a context in which the frequency stability can be predicted on the basis of measurable parameters or characteristics of the resonance cell. For the present purpose we will use a somewhat different formulation. In a straightforward linear circuit analysis, it is shown that the frequency stability of the hyperfine resonance line is transferred to the quartz oscillator (Vanier et al. 1979) and the resulting frequency stability of the system in the time domain can be written as (Vanier and Bernier 1981; *QPAFS* 1989):

$$\sigma(\tau) = \frac{KN}{\nu_0 I_{bg} q} \tau^{-1/2} \qquad (3.72)$$

where:

N is the noise spectral density observed at the photo detector originating from all sources

I_{bg} is the current developed in the photodetector by the background radiation

τ is the averaging time

K is a factor that depends on the modulation wave form and is of the order of 0.2

q is a quality factor defined as the ratio of the contrast C of the resonance line to its line width $\Delta\nu_{1/2}$

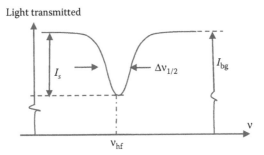

FIGURE 3.27   Representation of the various parameters used in the analysis of short-term frequency stability.

The exact meaning of those various parameters is made explicit in Figure 3.27. The contrast can thus be written as:

$$C = \frac{I_s}{I_{bg}}$$ (3.73)

and the quality factor as:

$$q = \frac{C}{\Delta v_{1/2}}$$ (3.74)

In the approach using a spectral lamp, the resulting pumping radiation consists of both D_1 and D_2 components and contains several hyperfine lines. The width of these lines is of the order of 1 GHz and, although hyperfine filtering reduces the intensity of the lines corresponding to the transitions $S_{1/2}$, $F = 2$ to $P_{1/2}$ and $P_{3/2}$, the radiation reaching the detector is intense and causes so-called shot noise whose spectral density is given by:

$$N^2 = 2eI_{bg}$$ (3.75)

where:
   e is the charge of the electron

In such a case, $\sigma(\tau)$ becomes proportional to the square root of I_{bg}. In standard operation, I_{bg} is large and this type of noise is generally predominant, well above thermal noise at the detector. Furthermore, the contrast of the hyperfine resonance line is small (a few tenths of a percent) due to the presence of optical lines that do not contribute to optical pumping or to the signal, and also due to the background radiation originating from the noble gas used to initiate and maintain the lamp plasma excitation. Assuming a hyperfine line width of the order of 250–500 Hz, the quality factor, as defined by Equation 3.74 is calculated to be ~10^{-5}. In practice, using a spectral lamp, the background light intensity may correspond to a current of ~100 μA, and the shot noise limited frequency stability is calculated to be approximately 0.5 to 1×10^{-11}. Instruments using either the integrated or the separated filter approach have a frequency stability in general agreement with this calculation, smaller units showing frequency instabilities above $10^{-11} \tau^{-1/2}$.

The above considerations show that in the case of shot noise, frequency stability is controlled by three parameters: line width, contrast, and background current. Consequently, a reduction in background radiation should result in better frequency stability in the shot noise limit. This consideration and possible simplification in design and construction have created an interest in the use of solid-state diode lasers instead of spectral lamps for implementing that type of clock. This is due to the fact that the laser spectrum is very narrow, <100 MHz, resulting in a more efficient optical pumping, leading to less background radiation and higher contrast of the hyperfine resonance line. In practice, as will be described below, a contrast larger than 10% is observed with laser optical pumping, due essentially to a reduction of the background current by about two orders of magnitude.

The analysis we developed in Chapter 2 can be applied directly to the case of ^{87}Rb, optically pumped with a laser tuned to the $S_{1/2} - P_{1/2}$ transition (D_1 radiation) as shown in Figure 3.28. As in the classical approach using a spectral lamp, a buffer gas is used for preventing relaxation of Rb atoms on the wall of the cell. This buffer gas is generally a mixture of a noble gas with N_2 as a constituent. The purpose of the mixture is to reduce considerably the dependence of the ^{87}Rb resonance frequency on temperature as will be explained below. On the other hand, N_2 is used specifically for quenching fluorescence radiation through collisions with Rb atoms in the excited state (Happer 1972). This process prevents optical pumping that would take place by means of scattered radiation, which would act as an added relaxation mechanism. The N–Rb atom collisions cause mixing of the $P_{1/2} - P_{3/2}$ states as well as decay of the Rb atoms from those states, resulting in a broadening of the optical resonance line at 794 nm. On the other hand, noble gas–Rb collisions introduce a certain degree of dephasing within the optical transition, which also causes optical broadening. In the analysis, it would seem natural to separate the two effects, excited state decay and dephasing. However, due to the complexity of the processes

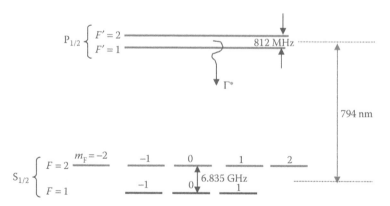

FIGURE 3.28 Ground state $S_{1/2}$ and first excited state $P_{1/2}$ energy manifolds of ^{87}Rb.

involved and the uncertainty on their relative importance in line broadening, we will assume that the resulting combined effect is a broadening of the excited state and we will simply associate this effect to a decay of the excited state. The broadening due to buffer gas collisions, as measured from absorption spectra, is about 20 MHz/Torr, added to the Doppler broadening, which is 530 MHz at a normal temperature of operation (~60°C). At a buffer gas pressure of 20 Torr, the optical line width is thus of the order of 930 MHz. For all practical purposes it is Lorentzian. This point was examined by Vanier (2005); in this work, the absorption spectrum under various light polarization conditions was examined. It should be noted that the spontaneous emission rate broadening is of the order of 5 MHz and is very small compared to the effect of buffer gas collisions. Consequently, we characterize the excited state decay rate by the parameter Γ^*. We evaluate its size from the observed line width to be about $2\text{--}5 \times 10^9$ s^{-1} at the pressures considered in the present analysis. The selection rules of this decay are not known. Consequently, it is assumed that the decay takes place at all Zeeman sublevels of the ground state with equal probability. With the buffer gas broadening considered, the splitting of the excited state is barely resolved. In practice, the laser is normally tuned to the transition $S_{1/2}$, $F = 1$ to $P_{1/2}$, $F = 1$. Consequently, in view of those characteristics, that is to say equal decay rate to all levels of the ground state and partially resolved hyperfine structure of the excited state, we can use a simple three-level model as was done in Chapter 2. The overlapping transition $S_{1/2}$, $F = 1$ to $P_{1/2}$, $F = 2$, does not affect much the analysis in the three-level model aside from altering slightly the calculated pumping rate and causing an added light shift. The alteration of the pumping rate can be taken care of by adjusting the calculated pumping rate while the light shift can be taken care of in a perturbation approach.

It should be mentioned that the use of unpolarized radiation prevents any population trapping in end Zeeman levels as is introduced in the case of optical pumping with circular polarization (Vanier et al. 2003a). On the other hand, the use of linear polarization could have an effect on the final distribution of population among the ground-state Zeeman levels. However, in view of the remarks made above in connection to random decay from the excited state, it is assumed that the radiation is unpolarized and that all Zeeman levels within one hyperfine level are equally populated.

Consequently, $\Delta v_{1/2}$ can be calculated at the exit of the cell by proper evaluation of the pumping rates and saturation factor. Following the discussion above on frequency stability, it appears that an important parameter is the quality factor q, proportional to the contrast. We thus extend the above analysis by making a numerical evaluation of the contrast that appears possible within the hypothesis made. This numerical calculation is done for several values of the absorption coefficient α as well as for several microwave field intensities in the cavity as a function of light intensity or pumping rate at the entrance of the cell. The length of the cell is assumed to be 2 cm. In the calculation, the contrast is obtained from Equation 3.73 in terms of the Rabi frequency $\omega_{1\mu m}$:

$$C = \frac{\left| \omega_{1\mu m}^2(z = L, \omega_M = \omega_{\mu'\mu}) - \omega_{1\mu m}^2(z = L, \omega_m - \omega_{\mu'\mu} = 100{,}000) \right|}{\omega_{1\mu m}^2(z = L, \omega_M - \omega_{\mu'\mu} = 100{,}000)} \tag{3.76}$$

The frequency difference imposed, $\omega_M - \omega'_{\mu\mu} = 100,000$, on the denominator guarantees that for the background radiation, the Rabi frequency is evaluated well outside microwave resonance.

Typical results are shown in Figures 3.29 and 3.30 for the contrast and the line width under various conditions representative of typical experimental situations. In those figures, the microwave Rabi frequency is fixed at a value of 1.41×10^3/s. Similar calculations were done for a fixed absorption coefficient $\alpha = 2 \times 10^{11}$/m/s but as a function of the microwave Rabi frequency with the pumping rate Γ_p as a parameter. The results are shown in Figures 3.31 and 3.32. Various conclusions can be drawn immediately from those figures. First, due to low spurious background radiation, the contrast can be large and reach a value of the order of 10 to 20% at high Rb densities (large α). This is more than an order of magnitude larger than with a lamp. As expected, the line width increases with light intensity, although non-linearly at low values of Γ_p. This is due to the non-linear nature of optical absorption at high Rb densities. The homogeneous model would give a line width that increases linearly with light intensity at all values of Γ_p. On the other hand, it is found that the contrast increases with microwave power applied, but tends to a limit. This saturation behaviour is due to the fact that above a certain value of the microwave intensity, the population of the two ground-state energy levels is equalized and the absorption coefficient becomes constant above a certain value of the Rabi frequency b. On the other hand, it is observed that the microwave excitation broadens the line considerably as made explicit in Figure 3.32 and, although the contrast may become large, there is obviously no increase of the quality factor q as defined by Equation 3.75 above. The line width increases linearly with the microwave field at high values of the Rabi frequency.

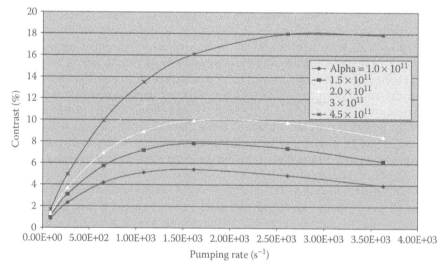

FIGURE 3.29 (Theory) Contrast calculated as a function of pumping rate for various values of the absorption coefficient. The microwave Rabi frequency, b, is assumed to be 1.41×10^3 s^{-1}. The points in the graphs originate from the software used in the calculation and help in identifying the various curves.

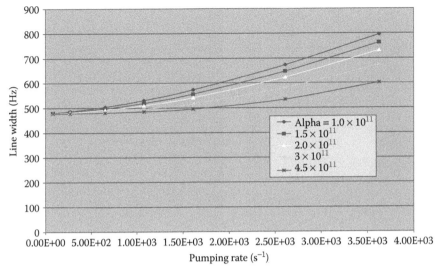

FIGURE 3.30 (Theory) Line width calculated as a function of pumping rate for various values of the absorption coefficient. The microwave Rabi frequency is assumed to be 1.41×10^3 s^{-1}.

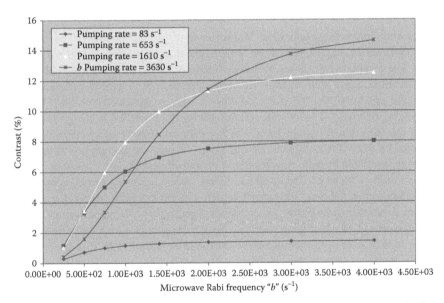

FIGURE 3.31 (Theory) Contrast as a function of the microwave Rabi frequency for four pumping rates. The absorption coefficient is set at $\alpha = 2 \times 10^{11}$ m^{-1} s^{-1}.

The other phenomenon that affects the operation of the frequency standard is the light shift given by Equations 2.43 and 2.44 of Chapter 2. This shift is large and affects the characteristics of the laser pumped frequency standard in an important manner. It displaces the position of the ground-state energy levels relative to each other, and consequently changes the transition frequency in an important way

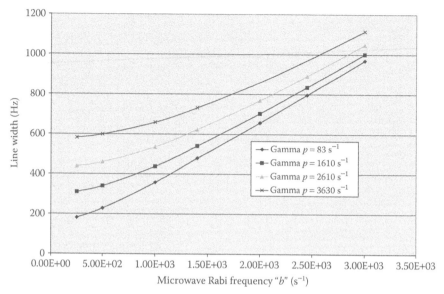

FIGURE 3.32   (Theory) Line width as a function of the microwave Rabi frequency for four values pumping rate Γ_p. The absorption coefficient has been set at $\alpha = 2 \times 10^{11}$ m^{-1} s^{-1}.

(Barrat and Cohen-Tannoudji 1961; Mathur et al. 1968; Vanier 1969). The analysis made above, however, is approximate in the sense that it considers only three energy levels. The complete calculation of this shift for a practical situation would be rather complex. First, the P state of the ^{87}Rb atom is formed of several hyperfine energy levels. On one hand, the $P_{3/2}$ state consists of 4 levels separated by 266, 153, and 70 MHz, respectively. The average wavelength for the transition to the ground state is 780 nm (D$_2$). On the other hand, the $P_{1/2}$ state consists of two hyperfine levels separated by 812 MHz and this transition has an average wavelength of 795 nm (D$_1$). Lasers are available for both wavelengths. The transition probability from the ground levels to each of these excited hyperfine levels varies according to rules connected to angular momentum matrix algebra (Clebsch–Gordan coefficients) and the absorption line is broadened by buffer gas collisions. Furthermore, the presence of Doppler broadening forces an averaging over velocities by means of a Maxwell distribution. Finally, radial intensity inhomogeneities of the laser beam, to be discussed below, alter the pumping rates across the beam and consequently alter the shift itself. The shift is also altered within the cell, the radiation being submitted to absorption along the propagation path.

Measurement results to be examined below are generally reported for very diverse experimental conditions regarding Rb density (temperature), buffer gas pressure, type of lasers, spectral width, and beam diameter. In view of all these considerations it is very difficult to develop an analysis that would cover experimental data at large. Furthermore, the purpose of the present text is to review the subject in the context of application to the implementation of a frequency standard. Consequently, the goal is not to verify in detail the exactness of the theory developed by means of comparison to experimental data, but to identify parameters controlling contrast line width and light shift and review proposed practical approaches leading to improved frequency stability. The analysis developed above should then be considered as a guide in the search for improvements, rather than a full explanation of experimental observations.

We have evaluated approximately the light shift for a specific simple situation. We have used the three-level model for a given buffer gas pressure that broadens the optical absorption line to ~1 GHz, including Doppler effect. This corresponds to a gas, such as nitrogen, at a pressure of about 3 kPa (25 Torr). In view of the important homogeneous broadening that takes place, it is expected that optical pumping is effective for all velocities. At that pressure, the shape of the optical absorption becomes approximately Lorentzian masking the Gaussian shape produced by Doppler broadening (530 MHz at 60°C). Consequently, we adjust the value of Γ^* in Equations 2.43 and 2.44 that give the width of the absorption line that is observed in practice (~1 GHz). This value is 6.28×10^9 s^{-1}. The calculation is thus made without averaging over velocities assuming then that all velocities are pumped due to the buffer gas homogeneous broadening. This approach would be less exact at low buffer gas pressures, since the line shape is a so-called Voigt profile originating from the convolution of a Lorentz with a Gaussian line. However, it appears to be a valid approximation in view of the many other effects that take place in the cell due to such causes as inhomogeneous pumping across the laser beam coupled to microwave saturation effects. In this context, it does not appear that a more exact approach is justified. This approach has also been used in the past to calculate the effect of optical pumping on the shape of absorption lines leading to very satisfying results (Vanier 2005).

The line broadening is thus assumed homogeneous and the calculation is made for three values of the pumping rate, $\Gamma_p = 2$, 3, and 4×10^3 s^{-1}, corresponding to values in Figure 3.30 that produce visible resonance line broadening. Values of the optical Rabi frequency, $\omega_{1\mu m}$, are calculated from these by means of Equations 2.41 or 2.42 and provide absolute values of the light shift. The results of that calculation are shown in Figure 3.33. As can be observed readily, the light shift in a laser pumped frequency standard is expected to be very important, being in the range of parts in 10^9 to 10^8. In a practical standard, the laser frequency and intensity, thus, need to be well-stabilized in order to achieve frequency stability in the 10^{-13}, as realized in existing standards using spectral lamps as pumping sources.

The theoretical results just presented regarding pumping rates, contrast, line width, and light shift are similar to those obtained by Camparo and Frueholz (1985) and by Mileti (Mileti 1995; Mileti and Thomann 1995) in a different mathematical context where the laser was essentially treated as a narrow optical pumping spectral source without coherence. In those calculations the pumping rates are introduced in a phenomenological way in the rate equations and the light shift is introduced as a perturbation. In the present case, however, the optical coherence introduced in the system by the laser is carried all through the analysis. This coherence leads to a very useful approach for calculating the absorption coefficient, the residual light intensity and the double resonance signal amplitude at the exit of the cell, through the solution of a first order differential equation obtained from Maxwell's equations. Furthermore, the present analysis leads directly and in a natural way to expressions

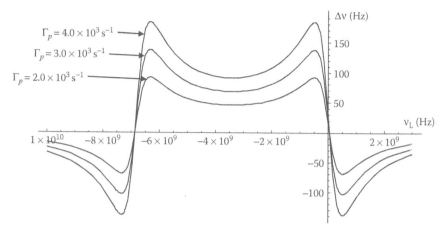

FIGURE 3.33 (Theory) Light shift expected in ^{87}Rb within the three level model for three values of the pumping rate corresponding to rates that produce visible broadening (see Figures 3.32 and 3.34). The parameter v_L used as x-axis is the laser frequency and has its origin at the frequency of the transition $S_{1/2}$, $F = 2$ to P.

for the line width, the light shifts, and the pumping rates and connects those results to the analysis concerning frequency stability. The results obtained in the present analysis for the light shift are also similar to those obtained by Mathur et al. using an operator formalism (Mathur et al. 1968). In the present calculation of the light shift, the hyperfine splitting of the excited state is completely neglected. This approach may be valid at high buffer gas pressures in which case the optical absorption lines are broadened and the excited state hyperfine splitting is not observable. In that case, the centre of the optical line appears as an average of the individual hyperfine absorption lines. At low pressure, the effect of the individual transitions can be taken care of by a perturbation approach in which the individual effects are added. In practice, however, the effect of the excited state hyperfine structure may be totally masked by various inhomogeneities of the laser beam to be discussed below.

3.4.2 EFFECT OF LASER RADIATION BEAM SHAPE

The analysis presented above does not take into account the shape of the laser radiation beam and the variation of the optical pumping rate in the radial direction. In general, the laser intensity is characterized by a Gaussian shape as a function of beam radius. This has important consequences since various regions of the beam effectively lead to different pumping rates. In practice, the photodetector, being large compared to the laser beam diameter, is exposed to the total beam and an average of the light intensity at the exit of the resonance cell is measured. The effect predicted earlier, in which the signal goes through a maximum versus pumping rate, the so-called *peaking* effect (Mileti 1995), is then masked. This is due to the fact that since different parts of the cell are exposed to different radiation intensities the resulting signal becomes an average over a range of pumping rates. The same remarks apply to the calculation of the line width and the light shift both varying with the radial distance. For this reason, these parameters are found in certain arrangements to be non-linear at high light intensities (Camparo et al. 1983). Due to the variation of the radiation intensity across the beam radius, the centre frequency of the resonance line is also a function of radial coordinate, due to a varying light shift. A measurement of the signal frequency at maximum is thus an average weighed by factors such as laser beam shape, buffer gas broadening, cavity field geometry, and microwave saturation. The measured light shift depends thus on actual experimental conditions.

It should be mentioned that this effect does not influence appreciably the final conclusion relative to the frequency stability of a frequency standard using a laser as optical pumping source. In practice, in the implementation of a laser pumped frequency standard, one would operate the system at a laser intensity that makes q optimum, which is in a region just below the maximum of contrast. The important parameters remain contrast and line width. The light shift has then to be addressed independently and its influence on frequency stability has to be evaluated for each experimental setup.

3.4.3 EXPECTATIONS RELATIVE TO SHORT-TERM FREQUENCY STABILITY

The results of the analysis just presented provide the background for a calculation of the expected frequency stability of a frequency standard such as one implemented according to Figure 3.26. These results make possible the evaluation of the quality factor defined through Equation 3.74 and consequently provide information to set the optimum point of operation for best frequency stability. For example, assuming an absorption coefficient $\alpha = 2 \times 10^{11}$/m/s, corresponding to a cell temperature of about 65°C, a contrast of 10% is obtained from Figure 3.31 at a pumping rate of 2000 s^{-1} and for a microwave Rabi frequency of 1.41×10^{3} s^{-1}. This high contrast is essentially due to the narrow width of the laser spectrum, which is more than an order of magnitude less than the spectral lamp, reducing background radiation. At that pumping rate, the hyperfine resonance line width, according to Figures 3.30 and 3.32, is about 575 Hz. In practice, using a vertical-cavity surface-emitting laser (VCSEL) in the same situation, a background current of the order of a few microamperes is detected. Taking only shot noise into consideration, one expects the frequency stability to be approximately

$$\sigma(\tau) \cong \text{a few} \times 10^{-14} \tau^{-1/2} \tag{3.77}$$

as calculated from Equations 3.72 and 3.75.

3.4.4 REVIEW OF EXPERIMENTAL RESULTS ON SIGNAL SIZE, LINE WIDTH, AND FREQUENCY STABILITY

Unfortunately, published experimental data often provide limited information on the conditions realized in practice. Line width is generally provided, but contrast is not a parameter that appears to have drawn attention in past publications. When contrast is reported, usually for a given set of optimum conditions, it is larger than 10% in qualitative agreement with the general prediction made above (Chantry et al. 1992; Mileti 1995).

It should be mentioned that in Chantry et al. (1992), ^{133}Cs was used as the reference alkali atom isotope. However, the system used for implementing a clock is the same as that used with ^{87}Rb and the three-level model developed above applies. The contrast is found to increase to a maximum value with microwave power with a line broadening in qualitative agreement with Equation 2.55. The signal amplitude at the photodetector is also often reported as a function of Rb density (cell temperature) and found to decrease at higher temperatures (Hashimoto et al. 1987; Matsuda et al. 1990). However, since absorption becomes important at higher temperatures, the background intensity also decreases and no conclusion can be drawn from this information, relative to contrast, quality factor q, and frequency stability, since information on background light intensity is lacking.

On the other hand, the effect of laser intensity variation across the beam on signal intensity was studied in some detail by Mileti (1995). It was made evident through expansion of the laser beam diameter by means of a telescope at the entrance of the cell, making the laser beam more homogeneous, and through the use of annular masks placed in front of the photodetector, hiding part of the laser beam. The arrangement made possible the detection of atoms submitted to approximately homogeneous optical pumping in the radial direction. The effect could be studied as a function of detected beam diameter and several conclusions were drawn relative to the observation of the *peaking* of the signal amplitude versus light intensity. The *peaking* effect predicted in Figures 3.29 and 3.31 could be observed in very specific conditions and a typical result is reproduced in Figure 3.34. In that figure, the laser intensity is given in terms of laser driving current and the signal size is given in μA. Consequently, the data cannot be compared directly to the theoretical graphs shown in Figure 3.29. Nevertheless, a *peaking* of the signal is observed and the general trend of the signal behaviour is observed.

Results obtained by Lewis and Feldman for line width as a function of laser radiation intensity are reproduced in Figure 3.35 (Lewis and Feldman 1981). An important broadening is observed. Assuming in first approximation that the broadening is linear with light intensity as would be obtained in the homogeneous model (Equations 2.55 and 2.50) and using information contained in Lewis and Feldman, 1981, one can relate power density used in the arrangement and pumping rate. At a laser radiation density of 250 μW/cm^2, calculation gives $\Gamma_p = 1800$ s^{-1}. This situates Lewis' graph in the upper portion of Figure 3.30 where the line width is nearly linear with pumping rate, making the linear assumption plausible in the present circumstances.

Results obtained by Camparo and Frueholz relative to contrast at relatively low temperature (37°C) are reproduced in Figure 3.36 (Camparo and Frueholz 1985). Results for line width are also reproduced in Figure 3.37. At that temperature the Rb density is rather low and, as pointed out by the authors, the cell is nearly optically thin. The absorption coefficient measured at that temperature is reported by the authors as $(\tau_D)^{-1} = 7.6$ cm. This coefficient is related to α by means of the relation, $(\tau_D)^{-1} = \Gamma^*/\alpha$. At the buffer gas pressure used (10 Torr, N$_2$), Γ^* is ~2 × 10^9 s^{-1}, and we obtain $\alpha = 2.63 \times 10^{10}$/m/s. The value calculated by means of Rb density (Equation 2.33) from the values used previously is 1.8×10^{10}/m/s. This difference may be due to the difference in density in the cell for a similar temperature, which as pointed out by several authors, may vary considerably depending on the type of

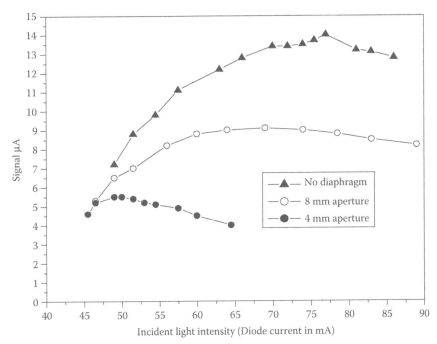

FIGURE 3.34 Typical result obtained by Mileti for the double resonance signal amplitude as a function of laser radiation intensity in a special experimental arrangement where the laser beam is made homogeneous in radial directions by expansion and portions of the detected transmission signal are selected by means of masks. (Data from Mileti, G., *Etude du Pompage Optique par Laser et par Lampe Spectrale dans les Horloges a Vapeur de Rubidium*, Thesis, Université de Neuchâtel, 1995.)

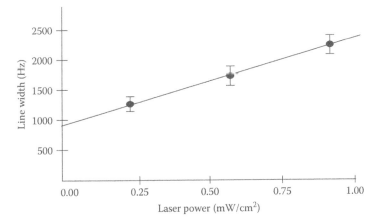

FIGURE 3.35 Hyperfine resonance line width reported by Lewis and Feldman as a function of laser light intensity. (Data from Lewis, L.L. and Feldman M., Optical pumping by lasers in atomic frequency standards. In *Proceedings of the Annual Symposium on Frequency Control* 612, 1981. Copyright 1981 IEEE.)

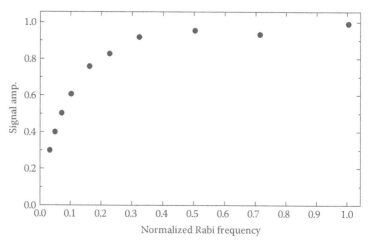

FIGURE 3.36 Normalized values of contrast versus normalized microwave Rabi frequency observed for the 0-0 hyperfine transition with laser optical pumping in a Rb cell at 37°C. (Reprinted with permission from Camparo, J.C. and Frueholz, R.P., *Phys. Rev. A*, 31, 144, 1985. Copyright 1985 by the American Physical Society.)

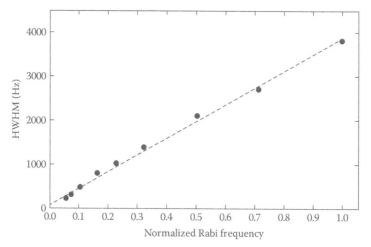

FIGURE 3.37 Half line width observed as a function of the normalized microwave Rabi frequency for the same situation as that giving the results of Figure 3.38. (Reprinted with permission from Camparo, J.C. and Frueholz, R.P., *Phys. Rev. A*, 31, 144, 1985. Copyright 1985 by the American Physical Society.)

glass used and the temperature cycling to which the cell has been exposed prior to measurements (Vanier 1968; Camparo et al. 2005).

The results presented by the authors on contrast are normalized, and a comparison to previous calculation can be made only on a qualitative basis. Nevertheless, it is readily observed that the general behaviour of the saturation observed in Figure 3.36 is the same as that calculated and shown in Figure 3.31. The same remark can be made for the data on line width. The large broadening observed in measurements of Camparo and Frueholz (1985) situates the data in the upper end of Figure 3.32 making the broadening linear with the microwave Rabi frequency as expected.

In some cases the *S/N* realized in practice is given for an optimum condition (Hashimoto and Ohtsu 1987, 1990; Chantry et al. 1992; Saburi et al. 1994). In that case, some information on the performance of the system may be extracted. To do so, Equation 3.64 can be transformed readily in terms of *S/N* through direct algebraic manipulations (Vanier and Bernier 1981). We obtain:

$$\sigma(\tau) = \frac{K}{Q_l S/N} \tau^{-1/2} \tag{3.78}$$

where:
K is about 0.2
Q_l is the resonance line Q

Equation 3.78 provides a simple way of evaluating the performance of the system through a direct measurement of two easily accessible parameters, line width and *S/N*. For example, in Hashimoto and Ohtsu (1987), using an edge-emitting diode, a *S/N* of 66 dB was reported with a line width of 570 Hz, leading to a predicted frequency stability of ~3 × 10⁻¹² $\tau^{-1/2}$. Saburi et al. (1994), using a DBR laser, with a narrow spectrum (~500 kHz), measured a *S/N* of 85 dB and the predicted frequency stability using Equation 3.78 is ~6 × 10⁻¹³ $\tau^{-1/2}$. In that case, however, the best measured frequency stability was 1 × 10⁻¹² $\tau^{-1/2}$. The difficulty with these reports arises in the absence of information on the actual background radiation level, which would provide a reference to calculate contrast, which, using Equation 3.72, would then permit an evaluation of the predicted frequency stability and possibly identify the origin of the noise.

Approaches have been proposed for increasing the *S/N* by concentrating all the atoms in the $F = 2$, $m_F = 0$ of the ground state of the Rb atom. This may be done, for example, by using a combination of polarized pulses of light and so-called π-RF pulses at the Zeeman frequency (Bhaskar 1995). In such an approach, however, the system operates in a pulse mode adding complexity to its practical implementation.

Unfortunately, the frequency stability predicted on the basis of shot noise alone is not generally realized in practice. A frequency stability of the order of 1 to 5 × 10⁻¹¹ $\tau^{-1/2}$ is reported when commercial laser diodes having large spectral widths (several tens of MHz) are used as optical pumping sources (Lewis and Feldman 1981; Ohtsu et al. 1985; Chantry et al. 1996). Better results are obtained when narrow band laser diodes, such as DBR, distributed feedback (DFB), and external cavity locked lasers are used (Saburi et al. 1994; Mileti et al. 1998; Ohuchi et al. 2000; Affolderbach and Mileti 2003a, b). Consequently, noise sources other than shot noise must be present when common diode lasers such as edge-emitting lasers and VCSELs are used as optical pumping sources. It is generally recognized that these noise sources originate in part from amplitude fluctuations (AM), inherent to the laser, and laser frequency fluctuations (FM) transformed into intensity noise by the resonance cell (Camparo and Coffer 1999). This will be discussed in Section 3.4.6.3, following a description of the various frequency shifts present in that type of optically pumped standard.

3.4.5 FREQUENCY SHIFTS

Optically pumped sealed cell frequency standards using a buffer gas and laser optical pumping are characterized by essentially all the same frequency shifts as those encountered when a spectral lamp is used. We have introduced above the most important one, the light shift. These shifts affect the accuracy of the standard. Furthermore, their fluctuation with time may affect frequency stability. We will review these shifts in the light of new experimental data obtained mostly with laser optical pumping.

3.4.5.1 Buffer Gas Shift

When the atoms are exposed to a travelling microwave field, a buffer gas is required to reduce Doppler broadening through Dicke effect (Dicke 1953). In the case studied here and represented in Figure 3.26, a cavity is used and the atoms are exposed to a microwave standing wave. They essentially stay in a field of the same phase during their time of interaction with the microwave radiation. In such a case, the buffer gas reduces relaxation of the Rb atoms upon collision with the cell inner surface by increasing the diffusion time. In practice, N_2 is used as one element of the buffer gas in order to quench the fluorescence emitted in the optical pumping process, which would cause random optical pumping equivalent to relaxation (Vanier 1968; Happer 1972).

Unfortunately, collisions between the Rb atoms and the buffer gas atoms cause a frequency shift. This shift is proportional to the density of the buffer gas or, in a sealed cell, to its pressure. Furthermore, this shift is temperature sensitive. The Rb hyperfine frequency is shifted by (*QPAFS* 1989):

$$\Delta v = P(\beta_{bg} + \delta_{bg}\Delta T + \gamma_{bg}\Delta T^2) \tag{3.79}$$

where:
   P is the buffer gas pressure
   β_{bg} is the pressure coefficient
   δ_{bg} is the linear temperature coefficient
   γ_{bg} is the quadratic temperature coefficient

These coefficients have been determined in several laboratories for many buffer gases and most alkali atoms. The pressure and linear temperature coefficients are both positive in the case of N_2. In practice, another buffer gas with a negative temperature coefficient is mixed with N_2 to minimize the temperature sensitivity of the resonance frequency at a given temperature (Missout and Vanier 1975). The gas most often used is Argon. The characteristics of those two gases are given in Table 3.5. The numbers have been obtained through an average of the most accurate published data (*QPAFS* 1989).

Equation 3.79 leads to a quadratic dependence of the frequency as a function of temperature since both δ_{bg} and γ_{bg} cannot be made equal to zero simultaneously. The ratio of the pressure of both gases at cell filling time has to be adjusted such as to position the maximum of the resulting curve at the desired temperature of operation. The behaviour just described has been reported experimentally by several authors in both cases of optical pumping with a spectral lamp and with a laser (Vanier et al. 1982a and b; Ohuchi et al. 2000; Affolderbach et al. 2006).

TABLE 3.5
Most Probable Pressure Shift and Temperature
Coefficients of ^{87}Rb in Buffer Gases such as Molecular
Nitrogen and Argon

	β_{bg} (Hz/Torr)	δ_{bg} (Hz/(°C·Torr))	γ_{bg} (Hz/(°C²·Torr))
Nitrogen (N$_2$)	546.9	0.55	−0.0015
Argon (A)	−59.7	−0.32	−0.00035

The residual frequency shift of such a mixture is of the order of 200 Hz/Torr. It is not possible in practice to set the total buffer gas pressure to a precision that would provide accuracy better than ~10^{-9}. For this reason, even with laser optical pumping, the Rb clock remains a secondary frequency standard and needs to be calibrated.

3.4.5.2 Magnetic Field Shift

As in the case of optical pumping with a spectral lamp, a magnetic field is required to provide an axis of quantization to the system and a reference axis for laser radiation. The magnetic field removes also Zeeman degeneracy in the ground state. The $m_F = 0$ sublevels are shifted quadratically causing a shift of the hyperfine frequency according to the equation (*QPAFS* 1989):

$$\Delta v_B(^{87}Rb) = 575.14 \times 10^2 B_0^2 \quad (\text{Hz}) \tag{3.80}$$

where:
  B_0 is the magnetic induction in tesla

This field is normally created by means of a solenoid inside a set of concentric magnetic shields. It is of the order of a few tens of μT and produces a shift of the order of 10 Hz. The stability of the current driving this field and the shielding factor of the enclosure must be compatible with the frequency stability desired. Generally, this requirement does not cause major problems.

3.4.5.3 Light Shift

In the three-level model used, only two transitions are considered and the analysis leads to a light shift given by Equations 2.43 and 2.44 and plotted in Figure 3.33 as a function of laser frequency with the laser intensity as parameter. As discussed above, it should be emphasized that the analysis made was based on several assumptions. The results obtained depend greatly on those assumptions and are to serve mainly as an indication of the size of the effect expected. In this section, we will summarize some of the experimental results reported in the literature and provide some indication on the importance of this shift in the implementation of a practical frequency standard. We will also examine proposals for reducing the effect.

In the analysis of the light shift, it is convenient to introduce two parameters characterizing typical experimental situations. One important parameter is the slope of the dispersion pattern at its centre, called the light shift coefficient β_{LS}. It is given by the sum of all light shifts that may be included in Equations 2.43 and 2.44. Close to one optical resonance line originating from one of the ground-state hyperfine levels, the shift from the other transition is small and the light shift, $\Delta\omega_{LS}$, may be written as:

$$\Delta\omega_{LS} = \beta_{LS}(\omega_L - \omega_M - \omega_{m\mu'}) \tag{3.81}$$

From Equation 2.44, it is readily shown that β_{LS} is a function of the laser power through the pumping rate Γ_p:

$$\beta_{LS} = \frac{d(\Delta\omega_{LS})}{d\omega_L} = \frac{\Gamma_p}{\Gamma^*} \tag{3.82}$$

On the other hand, for a given laser detuning $\Delta\omega_L$ from zero light shift, the resonant frequency is a function of the radiation intensity and it is convenient to define the intensity light shift coefficient α_{LS}. We write this shift as:

$$\Delta\omega_{LS} = \alpha_{LS}I \tag{3.83}$$

where α_{LS} close to the centre of the optical line may be written as:

$$\alpha_{LS} = \frac{\Delta\omega_L}{\Gamma^*} \tag{3.84}$$

It is readily shown by means of Equation 3.82 that α_{LS}, as defined, is directly related to β_{LS}. The coefficient α_{LS} is very important in practice, particularly when the laser frequency is locked to the maximum of the absorption line used for optical pumping. Such a locking does not always guarantee total independence of the hyperfine frequency against light intensity mainly due to asymmetry in the optical absorption line.

In practice, it is found that the light shift is not linear with light intensity. This is due to saturation effect and the non-homogeneity of the radiation intensity across the laser beam (Camparo et al. 1983). It is then more convenient to renormalize α_{LS} to a fractional change in light intensity at a given operating laser intensity, that is, fractional frequency shift per percent of light intensity change (Camparo 1996; Camparo et al. 2005). We call it $\alpha_{LS(\%)}$. This parameter is more useful than α_{LS} in evaluating the effect of laser intensity fluctuations on the frequency stability of the implemented frequency standard.

Several authors have reported on light shifts and on values of α_{LS} and β_{LS} in particular setups (Arditi and Picqué 1975; Lewis and Feldman 1981; Ohtsu et al. 1985; Hashimoto and Ohtsu 1987, 1990; Hashimoto et al. 1987; Hashimoto and Ohtsu 1989; Matsuda et al. 1990; Yamagushi et al. 1992; Deng et al. 1994; Saburi et al. 1994; Mileti et al. 1996; Ohuchi et al. 2000). When a laser is used without special arrangements for compensating for the light shift, the coefficient β_{LS} is found to be of the order 1 to 10×10^{-11}/MHz detuning of the laser from resonance depending on the laser power applied on the ensemble of atoms. These results are in general agreement with the calculation made above, which gives a light shift coefficient varying from 4.5 to 7.3 × 10⁻¹¹/MHz detuning for the three Rabi frequencies chosen in the particular example shown in Figure 3.33. This agreement is most satisfying in view of the kind of intuitive assumptions and approximation made in the analysis and the absence of information on the buffer gas pressure used in the papers cited. In the past, it was claimed that theory and experiments disagreed by more than an order of magnitude (*QPAFS* 1989). However, it appears that the conclusion was reached due to a low evaluation of the optical line width.

On the other hand, the value of α_{LS} obtained from published data spreads over a wide range. This is due to the size of laser detuning used in its determination. The size of α_{LS} can be evaluated either from Equation 3.84 or from Figure 3.33. For example, for a laser detuned by 250 MHz, that is at the laser tuning for maximum light shift in Figure 3.33, we obtain α_{LS} = 0.05 Hz per s⁻¹ (in terms of pumping rate). However, in the case that the laser frequency is locked close to the centre of the optical absorption line as is normally done, α_{LS} is much smaller by at least an order of magnitude.

A typical example of the phenomenon as measured in a laser pumped ⁸⁷Rb cell using radiation at 780 nm is shown in Figure 3.38 (Lewis and Feldman 1981). In this figure, the coefficient β_{LS} is evaluated to be 9.1 × 10⁻¹¹ per MHz for a power density of 250 μW/cm². An evaluation of $\alpha_{LS\%}$ can also be made from the data reported by the authors. At maximum light shift we obtain $\alpha_{LS\%}$ as 4.9 × 10⁻¹⁰/percent change in light intensity. In order to compare these results to those obtained in the previous

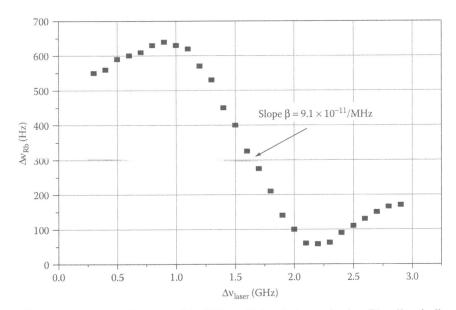

FIGURE 3.38 Light shift observed for 780 nm (D₂) optical pumping in a Rb cell optically pumped with a laser for a laser intensity of 250 μW/cm². (Reproduced with permission from Lewis, L.L. and Feldman, M., Optical pumping by lasers in atomic frequency standards. In *Proceedings of the Annual Symposium on Frequency Control* 612, 1981. Copyright 1981 IEEE.)

analysis, the pumping rate must be known. The authors provide information on the hyperfine resonance line broadening as a function of laser power density. From this information, assuming a saturation factor of the order of 1, providing a contrast of 10% as reported, we evaluate the pumping rate by means of Equation 2.55 to be 1800 s^{-1} at a power density of 250 μW/cm^2. At that pumping rate, our analysis gives $\beta_{LS} \cong 4 \times 10^{-11}$/MHz laser detuning and $\alpha_{LS\%} \cong 2 \times 10^{10}$/percent change in light intensity, values that fall in the range of those obtained in our calculation. These shifts are rather large and may affect the performance of the frequency standard through various mechanisms. When the laser is not tuned to the frequency where the light shift is zero, there can be direct conversion of laser intensity fluctuations into frequency instability of the clock.

Quasi-static laser frequency fluctuations that affect clock frequency stability in the long term are minimized by locking the frequency of the laser to the centre of the optical absorption line in the resonance cell. This is the simplest approach. However, the optical absorption line is broadened by the buffer gas (>750 MHz) and the frequency lock is generally weak. Furthermore, this optical line is a combination of several absorption lines corresponding to transitions from the ground state to the various hyperfine levels of the P state. In such a case, the light shift may not be zero and there may be a residual dependence of the frequency on the laser intensity. Another approach consists in locking the laser frequency to an external cell without buffer gas by means of saturated absorption of a chosen hyperfine line as described in Chapter 2. The saturated absorption line width is of the order of 5 MHz (Levi 1995; Levi et al. 1997; Ohuchi et al. 2000; Affolderbach et al. 2004). In that technique, an extra cell is required raising the complexity of the system. Furthermore, the laser frequency is not locked to the same frequency as that of the maximum absorption in the resonance cell. In such a case, there remains a shift dependent on light intensity, unless the laser beam frequency is translated by means such as an acousto-optic modulator.

Several proposals have been made for reducing the light shift. One approach proposed is to use a high buffer gas pressure in the resonance cell in order to broaden the optical resonance absorption line (Camparo 1996; Camparo et al. 2004, 2005). The added broadening reduces the slope of the dispersion curve at the centre of resonance and makes the clock frequency less sensitive to laser frequency and intensity fluctuations. There is a net advantage in using this approach, but at the expense of a weaker lock of the laser frequency to the absorption line. With a pressure of 100 Torr, the laser tuning sensitivity (β_{LS}) was reduced to ~6 $\times$ 10^{-13}/MHz laser detuning (Camparo et al. 2005). The sensitivity to laser intensity ($\alpha_{LS\%}$) was measured as ~10^{-11} per 1% change in light intensity for the laser locked to the broadened optical resonance line.

Another approach consists in introducing in the laser spectrum, sidebands that compensate the light shift of the main carrier (Affolderbach et al. 2003, 2005). This is similar to the technique used in the implementation of frequency standards using the CPT phenomenon (Levi et al. 2000) in order to avoid light shifts by selecting a proper index of modulation of the laser. The authors claim a complete cancellation of the light shift by a proper choice of the modulation index. Another similar approach uses two lasers (Deng 2000). This last technique has the added advantage of populating the $F = 2$, $m_F = 0$ level at the expense of the other Zeeman levels of the ground state.

An interesting technique has been proposed by Hashimoto and Ohtsu (Ohtsu et al. 1985) for tuning automatically the laser frequency to obtain zero light shift. It is based on the observation that when the laser frequency is detuned from exact optical resonance, the observed hyperfine resonance line becomes asymmetrical under special circumstances such as inhomogeneous broadening caused by a variation of laser intensity in the radial direction (Camparo et al. 1983). That property leads to the possible implementation of a system that measures the asymmetry, interprets it in terms of frequency, and controls the laser frequency such as to reduce the asymmetry to a minimum. In principle, the technique leads to zero light shift. The authors claim that using that approach, the long-term frequency stability of the clock they implemented was improved by a factor of 45 as compared to the case where the laser was not locked using that technique (Hashimoto and Ohtsu 1990). They report that the resulting observed frequency drift was reduced to 6.3×10^{-13}/hour. However, it may be mentioned that this drift corresponds to 7.6×10^{-11}/month and is larger than that observed in similar standards using a spectral lamp.

In a technique proposed by Camparo and Delcamp (Camparo and Delcamp 1995; Camparo 1996), a natural Rb cell is used and ground-state population inversion of the ^{87}Rb isotope is done by means of the fluorescence of ^{85}Rb optically pumped by a laser. The coincidence of optical lines allows optical pumping with specific properties regarding light shift. The authors claim that the frequency dependent light shift (β_{LS}) is reduced by at least an order of magnitude. However, the intensity dependent light shift (α_{LS}) is increased by an order of magnitude.

Other methods consist of pulsing the laser radiation and observing the hyperfine resonance signal in the dark (Alekseev et al. 1975; English et al. 1978). In principle, since there is no light present during the observation of the resonance signal, the light shift should disappear totally. Such an approach has been implemented in combination with the use of the Ramsey time separated pulse technique (Levi et al. 1997). The light shift was reduced to 3×10^{-13}/MHz. Another similar approach, proposed in the 1960s (Arditi and Carver 1964) with a spectral lamp makes use of stimulated emission in a cavity. In that case, the optical radiation is applied also as two successive Ramsey microwave pulses. Stimulated emission is observed after these microwave pulses, Ramsey fringes being observed after the second pulse. In such a case a high Q cavity is required for the observation of stimulated emission and the system is essentially an Rb maser below threshold. This approach was implemented recently by Godone et al. using intense laser radiation to destroy any ensemble residual microwave coherence between the optical pumping cycles (Godone et al. 2004a, 2006a). The results will be reviewed in Section 3.4.7.2.

Although these pulsing techniques appear to be rather effective, they add complexity to the system. There is also the possibility of the presence of a so-called *position shift* created by inhomogeneities in the ensemble (Risley and Busca 1978). Such an inhomogeneity could be created, for example, by a magnetic field gradient. In such a case, the average resonance frequency of the ensemble may depend on the light intensity, after the light has been turned off, since different regions of the resonance cell may be weighed differently, depending on the optical thickness of the ensemble.

A new method for elimination of the light shift by electronic means has been proposed by McGuyer et al. (2009). It is based on the observation that the out of phase signal detected at the lock-in amplifier used for locking the local oscillator frequency to the atomic resonance is a function of the light shift present in the detection system. That quadrature signal was thus used in another channel to lock to the system to zero light shift. It is claimed that when that type of feedback is introduced, long-term frequency stability is considerably improved up to averaging times of the order of 10^4 s.

The analysis of the effect of rapid fluctuations of the laser frequency and intensity on the clock behaviour requires knowledge of the spectral density of frequency and intensity fluctuations of the laser. This analysis is done in Section 3.4.6.1.

3.4.5.4 Spin-Exchange Frequency Shift

Spin-exchange interaction taking place in collisions between alkali atoms causes relaxation of the population and of the coherence that exists in the ground state. The relaxation has been taken care of above in introducing it phenomenologically in the rate equations by means of γ_1 and γ_2. These rates are given by (QPAFS 1989):

$$\gamma_1 = n\langle v_r \rangle \sigma_{se} \tag{3.85}$$

$$\gamma_2 = \frac{6I+1}{8I+4} n\langle v_r \rangle \sigma_{se} \tag{3.86}$$

where:
  n is the alkali atom density
  $\langle v_r \rangle$ is the relative velocity of the atoms
  σ_{se} is the collision cross section
  I is the nuclear spin

Values of σ_{se} as measured by various authors have been tabulated in QPAFS (1989). For a temperature of 60°C, the density of Rb is ~3 × 10^{11}/cm³ and $\langle v_r \rangle$ is 400 m/s. In the case of ^{87}Rb, $\sigma_{se} = 1.8 \times 10^{-14}$ cm², the rate γ_1 is 216 s⁻¹ and γ_2 is 135 s⁻¹. The spin-exchange contribution to the line width is thus 43 Hz. On the other hand, such collisions introduce also a phase shift in the magnetic moment of the atom and cause a frequency shift given by the relation:

$$\Delta\omega_{se} = \frac{1}{4} n\langle v_r \rangle \lambda\Delta \tag{3.87}$$

where:
  Δ is the fractional population difference between the two ground levels
  λ is the cross section for such a frequency shift and is calculated as 6.9 × 10^{-15} cm²
      (Mileti et al. 1992)

The frequency shift introduced thus depends on the population inversion. In the case of pumping from the lower level, $F = 1$, and total population inversion, the shift is of the order of –8 × 10^{-12}/K. In the situation where the lower level is populated, the shift is 1.3×10^{-11}/K. This shift is not negligible and it could very well cause a long-term degradation of frequency stability since it is directly proportional to density. It is well-known that the establishment of thermal equilibrium in a sealed cell is a very long process (Camparo et al. 2005). Consequently, it is possible that the phenomenon is in part responsible for long-term fluctuations and drift observed in Rb frequency standard in general (see discussion on recent spin-exchange calculation in Chapter 1, Section 1.1.3).

3.4.5.5 Microwave Power Shift

If the atomic ensemble is perfectly homogeneous regarding resonance frequency, simple logic leads to the conclusion that the observed resonance frequency should be independent of the microwave power used to observe the resonance signal. The observed frequency should also logically be independent of the shape of the field mode within the cavity. This is due to the fact that all parts of the ensemble have the same resonance frequency and interrogation of any part with different microwave intensity does not alter the observed average frequency. However, as was made explicit above, a residual light shift may be present in the ensemble under certain conditions of operation. This light shift varies continuously along the path of the light beam due to absorption as well as transversely since the light intensity may vary in the radial direction. Since the atoms are fixed in space due to the presence of the buffer gas, this causes inhomogeneous broadening of the resonance line. The measured resonance frequency becomes a function of the applied microwave power since different parts of the ensemble have different resonance frequency and are weighed differently due to the shape of the cavity mode and due to saturation effects. It should be realized that any source of inhomogeneity, such as a magnetic field gradient, would create the same effect (Risley et al. 1980). This effect is called the *power shift*. The centre frequency of the resonance line appears to be a function of the power applied. In some practical devices, the power shift may be of the order of parts in 10^{10}/dB of RF power fluctuation (*QPAFS* 1989).

This effect was studied using a spectral lamp as optical pumping source (Risley and Busca 1978). It was shown that the power shift was indeed due to gradients within the cell and the effect was called the *position shift*. Risley et al. (1980) have shown in more elaborate studies that in a wall coated cell without buffer gas, the atoms being free to move around the cell and to average possible resonance frequency gradients, the power shift disappears completely.

This effect is still present in the case of pumping with a laser and can in fact be amplified since the laser beam intensity varies strongly in the transverse direction. A three-dimensional model of the clock incorporating this effect was also made by Camparo and Frueholz (1989).

In a setup using a Cs cell with a mixture of Ar-N_2 at 39 Torr, a power shift of 3.8×10^{-10}/dB variation of microwave power in normal operating conditions was observed (Yamagushi et al. 1992). A similar result was obtained in the case of Rb (Camparo et al. 2005). In that case, the cell contained pure isotopic ^{87}Rb in a nitrogen buffer gas at a pressure of 100 Torr. The intent in using such a high buffer gas pressure was to reduce light shift as explained above. The atomic ensemble was pumped with a junction transverse strip (JTS) laser tuned to the D_1 wavelength. The laser was locked to the absorption line either from levels $F = 1$ or $F = 2$ of the ground state and a residual light shift of 10^{-11}/% light intensity was observed. A microwave power sensitivity of 5×10^{-10}/dB was measured. Mileti (1995) found in preliminary measurements a microwave power shift of 10^{-11}/dB when the laser was locked to the D_1 optical absorption line. A residual light shift of the order of 2×10^{-11}/% light intensity was also present.

These results appear to lead to the conclusion that, since a residual light shift exists even when the laser is locked to the centre of the optical absorption line, the resonance frequency of the atoms are dependent on their position within the cell. The microwave field, varying also with position, then causes a dependence of frequency on microwave power, various parts of the cell showing varying weights on averaging due to saturation of the signal. This conclusion is in agreement with the results found by Risley et al. (1980) in the case of spectral lamps.

Situations may be found in which the power shift caused by the residual inhomogeneous light shift may be cancelled by a similar opposing effect introduced by an intentionally applied magnetic field gradient. This effect was studied in some detail in a spectral lamp pumped standard and it was found that a microwave power-independent setting could be found for a given value of the magnetic field which causes cancellation of the inhomogeneous frequency shifts (Sarosy et al. 1992).

It should be mentioned that a residual power shift of 10^{-11}/dB as mentioned above requires the microwave generator to be stabilized to 10^{-2} dB to obtain a frequency stability of 10^{-13} in the long term. This may be a difficult requirement to achieve. However, a microwave power stabilizing technique such as that proposed by Camparo based on Rabi resonances could be used (Camparo 1998a, b; Coffer and Camparo 2000).

3.4.5.6 Cavity Pulling

The implementation of a Rb frequency standard with laser optical pumping does not alter the effect of cavity tuning on its frequency when compared to optical pumping with a spectral lamp. We recall that in both types of frequency standards the population difference between the ground-state hyperfine levels $F = 2$, $m_F = 0$, $F = 1$, $m_F = 0$ is the parameter monitored. In that case the clock frequency pulling is given by (QPAFS 1989):

$$\Delta\omega_{clock} = \frac{Q_{cav}}{Q_{at}} \frac{\alpha}{1+S} \Delta\omega_{cav} \tag{3.88}$$

where:
   Q_{cav} is the quality factor of the cavity
   Q_{at} is the atomic resonance quality factor
   S is the saturation factor of the resonance
   α is the measure of power emitted by the atoms relative to the power absorbed by the cavity

In a typical case, $\alpha \sim 10^{-2}$, $S \sim 2$, $(Q_{cav}/Q_{at}) \sim 2 \times 10^{-5}$, and the pulling factor is $\sim 7 \times 10^{-8}$. This is not negligible, as is often assumed in the case of passive standards, and special care must be taken in the design of the cavity and its temperature control in order to have cavity frequency stability compatible with the long-term frequency stability desired. For a required long-term frequency stability of ~$\pm 10^{-13}$, the cavity must stay tuned within ± 10 kHz. Typically a TE_{111} cavity made of copper may have a temperature coefficient (TC) of the order of 200 kHz/degree (Huang et al. 2001). This means that a temperature stability of ± 50 millidegrees over long periods is required to achieve a frequency stability of $\pm 10^{-13}$. Other designs, such as the magnetron cavity, have a TC that is an order of magnitude smaller. In the article cited, a magnetron cavity with eight electrodes had a TC of 28 kHz/K. The magnetron cavity referred to in Figure 1.E.2, with a slightly different design is reported to have a TC of 7 kHz/K (Affolderbach 2014, pers. comm.). These lower TCs reduce considerably the demand on temperature control.

It should also be mentioned that according to Equation 3.88, a change in cavity Q with time may have an important effect on long-term stability even if the cavity is perfectly tuned. This is due to the fact that a change in cavity Q will affect the microwave intensity at the resonance cell, which in turn could affect the residual power shift. This was studied in some detail by Coffer et al. who found that cycling of cavity temperature was affecting Rb distribution within the cell causing metallic deposits on the cell walls and causing degradation of cavity Q (Coffer et al. 2004). This is another effect that can affect the long-term frequency stability of the frequency standard.

3.4.6 IMPACT OF LASER NOISE AND INSTABILITY ON CLOCK FREQUENCY STABILITY

In practice, a wide variety of diode laser designs have been tested for implementing an optically pumped passive Rb standard. A description of such lasers and their characteristics was done in Chapter 2. For our purpose, those lasers can be divided into two general classes, edge-emitting lasers that emit coherent light parallel to the boundaries between the semiconductor layers and VCSELs that emit coherent light perpendicular to the boundaries between the layers. Since the power density required in the present application is less than a few hundreds of microwatts per cm^2, we will limit the discussion to low power lasers, that is to say those diode lasers that emit a total power of the order of a few milliwatts. The edge-emitting diode has typically a spectral emission line width of the order of 20 to 50 MHz while the spectral width of VCSEL is of the order of 50 to 100 MHz.

Unfortunately, as mentioned in Chapter 2, edge-emitting laser diodes have the undesirable characteristics of mode hopping when either temperature or driving current are altered to adjust wavelength and power. This characteristic makes them, in some instances, unusable since the wavelength desired cannot be realized in practice. Diodes with the appropriate wavelength are then selected. In general the yield is very low, 80% of the diodes with wavelength outside the range desired. On the other hand, VCSELs have been designed in general with an internal configuration that prevents mode hoping and, thus, are better adapted to the application reviewed. VCSELs have several other advantages over edge-emitting diodes. The VCSEL is cheaper to manufacture in quantity, is easier to test, and is more efficient requiring less electrical current to produce a given coherent energy output. VCSELs emit a narrow, more nearly circular beam than traditional edge emitters; this makes easier the manipulation of the laser beam such as coupling into an optical fibre. Other types of lasers, showing interesting characteristics regarding spectral line width were also described. Those are special constructions edge-emitting laser diodes such as DBR lasers, or DFB lasers. They have narrow emission spectra (less than a few MHz).

3.4.6.1   Spectral Width, Phase Noise, and Intensity Noise of Laser Diodes

As made explicit in Equation 3.72, the short-term frequency stability of laser-pumped frequency standards is determined by noise from a number of sources. When lasers are used, intensity noise and phase noise may affect the clock frequency stability through various mechanisms. Direct detection of intensity fluctuations at the photodetector, FM-AM conversion noise in the resonant cell, as well as FM conversion through the light shift may affect the frequency stability of the implemented frequency standard. Therefore, a characterization of phase noise and intensity noise of the lasers used is of prime interest. We will limit the study to those laser diodes that have been used and show promises in the implementation of the intensity optical pumping (IOP) double resonance frequency standard presently under study.

3.4.6.1.1   Laser Phase Noise

Phase fluctuations in laser originate from many sources such as spontaneous emission and mechanical instabilities of the structure. The first source is important in diode lasers and generally leads to white frequency noise. We call $S_\phi(f)$ (rad²/Hz) and $S_v(f)$ (Hz²/Hz) the phase and frequency spectral density of phase and of frequency fluctuations, respectively. They are related through the relation:

$$S_v(f) = f^2 S_\phi(f) \tag{3.89}$$

where:

f is a Fourier frequency

The laser emission spectrum is generally Lorentzian and its width originates primarily from those phase or frequency fluctuations. We have addressed this question in Chapter 2 by means of Equation 2.16. An approximate value of the noise spectral density in the case of white frequency noise may be obtained from the laser emission spectrum through the relation (Halford 1971):

$$\Delta v_{1/2} \cong 2\pi S_v(f) \tag{3.90}$$

This relation should be used with caution since it assumes that the laser frequency fluctuations are constant with frequency (white). As just mentioned, for DBR and DFB lasers the line width may be of the order of a few MHz or less and for VCSELs it may be 50–100 MHz. Consequently, the phase spectral density varies widely depending on the type of laser used particularly the quality of their internal cavity. As was shown n Chapter 2, it is possible to reduce the spectral width of those lasers to a few hundred kHz by means of special extended structures in which the laser is part of a resonator acting as a high Q cavity. In that case, the laser frequency fluctuation spectral density is considerably reduced.

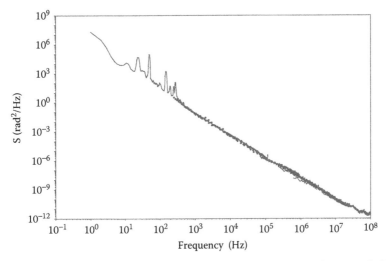

FIGURE 3.39 Spectral density of phase fluctuations of a DFB laser. (Data from Mandache, C., *Rapport de stage*, Mairie de Paris, Paris, France, unpublished.)

As an example, the phase spectral density of a DFB laser is shown in Figure 3.39 (Mandache 2006). The frequency noise spectrum is nearly white with $S_\nu(f) \sim 10^6\,\mathrm{Hz^2/Hz}$, which according to Equation 3.90 gives a spectral width of ~6.3 MHz. The measured width was of the order of 7 MHz. Mileti has reported a spectral density for a free running laser and for the same laser locked to an absorption line. In the case of the locked laser, the noise appeared to be nearly white. It was reduced by a factor of about 50 at a Fourier frequency of 300 Hz, when compared to the free running situation, and its measured frequency fluctuations spectral density was $16 \times 10^6\,\mathrm{Hz^2/Hz}$ (Mileti 1995).

3.4.6.1.2 Relative Intensity Noise

The noise spectral density of the photocurrent fluctuations at the output of a detector illuminated by an optical source may be written as:

$$\mathrm{PSD}_{\Delta I}(f) = 2eI_{\mathrm{ph}} + I_{\mathrm{ph}}^2\mathrm{RIN}(f) \qquad (3.91)$$

where:
I_{ph} is the average value of the output photocurrent
e is the electron charge

The first term on the right-hand side of Equation 3.91 is the shot noise introduced in Equation 3.75. The second term represents the power spectral density of the relative intensity fluctuations. These are characterized by the relative intensity noise (RIN) parameter of the source defined as:

$$\mathrm{RIN} = \frac{(\mathrm{PSD})_{\mathrm{in}}}{I_{\mathrm{ph}}^2} \qquad (3.92)$$

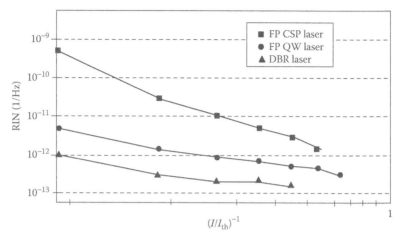

FIGURE 3.40   Relative Intensity Noise as a function of laser driving current at the frequency 1 kHz. FP stands for Fabry–Pérot construction and QW stands for Quantum Well. The FP_CSP laser is a single mode GaAlAs laser. DBR stands for Distributed Bragg Reflector. (Data from Sagna, N. et al., Noise measurement in single-mode GaAl As diode lasers. In *Proceedings of the European Forum on Time and Frequency* 521, 1992.)

and the RIN is generally a function of frequency. Laser diodes intensity noise has been the subject of numerous studies and reports have been written on its characterization for several types of lasers (Joindot 1982; Sagna et al. 1992; Mileti 1995; Coffer and Camparo 1998; Mandache 2006). In general, RIN is a complex function of frequency and laser driving current. A typical result of measurements made at 1 kHz is shown in Figure 3.40 for various lasers, FP and DBR types (Sagna et al. 1992). Mileti (1995) reports a RIN of 2.8×10^{-13}/Hz for an edge-emitting diode laser at a frequency of 300 Hz and a current about 70% above threshold. This is of the order of magnitude of the results shown in Figure 3.40 for the FP lasers. Results on a DBR laser show a RIN of the order of 10^{-13}/Hz at a driving current 50% above threshold. Saburi et al. (1994) also report a RIN of the order of 10^{-13}/Hz for a DBR laser having a spectral line width of 500 kHz. On the other hand, the RIN of VCSEL diodes is a complex function of frequency and driving current and is generally larger than the RIN of edge-emitting diodes.

It may be added that solid-state laser diodes may be characterized by the presence of a residual broadband spectrum spreading over several nanometres (Mileti 1995). To the authors knowledge, no work has been reported on its effect on the frequency stability of a laser pumped frequency standard. In principle it would add directly to the intensity noise at the detector, since that radiation is not absorbed by the cell.

3.4.6.2   Impact of Laser Noise on Clock Short-Term Frequency Stability

3.4.6.2.1   Laser Intensity Fluctuations

Laser intensity fluctuations have been characterized above by means of the relative intensity noise concept RIN. For an edge-emitting laser diode, the RIN may be of the order of 5×10^{-13}/Hz at a frequency of 200 Hz and depends slightly on the frequency at which it is determined (Mileti 1995). This noise, modulating the amplitude of laser radiation, is transmitted through the resonance cell, and is detected by the photodetector as noise directly added to the shot noise component. Using this value and the numbers used before for the hyperfine resonance line characteristics (contrast = 10%, $\Delta v_{1/2}$ = 500 Hz), Equation 3.72 predicts a frequency stability better than 10^{-13}. It is thus concluded that noise detected directly at the photodetector originating from laser intensity noise cannot be responsible for the measured frequency stability of 10^{-11} using such diodes. However, VCSELs are known to have much larger RIN and, if used in an experimental setup, may cause observable frequency instabilities in the short term.

Laser intensity fluctuations can also influence the frequency stability of the clock through the light shift. In the case that the laser tuning and locking is such as to leave a residual light shift, laser intensity fluctuations ΔI are converted into clock frequency fluctuations Δv_{LS} through the dispersion pattern of the light shift. We call S_{LS} the slope of this light shift against current at the detector:

$$S_{LS} = \frac{\Delta v_{LS}}{\Delta I} \qquad (3.93)$$

and the frequency instability of the frequency standard caused by this process may be written as:

$$\sigma(\tau) = \frac{1}{\sqrt{2}} \frac{(\text{RIN})^{1/2} \times I \times S_{LS}}{v_0} \tau^{-1/2} \qquad (3.94)$$

With a laser locked to the optical absorption line in the cell, the residual S_{LS} as mentioned above may be of the order of 5×10^{-11} per 1% change in light intensity at the photodetector. In that case with a RIN of 5×10^{-13}/Hz we obtain $\sigma(\tau) = 2.5 \times 10^{-16}$, a totally negligible contribution. Even if the RIN, as may happen in certain VCSELs, is increased by two orders of magnitude, the contribution would not be visible at the level of frequency stabilities observed presently.

3.4.6.2.2   Laser Frequency Fluctuations

In the short term, the frequency fluctuations of the laser may affect the clock frequency stability through several processes such as, the frequency-locked loop of the laser itself, the light shift and non-linear optical absorption in the cell.

In the first case, the process consists of a conversion of laser frequency noise through the discriminator pattern of the optical frequency-locked loop. If the servo system has a broad band, rejecting laser fluctuations at frequencies greater than the modulation frequency of the rf generator, the effect, in principle, should be negligible. However, higher frequency noise components may create beat notes between high frequency components and may lead to lower frequencies fluctuations, which appear as amplitude noise. This adds directly to the shot noise component. The process is rather complex, but a rough evaluation shows that the noise generated may be of the order of magnitude of shot noise (Mileti 1995).

In the second case, the laser frequency fluctuations are transformed into clock frequency fluctuations through the dispersion pattern of the light shift. The effect on the frequency fluctuations of the clock in the frequency domain, $S_y(f)_{fs}$, can be evaluated through the expression:

$$S_y(f)_{fs} = S_y(f)_{laser} \times \beta_{LS}^2 \tag{3.95}$$

where:

$S_y(f)_{laser}$ is the laser frequency fluctuations spectral density

β_{LS} is the light shift coefficient defined through Equation 3.82

The laser spectral density is a function of the optical servo loop bandwidth and may be of the order of several kHz/Hz$^{1/2}$. On the other hand, as reported above, β_{LS} may be of the order of $5 - 10 \times 10^{-11}$/MHz. For the purpose of calculation of an order of magnitude, we assume a $S_y(f)_{laser}$ for the DFB laser reported above, 10^6 Hz^2/Hz and a $\beta_{LS} = 1 \times 10^{-10}$/MHz or 0.68 Hz/MHz of laser detuning. Assuming white frequency noise, the calculated clock frequency stability is then:

$$\sigma_y(\tau) = 7 \times 10^{-14} \tau^{-1/2} \tag{3.96}$$

a negligible contribution. In the case of an edge-emitting laser locked to linear absorption in the cell mentioned earlier, the $S_y(f)_{laser}$ is 16×10^6 Hz^2/Hz and the effect on frequency stability is $\sim 3 \times 10^{-13}$.

In the third case, the laser frequency fluctuations (FM) are converted into AM by non-linear resonant absorption processes within the resonance cell (Camparo and Buell 1997; Coffer and Camparo 1998). When resonant laser light passes through the vapour, the laser intrinsic phase fluctuations induce random variations in the medium absorption cross section (Camparo 1998a, b, 2000; Camparo and Coffer 1999; Coffer et al. 2002). This phenomenon has feedback on the transmitted radiation and as a consequence laser phase noise (PM) is converted into transmitted laser intensity noise (AM). This process is nonlinear, so that in an optically thick vapour the conversion of PM to AM can increase an optical beam's relative intensity noise by orders of magnitude. In particular, it was found that the resultant RIN after traversal of a cell increased as the square root of the laser spectral width up to a value of the order of the dephasing rate of atoms in the absorption cell. However, it is also found that when the buffer gas dephasing time is much shorter than the field correlation time, or in other words the absorption line width is much larger than the laser spectral width, the PM to AM conversion becomes inefficient. An analysis based on the concepts of cross-sectional fluctuations was found to be in agreement with this observation (Coffer et al. 2002). Consequently, in order to reduce the effect of FM to AM conversion on clock's frequency stability, it appears that at low buffer gas pressures a narrow laser spectrum is preferable, while at high pressure a wider laser spectrum could be used.

Along this line of thought, several attempts have been made at reducing the effect of laser FM noise on the frequency stability of the laser optically pumped Rb standard. A first approach consists in using a laser with a narrow spectral width. In that case, with the laser frequency locked to an external cell, preferably using saturated absorption, clock frequency stability well below the 10^{-11} at 1 s has been obtained. For example, Saburi et al. report a measured frequency stability of $1 \times 10^{-12} \tau^{-1/2}$

using a DBR laser having a spectral line width claimed to be 500 kHz (Saburi et al. 1994). On the other hand, a frequency stability of 3×10^{-13} at an averaging time of 1 s was reported by Mileti et al. using a DBR laser with a spectral width of 3 MHz (Mileti et al. 1998). In that case, however, special care was taken at reducing noise originating from the interrogating microwave generator in order to avoid intermodulation effects to be discussed below. In those two cases (Saburi et al. 1994; Mileti et al. 1998) the cell contained a mixture of buffer gases, one constituent being N_2, but its pressure was not reported.

In an interesting approach, a clone cell was used to cancel passively the intensity noise generated by the clock resonance cell (Mileti et al. 1996). The technique relies on the correlation of the intensity noise generated in two identical cells through FM–AM conversion. The clone cell is not excited by microwave and serves only at creating an intensity noise correlated to the noise created in the clock cell. It appears that the system, although somewhat complex, is rather efficient in noise cancellation. Frequency stability of $5 \times 10^{-13} \tau^{-1/2}$ was reported for a clock using that technique and an extended cavity laser stabilized by means of saturated absorption in an extra external cell.

We have described in Chapter 2 the system developed by Affolderbach et al. for stabilizing a laser frequency through an extended cavity and saturated absorption (Affolderbach et al. 2004). The laser frequency stability obtained should in principle be an excellent optical pumping source for implementing a frequency standard whose frequency stability would not be affected by laser instability in the range of measurements extending to 10^5 s. In fact, the authors have integrated their external cavity diode laser (ECDL) saturated absorption stabilized laser in a frequency standard and obtained a clock short-term frequency stability of $3 \times 10^{-12} \tau^{-1/2}$. Using the Doppler stabilized laser, the authors found that the frequency stability was worst by an order of magnitude over most of the range of averaging times above 100 s (Affolderbach et al. 2004). It should be added that the saturated absorption approach has a drawback in the sense that the laser emission frequency is set by a different cell than the one used as clock reference and may be shifted from resonance. The effect may create a light shift that is dependent on intensity of the laser. It is possible to correct for that property by adjusting the buffer gas pressure in the clock cell in order to shift the optical absorption frequency by the proper amount, minimizing the light shift (Affolderbach et al. 2006). The application of that technique resulted in a unit with a frequency stability of $3 \times 10^{-12} \tau^{-1/2}$ in the range extending to $\tau = 10^4$ s.

In another approach, Camparo et al. (Camparo et al. 2004, 2005) used a cell containing ^{87}Rb with a N_2 buffer gas at a pressure of 100 Torr. The optical absorption line width was 1.6 GHz. Optical pumping was done by means of a JTS laser having a spectral width of 21 MHz and locked to the linear D_1 absorption line of the resonance cell. The results are reproduced in Figure 3.41. The clock frequency stability is:

$$\sigma(\tau) = 1.8 \times 10^{-12} \tau^{-1/2} + 1.1 \times 10^{-13} \tau^{1/2} \qquad (3.97)$$

in the range of measurements reported. Since a single resonance cell generates locking signals for both the laser wavelength and the crystal oscillator, the atomic clock has real potential for miniaturization. Actually, a similar approach was used with a

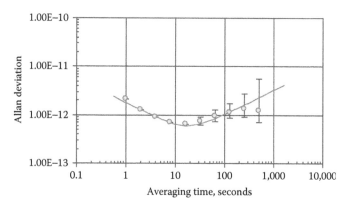

FIGURE 3.41 Frequency stability of a Rb frequency standard using a JTS (Junction Transverse Stripe) laser as optical pumping source. (Reproduced with permission from Camparo, J.C., *Personal communication*, 2007; Camparo, J.C. et al., Reducing PM-to-AM conversion and the light-shift in laser-pumped, vapor-cell atomic clocks. In *Proceedings of the IEEE International Frequency Control Symposium* 134, 2004. Copyright 2004 IEEE.)

VCSEL diode and an improvement in frequency stability by more than a factor two was obtained, the best stability being observed at the same averaging time and being of the order of 10^{-13} (Bablewski et al. 2011).

3.4.6.2.3 Intermodulation Effect

Another phenomenon, which appears to have limited the frequency stability of passive frequency standards, has been the so-called intermodulation effect (Kramer 1974). This effect is present in all types of passive atomic frequency standards and is not connected to the use of lasers for optical pumping. However, it is sufficiently important to be mentioned here. It consists of the so-called aliasing of high frequency noise present in the interrogating signal at even harmonics of the modulation frequency used in the servo loop, to the fundamental frequency in the interrogating signal. An analysis of the phenomenon was made by Audoin et al. in a quasi-static approach (Audoin et al. 1991). It was shown that the main effect was due to noise at the second harmonic of the fundamental modulation frequency f_M:

$$\sigma(\tau) = \frac{1}{2}\left[S_{yLO}(2f_M)\right]^{1/2}\tau^{-1/2} \tag{3.98}$$

A simple calculation shows that such an effect using a low quality quartz oscillator as local oscillator could be disastrous leading to a frequency stability in the 10^{-11} range. Verification of the effect was made in various studies using either spectral lamps or laser optical pumping and several approaches were proposed for minimizing the effect. One approach consisted in introducing notch filters at the second harmonic of the modulation frequency at the output of quartz crystal oscillator used as local oscillator (Szekely et al. 1994). A more direct approach was to reduce the level of noise in the quartz oscillator and following multiplication chain (Deng et al. 1997, 1998). It was also found, in particular, that the effect could be reduced

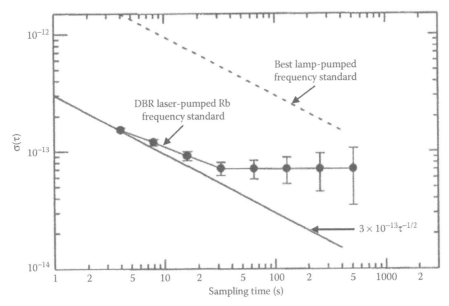

FIGURE 3.42 Frequency stability reported for passive Rb frequency standard using a DBR laser as pumping source locked to saturated absorption. In that case special care was taken to minimize intermodulation effects from the local oscillator. (Reproduced with permission from Mileti, G. et al., *IEEE J. Quantum Electron.*, 34, 233, 1998. Copyright 1998 IEEE.)

by using square wave modulation with wide band detection and demodulation at the synchronous detector (De Marchi et al. 1998). The question was addressed in more details using a dynamical analysis (Ortolano et al. 2000; Beverini et al. 2001). Results obtained in Mileti et al. (1998) in which the local oscillator characteristics were such as to minimize intermodulation effects are reproduced in Figure 3.42. The setup uses a DBR laser for optical pumping source, locked to a cross-over transition saturation absorption line in an external cell. It may be added that due to the somewhat pernicious consequences of the intermodulation phenomenon it is not evident that, without particular care, the frequency stability, as reported in several earlier publications for laser pumped passive frequency standards, was not affected by this intermodulation effect. It is obvious that special care needs to be taken in actual implementations of the local oscillator and modulation scheme if ultimate short-term frequency stability is the goal aimed for. Further experiments, under a well-controlled environment, have resulted in an improved short term frequency stability of $1.4 \times 10^{-13} \tau^{-1/2}$, with a somewhat similar behaviour in the medium term region (Bandi et al. 2014).

3.4.6.3 Medium- and Long-Term Frequency Stability

The question of medium- and long-term frequency stability remains a question to be addressed with some attention. In fact there are no a priori reasons to believe that the long-term frequency stability of a laser pumped frequency standard will be better than that of a standard using a spectral lamp as pumping source. There were hypothesis advanced in the past regarding the possibility that the light shift could be responsible for the long-term drift often observed in standards using a spectral lamp. The question was raised recently by Camparo, and the experimental data led to the conclusion that this was an unlikely cause (Camparo 2005). We have discussed that question in Chapter 2. However, pumping with a laser raises the question to another level. As was discussed above, the light shift may be of the order of several hundred Hz depending on the laser tuning. Consequently, the laser frequency needs to be stabilized with care. Results were reported above regarding such stabilizing systems and it was concluded that, although solutions exist such as using narrow saturated absorption resonance lines, long-term laser frequency stability has not been addressed in sufficient details for the particular application considered. Furthermore, in cases where the laser frequency is locked to linear absorption in the clock cell there may remain a light shift that depends on the intensity of the laser radiation. Studies have been made on the possibility of stabilizing both laser frequency and intensity through servo loops on the laser driving current. It was found that the approach led to practical difficulties, the two parameters not being independent (Tsuchida and Tako 1983). An approach recently used in the control of light shift in a frequency standard based on the CPT phenomenon may offer a solution to the problem. The technique uses an external LCD intensity control independent of laser driving current. The authors claim excellent results reducing the light shift medium-term effect ($100 \text{ s} < \tau < 10,000 \text{ s}$) by at least an order of magnitude (Shah et al. 2006).

The question of long-term drift in sealed cells passive frequency standards thus appears to remain unanswered even using laser optical pumping. A possible answer to the question may reside in the recent experimental results obtained by Camparo in the observation of the long time constants required to reach Rb density equilibrium in cells whose thermal equilibrium had been perturbed (Camparo et al. 2005). As outlined in Chapter 2, it was found that even after hundreds of days, thermal equilibrium was not yet reached under certain circumstances. In that case light absorption varies with time with obvious changes of intensity within the cell and consequences on light shift. Furthermore, changes in Rb density have direct effect on frequency through the spin-exchange shift (Micalizio et al. 2006). These combined effects may play an important role in the long-term drift as observed in sealed cell Rb frequency standards.

This question may be addressed by examining the actual dynamics of thermal equilibrium of various parts of the resonance cell. The cell is generally made of a central glass part several centimetres in dimensions and a small stem, millimetres in diameter, containing the Rb film. This stem is usually maintained at a temperature of a few degrees lower than the cell core itself. This is done to prevent Rb migration in the core of the cell itself, which would have an effect on the cavity Q. Experiments have been done on the influence on the cell transmission and resonance frequency of an abrupt change of temperature of the stem while the cell core temperature is kept constant (Bandi et al. 2014). As was mentioned earlier, it is found that equilibrium is reached only a very long time after the stem change in temperature. In fact, there are two effects observed independently due to a change in buffer gas density and a change in Rb density in the cell. The first effect is very rapid, the frequency of the cell changing abruptly. However, no change in light intensity transmitted is observed, which leads to the conclusion that the Rb density is not altered. This last observation reflects the fact that Rb density does not change rapidly in the cell. It is found, however, that the frequency varies smoothly over a long period with a time constant of the order of days, the change being opposite to the actual rapid change in frequency observed immediately after the stem change in temperature. These results appear to confirm the conclusion that medium- and long-term frequency fluctuations may come from fluctuations of equilibrium density in the cell itself leading to a frequency shift presumably caused by spin-exchange interaction.

3.4.7 OTHER APPROACHES USING LASER OPTICAL PUMPING WITH A SEALED CELL

3.4.7.1 Maser Approach

In the classical approach using a spectral lamp as an optical pumping source, it is possible to create a population inversion such as to obtain self-sustained stimulated emission (maser) when such an optically pumped ensemble of ^{87}Rb atoms is placed in a microwave cavity (Davidovits and Novick 1966). The main threshold conditions for self-sustained oscillation are that the cavity Q be sufficiently large and that a pumping rate compatible with the Rb density be achieved (Vanier 1968). Such a maser has raised considerable interest due to its excellent short-term frequency stability (a few in 10^{13} at 1 s) and its small size. Its properties were studied in detail (Têtu et al. 1973; Busca et al. 1975).

It is natural to ask if the cumbersome spectral lamp used in the implementation of that maser could not be replaced by a laser as in the passive approach described above. Attempts were made and self-sustained oscillation could be realized under certain conditions (Michaud et al. 1990, 1991; Deng et al. 1994). It was found that at the low buffer gas pressure used for maximum gain in the conventional approach, lasers having a narrow spectral width do not provide optical pumping efficient enough to reach oscillation threshold conditions. At low buffer gas pressures, groups of atoms may be off optical resonance due to Doppler shift reducing the number of atoms optically pumped to the upper level. This problem was addressed by modulating the laser frequency in order to cover the cell optical absorption spectrum (Michaud et al. 1990, 1991). At the time of writing, although this IOP Rb maser shows some interesting characteristics, there appears to be little activity on the development of the device.

3.4.7.2 Laser Pulsing Approach

In the early 1960s pulsed optical pumping was proposed as an avenue for addressing the light shift problem when a cell with buffer gas is used. The approach used the time domain Ramsey pulse technique (Ramsey 1956), microwave stimulated emission in a cavity being the measured parameter (Arditi and Carver 1964). In that early development, the population inversion was accomplished by means of a spectral lamp. In that case the difficulty in obtaining large population inversion with short light pulses limited considerably the population inversion and the S/N. Consequently the interest in such an approach was lost for sometime. However, the advent of solid-state diode lasers at the appropriate wavelength and delivering sufficient power raised again the interest in the approach providing more controllable and more efficient optical pumping (Godone et al. 2004a, 2005, 2006b). In the technique, a strong laser pulse is applied to the ensemble of atoms placed in a high Q microwave cavity resonant at the hyperfine frequency of the atoms. Two short microwave pulses are then applied in succession simulating in the time domain the same conditions on the ensemble as in the classical Ramsey technique applied at two space intervals on an atomic beam. The optical pumping pulse and the two microwave pulses provide a Ramsey cycle. State superposition represented classically by an inclination angle θ of the magnetization from the quantization axis results from application of those microwave pulses. Stimulated emission takes place after each pulse with an amplitude that depends on the tuning and power of the applied microwave field and the length of the pulse. The time between the pulses must be less than the decorrelation time caused by relaxation processes of the atoms in the cell. Coherent microwave stimulated emission is detected after the second pulse. Its amplitude as a function of microwave frequency is characterized by the presence of Ramsey fringes, of which the central one can be used, as in the beam approach, to lock the microwave frequency generator to the centre of the hyperfine transition. The advantage of using a laser is that it allows sufficient power for efficient optical pumping and for creating a total loss of microwave phase memory between the successive Ramsey cycles.

The technique just described operates essentially as a maser in the pulse mode, using detection of stimulated emission after the second pulse, but taking advantage of the narrowing of the resonance line by means of the Ramsey interference fringe approach in the time domain. It remains, however, a passive device since it is the amplitude the microwave energy that is detected and used as an indication that the microwave interrogation frequency is resonant with the clock transition. In principle, however, since the light intensity is reduced to zero after the optical pumping pulse, the light shift should also be reduced to zero since the microwave interrogation is done in the dark. The approach was studied in detail and refined to effectuate selective optical pumping such as to populate only one of the clock levels, increasing by the same fact the signal output for the same Rb density. A clock was implemented and a frequency stability of $1.2 \times 10^{-12} \tau^{-1/2}$ was observed over a range of averaging times $1 < \tau < 50,000$ s (Micalizio et al. 2009).

It is also possible to observe the resonance directly by means of a small amplitude laser pulse applied on the cell after the second microwave pulse. The laser pulse acts as a probe and, if appropriately tuned, it measures the population difference between the two clock levels. The probe does not introduce a light shift since it is applied when the microwave interrogating frequency is off. This is very much as in the case of the optically pumped beam approach. It is shown, however, that the central fringe observed by means of microwave stimulated emission is narrower than that observed by means of populations difference by a factor of 2. This property would in principle lead to the conclusion that the stimulated emission approach has an advantage over the population detection approach. However, the stimulated emission approach requires the use of a cavity with a moderate Q (5,000–10,000) for microwave detection as in a maser. In such a case, as described in some detail earlier, cavity pulling may have an influence on frequency stability. This is not the case when optical detection is used. A cavity with a low Q can then be used, reducing by the same fact cavity pulling. Furthermore, optical photons have energy higher than microwave photons by 4 to 5 orders of magnitude. Consequently, a higher S/N is expected in principle in optical detection as compared to microwave detection. Such an approach was also studied in detail and resulted in the most interesting results. Actually a frequency standard constructed on the basis of optical detection showed improved frequency stability over the microwave detection approach by nearly an order of magnitude. Its frequency stability was $1.6 \times 10^{-13} \, \tau^{-1/2}$ over a range of averaging times $1 \ll \tau \ll 10,000$ s (Micalizio et al. 2012a, 2012b).

It should be mentioned, however, that the use of a cell for the implementation of such a standard, aside from the cancellation of the light shift, leads to the same questioning regarding long-term frequency stability. Other effects, such as, temperature fluctuations affecting frequency stability through the buffer gas remain as a major cause of instability, as well as fluctuations of Rb density with time through possible chemical reaction affecting frequency stability through spin-exchange interactions. Due to those effects and possibly other unknown phenomena, the long-term frequency stability of such a standard appears to be limited to the 10^{-14} range (Micalizio et al. 2012c).

3.4.7.3 Wall-Coated Cell Approach

According to Equations 3.72 and 3.74, a reduction in line width is, in principle, a direct way of increasing frequency stability in the short term. In all practical implementations realized up until now, a buffer gas has been used to avoid relaxation caused by collisions of the Rb atoms with the wall of the containing cell. The buffer gas increases the diffusion time to the walls and at pressures above, say 20 Torr, relaxation is caused essentially by spin-exchange collisions between the alkali atoms and collisions of the atoms with the buffer gas molecules. The other broadening mechanisms caused by the rf interrogation signal, through the saturation parameter S, and the optical pumping itself, through the pumping rate Γ_p, contribute also to the resulting line width. Over the years, suggestions have been made on replacing the buffer gas by a non-relaxing wall coating (Singh et al. 1971; Robinson and Johnson 1982). Much work has been done on the effect of collisions on Zeeman coherences for alkali atoms colliding with surfaces made of long chains hydrocarbon (Bouchiat 1965). Such studies have been extended to the case of hyperfine coherence in cells coated with Paraflint© (Moore and Munger) (Vanier et al. 1974). A relaxation rate of 25 s^{-1} has been measured for ^{85}Rb at the hyperfine frequency of 3.0 GHz in a 6.6 cm diameter cell at 27°C. This corresponds to a line width of 8 Hz. The wall shift was measured and found to be 23 Hz at that temperature. For ^{87}Rb, similar results were obtained, the wall shift being of the order of 130 Hz in a 2.5 cm diameter cell at the same temperature (Vanier et al. 1981). Although these results, at first sight, appear most promising, it is important to recall that the line width in an operating device is also determined by several other parameters mentioned earlier. For example, at an operating temperature of 65°C, the spin-exchange broadening $(5/8 \, n\langle v_r\rangle\sigma)$ is of the order of 65 Hz. In order to optimize signal size and contrast, the optical pumping rate (Γ_p) and microwave Rabi frequency (b) are generally set at a point where line width is doubled in each case. The resulting line width, assuming that wall collision broadening is negligible compared to those other mechanisms, would be of the order of 260 Hz. Even in the presence of those added broadenings, there would still be a net gain in line width in using wall coating compared to buffer gas in which case the operating line width is generally above 500 Hz. It may still be advantageous to operate at lower temperatures where spin exchange is less important. Such an approach needs to be studied experimentally.

On the other hand, temperature dependence of the wall shift may be a handicap for long-term frequency stability. In the 2.5 cm cell mentioned above, a temperature coefficient of $10^{-10}/°C$ has been reported requiring a temperature stabilization to ±1 mdegree for a frequency stability of 10^{-13} (Vanier et al. 1981). Nothing is known either on the long-term stability of the wall shift. It should be mentioned that the light shift, although still present, should have a behaviour different from that observed in cells using a buffer gas. Due to the free motion of the atoms, this light shift should not produce inhomogeneous broadening and consequently the power shift observed in certain circumstances in cells using a buffer gas should be absent.

An actual frequency standard using a paraffin-coated cell and laser optical pumping was implemented (Bandi et al. 2012). They obtained a frequency stability of $3 \times 10^{-12} \tau^{-1/2}$ for averaging times between 1 and 100 s. It should be mentioned that the use of laser optical pumping should not alter the conclusions reached above, aside from an amplification of the effect of internal Ramsey interference line narrowing that has been proposed by Xiao for explaining the somewhat *peaky* aspect of the line shape observed when a narrow laser beam is used for optical pumping (Xiao et al. 2006).

3.5   CPT APPROACH

In Chapter 2, we have introduced the phenomenon of CPT, outlined the physics involved, and developed the basic mathematical context. In the present chapter (Section 3.1.2), we have described how the phenomenon could be exploited in an atomic beam to implement a frequency standard using the Ramsey space separated pulse interrogation technique. We have shown that the phenomenon made possible an approach in which the microwave cavity was not required, the atoms being excited in a coherent superposition state by the two laser radiation fields separated in frequency by the hyperfine frequency of the atom used. In the present section, we wish to describe a frequency standard based on the use of that phenomenon in sealed cells either in the passive or active modes.

3.5.1   SEALED CELL WITH A BUFFER GAS IN CONTINUOUS MODE: PASSIVE FREQUENCY STANDARD

An experimental arrangement using a sealed cell containing ^{87}Rb is shown in Figure 3.43. It should be mentioned that the same setup can also be implemented with Cs with appropriate laser wavelength and microwave frequency. The set up shown can be used either to observe the CPT phenomenon through the transmitted radiation or to implement a frequency standard by closing the feedback loop by

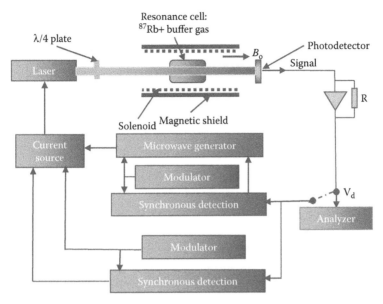

FIGURE 3.43 Experimental arrangement used either to observe the CPT phenomenon in transmission or implement a passive frequency standard.

means of appropriate modulation at a frequency ω_{mod} and synchronous detection. As shown in the figure, a second feedback loop is included in the system for locking the laser wavelength to the optical absorption in the cell. In that second loop, the modulation frequency of the laser current is different from that applied to the microwage generator. We assume that a buffer gas such as N_2 is used. As mentioned earlier, collisions of atoms with that molecule have the property of quenching the fluorescence. This is a desirable property since fluorescence photons emitted at random would cause optical pumping resulting in a loss of coherence in the ensemble. The use of such a gas prevents of course detection of the resonance CPT phenomenon by means of fluorescence. As is readily observed in Figure 3.43, no microwave field is applied to the atomic ensemble and no microwave cavity is required. This is an advantage of the approach over the classical IOP approach that uses the double microwave-optical resonance technique using a cavity as described previously.

The two approaches have been compared in some details in the same cell and conclusions have been reached regarding specific advantages of CPT over IOP in implementing a passive closed cell atomic frequency standard (Vanier et al. 2001, 2003). The microwave Rabi angular frequency b is thus set equal to zero in Equations 2.66 through 2.72. As mentioned earlier, the solution of this set of equations can be done exactly (Orriols 1979). However, as also mentioned, such a solution is rather complex and is not transparent to easy interpretation. It is best to do approximations in order to obtain some insight into the physical phenomena taking place and to evaluate the importance of the various parameters in the final implementation of a frequency standard.

3.5.1.1   Signal Amplitude and Line Width

3.5.1.1.1   Homogeneous Three-Level Model

In practice, the parameter measured in Figure 3.43 is the light intensity at the exit of the cell. Actually, the CPT phenomenon is observed as an increase of transparency of the cell, or a reduction in power absorbed, at exact resonance. We will recall the physics described in Chapter 2 and the main results obtained. The power absorbed by slice dz of the cell is given by:

$$\Delta P_{\text{abs}}(z) = n\hbar\omega_l \Gamma^* \rho_{mm} dz \tag{3.99}$$

where:

   n is the alkali metal density

The energy absorbed is given back either as fluorescence or to the buffer gas vibration modes when a quenching gas such as N_2 is used. Consequently, a measure of transparency is obtained from the change of the population of the excited state ρ_{mm} upon CPT resonance. The value of ρ_{mm} is given by Equation 2.66 in stationary state. In the case of low Rb density, the absorption is small and the system may be considered homogeneous. As an approximation, we may then assume that the density matrix elements are constant throughout the cell. If the sidebands at ω_1 and ω_2 created by modulation of the laser at the frequency ω_m have the same amplitude and the laser is exactly tuned to the optical transition, the population of the two ground levels remain equal, that is $\rho_{11} = \rho_{22}$, and straightforward algebra gives the excited state population as:

$$\rho_{mm} = \frac{\omega_R^2}{\Gamma^{*2}}(1 + 2\delta_{\mu\mu'}^r) \tag{3.100}$$

where:

   $\delta_{\mu\mu'}^r$ is the real part of the coherence created in the ground state by the CPT phenomenon and is given by:

$$\delta_{\mu\mu'}^r = -\frac{1}{2}\frac{(\gamma_2 + \omega_R^2/\Gamma^*)}{(\gamma_2 + \omega_R^2/\Gamma^*)^2 + \Omega_\mu^2} \tag{3.101}$$

with

$$\Omega_\mu = (\omega_1 - \omega_2) - \omega_{\mu'\mu} \tag{3.102}$$

The parameter γ_2 is the coherence relaxation rate, ω_R is the Rabi frequency caused by laser radiation and Γ^* is the excited state decay rate including buffer gas interaction. The transmission shows a sharp increase at resonance when $(\omega_1 - \omega_2) = \omega_{\mu'\mu}$. A typical experimental signal observed directly on an oscilloscope by means of laser frequency slow modulation is shown in Figure 3.44. This CPT transmission resonance signal has a Lorentzian shape and has all the properties of the ground-state hyperfine resonance as in the case of the double resonance technique used in standards based

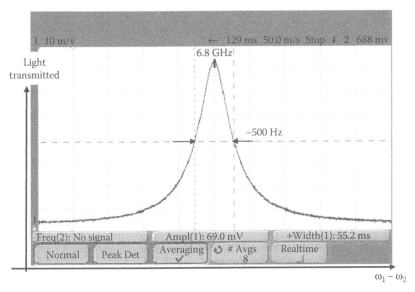

FIGURE 3.44 Experimental transmission CPT signal in a small cell in ^{87}Rb as observed directly by sweeping slowly the microwave modulation of the laser. In that particular recording made in an Ar/N$_2$ mixture with pressure ratio of ~1.5 and at a temperature of about 65°C, the line width is of the order of 500 Hz and the contrast is approximately 5%.

on intensity optical pumping described earlier. The width of the resonance line is given by:

$$\Delta\nu_{1/2} = \frac{(\gamma_2 + \omega_R^2/\Gamma^*)}{\pi} \qquad (3.103)$$

and is a function of the square of the Rabi frequency, ω_R^2, thus proportional to the laser radiation intensity.

3.5.1.1.2 Inhomogeneous Model with a Trap

The previous model, although providing the basic physics behind the CPT phenomenon for implementing a frequency standard, does not represent completely the experimental situation. Firstly, optical absorption is not negligible and the ensemble at normal temperatures of operation (~60°C for Rb) is optically thick to a certain extent. The system is no longer homogeneous. Furthermore, the system includes several energy levels and the transitions from the two selected levels μ and μ' of the ground state to level m are not closed. In particular, atoms excited to the P state under circular polarization (σ^+ or σ^-) and falling into either level $m_F = +2$ or $m_F = -2$ of the ground state are trapped in those levels for a time of the order of $1/\gamma_2$ and are lost for the CPT phenomenon involving the $m_F = 0$ levels. This phenomenon may be accounted for by simply introducing a fourth level into the three-level model as shown in Figure 3.45 (Vanier et al. 2003a).

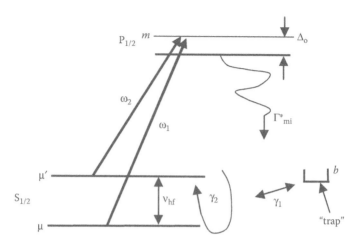

FIGURE 3.45 Four-level model used in the analysis to take into account the trapping property of the levels not involved in the Λ scheme and not excited by the laser radiation under circular polarization.

All Equations 2.66 through 2.72 stay the same except that we now have an added equation taking into account decay from level m to level b:

$$\frac{d}{dt}\rho_{bb} + \left(\gamma_1 - \frac{1}{3}\right)\rho_{bb} = \Gamma^*_{mb}\rho_{mm} \tag{3.104}$$

with

$$\rho_{\mu\mu} + \rho_{\mu'\mu'} + \rho_{bb} + \rho_{mm} = 1 \tag{3.105}$$

We may assume that, due to the large size of Γ^*, ρ_{mm} stays always very small and that the total equilibrium population of the ground-state levels is 1. In order to take into account the inhomogeneous character of the ensemble, it is required to obtain a relation between light intensity and position in the cell. This is done in the following way.

The coherent laser radiation field interacts with the ensemble and creates a polarization P_n. The field E_n and polarization P_n are connected by the relation (*QPAFS* 1989):

$$\frac{\partial^2 E_n}{\partial z^2} - \frac{1}{c^2}\frac{\partial^2 E_n}{\partial t^2} = \mu_0 \frac{\partial^2 P_n}{\partial t^2} \tag{3.106}$$

where:
  c is the speed of light
  μ_0 is the permeability of free space

On the other hand, the polarization is given by:

$$P = nTr(\rho P_{op}) \tag{3.107}$$

where:

n is the Rb density

P_{op} is the equivalent quantum mechanical operator of the classical electrical polarization

In that expression Tr, means the trace of the product of the two matrices ρ and P_{op}. After some algebra, assuming a stationary state and making the adiabatic approximation one obtains:

$$\frac{\partial E_n}{\partial z} = \left(\frac{n\omega_n d_{ij}}{c\varepsilon_o} \right) \mathrm{Im}\,\delta_{ij}(z) \tag{3.108}$$

where:

ω_n is the laser sideband angular frequency

ε_o is the permittivity of free space

d_{ij} is the Rb atom electric dipole moment for transition i to j and it contains the appropriate information on the transition probability of the transitions involved

$\mathrm{Im}\,\delta_{ij}$ is the imaginary part of the optical coherence (Equations 2.69 and 2.70) created in the system by the laser radiation. The problem thus reduces to that of evaluating the value of δ_{ij} in stationary state. To simplify the calculation, we assume that the two radiation fields have equal intensities. Since we do not know the dynamics of the quenching by the buffer gas, we also assume that the decay rates from the excited state to all levels of the ground state are equal:

$$\Gamma^*_{m\mu} = \Gamma^*_{m\mu'} = \Gamma^*_{mb} = \frac{\Gamma^*}{3} \tag{3.109}$$

Since the fractional population ρ_{mm} of the excited state is always small ($<10^{-6}$) relative to that of the ground levels, we assume that in equilibrium, in the absence of radiation, the ground-state levels have equal populations:

$$\rho_{\mu'\mu'}(eq.) = \rho_{\mu\mu}(eq.) = \rho_{bb}(eq.) = \frac{1}{3} \tag{3.110}$$

Using Equations 2.57 and 2.58, Equation 3.108 can be transformed readily into:

$$\frac{\partial \omega_R}{\partial z} = \alpha\,\mathrm{Im}\,\delta_{\mu n} \tag{3.111}$$

where a is the absorption coefficient and is given by:

$$\alpha = \left(\frac{\omega}{c\varepsilon_o \hbar} d_{\mu n}^2 \right) n \tag{3.112}$$

The imaginary part of $\delta_{\mu m}$ is obtained from the rate equations at equilibrium. We assume that the laser is tuned to the optical transition and that the populations of the two ground levels contributing to the CPT phenomenon remain equal. Making the adiabatic approximation in which one assumes that the excited state population and coherence follow the slow evolution of the ground state, one obtains:

$$\text{Im}\,\delta_{\mu m} = -\frac{\omega_R}{\Gamma^*}\left(\frac{1}{3} - \frac{(2/9)(\Gamma_p/\gamma_1)}{1+(2/3)(\Gamma_p/\gamma_1)} + \delta'_{\mu\mu'}\right) \tag{3.113}$$

where

$$\delta_{\mu\mu'} = \frac{-(2/3)\Gamma_p(\gamma_2 + 2\Gamma_p) + (4/9)\left[\Gamma_p^2(\gamma_2 + 2\Gamma_p)/\gamma_1(1+2\Gamma_p/3\gamma_1)\right]}{(\gamma_2 + 2\Gamma_p)^2 + (\omega_{12} - \omega_{\mu'\mu})^2}$$

$$+ i\frac{\left\{(2/3)\Gamma_p - (4/9)\left[\Gamma_p^2/\gamma_1(1+2\Gamma_p/3\gamma_1)\right]\right\}(\omega_{12} - \omega_{\mu'\mu})}{(\gamma_2 + 2\Gamma_p)^2 + (\omega_{12} - \omega_{\mu'\mu})^2} \tag{3.114}$$

and

$$\rho_{bb} = \frac{1}{3} + \frac{(4/9)(\Gamma_p/\gamma_1)}{1+(2/3)(\Gamma_p/\gamma_1)} \tag{3.115}$$

To simplify writing, we have defined the pumping rate, function of position in the cell, as

$$\Gamma_p(z) = \frac{\omega_R^2(z)}{2\Gamma^*} \tag{3.116}$$

In order to obtain the value of the radiation intensity at the exit of the cell, Equation 3.111 needs to be solved with $\text{Im}\,\delta_{\mu m}$ and $\delta'_{\mu\mu'}$ given by Equations 3.113 and 3.114. Unfortunately, due to the complexity of the right-hand side, Equation 3.111 cannot be solved analytically and a numerical approach is required in which $\omega_R^2(z)$ is evaluated at the exit of the cell, that is $z = L$.

In view of the form of $\delta_{\mu\mu'}$, the resonance line shape is assumed to be Lorentzian with a width given by:

$$\Delta\nu_{1/2} = (1/\pi)\left(\gamma_2 + \frac{\omega_R^2(z=L)}{\Gamma^*}\right) \tag{3.117}$$

where:
γ_2 is the relaxation rate in the absence of laser radiation or $\omega_R = 0$

γ_2 includes several mechanisms such as relaxation by diffusion to the cell walls, relaxation by collision with the buffer gas molecules, and spin-exchange interaction between Rb atoms as described in the case of the IOP frequency standard.

3.5.1.2   Practical Implementation and Its Characteristics

3.5.1.2.1   Frequency Stability

In the implementation of a passive frequency standard based on the CPT phenomenon, the frequency of the microwave generator used to modulate the laser frequency is locked to the centre of the hyperfine resonance line. This may be done by modulating the frequency of the microwave generator at a low frequency and using synchronous detection to create an error signal. If the amplitude of this modulation is of the order of one half the resonance line width, the short-term frequency stability in the limit of shot noise is given by Equations 3.72 and 3.75, which can be written as (Vanier and Bernier 1981; Vanier 2002; Vanier et al. 2003b):

$$\sigma(\tau) = \frac{K}{4\nu_{hf}} \sqrt{\frac{e}{I_{bg}}} \frac{1}{q} \tau^{-1/2} \tag{3.118}$$

where:

K is a constant that depends on the type of modulation used and is of the order of 0.2

ν_{hf} is the hyperfine frequency

e is the charge of the electron

I_{bg} is the background current created by the residual transmitted radiation reaching the photodetector

τ is the averaging time

q is a quality figure defined earlier as the ratio of the contrast C to the line width $\Delta\nu_{1/2}$:

$$q = \frac{C}{\Delta\nu_{1/2}} \tag{3.119}$$

The contrast C is defined as the CPT signal intensity divided by the background intensity. To obtain best frequency stability, it is thus important to maximize contrast and to minimize line width. In practice, as we have outlined in the case of the double resonance frequency standard using laser optical pumping, the laser spectrum is affected by amplitude and frequency fluctuations, which are additional sources of noise affecting frequency stability, and the limit of shot noise given above is not reached. For example, the amplitude noise, characterized by the relative intensity noise parameter, RIN (Sagna et al. 1992), adds directly to shot noise. For amplitude noise, Equation 3.118 becomes:

$$\sigma(\tau) \approx \frac{1}{\sqrt{2}} \frac{(RIN)^{1/2}}{4q\nu_{hf}} \tau^{-1/2} \tag{3.120}$$

In such a case, it is also important to maximize the quality figure q. We recall that laser frequency fluctuations are transformed into amplitude fluctuations through various resonance mechanisms in the atomic ensemble and contribute to additional noise thus affecting frequency stability (Camparo and Buell 1997; Coffer et al. 2002).

The signal amplitude is a function of Rb density and pumping rate. A typical experimental result is shown in Figure 3.46 for a cell containing a buffer gas mixture of Ar and N_2 in a pressure ratio of 1.4 at a total pressure of 10.5 Torr. The solid curve through the experimental points is a plot of a numerical integration of Equation 3.111 using the four-level model developed above. The free parameter used in this plot is the maximum contrast observed. It is seen that the contrast is limited to a maximum value of the order of 5%. This is due to the fact that in ^{87}Rb only two levels out of eight in the ground state contribute to the CPT phenomenon, the other levels contributing individually to optical absorption. Furthermore, in modulating the laser through the bias current, many sidebands are produced which contribute to the background light intensity. It is possible to obtain a larger contrast by raising the cell temperature, thereby increasing Rb density and optical absorption. The contrast maximum is then observed at a larger light intensity. An undesired added side effect, however, is an increase in line width due to increased laser radiation broadening and spin-exchange interaction. In practice, the optimum operating condition giving a maximum quality figure q is determined experimentally.

The behaviour of the line width is also shown in Figure 3.46 as a function of light intensity. The results confirm the linear dependence of the line width on light intensity. It is used to calibrate the system in terms of Rabi frequency by means of Equation 3.117. The four-level model was further verified by applying it to an ensemble at several other temperatures. It was found to agree well with the experimental data (Vanier et al. 2003a).

It may be pointed out that the standard three-level model without trap predicts a contrast that increases with Rb density and laser radiation intensity. This is not in agreement with the experimental data that shows a maximum in response to laser intensity. The maximum is caused by the fact that the trapping of atoms in level $m_F = 2$ becomes important at radiation intensities larger than the relaxation rate in the ground state.

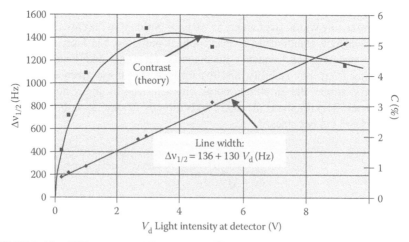

FIGURE 3.46 CPT contrast and line width in a ^{87}Rb cell containing an Ar-N_2 buffer gas mixture and operated at 75° C. The points are experimental and the solid curves are theoretical.

FIGURE 3.47 Commercial passive Rb frequency standard based on CPT. The system is totally autonomous, using a digital electronics approach for accomplishing the wavelength and frequency lock of the laser and microwave source to the atomic resonances. Its volume is of the order of 125 cm^3. (Reproduced with permission from Vanier, J. et al., *IEEE Trans. Instrum. Meas.*, 54, 2531, 2005. Copyright 2005 IEEE.)

A totally autonomous frequency standard using the basic CPT approach is shown in Figure 3.47 (Vanier et al. 2004, 2005). The alkali metal atom used is ^{87}Rb and the buffer gas is a mixture of Ar and N$_2$ in a pressure ratio of about 1.5. The cell is a glass enclosure of a few cm^3 in volume and the optical radiation source is a VCSEL modulated in frequency at 3.4 GHz. The contrast observed at the optimum operating point is of the order of 5% and the line width is somewhat less than 500 Hz. The electronic servo systems used are made entirely of digital electronics. The unit has a volume of 125 cm^3. The frequency stability obtained is of the order $3 \times 10^{-11}\tau^{-1/2}$ and reaches a level below 10^{-12} at an averaging time of the order of 2000 s as shown in Figure 3.48. It is one of the best CPT frequency standards that has been realized as a complete self-locking and sustaining unit.

3.5.1.2.2 Frequency Shifts

As in the standard IOP double resonance approach, there are many frequency shifts that are present in such an implementation. These include magnetic field shifts, buffer gas shifts and light shifts. Although they are well-known and do not create a major problem in the operation of the system, instability in these shifts may affect frequency stability in the medium- and long-term regions of averaging times. Regarding magnetic field and buffer gas shifts, the reader is referred to the discussion made in relation to the implementation of double resonance Rb IOP frequency standards.

As discussed previously, the light shift is a frequency shift that is important in the classical IOP approach using laser optical pumping. Fortunately, this shift appears to be well under control in CPT. In the case of frequency modulated lasers, this is due to the fact that for equal amplitudes, the first sidebands radiation fields displace both ground-state levels by the same amount. On the other hand, the total light shift

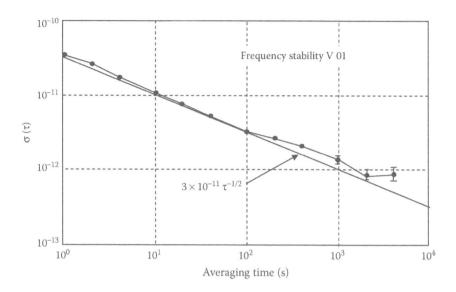

FIGURE 3.48 Frequency stability of the CPT frequency standard shown in Figure 3.47, using a small cell (a few cm³) and a VCSEL as the optical pumping source. (Reproduced with permission from Vanier, J. et al., *IEEE Trans. Instrum. Meas.*, 54, 2531, 2005. Copyright 2005 IEEE.)

introduced by all sidebands disappears for certain modulation conditions. This question has been studied in some detail (Vanier et al. 1999; Levi et al. 2000; Zhu and Cutler 2000). The total sidebands light shift is given by:

$$\frac{\Delta\omega_{LS}}{\omega_{\mu\mu'}} = \left(\frac{\omega_R}{\omega_{\mu\mu'}}\right)^2 \left[\Theta(m) + \xi(m)\left(\frac{\Delta_o}{\omega_{\mu\mu'}}\right)^2\right] \qquad (3.121)$$

where it is recalled that Δ_o is the laser detuning defined in Figure 3.45. The coefficients $\Theta(m)$ and $\xi(m)$ are given by the following expressions:

$$\Theta(m) = J_o^2(m) + \frac{1}{2}J_{p/2}^2(m) - 2\sum_{n=1\neq p/2}^{\infty} J_n^2(m)\left[\frac{p^2}{(2n)^2 - p^2}\right] \qquad (3.122)$$

$$\xi(m) = 4J_o^2(m) - 8\sum_{n=1\neq p/2}^{\infty} J_n^2(m)\frac{12n^2 + p^2}{\left[(2n)^2 - p^2\right]^3}p^4 \qquad (3.123)$$

In these expressions, J_i is a Bessel function and p is an even integer defined as the ratio of the hyperfine frequency $\omega_{\mu\mu'}$ to the laser modulation frequency ω_m. When p is set equal to 2, the two first sidebands J_{1+} and J_{1-} on each side of the carrier are used in the CPT-Λ scheme and in the case of ^{87}Rb, ω_m is $2\pi \times 3.41 \times 10^9$ s^{-1}. The two functions are plotted in Figure 3.49 as a function of the index of modulation m.

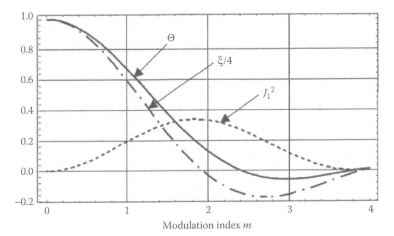

FIGURE 3.49 Variation of the light shift coefficients with modulation index in the case the first sidebands J_{1+} and J_{1-} are used to excite the CPT phenomenon ($p = 2$). (With kind permission from Springer Science+Business Media: *Appl. Phys. B*, Atomic clocks based on coherent population trapping: A review. 81, 2005, 421, Vanier, J.)

As is readily observed, the main component of the light shift, called the power shift and given by the coefficient $\Theta(m)$, vanishes at a laser modulation index, $m = 2.4$. This is due to the fact that the sidebands created in the modulation process create light shifts that compensate each other. A typical experimental result is shown in Figure 3.50 for the unit just illustrated in Figure 3.47 (Vanier et al. 2004, 2005). These results show that the clock frequency becomes independent of laser intensity for a modulation index of ~2.4 as predicted. On the other hand, the quadratic part of the light shift given by coefficient $\xi(m)$ is of the order of 10^{-14}/MHz of laser detuning from optical resonance and is negligible in most circumstances (Levi et al. 2000).

3.5.1.2.3 Line Shape of Resonance Signal
In the analysis just presented, the amplitude of the two sidebands J_{1+} and J_{1-} were assumed equal. Furthermore it was assumed that the laser is tuned exactly to the optical transition. Studies of the shape and of the centre frequency of the hyperfine resonance as observed in transmission have been made for the case when these two conditions are not satisfied (Levi et al. 2000). The problem was analyzed for the case of a three-level model in an optically thin sample. A numerical solution of the rate equations showed that when the two conditions are not satisfied, the line shape is no longer Lorentzian, as is shown in Figure 3.51a. However, the minimum of the resonance line is not displaced in the process. This effect is observed experimentally as is shown in Figure 3.51b.

The question of the effect of the laser light beam transverse variation on the shape of the CPT resonance line was also addressed (Levi et al. 2000; Taichenachev et al. 2004). The line shape is altered because the pumping rate $\omega_R^2/2\Gamma^*$ is a function of radial position across the laser beam. In the case of low pumping rates, it is found that the line shape remains Lorentzian while in the case of large pumping rates, the line shape becomes sharper than a Lorentzian, although it remains symmetrical.

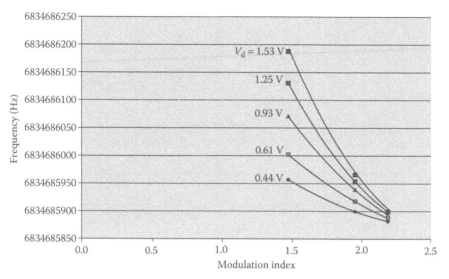

FIGURE 3.50 Light shift observed in the compact unit described in the text as a function of the modulation index of the laser for various radiation intensities. (Reproduced with permission from Vanier, J. et al., Practical realization of a passive coherent population trapping frequency standard. In *Proceedings of the IEEE International Ultrasonics, Ferroelectrics, and Frequency Control Joint 50th Anniversary Conference* 92, 2004. Copyright 2004 IEEE; Vanier, J. et al., *IEEE Trans. Instrum. Meas.*, 54, 2531, 2005. Copyright 2005 IEEE.) The numbers associated with the various curves are light intensities as measured at the photodetector in Figure 3.43.

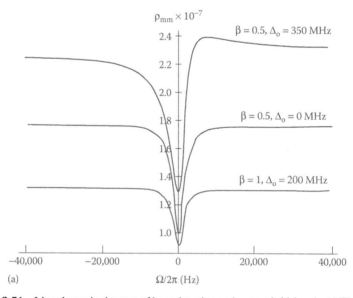

FIGURE 3.51 Line shapes in the case of laser detuning and unequal sidebands. (a) Theoretical; β stands for the ratio ω_{R1}/ω_{R2} and Δ_o is the laser central frequency detuning defined in Figure 3.45.

(Continued)

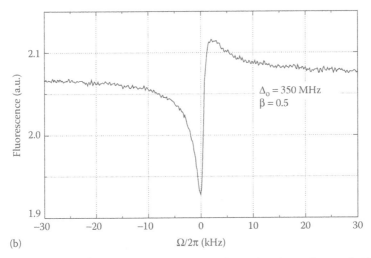

FIGURE 3.51 (Continued) Line shapes in the case of laser detuning and unequal sidebands. (b) Experimental. (Data from Godone, A. et al., *Coherent Population Trapping Maser*, CLUT, Torino, Italy, 2002. Copyright 2002 CLUT.)

It is interesting to point out that the four-level model with a trap, used above to explain the behaviour of the CPT transmission resonance line contrast in ^{87}Rb, may also be used to analyze the shape of the absorption spectrum for the case of an unmodulated laser radiation field. In practice, it is found that the recorded shape of the optical absorption for the D1 line is a function of polarization and light intensity. This is due to the fact that upon slow sweeping of the laser wavelength across the absorption line, optical pumping takes place, with various energy levels acting as traps. An analysis similar to that done above but with monochromatic radiation was done for ^{87}Rb. The results are reported in Appendix 3.C and are found to be in agreement with the observed absorption spectra. These results validate to some extent the four-level model used above to explain the contrast observed in the CPT transmission resonance line.

Aside from the early work reported on ^{87}Rb and ^{133}Cs (Cyr et al. 1993; Levi et al. 1997), work has also been done using ^{85}Rb (Lindvall et al. 2001; Merimaa et al. 2003). In that last case, the hyperfine frequency is 3.035 GHz. An edge emitter diode was used, modulated at that frequency. The carrier J_0 and one of the first sidebands J_1 were used to generate the two laser frequencies identified as ω_1 and ω_2 in Figure 3.45. Argon and neon were used as buffer gases. The authors have obtained a CPT hyperfine resonance line width at room temperature of the order of 20 Hz for zero light intensity. This compares well to a line width of 10 Hz measured in the dark in such buffer gases at room temperature by means of the pulsed stimulated emission technique (Vanier et al. 1974). In their case, a frequency generator locked to the CPT resonance line gave a frequency stability of $3.5 \times 10^{-11}\tau^{-1/2}$ for $1\,\text{s} < \tau < 2000\,\text{s}$.

3.5.1.2.4 Other Approaches

Cs has also been used in bench setups to demonstrate the possibility of making a small optical package (Levi et al. 1997; Kitching et al. 2000, 2001). The implementation of an optical package using a silicon substrate as the container has been reported (Knappe et al. 2004). The intention is the implementation of an integrated unit including control electronics on the same substrate. Since the cell itself is very small (~mm³), diffusion to the walls causes a large CPT resonance line broadening (~7 kHz) and reduces the contrast considerably (<1%) relative to cells having a volume in the cm³ range. It may be mentioned that when the necessary hardware consisting of temperature control, isolation, solenoid, and required magnetic shields are taken into account, not much is gained in volume reduction in such packages and the cost on frequency stability is important. In fact, the frequency stability obtained using such small cells and a VCSEL for optical pumping in a bench setup is of the order of $2 \times 10^{-10} \tau^{-1/2}$ for 1 s < τ < 30 s. A rather important linear frequency drift is observed above this averaging time. However, it was shown that the use of DFB laser for optical pumping improves the overall frequency stability to 3.8×10^{-11} at 1 s averaging time reaching 10^{-12} at 1000 s (Boudot et al. 2012). Other work made towards the implementation of small cells has also been reported (Zhu et al. 2004; Lutwak et al. 2009).

Studies have been initiated using amplitude modulation of the laser to generate the field required to create the CPT phenomenon (W. Happer, *pers. comm.*). The goal is to use very high buffer gas pressures to inhibit diffusion to the cell walls and thus making possible narrow resonance lines even in very small cells. The idea is based on the concept that CPT behaves differently when amplitude modulation of the laser is used instead of frequency modulation to create the sidebands. With amplitude modulation it appears that the excitation of the CPT phenomenon is possible with broad overlapping optical lines (large buffer gas pressures) (Jau et al. 2004). In the case of frequency modulation, the observation of the CPT signal requires that the optical lines be resolved making the signal amplitude a function of the buffer gas pressure.

3.5.1.2.5   Reflections on Frequency Stability

The expected frequency stability may be calculated with Equation 3.118 in the limit of shot noise. The quality figure q is obtained from the contrast and the line width as calculated above. In Vanier et al. (2004), for a temperature of the order of 65 to 70°C, a q figure of the order of 1.5×10^{-4} for a background intensity $I_{bg} = 10 \times 10^{-6}$ A is obtained. The resulting expected calculated frequency stability is of the order $7 \times 10^{-14} \tau^{-1/2}$.

Unfortunately such frequency stability is not observed experimentally as reported above. A frequency stability of the order of 3 to $5 \times 10^{-11} \tau^{-1/2}$ has been a common observation in most devices, two orders of magnitude less than that expected from the limit of shot noise (Merimaa et al. 2003; Vanier et al. 2003b, 2004). This behaviour is similar to the case of the classical passive IOP frequency standard using a solid-state diode laser for optical pumping source as was reported earlier.

It is believed that this behaviour is due to the inherent AM and FM noise imbedded in the laser radiation as in the case of the IOP approach. The AM noise appears directly at the photodetector as intensity noise and adds directly to shot noise. That noise contributes to the order of several parts in 10^{-13} to the frequency instability and of course depends on the RIN of the laser used. On the other hand, it is believed that the laser FM noise is converted into AM fluctuations by the nonlinear optical resonance absorption within the ensemble as we have discussed earlier. It also adds directly to shot noise at the detector. We have also mentioned that experiments have been made in order to compensate this effect by either reducing the spectral width of the laser spectrum or by direct compensation by means of a clone resonance cell in the IOP approach. Relative success was obtained in that last case and although the technique is somewhat complex, it appears to provide an avenue for reducing the effect.

3.5.2   Active Approach in a Cell: The CPT Maser

Another approach using CPT for implementing a frequency standard consists in exploiting directly the hyperfine coherence created in the ground state as in a maser. This coherence creates in turn an oscillating magnetization at the same frequency (Vanier et al. 1998; Godone et al. 1999). When the ensemble is placed in a cavity, the magnetization excites the cavity mode and creates an oscillating magnetic field. This oscillating magnetic field reacts back on the atomic ensemble and causes stimulated emission in the same manner as in a maser. This CPT maser has very interesting characteristics such as absence of threshold relative to pumping rate, alkali atom density and cavity Q, a great improvement over the intensity pumped Rb maser which has critical threshold conditions relative to these parameters (Vanier 1968). It may be mentioned that these threshold conditions have prevented the practical realization of Cs maser using IOP (Vanier and Strumia 1976), while the same Cs maser was readily implemented using CPT (Vanier et al. 1998).

The analysis of the CPT maser is more complex than that of the passive approach in view of the interaction of the atoms with the microwave field in the cavity. The analysis must take into account the phase relationship ϕ between the microwave field and the oscillating magnetization. Furthermore, the phase of the optical fields is coupled to the hyperfine coherence. An analysis taking into account the interrelation between these phases has been developed in the case of a closed three-level system (Godone et al. 2000). A more exact analysis of the CPT maser characteristics including optical pumping in an open system is more complex. However, the most important characteristics of the maser may be obtained from approximate models that provide some insight in the physical phenomena taking place. For this reason, we will outline a few approaches that make explicit the fundamental characteristics of the maser under specific conditions of operation as a frequency standard.

3.5.2.1   Basic CPT Maser Theory

The experimental setup used in the analysis is that illustrated in Figure 3.52 with the cavity tuned close to the hyperfine frequency. In a maser, the energy given by the atomic ensemble is dissipated in the cavity walls and coupling loop. An analysis using a so-called self-consistent approach is given in Appendix 3.D. It is shown that the power delivered by the atoms is given by the expression:

$$P_{\mathrm{at}} = \frac{1}{2} \frac{N\hbar\omega k}{\left[1 + 4Q_L^2 \left(\Delta\omega_c / \omega_{\mu'\mu}\right)^2\right]} \left|2\delta_{\mu'\mu}\right|^2 \tag{3.124}$$

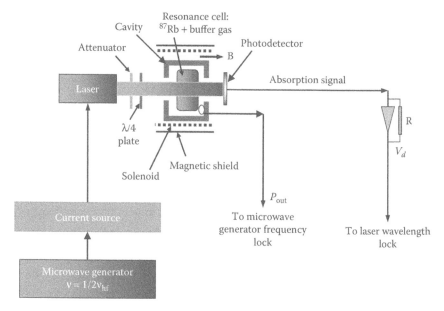

FIGURE 3.52   Block diagram of the CPT maser using ^{87}Rb.

where k is given by Equation 3.D.3. An important relation is also obtained between the Rabi frequency, a measure of the microwave field intensity in the cavity, and the coherence existing in the ensemble:

$$\langle b \rangle = 2k\left|\delta_{\mu'\mu}\right| \tag{3.125}$$

where the brackets <> mean average over the length of the cell and the double bar | | symbol means absolute value. On the other hand, the phase between the field and the magnetization in the ensemble is given by:

$$\phi = \frac{\pi}{2} + \tan^{-1} 2Q_L \frac{\Delta\omega_c}{\omega} - \tan^{-1} \frac{\delta^i_{\mu'\mu}}{\delta^r_{\mu'\mu}} \tag{3.126}$$

making explicit the phase quadrature, close to $\pi/2$, between these two physical quantities. The problem is thus one of evaluating the coherence $\delta_{\mu'\mu}$. This is done from the rate equations. Several approximations adapted to the experimental situation at hands can be made in such a process and they are outlined in the following paragraphs.

3.5.2.1.1   Homogeneous Model

A simple approach for obtaining an expression for the maser power output consists in assuming that the ensemble is optically thin and in replacing $\delta_{\mu'\mu}$ in Equation 3.124 by the value found in Equation 3.114. This approximation is valid only at low temperatures. In doing so one entirely neglects the feedback of the microwave field on the atomic ensemble coherence and on the population of the levels. This is actually rather approximate, but the exercise nevertheless provides some insight into the behaviour of the maser. In that exercise, it is readily observed that the maser does not have a threshold relative to cavity Q, density, line width, or pumping rate.

A more realistic model consists in solving the set of rate Equations 2.66 through 2.72 with the presence of the microwave Rabi frequency in the equations and taking into account its coupling to the coherence by means of Equation 3.125. The system may then be solved taking into account the phase of the microwave radiation given by Equation 3.126. The results show again that the maser does not have a threshold relative to the parameters mentioned above (Godone et al. 2000).

3.5.2.1.2 Inhomogeneous Model

In practice, however, as in the case of the passive approach, the system becomes optically thick at normal temperatures of operation and the ensemble cannot be considered homogeneous. In such a case, the amplitude of the various parameters varies along the length of the cell and the set of Equations 2.66 through 2.72 is valid only locally, that is for a thin slice of the medium. Furthermore, the phase of the laser sidebands, when taken into account, is coupled to the microwave field and varies along the length of the cell. The problem has been solved in the case of a closed three-level system by dividing the cell into small slices and by means of a numerical integration over the cell length of the Rabi frequency b, which is a function of distance in the cell and of sidebands phase (Godone et al. 2002a, b).

It is found in particular that the calculated maser power is a function of the ratio of cell length to microwave wavelength. In fact, for low densities, and a cell whose length L is equal to $\lambda_{\mu w}$, the power output is equal to zero. This effect results from the fact that the elementary magnetizations in each slice emit with their own phase. The phase of the microwave radiation created in the second half of the ensemble is opposite to that of the radiation created in the first half. However, at high densities, the system is not homogeneous and the phase cancellation from one part of the cavity to the other is not exact: the part of the cell at the entrance of the cavity contributes more than the part at the exit of the cavity since radiation is strongly absorbed. Nevertheless, the power calculated still decreases for cell lengths larger than $\sim\lambda/2$.

The analysis can also be developed for the four-level model developed above in the case of the passive approach. In that case, in order to simplify the analysis, the cell dimension is assumed less than a half wavelength and the variation of phase is neglected along the length of the cell. The solution consists in dividing the cell into thin slices and solving Equations 2.66 through 2.72 for each of these slices as was done in the case of the three-level model. The value of the Rabi frequency is calculated at each slice by means of the Equation 2.32 with α defined by Equation 2.33 as in the case of the passive approach. The problem is reduced to one of solving the resulting set of equations that includes the self-consistent condition in which the power emitted by the ensemble is equal to that lost in the cavity. Those equations are:

$$b = -2\frac{k}{L}\int_0^L \delta_{\mu\mu'}^r(z)dz \qquad (3.127)$$

$$\frac{\partial}{\partial z}\Gamma_p = -\alpha\frac{2\Gamma_p}{3\Gamma^*}\left\{1 - \frac{(2/3)(\Gamma_p/\gamma_1)}{1+(2/3)(\Gamma_p/\gamma_1)} + 3\delta^r_{\mu\mu'}\right\} \qquad (3.128)$$

$$\left(\gamma_1 + 2\Gamma_p + \frac{b^2}{\gamma_1 + 2\Gamma_p}\right)\delta^r_{\mu\mu'} = -\Gamma_p\left\{\frac{2/3}{1+(2/3)\Gamma_p/\gamma_1}\right\} \qquad (3.129)$$

where the pumping rate is given by:

$$\Gamma_p = \frac{\omega^2_R}{2\Gamma^*} \qquad (3.130)$$

The power emitted by the atomic ensemble is given by:

$$P_{at} = \frac{kN_a}{2L^2}\hbar\omega_{12}\left[2\int_0^L \delta^r_{\mu\mu'}(z)dz\right]^2 \qquad (3.131)$$

These equations are solved through a recursive method in which the Rabi frequency b is first given a value. Integration is then done numerically for a given value of the pumping rate Γ_p at the entrance of the cell. The solution provides a new value of b, which is then used as the initial parameter for a new integration. The process converges rapidly (Godone et al. 2002). The result of this exercise is shown as the solid lines in Figure 3.53 as a function of the pumping rate Γ_p for three values of α corresponding to the temperature indicated.

Experimental results obtained in similar conditions are shown as points on the same graph (Godone et al. 2002). It should be pointed out, as done previously, that in general the density of atoms in the Rb ensemble is not well-known and depends to a large extent on the previous temperature cycling of the cell used or its history (Gibbs 1965; Vanier 1968). The experimental value of the density is usually smaller than that obtained from tables. For this reason, the free parameter used in the adjustment of the theoretical results to the experimental data is the power output, function of density. It has been adjusted to match approximately the maser power at the highest temperature. It is observed that this procedure nevertheless provides a semi-quantitative agreement with the experimental data.

3.5.2.2 Frequency Stability

The CPT maser may be considered to be a hybrid device, the atoms emitting energy, but at a frequency given by the difference of the two laser sidebands, $\omega_1 - \omega_2$. In that case, the frequency of the microwave generator used to modulate the laser needs to be locked to the frequency of the maser emission line maximum by synchronous detection as in the passive case. It is thus a frequency-lock system similar to that used in the case of a passive maser.

Considering only thermal noise, the short-term frequency stability limit of the CPT maser may be written as (Godone et al. 2002, 2004):

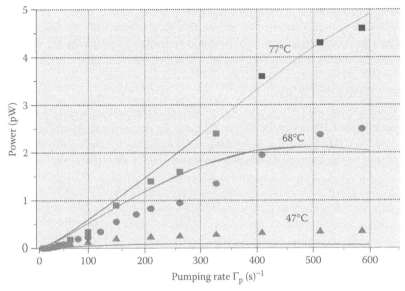

FIGURE 3.53 Power output of the CPT ^{87}Rb maser. Continuous solid lines are obtained from the four level-model developed in the text while the points are experimental. (Data from Godone, A. et al., *Coherent Population Trapping Maser*, CLUT, Torino, Italy, 2002. Copyright 2002 CLUT.)

$$\sigma(\tau) \approx \sqrt{\frac{Fk_BT}{2P_o}} \frac{1}{Q_a} \tau^{-1/2} \qquad (3.132)$$

where:

    T is the temperature of the cell-cavity arrangement
    P_o is the power output of the maser
    Q_a is the atomic line quality factor
    F is the noise figure of the amplifier in the first stage of the receiver

Since the output frequency is the difference frequency of the correlated sidebands, laser frequency noise should not affect the CPT maser output frequency directly. However, these fluctuations may affect the maser frequency stability through the light shift that will be examined below. As in the case of the passive CPT approach, it is important to minimize the maser line width to obtain the largest possible line Q in order to maximize frequency stability. Similarly, higher emission power improves frequency stability. However, since higher power generally leads to a greater line width, a trade off must be made. In a typical situation, around 60°C with a pumping rate of 300 s^{-1}, the power delivered by the atoms to the cavity is typically 1×10^{-12} W. The line width is of the order of 175 Hz providing a line Q_a of 3.9×10^7. With a receiver noise figure of 1.2, the expected frequency stability is then of the order of $1 \times 10^{-12} \tau^{-1/2}$. An experimental setup in which a crystal oscillator was frequency locked to the maser power output maximum gave, under conditions similar to those just described, a short-term frequency stability of the order of $3 \times 10^{-12}\tau^{-1/2}$ (Godone et al. 2004). It is mentioned by the authors that their implementation was not optimized and that better frequency stability could be realized through the use of a better quartz crystal oscillator in the frequency-locked loop (Godone et al. 2002).

3.5.2.3 Frequency Shifts

All the shifts mentioned in the case of the CPT passive frequency standard approach are present in the CPT maser. Furthermore, there are frequency shifts that are particular to the maser. In particular, a detuning of the laser by Δ_o coupled to a difference in amplitude of the laser sidebands produces a frequency shift given by:

$$\Delta\omega_{LS} = -\frac{(1-\beta^2)\omega_{R1}^2}{4}\left[\frac{\Delta_0}{\left(\Gamma^*/2\right)^2 + \Delta_0^2}\right] \qquad (3.133)$$

where:
β is the ratio of the sidebands amplitude

This is totally different from the case of the passive CPT standard where such a condition, laser detuning and difference in sidebands, produces a distortion of the resonance line without displacement of its maximum. This is due to the fact that in the two approaches, different physical observables are measured. In the case of the passive approach, the optical coherence is the measured parameter through the absorption of the medium, while in the case of the maser it is the ground-state coherence that is detected and measured directly.

On the other hand, the interaction of the atomic ensemble with the cavity microwave field creates added frequency shifts such as the cavity pulling, a microwave power shift and a propagation shift.

The cavity pulling is well-known and is common to all masers. In the case of the CPT maser, which is a hybrid between an oscillating maser and a passive maser, the shift is a function of the importance of the microwave feedback on the atomic ensemble. It is calculated from the phase introduced by the detuning through Equation 3.114 its effect on the frequency of the oscillating magnetization. The frequency shift is:

$$\Delta v = \frac{Q_L}{Q_a}\Delta v_c (S-1) \qquad (3.134)$$

where:
Q_L is the cavity loaded quality factor
Q_a is the atomic line quality factor
Δv_c is the cavity detuning
S is the ratio of the total line width including microwave interaction to the line width including all causes except the microwave interaction

This shift creates a stringent demand on the frequency stability of the cavity resonance frequency. For a cavity Q_L of 10,000, and an atomic line Q_a of 5×10^7, the cavity must be stable to within 3 to 4 Hz to obtain a maser fractional frequency stability of the order of ~10^{-13}.

The microwave power shift originates from the fact that upon stimulated microwave emission the populations of the ground levels are no longer equal. This creates an asymmetry in the system and an effect similar to the linear light shift mentioned above (Equation 3.133) is introduced. It adds directly to the light shift calculated previously for the effect of sidebands of different amplitudes.

Finally, an important effect predicted by the analysis is the introduction of a frequency shift associated with the phase of the sidebands radiation. The shift has been called a propagation shift and becomes rather important in the case of high densities. It has the effect of altering the value of the modulation index for which the power light shift becomes zero. It can thus be cancelled in the same way as the sideband light shift by a proper adjustment of the modulation index (Godone et al. 2002a and b, 2004b). Figure 3.54 is an illustration of the frequency stability that has been measured for a CPT-^{87}Rb maser (Godone et al. 2004).

In the setup studied, it is possible that the behaviour of the frequency stability observed for τ > ~200 s is due to fluctuations of the frequency shifts outlined above. However, it should be mentioned that in the particular setup reported, the maser cavity was made of copper with a temperature sensitivity of more than 100 kHz/degree (Godone et al. 2002a). In that case, environmental fluctuations may have a direct effect on the cavity tuning and cause long-term random fluctuations or frequency drifts that will be reflected by an increasing $\sigma(\tau)$ as the one observed.

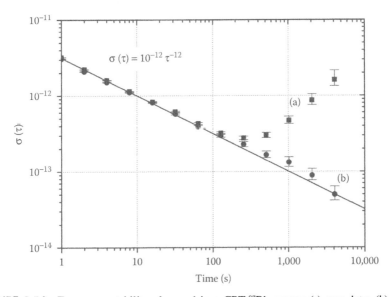

FIGURE 3.54 Frequency stability observed in a CPT-^{87}Rb maser: (a) raw data; (b) drift removed. (Reprinted with permission from Godone, A. et al., *Phys. Rev. A*, 70, 012508, 2004b. Copyright 2004 by the American Physical Society.)

3.5.3 TECHNIQUES FOR IMPROVING *S/N* RATIO IN THE PASSIVE IOP AND CPT CLOCK APPROACH

We have described above the basic physics for implementing either an IOP or a CPT frequency standard. The essential was given as developed in the early 2000. Intense development followed to improve standards either in size or frequency stability.

As we have seen, in the observation of CPT, selection rules forces the use of circularly polarized radiation and, for ^{87}Rb for example, the atoms end up upon optical pumping in end levels such as $m_F = +1$ and $+2$. The contrast of the 0-0 transition is limited to about 5% by the accumulation of atoms in those so-called trapping levels that do not take part of the CPT phenomenon. In the case of IOP, no specific polarization is used in inverting the population. When radiation corresponding to the transition $\left| S_{1/2}, F = 2 \right\rangle \rightarrow \left| P \right\rangle$ is used for optical pumping of the ^{87}Rb ground state, for example, the atoms are distributed or *diluted* among all sublevels of the lower ground state, in the $F = 1$ level. Consequently in both cases the observation of the field independent resonance $\left| F = 2, m_F = 0 \right\rangle \rightarrow \left| F = 1, m_F = 0 \right\rangle$ is characterized by a small contrast. An intense background radiation is present that contributes to shot noise, limiting *S/N* and in turn increasing frequency instability in the short term.

In order to increase *S/N*, several techniques have been proposed. One consists in using a transition between those end levels, in which atoms tend to accumulate, resulting in an increase in signal amplitude, through the simple increase of atoms contributing to the resonance phenomenon. One added advantage resides in a reduction of spin-exchange interaction that vanishes when the polarization is a maximum (Jau et al. 2003, 2004). However, these end transitions are field dependent in first order, a property that adds a difficulty in the practical implementation of a frequency standard. It is then necessary to stabilize the applied magnetic field, which can be done by means of a Zeeman transition between field dependent levels within a given F level.

In the case of CPT, so-called push–pull optical pumping (PPOP) has been proposed, in which the light polarization is switched between right and left circular polarization at the rate $2/T_{00}$, where T_{00} is the period of the 0-0 transition (Jau et al. 2004). Short pulses of right circular polarization (RCP) alternate with pulses of left circular polarization (LCP). The atoms accumulate in the dark superposition state and absorb little light from the radiation of either polarization. This technique thus avoids the trapping of atoms in levels that do not contribute to the CPT phenomenon. The 0-0 transition is enhanced. A similar approach uses optical pumping in which two L schemes are excited at the same time with two radiation fields having perpendicular linear polarization, or the so-called lin ⊥ lin polarization approach (Zanon et al. 2005a). It is shown that these two approaches essentially amount to the same physics (Liu et al. 2013a). Other approaches have also been suggested in which counter propagating waves (Kargapoltsev et al. 2004; Taichenachev et al. 2004) are used. All these techniques make possible an increase in contrast and may be useful when small cells are used.

The continuous wave (CW) operation of the CPT clock however suffers from broadening of the resonance line by the pumping radiation and furthermore from the intense light shift that is present if the modulation amplitude of the laser is not adjusted properly.

This last effect is rather important in the case that the laser fields creating the CPT phenomenon originates from two lasers slaved to each other with a difference frequency corresponding to the hyperfine transition. A solution to that problem resides in the use of pulsed optical pumping in a time domain Ramsey scheme (Zanon et al. 2005a) exciting at the same time two Λ pumping schemes with lin $\perp$ lin radiation fields. In such an approach, a first pulse puts the atomic ensemble in a 0-0 superposition state. The ensemble is then let to evolve freely in that superposition state for a time T when a probing pulse is applied to the ensemble to determine its state of coherence. The interaction of this second pulse gives rise to Ramsey fringes with widths of the order of 1/2T. Typical values reported for the length of pulses are 2 ms for the first pulse, T = 3 ms as free time and 25 μs for the probe (Liu et al. 2013b). A most promising result was obtained upon optimizing several parameters of a laboratory system such as buffer gas pressure and temperature: frequency stability of $3.2 \times 10^{-13}\ \tau^{-1/2}$ was measured up to $\tau = 3000$ s (Danet et al. 2014; Kozlova et al. 2014). This is effectively the best result obtained on a frequency standard based on the passive approach in using CPT.

3.5.4 CPT IN LASER-COOLED ENSEMBLE FOR REALIZING A FREQUENCY STANDARD

Finally, in view of the introduction of laser cooling of atomic ensemble reaching very low temperatures and reducing Doppler effect, the question is raised as to the use of CPT in such cold ensembles. A first proposal was made by Zanon et al. (2003) using a molasses and pulsed CPT, simulating the time sequence used in a fountain but without launching the atoms. A complete laboratory system was developed along that idea by Esnault et al. (2013). The system operates in a sequence as just mentioned. Cooling lasers are on for a cycle of 45 ms creating the MOT containing the order of 10^6 Cs atoms. The atoms can be recycled with a recovery efficiency of about 80% at the end of the total clock sequence. After cooling, the atoms are excited by means of a double Λ CPT scheme as illustrated in Figure 3.55.

The two bichromatic beams are polarized linearly and sent on the free falling atomic ensemble from opposite directions in a lin ‖ lin configuration avoiding Doppler effect. The ensemble is first pumped into a superposition state by means of a Ramsey CPT pulse 400 μs long. The system is then left to evolve freely for a time

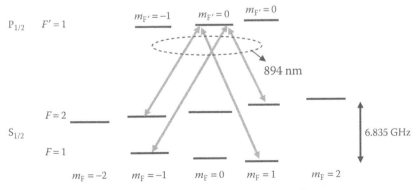

FIGURE 3.55 Double Λ scheme used in the CPT excitation of ^{133}Cs in a MOT.

T_R after which the state of the system is detected with a probe laser beam under the form of a pulse 50 μs long. For a T_R of 8 ms a central fringe line width of 62.5 Hz was observed. It should be noticed that the two Λ systems probed have first order magnetic field shifts that are equal but opposite in sign and are very sensitive to magnetic field gradients. The frequency stability of the laboratory setup, operated without magnetic shielding was characterized by a frequency stability of 4×10^{-11} $\tau^{-1/2}$, a result that is probably not representative of the capabilities of the technique.

3.6   LASER-COOLED MICROWAVE ION CLOCKS

We have outlined in Chapter 1 some recent developments on ion frequency standards. The outline was essentially limited to Yb+ and Hg+. In that last case we outlined the approach using a linear trap which offers very promising avenues for implementing a small size passive frequency standard in the microwave range with extremely good frequency stability. However, there has been further work on the possibility of using other ions and laser cooling for that purpose and we wish to examine briefly some of those avenues in this section.

The ions of greatest interest are listed in Table 3.6. Those are ions with odd isotopic numbers with a ground-state hyperfine splitting given in the table. They can thus be used, in principle as ^{199}Hg+, to implement a frequency standard in the microwave range if that ground-state hyperfine splitting gives rise to an accessible frequency.

We may mention immediately that intensive research is also underway at a number of laboratories worldwide with the intent of implementing an optical frequency standard using some of those ions, in particular using traps with a single cold ion. We will examine those in the next chapter, which will be concerned with the progress accomplished in the practical realization of optical frequency standards.

TABLE 3.6
List of Some Ions with a Ground-State Hyperfine Splitting That Shows Interest in the Implementation of an Atomic Frequency Standard

Ion	I	Ground-State Hyperfine Frequency (Hz)	References
^{9}Be+	3/2	1,250,017,674.10(1)	Bollinger et al. 1983b
^{43}Ca+	7/2	3,255,608,29	Steane 1997
^{111}Cd+	1/2	14,530,507,349.9(1.1)	Zhang et al. 2012
^{113}Cd+	1/2	15,199,862,855.0125(87)	Zhang et al. 2012
^{135}Ba+	3/2	7,183,340,234.35(0.47)	Becker et al. 1981
^{137}Ba+	3/2	8,037,741,667.694(360)	Blatt and Werth 1982
^{171}Yb+	1/2	12,642,812,118.4685(10)	Warrington et al. 2002; Schwindt et al. 2009
^{199}Hg+	1/2	40,507,347,996.84159(44)	Prestage et al. 2005
^{201}Hg+	3/2	29,954,365,821.1(2)	Taghavi-Larigani et al. 2009

Source: BIPM. 2012. Bureau International des Poids et Mesures, Comité International des Poids et Mesures 101e session du CIPM – Annexe 8 203.

The introduction of laser cooling was one of the most significant steps in the use of traps for the development of improved frequency standards using ions. Laser cooling can be applied to ions in both Paul and Penning traps (Itano and Wineland 1982). Sometimes, due to inaccessible wavelengths with a laser it is preferable to use an indirect approach to effectuate the cooling. In that case, a second species of ions is introduced in the trap and laser cooled. Those cold ions then cool the desired ions by means of collisions in a process known as *sympathetic cooling* (Larson et al. 1986).

In the present section, we will concentrate on the development of laser-cooled ion standards in the microwave range for some selected ions not already covered in Chapter 1. We will not describe actual setups such as traps and feedback loops used to lock a reference oscillator to the hyperfine transition. We will rather concentrate on the physics involved in the utilization of such ions and results obtained regarding frequency stability and accuracy.

3.6.1 ^{9}Be$^+$ 303 MHz Radio-Frequency Standard

The ^{9}Be$^+$ ion has a nuclear spin $I = 3/2$. The ground state $S_{1/2}$ consists of two hyperfine levels, $F = 2$ and $F = 1$. The Zeeman energy levels variation with magnetic induction B is shown in Figure 3.56.

In the early 1980s, Bollinger et al. succeeded in storing approximately 300 ^{9}Be$^+$ ions in a Penning trap without laser cooling (Bollinger et al. 1983a). In such a trap, ^{9}Be$^+$ ions are confined by means of static magnetic and electric fields and may be stored for hours. Unfortunately, the large magnetic field required in such a trap does not permit the operation of the frequency standard at the hyperfine transition frequency of 1.25 GHz at low field. However, at a magnetic induction $B_1 = 0.8194$ T, the transition between Zeeman sub levels $|F = 1, -3/2, 1/2\rangle$ and $|F = 1, -1/2, 1/2\rangle$ at a frequency

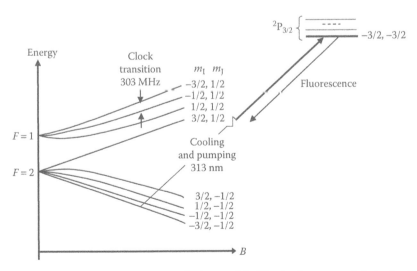

FIGURE 3.56   Ground state and first-excited state of the ^{9}Be$^+$ ion in a magnetic induction B of interest in the implementation of a microwave frequency standard at the rf frequency of 303 MHz.

$\nu_1 = 303$ MHz, as shown in the figure, can be used as the clock transition. The frequency depends quadratically on the magnetic induction as:

$$\frac{\Delta \nu_1}{\nu_1} = -0.017 \left(\frac{\Delta B_o}{B_o} \right)^2 \qquad\qquad (3.135)$$

In the approach used at that time, the ions are first cooled and optically pumped by means of a 313 nm narrowband source tuned to the $^2S_{1/2} \left| F = 2, -3/2, -1/2 \right\rangle \rightarrow {}^2P_{3/2} \left| -3/2, -3/2 \right\rangle$ transition. The 313 nm radiation is obtained by generating a second harmonic of the output of a single mode cw dye laser. The optical pumping process is somewhat unusual in that the other weak transitions are pumped by the 313 nm radiation on their Lorentzian wings and the atoms, due to different states transition probabilities, are finally optically pumped into the $^2S_{1/2} (-3/2, -1/2)$ state although atoms in that state are driven to the excited state at the highest rate (Itano and Wineland 1982). A mixing of the $\left| F = 1, -3/2, 1/2 \right\rangle$ and $\left| F = 2, -3/2, -1/2 \right\rangle$ ground states is done by means of radiation at 23.9 GHz, transferring half the population of the optically pumped state $\left| F = 2, -3/2, -1/2 \right\rangle$ to state $\left| F = 1, -3/2, 1/2 \right\rangle$. This process is essentially an electron spin flip. The detection of the clock transition frequency ν_1 is then done by exciting the $\left| F = 1, -3/2, 1/2 \right\rangle \rightarrow \left| F = 1, -1/2, 1/2 \right\rangle$ transition and then measuring the effect on the fluorescence created by the cooling radiation. The optical preparation, cooling and pumping is pulsed, being turned off during interrogation at the clock frequency ν_1. The interrogation is done by means of short Ramsey pulses separated by 19 s. The central Ramsey fringe has a width of 25 mHz giving a line Q of 1.2×10^{10}. In the system developed, the main residual frequency shift was the second order Doppler effect evaluated to be -3.8×10^{-13} with an uncertainty of 9×10^{-14}. Using a passive H maser as reference, the frequency was measured to be $\nu_1 = 303,016,377.265070(57)$ Hz. A fractional frequency stability of $\sigma_y(\tau) \simeq 2 \times 10^{-11} \tau^{-1/2}$ for 400 s $< \tau < 3200$ s was measured (Bollinger et al. 1985).

Unfortunately, in order to eliminate the Stark effect (light shift) caused by the cooling radiation at 313 nm, the laser needs to be turned off during interrogation of the clock transition at 303 MHz. During that period of interrogation, the atomic ensemble heats up to temperatures of the order of 20 to 30 K causing an increase of the second order Doppler shift, which limits the accuracy of the clock (Bollinger et al. 1985). For this reason, it was proposed to continuously cool the ions by means of so-called sympathetic cooling using an ensemble of Mg$^+$ ions loaded in the trap and laser cooled. The Be$^+$ ions ensemble cooled in a first step by means of a pulse of 313 nm radiation are then cooled continuously to about 250 mK by means of Coulomb interaction with the cooled Mg$^+$ ensemble. The system implemented is essentially the same as with a single species but somewhat more complex relative to detection of the clock transition. In the system developed, the Penning trap stores at the same time 5,000 to 10,000 ^{9}Be$^+$ and 50,000 to 150,000 ^{26}Mg$^+$ ions. In this approach, the first stage of cooling and pumping of the ^{9}Be$^+$ ions are done on the transition $^2S_{1/2} \left| F = 2, 3/2, 1/2 \right\rangle \rightarrow {}^2P_{3/2} \left| 3/2, 3/2 \right\rangle$ and atoms are pumped by the same mechanism described above into state $^2S_{1/2} \left| F = 2, 3/2, 1/2 \right\rangle$. After the 313 nm cooling radiation is turned off, atoms are transferred by means of successive rf π pulses to the state $\left| F = 1, -1/2, 1/2 \right\rangle$ by means of state $\left| F = 1, 1/2, 1/2 \right\rangle$. The frequency of the transitions involved is 321 and 311 MHz, respectively. The Ramsey method of separated oscillatory field is then used to interrogate the 303 MHz clock transition with two pulses 1 s long and separated by 100 s. The resonance effect is then detected by measuring the number of atoms in state $\left|F = 1, -1/2, 1/2\right\rangle$. This is done by a similar process as the one used to populate state $\left|F = 1, -1/2, 1/2\right\rangle$ by reversing the order of the two rf π pulses used previously and repopulating state $\left|F = 2, 3/2, 1/2\right\rangle$. The 313 cooling radiation is then turned back on and the population in that state is then detected as a decrease of the Be^+ fluorescence when resonance at the clock transition is achieved (Bollinger et al. 1991).

The use of continuous sympathetic cooling by means of Mg^+ permits long interrogation time leading to a line Q of 6.2×10^{10} for 100 s free precession between Ramsey pulses. The frequency stability of a complete system using a passive H maser as reference oscillator was measured as $3 \times 10^{-12} \tau^{-1/2}$ for 10^3 s $< \tau < 10^4$ s. Unfortunately, although a low ion temperature resulted from the technique of sympathetic cooling leading to a small second order Doppler shift evaluated to be of the order of -1.2×10^{-14}, it appeared that a pressure shift cause by residual gases of unknown origins in the vacuum system limited the accuracy of the laboratory system to 1×10^{-13}. Nevertheless it appears that the success obtained with the sophisticated approach used in the implementation of such a standard allowed hopes in developing further the techniques experimented and in using them in many other cases requiring pulsing approaches.

Along this line of research, it is of value to mention that recently, Cozijn et al. (2013) have demonstrated laser cooling of trapped beryllium ions at 313 nm using a so-called ridge waveguide diode laser chip operating at 626 nm, which simplifies the system. Second harmonic radiation was generated by means of an external cavity containing a nonlinear crystal. That radiation was used for cooling the Be^+ ions. They detected 600 ions, stored in a linear Paul trap, cooled to a temperature of about 10 mK. The system could then be used for developing a clock operating in a low magnetic field on the hyperfine transition at 1.25 GHz.

3.6.2 $^{113}Cd^+$ AND $^{111}Cd^+$ ION TRAP

Studies of $^{113}Cd^+$ and $^{111}Cd^+$ trapped ions atomic clock were initiated at Tsinghua University and led to interesting results on the use of those ions for implementing a frequency standard (Zhang et al. 2012). The main advantage of those ions, both having a nuclear spin $I = 1/2$ lies in their simple energy level structure, as shown in Figure 3.57.

A linear quadrupole Paul trap was used and a single laser was found sufficient to realize laser cooling. The laser radiation at 214.5 nm frequency is produced by means of a tunable diode laser with its frequency quadrupled and is used for cooling, pumping as well as detection. The clock frequency is interrogated by means of an applied microwave field generated by means of a horn. Pulsed Ramsey technique in the time domain was used with pulses length $\tau = 5$ ms and time $T = 100$ ms between pulses. The effect of the Ramsey pulses interaction on the population of the ions in the $\left|^2S_{1/2}, F = 1\right\rangle$ is measured through the emitted fluorescence and detected by means of a photomultiplier tube. The frequency of the clock transition for both isotopes was measured against a Cs clock calibrated with an accuracy of 5×10^{-13}. The results are

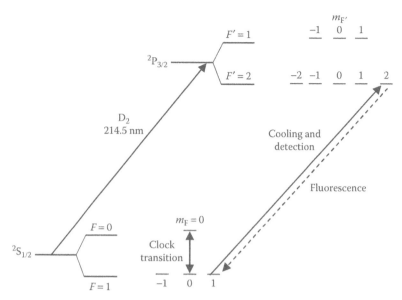

FIGURE 3.57 ^{111}Cd$^+$ and ^{113}Cd$^+$ relevant energy levels and transitions for implementing a microwave frequency standard with a clock transition frequency of either 15.2 GHz or 14.5 GHz.

$$\nu_{hs}\ (^{113}Cd^+) = 15,199,862,855.0125(87)\ Hz.$$
$$\nu_{hs}\ (^{111}Cd^+) = 14,530,507,349.9(1.1)\ Hz.$$

Based on the results obtained, work was initiated on the implementation of a transportable frequency standard using ^{113}Cd$^+$ (Zhang et al. 2014). The clock's measured frequency stability was better than $1.7 \times 10^{-12}\ \tau^{-1/2}$ for averaging times above 100 s thus reaching 2.3×10^{-14} at 4000 s.

3.6.3 ^{171}Yb$^+$ Laser-Cooled Microwave Frequency Standard

The ^{171}Yb$^+$ ion has raised considerable interest for the implementation of a microwave frequency standard. The advantages of this ion originate from low Doppler shifts associated with its large mass, its simple ground-state levels structure, the ion having a nuclear spin $I = 1/2$, and its relatively large hyperfine transition frequency, 12.6 GHz. The lower manifolds of energy levels of interest are shown in Figure 3.58. The optical transition at $S_{1/2} \rightarrow P_{1/2}$, $\lambda = 369.5$ nm used for cooling and detection can be accessed by frequency-doubled lasers. Atoms that decay to the $D_{3/2}$ state are re-pumped to the cooling cycle by means of radiation generated by a diode laser at 935 nm. A microwave frequency standard based on that ion has been developed at National Measurement Institute, Australia (Warrington et al. 2002; Park et al. 2007). A linear Paul trap was used to store the ions. In their case, laser cooling was accomplished by means of a frequency-doubled titanium-doped sapphire laser. That radiation was also used to prepare and probe the population of the ground state.

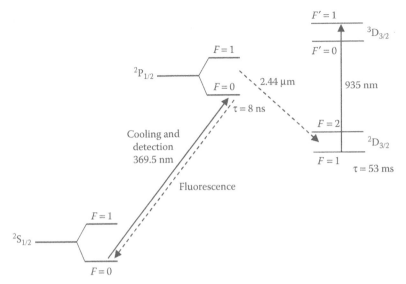

FIGURE 3.58 Lower manifolds of energy levels of ^{171}Yb$^+$ ions of interest in implementing a microwave frequency standard.

The system used Ramsey's technique of separated pulses in the time domain with microwave pulses length $\tau = 400$ ms and time between pulses $T = 10$ s. The clock transition frequency was measured with an accuracy of 8 parts in 10^{14}, limited by the homogeneity of the magnetic field.

A small clock using that ion was also realized at Sandia National Laboratories (Schwindt et al. 2009). It was implemented as a highly miniaturized trap. The package contains the ion trap, a Yb source, and a getter pump. The Yb source is constructed from a micro-machined silicon hotplate that has Yb evaporated onto the surface. The ion trap is a linear rf Paul trap in an octupole configuration having dimensions of a few millimetres. The low-power light sources for state detection at 369 nm and for photoionization at 399 nm are provided by a frequency doubled vertical external cavity surface-emitting laser (VECSEL). The long-term frequency stability was of the order of 10^{-14} for an averaging time of one month.

APPENDIX 3.A: FREQUENCY STABILITY OF AN ATOMIC FOUNTAIN

Analysis of the frequency stability of the classical room temperature frequency standards was done in detail in Volume 2 of *QPAFS* (1989). We wish to repeat the exercise in the case of the fountain, which is different since it operates in a pulse mode. The frequency lock of a reference oscillator that provides the microwave interrogation signal in the fountain is done in a similar way as in the case of the ^{199}Hg$^+$ standard using a Paul trap. In that case, optical pumping is used to invert the population and resonance is detected by means of the fluorescence that is altered when transitions are excited at the hyperfine frequency. The interrogation microwave radiation is provided by a quartz crystal oscillator multiplied to the hyperfine frequency of the mercury ion (40 GHz). The frequency of the oscillator is locked to the atomic resonance in a digital process. The interrogation frequency is switched alternatively to each side of the resonance line by $\pm\omega_m$, approximately one half the line width. The fluorescence photons emitted by the ensemble of ions are then counted for both frequencies. The difference in photons number for $\pm\omega_m$ is a measure of the detuning of the average interrogating frequency from the centre of the resonance line. This signal is then used to lock the average frequency of the reference quartz crystal oscillator to the centre of the resonance line. The frequency stability analysis is done on the basis that the counting of N photons is characterized by fluctuations ΔN resulting in noise at the detector.

The same kind of analysis can be done for the case of the fountain, the main difference being that the fountain is not continuous and the resonance line is observed through the Ramsey interference technique creating fringes. We provide in Figure 3.A.1 a graphic representation defining the various parameters used in the frequency locking technique.

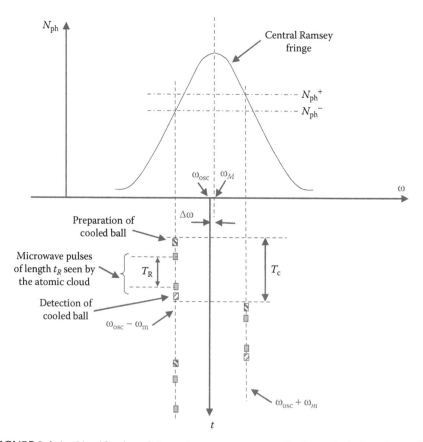

FIGURE 3.A.1 Identification of the various parameters used in the analysis done for establishing the frequency stability of the fountain.

The central Ramsey fringes has a shape given by Equation 1.1 in which we set the phase Φ equal to zero and assume the power of the applied microwave field to be adjusted such that $b\tau = \pi/2$, an optimum value making $\sin b\tau = 1$. The number of fluorescence photons counted by the detection system for one ball traversing the system is thus;

$$N_{ph} = \frac{1}{2} N_{at} n_{ph} \varepsilon_c \varepsilon_d \left[1 + \cos(\Omega_o T_R) \right] \qquad (3.A.1)$$

where:

N_{at} is number of atoms in the ball traversing the cavity and evaluated by means of fluorescence photons counted at the detector

n_{ph} is the number of fluorescence photons emitted per atoms

ε_c is the collection efficiency of the optical system

ε_d is the photodetector efficiency

These parameters can be condensed into a single parameter called $n_d = n_{ph} \varepsilon_c \varepsilon_d$. We note that at resonance, at the maximum of the fringe we have:

$$N_{ph,M} = N_{at} n_{ph} \varepsilon_c \varepsilon_d \qquad (3.A.2)$$

The frequency of the interrogating oscillator is modulated at frequency ω_m, approximately one-half the fringe line width. Its frequency is:

$$\omega_{\pm} = \omega_M + \Delta\omega \pm \omega_m \qquad (3.A.3)$$

where:

ω_M is the frequency of the fringe maximum

$\Delta\omega$ is the small detuning of the oscillator from the fringe maximum

The passage of the cloud of atoms in the cavity during their free flight may be represented as the application of two successive microwave pulses of length t_R, on each side of the resonance line. They are separated by a time T_R, which is much larger than t_R. The number of photons counted on each side of the fringe is:

$$N_{ph}^{+} = N_{ph}(\omega_{osc} + \omega_m) + \delta N_{ph}^{+} \qquad (3.A.4)$$

$$N_{ph}^{-} = N_{ph}(\omega_{osc} - \omega_m) + \delta N_{ph}^{-} \qquad (3.A.5)$$

where:

δN_{ph}^{+} and δN_{ph}^{-} are noise components in the detected photon flux

In practice the servo loop makes the average interrogating oscillator frequency very close to the hyperfine frequency and $\Delta\omega$ is very small. Due to the symmetry of the fringe, the difference in incident fluorescence photons counted is essentially given by

$$N_{ph}^{-} - N_{ph}^{+} = \delta N_{ph}^{-} - \delta N_{ph}^{+} \qquad (3.A.6)$$

and for $\Delta\omega$ very small we may write:

$$\frac{\partial N_{ph}}{\partial \omega}\bigg|_{\omega_{hfs}-\omega_m} = \frac{1}{2}\frac{\delta N_{ph}^- - \delta N_{ph}^+}{\Delta\omega} \qquad (3.A.7)$$

The factor 2 is introduced because a small change in $\Delta\omega$ of the interrogating frequency produces a change of photon flux of $2\Delta I$, each side of the fringe having opposite slopes. Fluctuations in fluorescence photons δN_{ph}^+ and δN_{ph}^- may then be interpreted as causing small frequency changes or fluctuations $\Delta\omega$ of the resonant frequency. Equation 3.A.7 may then be written as:

$$\Delta\omega = \frac{1}{2}\frac{\delta N_{ph}^+ - \delta N_{ph}^-}{\left[\partial N_{ph}/\partial \omega\right]_{\omega_M \pm \omega_m}} \qquad (3.A.8)$$

The fluctuations $\delta N_{ph}^\pm$ being observed at different times are not correlated. Since they are added in the counting, they simply add up. This equation may be written as:

$$\frac{\left(\Delta\omega\right)^2}{\omega_M^2} = \frac{1}{2}\frac{\left(\delta N_{ph}\right)^2}{\left[\left(\partial N_{ph}/\partial \omega\right)_{\omega_M \pm \omega_m}^2\right]\omega_M^2} \qquad (3.A.9)$$

In order to continue the calculation we need the derivative of the photon flux reaching the detector region as a function of applied frequency. This is readily calculated from the fringe shape given by Equation 3.A.1:

$$\frac{\partial N_{ph}}{\partial \omega}\bigg|_{\omega_{hfs}\pm\omega_m} = \frac{1}{2}N_{ph,M}\,n_d T_R \qquad (3.A.10)$$

where:

$N_{ph,M}$ is the number of photons detected at the maximum of the fringe given by Equation 3.A.2

T_R is the time between the two transits (up and down) of the atomic ball in the cavity

Equation 3.A.9 may thus be written as:

$$\frac{\left(\delta\omega\right)^2}{\omega_M^2} = 2\frac{\left(\delta N_{ph}\right)^2}{\pi^2 N_{ph,M}^2 Q_l^2} \qquad (3.A.11)$$

where:
Q_l is the line Q of the central fringe used in the servo-locking system

We must realize that the system is pulsed, atomic balls being prepared and sent up at a rate of the order of one per second. On the other hand, the system requires at least two balls to obtain an error signal driving the frequency-locked servo. Consequently, the analysis must be understood as being valid for averaging periods longer than that pulsation period. Fluctuations on both sides of Equation 3.A.11 can be interpreted as root mean square values, and defining $y = \delta\omega/\omega$, we may write the fluctuations in terms of spectral densities and we obtain the general expression:

$$S_y(f) = 2\frac{S_{\delta N}(f)}{\pi^2 N_{ph,M}^2 Q_l^2} \tag{3.A.12}$$

We thus need to evaluate $(\delta N)^2$ for the various types of noise enumerated in the main text. On the other hand, the physical processes generating noise in the fountain are white frequency noise and in that case the two-sample variance can be written as:

$$\sigma_y^2(\tau) = \frac{1}{2}h_o\tau^{-1} \tag{3.A.13}$$

where:
h_o, assuming $S_{\delta N}$ independent of frequency, is S_y

We examine various types of noise present in the fountain.

3.A.1 SHOT NOISE

Fluorescence photons emitted by the atoms in the detection process originates from spontaneous emission, a random process. The number of emitted photons per atom, n_{ph}, is random and fluctuates with time. This is a statistical process and is generally called shot noise. The spectral density of the fluctuations in the photons number at the detector is then equal to $2N_{at}n_d$ (QPAFS 1989). However, after the second cavity, when the frequency of the interrogating radiation is tuned to $\pm 1/2$ the fringe linewidth, the atoms are found in a superposition state and only half of the total number of atoms in the ball make transitions and thus contribute to noise. The spectral density of noise is thus $S_{\delta N} = N_{at}n_d$. At the maximum of the fringe, where the interrogating microwave is tuned to atomic resonance, the atoms have made a complete transition and are found all in the same state. In that case, the number of photons that are detected is $N_{ph,M} = N_{at}n_d$.

The fountain operates in a discontinuous way by sending atomic clouds or balls of atoms up in a free flight and the actual number of photons counted is a function the cycling process taking place. As shown in Figure 3.A.1, T_c is defined as the cycle time of the cloud, including cooling, trapping, launching, preparation, interrogation, and detection. The analysis just made makes sense for averaging times τ longer that T_c and the average number of atoms reaching the detector per cycle produces a number of photons equal to $N_{at}n_d/T_c$. At the top of the fringe, this is effectively the number of atoms detected as having made a complete transition and we have $N_{ph,M} = N_{at}n_d/T_c$. When these considerations are taken into account, Equation 3.A.12 becomes

$$S_y(f) = 2\frac{T_c}{\pi^2 N_{at}n_d Q_l^2} = h_o \tag{3.A.14}$$

This is white frequency noise and the two sample variance $\sigma_y^2(\tau)$ of shot noise becomes:

$$\sigma_{ySN}^2(\tau) = \frac{1}{\pi^2} \frac{1}{N_{at}n_d} \frac{1}{Q_l^2} \frac{T_c}{\tau} \qquad (3.A.15)$$

an expression often quoted in the literature under various forms.

3.A.2 QUANTUM PROJECTION NOISE

As described in the main text, this type of noise originates from the probabilistic nature of quantum mechanics. It is present in the fountain because the interrogation radiation is tuned at $\pm 1/2$ a line width from resonance where the atoms are in a superposition state. Let us analyze the behaviour of one atom in such a state and exposed to an electromagnetic field. Assume this atom has an energy level structure as in Figure 3.A.2. The two lower energy levels are Zeeman sub-states called $|u_1\rangle$ and $|u_2\rangle$ with energy E_1 and E_2. An electromagnetic field with a frequency close to resonance is applied to the atom. It may excite transitions between those levels and is characterized by a Rabi frequency b. The state of the atom is described by:

$$\Psi = c_1|u_1\rangle + c_2|u_2\rangle \qquad (3.A.16)$$

A laser can be used as a probe to verify in which of the two ground levels the atom is. This is done by means of the fluorescence as in the case of the fountain. We assume that the transition $|u_2\rangle \rightarrow |u_3\rangle$ is cyclic and the atom falls in level E_2 when decaying by spontaneous emission. With the detection laser off, such a system can be solved exactly for the coefficient c_1 and c_2 of the wavefunction (*QPAFS* 1989). Assuming as initial condition, $c_2 = 0$ and $c_1 = 1$ at $t = 0$ one obtains:

$$c_2 = -i\frac{2b}{\Omega}\sin\frac{\Omega}{2}t\,e^{-i(\Omega_o/2)t} \qquad (3.A.17)$$

$$c_1 = \cos\frac{\Omega}{2}t - i\frac{\Omega_o}{\Omega}\sin\frac{\Omega}{2}t\,e^{i(\Omega_o/2)t} \qquad (3.A.18)$$

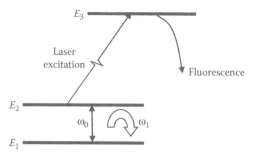

FIGURE 3.A.2 Representation of the Zeeman states of an atom exposed to a radiation field creating transitions with characteristic Rabi frequency b.

where:

$$\Omega = \left[\left(\omega - \omega_o \right)^2 + \left(b \right)^2 \right]^{1/2}$$
(3.A.19)

and

$$\Omega_o = \left(\omega - \omega_o \right)$$
(3.A.20)

Thus the probability of finding the atom in either state is given by:

$$p_2 = |c_2|^2 = \left(\frac{\omega_1}{\Omega} \right)^2 \sin^2 \frac{\Omega}{2} t$$
(3.A.21)

$$p_1 = |c_1|^2 = \cos^2 \frac{\Omega}{2} t - \left(\frac{\Omega_o}{\Omega} \right)^2 \sin^2 \frac{\Omega}{2} t$$
(3.A.22)

This solution is normalized and gives

$$|c_1|^2 + |c_2|^2 = 1$$
(3.A.23)

On the other hand, at resonance $\omega = \omega_o$ and

$$p_2 = |c_2|^2 = \sin^2 \frac{1}{2} bt$$
(3.A.24)

and the probability of finding the atom in level 2 is unity for $bt = \pi$. Thus if the microwave radiation is applied for a time t equal to b/π the atom is found with probability equal to 1 or certainty in level E_2. In magnetic resonance this is called a π pulse. If $bt = \pi/2$, then $p_1 = p_2 = 1/2$ and there is equal probability of finding the atom in either state. If the atom has been placed in state E_2 by means of a π pulse, the detecting laser will excite it to state $|u_3\rangle$ or level E_3. The atom will then decay spontaneously to level 2 and emit a photon. The outcome of the measurement can be predicted with certainty. In the second case, when a $\pi/2$ pulse is applied, the excitation of the atom by the laser cannot be predicted with certainty. It is characterized by a probability. Actually, we can say that when the laser radiation interacts with the atom, the atom state vector is projected into one of the two states and the process is random. This is a feature of quantum mechanics that we cannot escape. The process is also called *wavefunction collapse*. As in all random processes, fluctuations are created in the number of atoms being detected by means of fluorescence photons and they appear as noise in the detection process. The phenomenon has been called *quantum projection noise* (Itano et al. 1993). It contributes to the spectral density of noise in Equation 3.A.12 above. Our reasoning has been done in the context where one single atom is used. However, the process can be applied to a large number of atoms, N_{at}, that are not correlated. The situation is similar to that in which the single atom is excited N_{at} times.

In the case of the fountain, the situation is very similar. At the maximum of the fringe the atoms are in one of the two states, depending on the excitation process.

They are found with certainty in one of those two states and there are no fluctuations. However, in order to provide an error signal for the servo system we interrogate the atoms at half the line width of the Ramsey fringe. According to Equation 3.A.16, there is equal probability of finding the atoms in either hyperfine states of the ground state. In that case, quantum projection noise is a maximum.

Let us analyze the situation with a single atom and calculate the variance of the random quantum projection process. We define a projection operator $P_2 = |u_2\rangle\langle u_2|$. Its expectation value is readily calculated from Equation 3.A.16 as $\langle P_2 \rangle = |C_2|^2 = p_2$, which is the probability of finding the atom in state $|u_2\rangle$. It is also readily shown that the application of P_2 twice leads to

$$P_2^2 = \left(|u_2\rangle\langle u_2|\right)\left(|u_2\rangle\langle u_2|\right) = \left(|u_2\rangle\langle u_2|\right) = P_2 \tag{3.A.25}$$

The variance of the quantum mechanical projection, corresponding to a measurement, may be calculated as follows:

$$\left(\Delta P_2\right)^2 = \left\langle \left(P_2 - \langle P_2 \rangle\right)^2 \right\rangle \tag{3.A.26}$$

$$\left(\Delta P_2\right)^2 = \left\langle P_2^2 - 2\langle P_2 \rangle P_2 + \langle P_2 \rangle^2 \right\rangle \tag{3.A.27}$$

Using property given by Equation 3.A.25 we obtain:

$$\left(\Delta P_2\right)^2 = \langle P_2 \rangle - \langle P_2 \rangle^2 \tag{3.A.28}$$

or in terms of state probability;

$$\left(\Delta P_2\right)^2 = p_2\left(1 - p_2\right) \tag{3.A.29}$$

The fluctuations in the number of atoms making the transition due to quantum projection noise is thus:

$$\left(\Delta N_{QN}\right)^2 = N_{at}\, p_2\left(1 - p_2\right) \tag{3.A.30}$$

This is 0 for $p_2 = 1$, the atoms being all in state $|u_2\rangle$ or $p_2 = 0$ all atoms being in state $|u_1\rangle$. In the case $p_2 = 1/2$, this gives quantum mechanical fluctuations equal to $N_{at}/4$. These are fluctuations in the number of atoms that have made the transition. This is reflected in fluctuations in the number of photons as observed at the detector, which is then given by:

$$\Delta N_{ph,QN}^2 = \frac{N_{at}\, n_d^2}{4} \tag{3.A.31}$$

Using the same approach as in the case of shot noise we interpret these fluctuations as contributing to the spectral density of the noise affecting the frequency stability of the fountain. We average over on cycle and obtain:

$$\sigma_{yQN}^2(\tau) = \frac{1}{4}\frac{1}{\pi^2}\frac{1}{N_{at}}\frac{1}{Q_l^2}\frac{T_c}{\tau} \qquad (3.A.32)$$

APPENDIX 3.B: COLD COLLISIONS AND SCATTERING LENGTH

In *QPAFS*, Volume 1 (1989), we have presented an analysis of the effect of collisions in the context of spin-exchange interactions between alkali and H atoms. The
analysis was presented in both approaches, semi-classical and quantum mechanical.
An important concept introduced in the quantum mechanical approach is that the
collision of two atoms could be represented as the diffusion of a fictitious particle
with kinetic energy

$$E = \frac{\hbar^2 k^2}{2\mu} \qquad (3.B.1)$$

where:
  $\vec{k}$ is the wave vector of the fictitious particle of reduced mass μ having momentum
  $\vec{p} = \hbar \vec{k} = \mu \vec{v}$

This particle evolves in an effective potential given by:

$$V_{s,t}^{eff} = U_{s,t}(R) + \frac{\hbar^2}{2\mu}\frac{l(l+1)\hbar^2}{R^2} \qquad (3.B.2)$$

where:
  μ is the reduced mass of the two colliding particles
  R is the distance between the two atoms
  $U_{s,t}$ is the potential of interaction of the two particles forming either a singlet or
      a triplet state

The second term of the expression acts as a centrifugal barrier created by the relative motion of the two atoms with l being the order of the Legendre polynomial P_l,
which is a solution of Schrödinger's equation for the two possible singlet and triplet
potentials leading to partial waves identified by the quantum number l. The analysis
shows that the collision creates a phase shift η_l in the radial part of the wavefunction,
a phase shift that is proportional to the potential difference $(U_s - U_t)$. The collision
cross sections in Equation 3.51 are defined as the flux of atoms per unit area in a
given direction divided by the incident flux. Calculation then shows that:

$$\sigma = \frac{\pi}{k^2}\sum_l (2l+1)\sin^2\eta_l \qquad (3.B.3)$$

and

$$\lambda = \frac{\pi}{k^2}\sum_l (2l+1)\sin 2\eta_l \qquad (3.B.4)$$

In order to determine the actual cross sections, knowledge of the phase shift η_l is thus required. This phase shift is function of the form of the two potentials created in the formation of the triplet and singlet states upon collision. However in the fountain, the temperature of the atoms in the ball is very low and only S waves ($l = 0$) contribute to the cross sections. In that context, a parameter called the diffusion length a is defined as:

$$a = \lim_{k \to 0} -\frac{\tan \eta_0(k)}{k} \tag{3.B.5}$$

Equation 3.51 can be written as:

$$\frac{d}{dt}\rho_{\alpha\beta} = -(\Gamma + i\Delta v)\rho_{\alpha\beta} \tag{3.B.6}$$

Γ being the broadening of the resonance line, while Δv is the frequency shift, functions of the cross sections σ and λ. These cross sections can be expressed in terms of that diffusion length and a calculation assuming that the atoms are distributed among all levels j of the ground state then shows that the broadening and the frequency shift are given by:

$$\Gamma = -\frac{4\pi\hbar}{m}\sum_j n_j(1+\delta_{\alpha j})(1+\delta_{\beta j}) \operatorname{Im}\left\{a_{\alpha j} + a_{\beta j}\right\} \tag{3.B.7}$$

$$\Delta v = -\frac{2\hbar}{m}\sum_j n_j(1+\delta_{\alpha j})(1+\delta_{\beta j}) \operatorname{Re}\left\{a_{\alpha j} - a_{\beta j}\right\} \tag{3.B.8}$$

where:
   n_j is the density of atoms in level j

The problem is thus one of calculating the diffusion length from the potentials. Calculations then show that for a temperature of 1 μK and a density of $10^9/\text{cm}^3$, the shift in Cs is of the order of -17×10^{-13} while in Rb it is of the order of 1.2×10^{-13}, making clear the advantage of the Rb fountain relative to the Cs fountain in that regard (Kokkelmans et al. 1997).

APPENDIX 3.C: OPTICAL ABSORPTION OF POLARIZED LASER RADIATION INCLUDING OPTICAL PUMPING

This appendix concerns optical absorption spectra of the cell used in the CPT passive approach experiments described in the main text. The level structure of the lower manifold of ^{87}Rb pertinent to the present discussion is shown in Figure 3.C.1. The absorption spectra under various conditions are shown in Figure 3.C.2. The spectra are obtained for two light intensities and two polarizations, linear polarization propagating in the direction of the magnetic field, called σ, and circular polarization, σ^+ (or σ^-). As is clearly observed in that figure there is a drastic change in

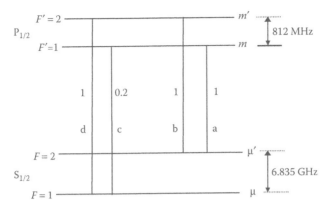

FIGURE 3.C.1 Lower levels $P_{1/2}$ and $S_{1/2}$ of the ^{87}Rb atom. The numbers in circles are transition probabilities summed over the Zeeman sublevels of each manifold.

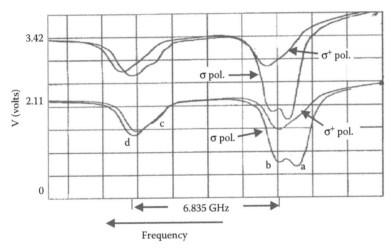

FIGURE 3.C.2 (Experimental) Absorption spectrum of a ^{87}Rb cell under linear polarization (σ) and circular polarization (σ^+). The laser is not modulated. The absorption lines are identified by a, b, c, and d corresponding to the transitions shown in Figure 3.C.1. The absorption spectrum is shown for two light intensities of 21 μW (2.11 V) and 34 μW (3.42 V) incident on the cell.

behaviour in passing from linear polarization to circular polarization. Absorption through transition a is considerably reduced under circular polarization. Transition probability between the various Zeeman sublevels cannot explain alone this behaviour. In fact a summation over all these levels considering appropriate transitions gives equal transition probabilities whatever the polarization. The effect can only be explained through a detailed examination of the dynamics of the transitions causing optical pumping that takes place in the presence of a buffer gas. This is shown in Figure 3.C.3 for the case of σ^+ polarization. As is observed, for transition a levels $F = 2$, $m_F = 1$ and 2 are not connected by the radiation. For transition b,

σ⁺ polarization

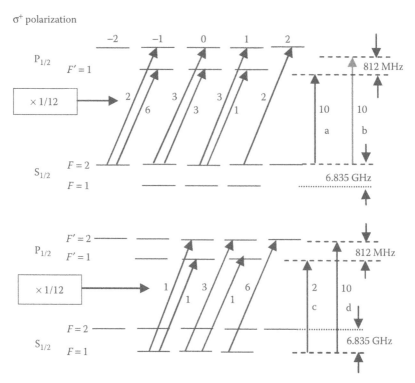

FIGURE 3.C.3   Manifold of ⁸⁷Rb lower energy levels used to illustrate the characteristics of optical absorption. The numbers attached to the arrow are transition probabilities.

only level $F = 2$, $m_F = 2$ is not connected. Atoms excited to the P state decay to all levels equally due to collisions with the buffer gas and are trapped in these levels. This effect is essentially an optical pumping process (Vanier et al. 1982b). Decay from the excited state takes place at rate Γ^* (~10^9 s⁻¹), while equilibrium in the ground state is re-established through relaxation at rate γ_1 (~10^3 s⁻¹). There are thus a large number of atoms trapped in these levels and they no longer contribute to absorption.

The situation for transitions c and d originating from level $F = 1$ is different, as is seen in Figure 3.C.3. Only level $F = 1$, $m_F = 1$ can act as a trap for transition c. However, in all cases the other hyperfine level not involved in the transition can also act as a trap. The only difference between the two cases is related to the number of trapping levels involved. In the case of σ polarization the situation is radically different. A similar representation as that shown in Figure 3.C.3 shows that transitions are possible from all Zeeman sublevels and there are no trapping levels as in the previous case. Optical pumping is possible only to the other hyperfine level not involved in the transition.

An analysis can be made of this effect using the tools developed in the present text. Since there is only one radiation field present there is no coherence introduced in the ground state. Solving Equation 3.111 through numerical integration, we can

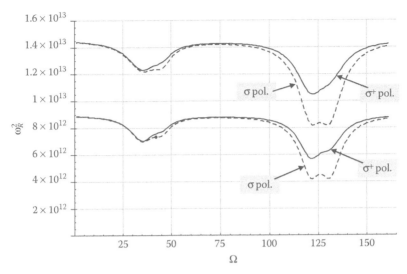

FIGURE 3.C.4 Calculated absorption spectrum (D1) by a cell containing ^{87}Rb and a buffer gas for two values of the intensity of a monochromatic laser radiation corresponding approximately to those used in the case of Figure 3.C.2. The constants assumed in the calculation are $\alpha = 2.1 \times 10^{11}$ m^{-1} s^{-1} and $\Gamma^* = 2 \times 10^9$ s^{-1}.

calculate readily the intensity of the transmitted radiation (square of the amplitude of the optical Rabi frequency) at the exit of the cell. The required expression of the optical coherence is calculated from rate equations similar to the set 2.66 through 2.71. In the calculation we assume the appropriate number of trap levels as determined from Figure 3.C.3 for σ^+ polarization, and from a similar one for σ polarization. We use tabulated values of transition probabilities. This is done for two intensities approximating the experimental situation shown in Figure 3.C.2. The result of this exercise is shown in Figure 3.C.4.

These results can be compared to the actual measured absorption spectrum of laser radiation shown in Figure 3.C.2. It appears that the model developed based on optical pumping explains rather well at least qualitatively the experimental observations. The net effect is a transfer of population from absorbing levels to other levels not coupled by the radiation, decreasing considerably the absorption and affecting the shape of the overlapping absorption lines. The present results validate to a certain extent the four-level model that was used in the calculation outlined above in connection the CPT hyperfine resonance transmission contrast, whose calculation was developed in the main text.

APPENDIX 3.D: BASIC CPT MASER THEORY

In a maser, the energy given by the atomic ensemble is dissipated in the cavity walls and coupling loop. In a self-consistent approach, for continuous oscillation, it is required that the power emitted by the atoms equals the power dissipated in the cavity. The power dissipated in the cavity is given by (Collin 1991):

$$P_{\text{cav}} = \frac{\mu_o \omega}{2Q_L} \int_{V_c} |H(r)|^2 dV \qquad (3.D.1)$$

where:

Q_L is the loaded quality factor of the cavity

μ_o is the permeability of free space

ω is the angular frequency of the oscillating magnetic field $H(r)$ in that cavity

This expression can be written in terms of the Rabi angular frequency b as:

$$P_{\text{diss}} = \frac{1}{2} \frac{N\hbar\omega}{k} \mu_o \langle b \rangle^2_{\text{bulb}} \qquad (3.D.2)$$

where k is defined as:

$$k = \frac{NQ_L \eta' \mu_B^2 \mu_o}{\hbar V_{\text{bulb}}} \qquad (3.D.3)$$

and η' is the filling factor equal to:

$$\eta' = \frac{V_{\text{bulb}} \langle H_z(r) \rangle^2_{\text{bulb}}}{V_c \langle H^2(r) \rangle_c} \qquad (3.D.4)$$

The classical magnetization is coupled to the rf field through Maxwell's equations. In a cavity, the relation is (*QPFAS* 1989):

$$\ddot{\vec{H}}(r,t) + \left(\frac{\omega_c}{Q_L}\right)\dot{\vec{H}}(r,t) + \omega_c^2 \vec{H}(r,t) = \vec{H}_c(r) \int_{V_c} \vec{H}_c(r) \bullet \ddot{\vec{M}}(r,t) dv \qquad (3.D.5)$$

where:

ω_c is the cavity angular resonance frequency

V_c is the cavity volume

$\vec{H}_c(r)$ is the orthonormal cavity field mode

$\vec{M}(r,t)$ is the oscillating magnetization created by the coherence in the ensemble

It is given by:

$$\langle M_z \rangle = Tr(\rho M_{\text{op}}) \qquad (3.D.6)$$

Here, M_{op} is the equivalent quantum mechanical operator of the classical magnetization. The result is:

$$\langle M_z \rangle dv = -\frac{1}{2} n\mu_B (\rho_{\mu'\mu} + \rho_{\mu\mu'}) dv \qquad (3.D.7)$$

where:

n is the Rb density

We write H and M in complex form:

$$\bar{H}(r,t) = \left[H^{+*}(r)e^{-i\omega t} + H^{+}(r)e^{i\omega t} \right] \bar{z} \qquad (3.D.8)$$

$$\bar{M}(r,t) = \left[M^{+*}(r)e^{-i\omega t} + M^{+}(r)e^{i\omega t} \right] \bar{z} \qquad (3.D.9)$$

where:

H^{+} and M^{+} are complex amplitudes of the field and of the magnetization, respectively

We replace these expressions in Equation 3.D.5 and, keeping only the resonant component as in the rotating wave approximation used before, we obtain:

$$\left| \bar{H} \right| e^{-i\phi} = \frac{-iQ_L}{1 + 2iQ_L(\Delta\omega_c/\omega)} \bar{H}_c(r) \int_{V_c} \bar{H}_c(r) \cdot \left(\frac{1}{2} \right) n\mu_B (\delta_{24}^r + i\delta_{24}^i) dv \qquad (3.D.10)$$

where the off diagonal density matrix element has been written explicitly in complex form. Simple algebraic manipulations show that the phase of the field is given by

$$\phi = \frac{\pi}{2} + \tan^{-1} 2Q_L \frac{\Delta\omega_c}{\omega} - \tan^{-1} \frac{\delta_{24}^i}{\delta_{24}^r} \qquad (3.D.11)$$

making explicit the phase quadrature, close to $\pi/2$, between the field and the magnetization.

On the other hand, the power given by the atoms can also be obtained from Equation 3.D.1 by realizing that the value of H can be written in terms of the magnetization by means of Equation 3.D.10. One obtains:

$$P_{at} = \frac{1}{2} \frac{N\hbar\omega k}{\left[1 + 4Q_L^2 \left(\Delta\omega_c/\omega_{\mu'\mu} \right)^2 \right]} \left| 2\delta_{\mu'\mu} \right|^2 \qquad (3.D.12)$$

The energy given by the ensemble compensates for the losses in the cavity. Equating Equations 3.D.2 and 3.D.12, and assuming that $2Q_L(\Delta\omega_c/\omega)$ is much smaller than 1, one obtains the interesting relation between the absolute value of the coherence $\delta_{\mu'\mu}$ and the Rabi angular frequency averaged over the interacting atomic ensemble:

$$\langle b \rangle = 2k \left| \delta_{\mu'\mu} \right| \qquad (3.D.13)$$

Equations 3.D.11 through 3.D.13 are those used in the main text.

4 Optical Frequency Standards

Based on our discussion in Chapters 1 and 3, it appears that the field of atomic frequency standards is a rather mature field achieving precision and accuracy not reached in other fields of physics. However, in science, the challenge is always to attempt to do better. This is often driven by curiosity, but also with the hope of accomplishing greater precision and accuracy in measurement which may lead to verification, to higher levels, of predictions made by theories in parallel fields. Agreement with predictions is searched for and generally obtained. When this is so, our confidence in our theoretical model of the functioning of the universe is increased. However, very often, the results obtained in accurate and precise measurements do not agree with theoretical predictions and leads to a questioning of the basic physics that underlie our understanding of the accepted physical laws. This has happened quite often in the past. We may cite the expansion of the universe. Based on simple laws of attraction or relativistic curvature of the space–time canvas, the expansion should slow down. However, precise measurements on the electromagnetic radiation spectrum emitted by far-off galaxies have shown that the expansion actually accelerates with the passing of the millennia. This is totally contrary to our basic understanding of the physics involved and we do not yet have an explanation of the phenomenon (Vanier 2011). There are many other phenomena that have raised questions and these can often be answered by means of better measurements. For example, all our basic physical laws are based on the arbitrary introduction of parameters that we call fundamental constants. The actual stability with time of those constants is being questioned. Were they created at the same time as the original phenomenon that gave birth to our universe, the so-called big bang? Is it not possible that these constants have changed since the early days? As we will show later, we can answer to that kind of question to a certain level by means of atomic clocks. And this level depends on the precision reached with such atomic clocks. If it were only for that reason, we would be largely justified to attempt to develop still better clocks and reach a higher level of precision. There are, however, many other fields, where atomic clocks are used, fields that would profit from better accuracy and frequency stability of those clocks. We will outline them later. For the moment, following that short introduction on the need for better clocks, we will describe a new avenue in fulfilling that goal.

From the various analyzes and discussions made in Chapters 1 and 3 on atomic frequency standards that have been realized in the microwave range of the electromagnetic spectrum, it should be evident that there are not too many ways of improving the frequency stability and accuracy of those particular standards. Many approaches have been presented in the Chapter 1 culminating in the Rb and Cs fountains having best characteristics. The precision or frequency stability as well as the accuracy achieved in those fountains stand as the best obtained for standards in the microwave range. They appear as being limited to:

Frequency stability: $\sigma(\tau) \approx$ a few $10^{-14} \tau^{-1/2}$
Accuracy: $\Delta v / v \approx$ a few 10^{-16}

It should be noted that these characteristics are not totally independent of each other. It is not possible to determine a resonance frequency in a given atom by averaging several measurements with greater accuracy than is possible with the frequency stability realized at the averaging time at which the measurement is made. Consequently, before talking about accuracy, it is important to ensure that the frequency stability of the device is at least in the same range as the accuracy that is desired.

In order to identify possibilities of improving on those characteristics, it is best to examine first what is at the origin of those limits. Regarding frequency stability, we have derived in the case of the fountain an equation that says in a somewhat general way that the limit is determined in great part by quantum noise:

$$\sigma_{yQN}(\tau) = \frac{1}{2} \frac{1}{\pi} \frac{1}{\sqrt{N_{at}}} \frac{1}{Q_l} \sqrt{\frac{T_c}{\tau}} \qquad (4.1)$$

where:
   N_{at} is the number of atom contributing to the atomic resonance signal
   Q_l is the line Q
   T_c is the cycling time
   τ is the averaging time

If the frequency stability is limited by shot noise, a similar expression is obtained where N_{at} needs to be multiplied by the fraction of detected atoms contributing to the signal. The experimental approach used may limit the number of atoms that it is possible to trap and laser cool and consequently not much gain can be made in that direction. Even if it is possible to increase the number of atoms, a lowering in accuracy may result since a larger number of atoms may introduce a frequency shift caused by atom–atom interaction that is often difficult to evaluate as was shown in the case of the Cs fountain.

In principle, the line Q can be increased enormously since it is the ratio of resonant frequency to resonance linewidth. Thus, reducing the linewidth could be an approach. However, this approach may not be possible in the microwave range. In the case of the fountain, for example, the free time of the atomic ball between its passages in the microwave cavity could be increased by using a larger launching speed. The time during which the ball is in free flight would be increased and the line width would then be reduced. However, the expansion of the atomic ball in its path due to a residual speed of the atoms would decrease the number of atoms penetrating the microwave cavity on its second passage. This would of course reduce the signal (number of atoms N_{at}) and thus the short-term frequency stability. It appears that a cycling period of ~1 s is optimum, resulting in a line width of ~1 Hz and a line Q of ~10^{10}.

On the other hand, if an atomic resonance at a frequency of 400 THz, for example, were used and if its resonance line had the same width as in the case of the fountain, a line Q of ~4 × 10^{14} would result. This would give a line Q ~ 40,000 times larger than in the case of the fountain. The resulting frequency stability, $\sigma(\tau)$, according to Equation 4.1 would then increase by a large factor. However, this is possible only at the condition that a laser emitting radiation, with high spectral purity, is available. A frequency of 400 THz is in the optical range. This is the avenue that has been taken in the development of new atomic frequency standards in the last few decades.

4.1 EARLY APPROACH USING ABSORPTION CELLS

Early in the development of lasers at optical frequencies, it was realized that their mode of operation was radically different from other atomic oscillators, such as H and Rb masers. In those last cases, the frequency is determined primarily by the hyperfine resonance frequency of the atomic ensemble, which is largely independent of external perturbations. The frequency of the microwave cavity that allows oscillation does influence slightly the output frequency of the masers, but its effect is attenuated by a factor of the order of 10^5, because the Q of the resonance line is much larger than the cavity Q. In the case of lasers, operating at optical frequencies, the situation is reversed. The frequency is primarily determined by a high finesse optical cavity (Fabry–Perot cavity), which is required for oscillation and which selects a given frequency out of the possible Doppler broadened emission spectrum of the atomic ensemble providing the gain for oscillation. Consequently, the output frequency is a direct linear function of the tuning of the FP cavity. The laser is not by itself an accurate or stable frequency standard since its frequency depends on a mechanical structure.

We have described in *QPAFS*, Volume 2 (1989), and in Chapter 2 various methods of frequency stabilization of a laser cavity. One has to make use of an atomic reference and use the laser beam as interrogating radiation to be stabilized on a particular narrow resonance of the atomic reference. We need, however, in all cases to use narrow resonance lines and, in particular, to get rid of Doppler effect. One technique consists in placing a cell containing an atomic ensemble resonating at a given frequency within the range of oscillation of the laser and providing a narrow optical line that can be used as a reference. This is generally done by means of the technique of saturated absorption, cancelling first-order Doppler effect and leading to a line width of the order of one MHz or so, actually the natural line width of the transition used. Consequently, a line Q of the order of 10^8 is obtained. Another technique that has also been developed consists in using an atomic beam combined with Ramsey's interrogation technique, using two retro-reflected laser beams creating a standing wave. The technique is complicated by the fact that the atomic beam has necessarily a width larger than the wavelength of the radiation emitted by the laser to be stabilized. This complication can be addressed by using several laser beams either as standing or travelling waves or still, in the case of long wavelengths, using masks to orient in space parts of the reference atomic beam in regions where the laser beams have the same phase.

Resonance line widths in the range of 5 kHz are observed leading to a line Q at best of the order of 10^{10}.

Doppler effect can also be cancelled by using counter-propagating waves exciting so-called two photon Doppler-free transitions. That last technique was applied to mercury ions, $^{198}Hg^+$, in a Paul trap at room temperature exciting so-called quadrupole transitions $^2S_{1/2} \rightarrow {}^2D_{5/2}$ at 281.5 nm (Bergquist et al. 1985). The $^2D_{5/2}$ state is metastable and its lifetime is about 0.1 s. A line Q larger than 10^{14} thus appears to be feasible. The characteristics of the transition were measured by monitoring the fluorescence of the resonance transition $^2S_{1/2} \rightarrow {}^2P_{1/2}$ excited by coherent radiation at 194 nm. This is the so-called double resonance technique used extensively in microwave frequency standards, which we have described earlier. Unfortunately, the line width measured was of the order of 420 kHz limited mainly by the spectral width of the lasers used in the experiment. Thus, that technique shows little gain when compared to the microwave standards that use sources for interrogation with very narrow spectral lines. However, it provided the basis for future development using lasers with better characteristics, laser cooling, trapping, and the technique of shelving to be described below.

The approaches just mentioned have been described in some detail in *QPAFS*, Volume 2 (1989), and the reader is referred to Section 8.10 concerning the subject. Unfortunately, the full advantages attached to the use of a higher frequency and larger resonance line Q could not be totally exploited in those approaches due to the characteristics of the lasers available, the complexity of the devices developed, and their sensitivity to environmental perturbations. Frequency stabilities realized with those systems were of the same order of magnitude as those obtained in microwave standards. In order to obtain better frequency stability, it is thus clear that a totally new approach needed to be developed.

As we said several times, Doppler effect is the phenomenon that must be cancelled at all costs or at least reduced to a level that is negligible compared to other perturbations. Consequently, approaches in which Doppler effect at optical wavelength is inhibited needed to be developed and used. Furthermore, narrow lines in the optical range needed to be identified and exploited. This requires the use of atoms with a pair of levels having a very long lifetime in order to achieve narrow line widths. We will now address that question.

The amount of work that has been dedicated to this subfield in the last few decades is tremendous and the success that has resulted from that work is astonishing. A frequency stability reaching the level of 10^{-18} at an averaging time of 10,000 s has been realized and an accuracy close to that value has been accomplished. Such results open the door to new experiments in basic physics, which were not even dreamed of a few decades ago.

The number of scientific and review articles that have been published on the subject is extremely large. We will not detail all the optical clock systems that have been constructed and their specific particularities. We will rather outline the basic principles involved, describe a few basic systems and provide a summary of the results obtained with most interesting ions and atoms in particular experimental configurations.

4.2 SOME BASIC IDEAS

In the microwave range we have found that one way of getting rid of Doppler effect is by means of limiting the motion of atoms during the time of interaction with the interrogating radiation to regions smaller than a wavelength, that is to say using the so-called Lamb–Dicke technique. The condition can be readily fulfilled in a buffer gas or a mechanical structure of small dimensions. The same mechanical structure or buffer gas approach at optical wavelength is not possible due to the shorter wavelengths involved and the strong interaction with a buffer gas that broadens the resonance lines and shifts their frequency. However, the coupling of techniques such as atom laser cooling and trapping open new avenues that make possible the observation of narrow lines totally unaffected by Doppler effect.

A first condition of course is that we must find atoms for which laser cooling is possible. This means that the atom chosen must have a transition from the ground state to an excited state with strong transition probability, possibly a cycling transition to avoid as much as possible optical pumping into a state not connected to the cooling laser radiation. Lasers must be available at the frequency of that transition. Another condition is that a transition from the ground state with a narrow line width must also exist in the energy level manifold of the atom chosen to provide the clock reference frequency. Such combined conditions exist for an atom with an energy level structure such as the one shown in Figure 4.1. We call it a V energy level structure, as opposed to the Λ structure needed for the observation of coherent population trapping described earlier. As was mentioned above, such a combination is found in $^{198}\mathrm{Hg}^+$. It is also found in several other ions and atoms. The idea behind the choice of such a configuration is the coupling of the clock transition to the cooling transition. In a set up using such an atom or ion, cooling is accomplished by means of the strong transition $|1\rangle \rightarrow |2\rangle$ and fluorescence from spontaneous decay of excited state $|2\rangle$ is monitored. If the clock transition $|1\rangle \rightarrow |3\rangle$ is excited, atoms in ground state $|1\rangle$ are driven to state $|3\rangle$ affecting the level of fluorescence. They stay there for a given time, depending on their lifetime in that state. Consequently the clock transition can be monitored directly by means of the level of fluorescence originating from transitions $|2\rangle \rightarrow |1\rangle$.

Of course, state $|3\rangle$ must be a long-lived state in order to have a narrow clock transition. This is obtained by using an ion or an atom for which state $|3\rangle$ is a metastable

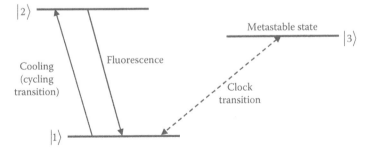

FIGURE 4.1 Basic energy level configuration of an atom or ion that is used in the implementation of an optical frequency standard.

state. A transition to such metastable state is, quantum mechanically, strictly forbidden. However, state mixing from nearby states, or from an applied magnetic field, affects the internal symmetry of such metastable states and a weak transition probability exists. In such cases, the width of the clock resonance line is then of the order of 1 Hz or less giving a line Q larger than 10^{14}, a factor of about 10^4 larger than the line Q's possible in the microwave range.

There are several ways of implementing an optical frequency standard using that approach. An elementary technique would consist of using an ensemble of ions or atoms cooled in a magneto-optical trapping (MOT). Another technique would use the approach proposed by Dehmelt (1975) in storing a single ion in a Paul trap. In that last case, collisions between ions would be avoided and, as will be shown below, cooling makes Lamb–Dicke technique very effective. The fluorescence detected would be that emitted by the single ion at the rate of the order $10^6 S^{-1}$ or higher, if state $|2\rangle$ is, for example, a P state and is driven by the cooling lasers at that rate. Another technique could be the storage of an ensemble of atoms in a periodic trap such as a lattice formed by means of an optical radiation field in the form of a standing wave as described in Chapter 2, the dimension of the wells satisfying the Lamb–Dicke condition for avoiding Doppler effect. These three techniques have been implemented in recent years with the last one using an optical lattice leading to frequency standards with accuracy better than that of the Cs fountain and frequency stability reaching a few in 10^{-18} for an averaging time of 10,000 s.

The basic idea in the practical implementation of such optical clocks is illustrated in Figure 4.2. The main difference from the implementation of a microwave frequency standard is in the detection of the clock resonance line, which is now at optical frequencies and is done by means of the so-called shelving technique. We will see below how this is done. We have shown also in the figure the presence of a frequency divider, called an optical comb. The presence of this device is required because there is no way of counting directly optical frequencies by means of standard electronic devices as is done in the microwave range, these devices having

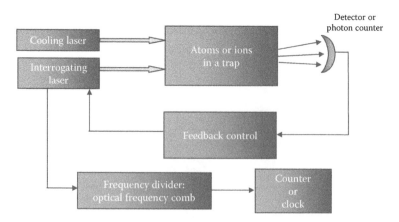

FIGURE 4.2   Block diagram making explicit the basic ideas behind the implementation of an optical frequency standard.

limited speeds. The optical frequency comb was invented recently and is now a standard tool for connecting optical and microwave frequencies together. We will describe its functioning in Section 4.6.1.

4.3 MOT APPROACH

In the MOT approach, an ensemble of atoms is cooled by the technique described earlier in Chapter 2, which consists of using an arrangement where six laser radiation beams tuned to a wavelength slightly lower than the transition $|1\rangle \rightarrow |2\rangle$ is shown in Figure 4.1. A molasses is formed and by applying a magnetic field gradient in a quadrupole configuration, a MOT is created. In the case of even isotope alkaline earth atoms for which there is absence of ground-state hyperfine structure, the nuclear spin I being 0, cooling is done only to the Doppler limit. In practice a temperature of the order of 1 mK is obtained in a ball of the order of 1 cm or so in radius. At that temperature, a residual atomic velocity of the order of 1 m/s persists. At optical frequencies this corresponds to a frequency shift of the order of MHz with a Doppler broadening of that order of magnitude. It is possible to reduce the temperature by using a second stage of cooling (Castin et al. 1989). Although in some configuration, cooling by means of a second stage can then be done to the recoil limit (Vogel et al. 1999), various frequency shifts introduce an uncertainty in frequency determination to the 10^{-13} to 10^{-14} level. The technique has nevertheless been used successfully in the preliminary determination of the frequency of the transition of the $^1S_0 \rightarrow {}^3P_0$ of mercury atom (Hg). The result was used for implementing an optical lattice type of optical frequency standard to be described later (Petersen et al. 2008). Similar measurements were done on the Strontium (Sr) atom using a MOT approach (Courtillot et al. 2003). On the other hand, a functional frequency standard using a free expanding MOT approach has been implemented using Ca (Oates et al. 1999, 2006). The lower energy levels of Ca are shown in Figure 4.3.

The system works as follows: It uses essentially a free expanding MOT that is created in the way described in Chapter 2. Atoms are loaded in the MOT from a Zeeman slowed beam and cooling is done on the transition $^1S_0 \rightarrow {}^1P_1$ at 423 nm. Roughly about 10^7 atoms are loaded in 3 ms in a volume of about 1 mm^3 and are

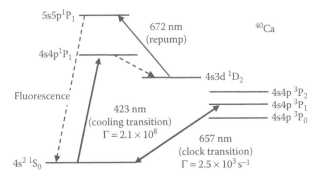

FIGURE 4.3 Lower manifolds of energy levels of interest of the ^{40}Ca isotope used as the heart of a frequency standard in the optical range.

trapped at a temperature of the order of 2 mK. Atoms that decay to the D state during cooling may be pumped back to the cooling cycle by means of radiation at 672 nm. Temperature can also be reduced by means of second-stage cooling techniques (Binnewies et al. 2001; Curtis et al. 2001). The MOT is interrogated by means of a probe radiation field consisting of Bordé–Ramsey pulses in which two pairs of pulses separated by a period T with a frequency near the clock transition at 657 nm are sent on the atomic ensemble from opposite directions. The dark period between the pairs is 3 µs. The probing radiation originates from a laser stabilized on an ultra-low expansion (ULE) cavity with a finesse of the order of 200,000. The line width of the radiation using cavity drift compensation may be of the order of 1–2 Hz when measured on a short timescale of 4 s. During interrogation of the clock transition, all lasers are turned off and the MOT is free expanding. The signal is detected by means of the effect of transitions $^1S_0 \rightarrow {}^3P_1$ on the fluorescence at 423 nm. The timescale involved in the cooling-interrogation-detection cycle is of the order of 3 ms, more than 100 times shorter than in the case of the fountain. Due to that short period, the atoms of the original MOT can be reused for the following cycle. Furthermore, such a short time cycle helps in reducing the effect of drifts of the high finesse cavity used for stabilizing the clock laser and reduces the so-called Dicke effect that affect frequency stability of the system when the interrogation cycle contains long dead times.

Such a clock was implemented as a complete system and was compared to an optical lattice clock based on the Yb atom to be described below (Oates et al. 2006). The comparison was done by means of an optical frequency comb also to be described below. The frequency of that comb was locked to the Yb lattice clock. Such a comparison and measurement showed a frequency stability of $5 \times 10^{-15} \tau^{-1/2}$ for the combination. It may be mentioned that Mg was also proposed and used for implementing an optical clock using a similar free expanding MOT approach (Keupp et al. 2005).

However, in the light of parallel developments, it appeared that the two other approaches, using single ion storage by means of a Paul trap and atom trapping by means of an optical lattice showed greater promises regarding accuracy and frequency stability. We will now describe those approaches.

4.4   SINGLE ION OPTICAL CLOCKS

4.4.1   THE CONCEPT

The use of ions in a Paul trap has been rather successful in the implementation of atomic microwave standards as described previously. Stability and accuracy in the range of 10^{-13} were realized. Still better results were obtained with a linear trap. However, frequency stability and accuracy in the microwave range using an ensemble of ions in a trap appears to be limited by the strong Coulomb interaction taking place between the ions themselves and the low quality factor possible for the clock transition at that frequency. It appears, however, that the use of ions at optical frequencies offers another avenue with promises regarding line Q and little perturbing interaction from the outside world.

One main idea is to get rid of ion-ion interaction by using a single laser-cooled ion in a Paul trap. In such a situation, the ion can almost be considered at rest isolated from the environment. In order to implement such an optical frequency standard, ions with an energy level structure as shown in Figure 4.1 are preferred. In the first proposal, a Tl$^+$ ion was suggested as candidate (Dehmelt 1973). Many other ions can be used. One main difficulty of course is the availability of lasers at the appropriate wavelength in order to cool and interrogate the ion at their resonance frequency. Since the clock transition is very narrow, an interrogating laser with a very narrow spectral width is required. In the microwave, that condition was realized through the use of high quality quartz crystal oscillators, with frequency multiplied to the desired value. Quartz crystal oscillators are known to have very high spectral purity. In extreme cases even cryogenic oscillators could be used. In the case of optical frequencies, in order to observe the very narrow line of the clock transition, lasers need to be stabilized to a cavity with very high finesse. The goal is to reduce their spectral width as was mentioned above in the case of the MOT implementation of an optical frequency standard.

A list of ions of interest is given in Table 4.1 along with the appropriate wavelengths and natural line width of the possible clock transitions. These ions belong to various columns of the table of elements, such as 2a and 3a and even from the rare earth group. They are chosen mainly because their energy level structure satisfies the condition made explicit in Figure 4.1.

In order to illustrate the operation of such a clock, we choose the example of an ion such as ^{199}Hg$^+$ which was used early in the development of such standards (Tanaka et al. 2003). The ion has a nuclear spin 1/2 and the energy level structure of its lower states of interest is shown in Figure 4.4. Due to its nuclear spin, such an ion has a hyperfine structure. It is the one that was studied in Chapter 1 in the description of a microwave standard in a Paul trap and in a linear trap exploiting the hyperfine resonance frequency in its ground state at 40 GHz. That ion has a lifetime of about 2 ns in state $P_{1/2}$ while in state $D_{5/2}$ it has a lifetime of about 90 ms. State $^2D_{5/2}$ is a metastable state and the transition $\left|^2S_{1/2}, F=0, m_F=0\right\rangle \to \left|^2D_{5/2}, F=2, m_F=0\right\rangle$ is called a quadrupole transition. The natural line width of such a transition is thus of the order of 2 Hz. The transition frequency being 1.06×10^{15} Hz, the theoretical line Q is 5×10^{14}.

The ion, of charge q and mass m, may be stored in a miniature radio-frequency (rf) Paul trap of the type illustrated in Figure 4.5 with a ring electrode internal radius r_0 and the distance z_0 between the end caps of the order of mm or less.

The rf trapping frequency may be of the order of 20 MHz with peak voltage of about 700 V. The whole structure may be cooled at cryogenic temperatures (liquid He) in order to improve vacuum and avoid collisions with background residual gas. In such an arrangement, ions may be created from a vapour background by means of a weak electron beam or still in some cases by photo-ionization, of which one single ion is captured by the trap. The trap has a secular frequency between 1.2 and 1.5 MHz and in general an ion can be maintained in place for long periods exceeding hundreds of days. The cyclic transition $\left|^2S_{1/2}, F=1\right\rangle \to \left|^2P_{1/2}, F=0\right\rangle$ at 194 nm is used for cooling. This frequency is obtained by means of multiplication from longer wavelength lasers, in a particular case, by so-called sum-frequency generation using a beta-barium borate (BBO) crystal (Tanaka et al. 2003). Due to off resonance interaction, the ion may be excited to level $\left|^2S_{1/2}, F=0\right\rangle$ resulting in optical pumping

TABLE 4.1

Characteristics of Ions of Interest That Can Be Used for Implementing a Double Resonance Optical Frequency Standard

Ion	Nuclear Spin I	Cooling Transition and Wavelength	Clock Transition and Wavelength	Clock Frequency (Hz)	Natural Line Width Lifetime	References
^{27}Al$^+$	5/2	$^1S_0 - {}^1P_1$, 167 nm	$^1S_0 - {}^3P_0$, 267 nm	1,121,015,393,207,851(6)	0.008 Hz, 20 s	Rosenband et al. 2006, 2007
^{40}Ca$^+$	0	4s $^2S_{1/2}$ – 4p $^2P_{1/2}$, 397 nm	4s $^2S_{1/2}$ – 3d $^2D_{5/2}$, 729 nm	411,042,129,776,393.0(1.6) 411,042,129,776,393.2(1.0) 411,042,129,776,398.4(1.2)	0.2 Hz	Gao 2013 Chwalla et al. 2008, 2009 Matsubara et al. 2012
^{43}Ca$^+$	7/2	4s $^2S_{1/2}$ – 4p $^2P_{1/2}$, 397 nm	4s $^2S_{1/2}$ – 3d $^2D_{5/2}$, 729 nm	411×10^{12}		Champenois et al. 2004 Kajita et al. 2005
^{87}Sr$^+$	9/2	$^2S_{1/2} - {}^2P_{1/2}$, 422 nm	$^2S_{1/2} - {}^2D_{5/2}$, 674 nm	$444{,}781{,}083.91(4)\ 10^6$		Barwood et al. 2003
^{88}Sr$^+$ quadrupole	0	$^2S_{1/2} - {}^2P_{1/2}$, 422 nm	$^2S_{1/2} - {}^2D_{5/2}$, 674 nm	444,779,044,095,510(50) 444,779,044,095,484.6 (1.5) 444,779,044,095,484.6	0.4 Hz	Dubé et al. 2013 Margolis et al. 2004, 2006 Wallin et al. 2013
^{115}In$^+$	9/2	$^1S_0 - {}^3P_1$, 230.6 nm	$^1S_0 - {}^3P_0$, 236.5 nm	$1{,}267{,}402{,}452{,}899.92(0.23) \times 10^3$	0.8 Hz	Sherman et al. 2005 von Zanthier et al. 2000 Wang et al. 2007
^{138}Ba	0	$^2S_{1/2} - {}^2P_{1/2}$, 493 nm	$5d^2D_{3/2} - 5d\ {}^2D_{5/2}$, 12.48 μm	$24{,}012{,}048{,}319(1) \times 10^3$		Madej et al. 1993 Whitford et al. 1994
^{171}Yb$^+$ quad	1/2	$^2S_{1/2} - {}^2P_{1/2}$, 370 nm	$^2S_{1/2} - {}^2D_{3/2}$, 436 nm	688,358,979,309,307.6(1.4)	10^{-9} Hz	Peik et al. 2007
^{171}Yb$^+$ oct	1/2	$^2S_{1/2} - {}^2P_{1/2}$, 370 nm	$^2S_{1/2} \rightarrow {}^2F_{7/2}$ nm	642,121,496,772,300(0.6)	3.1 s	Hosaka et al. 2005
^{198}Hg$^+$	0	$^2S_{1/2} - {}^2P_{1/2}$, 194 nm	$^2S_{1/2} - {}^2D_{5/2}$, 282 nm	1,064,721,609,899,144.94(97)		Tanaka et al. 2003

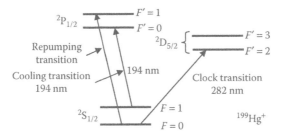

FIGURE 4.4 Lower energy levels manifold of interest in the implementation of a $^{199}\text{Hg}^+$ ion optical frequency standard in the optical range. The lifetime of the ion in state $^2\text{D}_{3/2}$ is ~90 ms while in state $^2\text{P}_{1/2}$ it is ~2 ns.

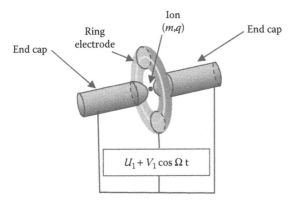

FIGURE 4.5 Miniature rf trap composed of two cylindrical end caps and one ring. It is used in ion trapping for implementing a single ion optical frequency standard.

of the ion to that state. This decreases the efficiency of cooling, the cooling laser being tuned to the other transition. In that case optical repumping at the transition $\left|^2\text{S}_{1/2}, F = 0\right\rangle \rightarrow \left|^2\text{P}_{1/2}, F = 1\right\rangle$ also at 194 nm, but displaced by the sum of the hyperfine frequencies of the excited and ground states, is done by means of another laser beam. Cooling is done to the Doppler limit of the order of a few mK. As mentioned above, the laser radiation needed to excite the clock transition and take advantage of its high Q must have a rather pure spectrum. In early developments of this type of frequency standard, the laser radiation at 282 nm for exciting the clock transition was generated by frequency doubling from a dye laser-emitting radiation at 563 nm. The laser was locked to a high finesse FP cavity isolated from vibrations. In such an arrangement, the spectrum of the resulting radiation may have a width less than 0.2 Hz. A small magnetic field is applied and, in the case of $^{199}\text{Hg}^+$, due to the presence of the nuclear spin $I = 1/2$, a transition involving levels with $m_F = 0$, independent of the magnetic field in first order, can be used as the clock transition.

The detection of the resonance line and the locking of the frequency of the interrogating laser is done as illustrated in Figure 4.6. The signal detected is that of the fluorescence at 194 nm, which is collected and counted as number of photons per second. A fluorescence signal indicates that the ion is in the $\text{S}_{1/2}$ ground state. Absence of

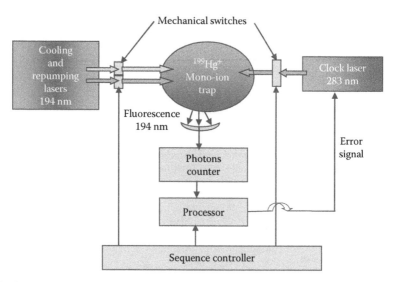

FIGURE 4.6 Diagram illustrating the basic principle of the mono-ion optical frequency standard. This diagram may be used to illustrate the functioning of that type of standard using other ions. In some cases repumping may have to be done at a wavelength that depends on the level structure of the ion used.

fluorescence indicates that the ion has been excited to the D state by the interrogating laser radiation. In view of the strong interactions that take place during cooling, light shifts may be present. Consequently, the system must be operated in a pulse mode in which the cooling and repumping lasers are blocked by mechanical shutters during interrogation. First the ion is cooled by means of the laser radiation at 194 nm. Repumping is accomplished at the same time and an ion temperature of the order of a few mK, the Doppler limit, is obtained. The repumping laser is then blocked and atoms are pumped into level $\left|^{2}S_{1/2}, F = 0\right\rangle$ by the cooling laser. The cooling laser is then blocked, ending the preparation period, and the laser interrogating radiation at 282 nm is applied for a short time, say 20 ms, exciting the clock transition $\left|^{2}S_{1/2}, F = 0, m_{F} = 0\right\rangle \rightarrow \left|^{2}D_{5/2}, F = 2, m_{F} = 0\right\rangle$. If a transition takes place during that time, the ion is said to be shelved in the metastable $^{2}D_{5/2}$ state. The lifetime of the ion in that state is of the order of 90 ms corresponding to a line width of the order of a few hertz. The choice of the length of the interrogation pulse is a function of this lifetime and actually determines the width of the resonance line observed, as was mentioned above. The longer is the pulse, the narrower is the width. The pulse is called a Rabi pulse to differentiate it from the Ramsey approach in which two pulses are applied in succession and fringes are observed. After the 20 ms pulse, the laser radiation at 194 nm is turned on again. Fluorescence at 194 nm is monitored and is used as the indication that a transition has taken place or not. Absence of fluorescence at 194 nm is the sign that the transition has taken place. After a certain time, dictated by the lifetime of the ion in state $D_{5/2}$, the ion returns to the ground state and fluorescence radiation is detected again. In some cases it may be desired to clean out that state by an extra radiation field pumping to the P state in order to start another cycle. These steps constitute a cycle as in the case of the fountain described in Chapter 3, which thus includes cooling, state preparation, interrogation, and detection.

For obtaining an error signal for frequency locking, the laser generating the 282 interrogating radiation, the cycle is repeated for the laser detuned to 1/2 a line width from resonance, alternatively on each side of the resonance line. As in the case of Cs and Rb fountains, that cycle is repeated several times providing some averaging. Subtraction of the photons count made on each side of the resonance line gives an indication of the proper tuning of the interrogation frequency from which an error signal is constructed numerically.

In the description just made, the interrogation is done by means of a Rabi pulse as mentioned. If the interrogation time lasts 40 ms, the Rabi resonance line width is about 20 Hz. However, the interrogation can also be done using the technique of Ramsey pulses providing fringes with a much narrower width than the Rabi single pulse technique just described (Letchumann et al. 2004). It is worth repeating that the quality of the standard that results from a given implementation depends largely on the quality of the original laser that is used for generating the interrogating radiation. The implementation of an optical frequency standard deals with atomic transitions at frequencies in the 10^{14}–10^{15} Hz and line widths of the order of 1–10 Hz. Consequently, lasers with spectral purity compatible with the requirement for observing such resonance lines must be available. In general, lasers of such a quality do not exist. Furthermore, in many cases, lasers are not available at the frequency required, particularly in the case of Hg^+ where the resonance lines fall in the ultraviolet (UV). In such a situation, the frequency of a lower frequency laser has to be multiplied or still altered by some means and its stability may be degraded. Those lasers in general thus need to be filtered somehow just to make possible the observation of such narrow resonance lines and furthermore need to be stabilized in order to stay within tuning range of the servo system used to lock them in frequency to the resonance line as was described above in the case of Hg^+. We have mentioned above the use of high finesse cavities to realize narrow spectral width of the laser spectrum. We have also described such systems in Chapter 2. They are used extensively in practice with refinements required by the particular application studied, such as isolation from environmental fluctuations and vibration.

This basic technique of using a single ion in a trap and the technique of shelving has been used extensively in implementing optical frequency standards with the various ions listed in Table 4.1. We will now give an outline of the physics behind the use of some of the ions listed. Reviews of that nature have been done previously by various authors (Madej and Bernard 2001; Gill et al. 2004; Poli et al. 2014). We will not give the details of specific implementations but will concentrate essentially on the quantum physics involved.

4.4.2 OUTLINE OF PARTICULAR IMPLEMENTATIONS WITH INDIVIDUAL IONS

4.4.2.1 $^{27}Al^+$ ($I = 5/2$)

^{27}Al is the stable isotope of aluminium. The lower manifolds of energy levels of interest in implementing a $^{27}Al^+$ single ion optical frequency standard are shown in Figure 4.7a. $^{27}Al^+$ has a nuclear spin I of 5/2 and thus its energy levels are split by

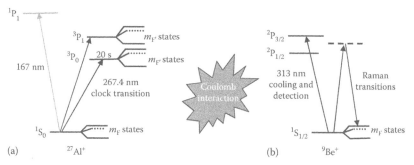

FIGURE 4.7   Energy levels of ^{27}Al$^+$ and ^{9}Be$^+$ of interest in the implementation of an Al$^+$ optical frequency standard.

hyperfine interaction. Also shown in the figure are the lower energy levels of the ^{9}Be$^+$ required as will be seen below for the practical realization of an optical clock using the ^{27}Al$^+$ ion. The clock transition $\left|^1S_0\right\rangle \rightarrow \left|^3P_0\right\rangle$ has a wavelength of 267.4 nm corresponding to a frequency of 1.121×10^{15} Hz. The interest in that transition originates from its narrow width of 8 mHz, originating from the lifetime, 20 s, of the P_0 state (Yu et al. 1992). The transition frequency also has a low sensitivity to electromagnetic perturbations. In fact it has the smallest sensitivity to Black Body radiation (BBR) among atomic elements currently under investigation for the implementation of an optical clock (Rosenband et al. 2006). Unfortunately, the cooling transition wavelength of ^{27}Al$^+$ is 167 nm and is not accessible with available lasers, being in the deep UV. To overcome the difficulty, another ion, in occurrence ^{9}Be$^+$, whose energy levels of interest are shown in Figure 4.7b, is used to cool the Al$^+$ ion by means of Coulomb interaction, both ions being stored in the same trap. In practice, the two ions form a two-ion pseudo-crystal structure, the Be$^+$ ion motion being coupled to the Al$^+$ ion by Coulomb interaction. The Be$^+$ ion is cooled to the Doppler limit by laser cooling at the accessible wavelength of 313 nm. The interaction between the two ions causes the Al$^+$ ion to be cooled at the same time by so-called sympathetic cooling (Larson et al. 1986). In practice, both ^{9}Be$^+$ and ^{25}Mg$^+$ have been used for that purpose with Mg$^+$ being more efficient because its mass has a value similar to that of Al.

Another basic idea behind the implementation of a ^{27}Al$^+$ clock is to use the Be$^+$ (or Mg$^+$) ion fluorescence upon cooling as a tool to detect the quantum state of the Al$^+$ ion. That ion (Be$^+$ or Mg$^+$) is called the *logic ion*, being the one on which the state of the Al$^+$ ion is mapped by means of excitation of appropriate transitions (Raman Sidebands pulses) that transfer superposition states from the Al$^+$ ion to the logic ion (Schmidt et al. 2005). The Al$^+$ ion is called the spectroscopic ion, being the one on which high resolution measurements are done, the particular transition $\left|^1S_0\right\rangle \rightarrow \left|^3P_0\right\rangle$ being used as the clock transition. This clock has become known as a *quantum logic clock.*

It should be mentioned that the implementation of such a clock is rather complex requiring several dye and fibre lasers, doubled and even quadrupled in frequency, in order to reach the various wavelengths required to do cooling, interrogating, and mapping of the information from the Al$^+$ to the Be$^+$ ion. Furthermore, since the ground state of ^{27}Al$^+$ is an S_0 state, there are no first-order magnetic field-independent transitions available for the clock transition. The clock operates on a so-called virtual field-independent transition created in the following way. The clock interrogation radiation at 267 nm is alternated between two frequencies that probe two field-dependent transitions, $\left|{}^1S_0, m_F = 5/2\right\rangle \to \left|{}^3P_0, m_F = 5/2\right\rangle$ and $\left|{}^1S_0, m_F = -5/2\right\rangle \to \left|{}^3P_0, m_F = -5/2\right\rangle$ (Bernard et al. 1998; Rosenband et al. 2007). The average of the two frequencies is the frequency of a transition that would exist between $m_F = 0$ pseudo-levels. The difference between the two frequencies is at the same time a measure of the magnetic field. There is a residual quadratic field dependence that is evaluated by means of second-order perturbation theory. The correction is small and does not affect substantially the accuracy of the clock. The trap, however, has to be well-shielded magnetically to avoid perturbations of the field-dependent levels from which the pseudo field-independent transition frequency is obtained.

Two ${}^{27}Al^+$ optical clocks using a linear Paul trap were constructed at NIST, Gaithersburg, Maryland, using Mg^+ and Be^+ as logic ions, respectively (Chou et al. 2010). The results are the following: frequency inaccuracy—8.6×10^{-18}; frequency instability—$2.8 \times 10^{-15}\, \tau^{-1/2}$; measurement uncertainty—7×10^{-18}; and frequency difference—1.8×10^{-17} between the two clocks.

A similar approach was used at Huazhong University of Science and Technology in China (Deng et al. 2013). The final goal for this ion clock is an accuracy level better than 10^{-17} and a frequency stability of $10^{-16} \times \tau^{-1/2}$. It appears that the goal has been fulfilled.

4.4.2.2 ${}^{40}Ca^+$ ($I = 0$) and ${}^{43}Ca^+$ ($I = 7/2$)

The calcium ion is one of the first ions that was probed with high resolution in the optical domain to determine its basic characteristics such as lifetime of excited states and cooling possibilities (see, e.g., Arbes et al. 1993; Nägerl et al. 1998). A comprehensive study of the properties of the ion was done to evaluate its characteristics as a possible frequency standard in the optical range (Champenois et al. 2001, 2004). More recently, several groups made a large number of experiments based on Ca^+ ions (see, e.g., Matsubara et al. 2004, 2008, 2012; Kajita et al. 2005; Degenhardt et al. 2005; Wilpers et al. 2006; Arora et al. 2007; Gao 2013; M. Kajita 2014, pers. comm.). Two isotopes have attracted interest, ${}^{40}Ca$ (97% abundance) and ${}^{43}Ca$ (0.14% abundance). The energy level manifolds of interest for those isotopes are shown in Figure 4.8. ${}^{40}Ca$ has no nuclear spin while ${}^{43}Ca$ has a nuclear spin equal to 7/2.

Laser radiation at 397 nm on the $\left|{}^2S_{1/2}\right\rangle \to \left|{}^2P_{1/2}\right\rangle$ transition is used for Doppler cooling and state detection. The clock transition is the $\left|{}^2S_{1/2}\right\rangle \to \left|{}^2D_{5/2}\right\rangle$ quadrupole transition at 729 nm. It has a natural line width of 0.2 Hz. A repumping laser at 866 nm tuned to the $\left|{}^2P_{1/2}\right\rangle \to \left|{}^2D_{3/2}\right\rangle$ transition is required. An advantage of the Ca^+ ion is that all transitions may be driven by means of laser diodes. In practice, as mentioned above for dye lasers, those lasers need to be well-stabilized, for example, by means of an external Zerodur FP reference cavity and a Pound–Drever–Hall stabilizer.

In a particular realization, a single ${}^{40}Ca^+$ ion is trapped and laser cooled in a miniature Paul trap (ring electrode and end caps with dimensions ~mm). In a typical

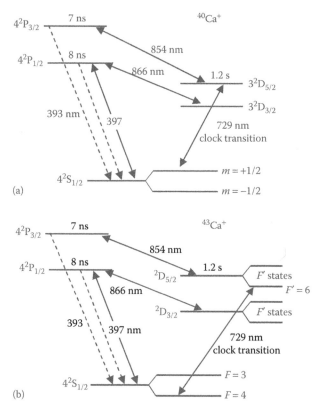

FIGURE 4.8 Manifolds of energy levels of interest in the implementation of (a) ^{40}Ca$^+$ and (b) ^{43}Ca$^+$ single ion frequency standard of interest in the optical range. The diagram makes explicit the presence of the nuclear spin in the case of the odd isotope ^{43}Ca.

configuration, such an ion could be maintained in the trap for more than 15 days (Gao 2013). After corrections are made for all biases to be examined later, the frequency of the clock transition was determined to be 411,042,129,776,393 Hz, with an uncertainty of the order of 4×10^{-15}. In the case of this even isotope with $I = 0$, no field-independent lines are present and the magnetic field bias is calculated by means of measurements on two field-dependent lines as described above in the case of Al$^+$. Similar measurements were done by Matsubara et al. (2012) and were approximately 5 Hz higher than the result just reported.

On the other hand, an optical frequency standard based on the $\left|^2S_{1/2}\right\rangle \rightarrow \left|^2D_{5/2}\right\rangle$ transition in ^{43}Ca$^+$ can also be realized (Kajita et al. 2005). A main advantage would be the use of a $\Delta m_F = 0$, $\left|^2S_{1/2}\right\rangle \rightarrow \left|^2D_{5/2}\right\rangle$, transition that is independent of the magnetic field in first order. However, the Zeeman second-order shift is large and furthermore, due to the large nuclear spin, there are many hyperfine components in the $\left|^2S_{1/2}\right\rangle \rightarrow \left|^2D_{5/2}\right\rangle$ transition. The elimination of the electric quadrupole shift to be discussed below is also difficult to accomplish (M. Kajita 2014, pers. comm.).

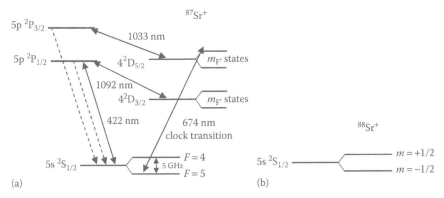

FIGURE 4.9   (a) Lower manifolds of energy levels of ^{87}Sr$^+$ of interest in implementing an optical frequency standard and (b) ground state of ^{88}Sr$^+$ with degeneracy removed by a magnetic field.

4.4.2.3 ^{87}Sr$^+$ (I =9/2) and ^{88}Sr$^+$ (I =0)

Natural abundance of ^{87}Sr is 7% and that of ^{88}Sr is 82.6%. The lower energy levels of interest of both isotopes are shown in Figure 4.9.

The clock transition is the $^2S_{1/2} \rightarrow {}^2D_{5/2}$ quadrupole transition with a wavelength of 674 nm. Its natural line width is 0.4 Hz. Cooling is done by means of the $^2S_{1/2} \rightarrow {}^2P_{1/2}$ transition at 422 nm. In the case of ^{87}Sr$^+$, there is no requirement for repumping from the ground state since the nuclear spin I being 0, there are no hyperfine levels in the ground state. However, there can be decay from the P state to the D state (branching). Repumping at 1033 and 1092 nm is needed to drive back to the cooling cycle the ion that has decayed to the D states as well as to recover the ion in the ground state when a clock transition has taken place. Those transitions are illustrated in Figure 4.9. In the case of ^{88}Sr$^+$, it is noted that the energy levels involved in the clock transition are field dependent in first order and a virtual transition independent of the field in first order is created by means of the scheme described above in the case of the Al$^+$ ion clock by alternating the interrogation frequency between pairs of field-dependent levels and averaging the results. The frequency stability of such an optical frequency standard for 40 ms probe pulses was evaluated to be $1.6 \times 10^{-14}\,\tau^{-1/2}$ for averaging times 30 s < τ < 5000 s (Barwood et al. 2012). The accuracy of the clock transition based on the determination of biases to be examined below was evaluated to be at the level of 2×10^{-17} (Madej et al. 2012). The absolute frequency of the clock transition was measured against the SI unit as:

$$\nu\left({}^{88}\mathrm{Sr}^+\right)=444,779,044,095,485.5\left(0.9\right),\left(2\times10^{-15}\right)\mathrm{Hz}$$

This is to be compared to the result 444,779,044,095,484.6(1.5), (3×10^{-15}) Hz obtained by Margolis et al. (2004).

^{87}Sr$^+$ was studied by Barwood et al. in connection to the implementation of an optical frequency standard similar to the one using ^{88}Sr$^+$ (Barwood et al. 2001, 2003). The difficulty with this isotope is due to the nuclear spin introducing energy levels and the presence of optical pumping of the ion into a dark state when laser cooling is effectuated and interrogation takes place. This problem is resolved by means of cooling using π polarization radiation resonant with both ground state $F = 3$ and $F = 4$ and modulating the cooling laser polarization and repumping radiation (Boshier et al. 2000; Barwood et al. 2001). One advantage in the use of that ion results in the possibility of using a clock transition between field-independent levels in first order that is levels for which $m_F = 0$. Hyperfine splitting constants were first determined and a calculation has been made using second-order perturbation theory of the actual second-order magnetic field coefficient originating from state $D_{5/2}$. The result is that level $^2D_{5/2}$, $F' = 7$ has the lowest quadratic coefficient and consequently the preferred transitions for clock application is the $\left|^2S_{1/2}, F = 5, m_F = 0\right\rangle \rightarrow \left|^2D_{5/2}, F = 7, m_F = 0\right\rangle$, the condition being for the transition to be allowed as a quadrupole transition, which requires that $F + F'$ is an even number (Barwood et al. 2003). The isotopic shift between the 674 nm quadrupole clock transitions of both isotopes $^{88}Sr^+$ and $^{87}Sr^+$ has been measured as 247.99(4) MHz (Barwood et al. 2003) and is an interesting parameter in basic physics studies concerned with the evolution of fundamental constants with time to be discussed in Chapter 5.

4.4.2.4 $^{115}In^+$ ($I = 9/2$)

There are many indium isotopes, but most have a very short lifetime. The isotope ^{115}In is stable with an abundance of 95.7% and has attracted interest for implementing a single ion optical frequency standard. Its nuclear spin is 9/2. Its lower energy levels structure is similar to that of $^{27}Al^+$ and is shown in Figure 4.10.

As for the case of a few other ions, the interest in that ion resides mainly in the fact that all transitions of interest can be excited using solid-state laser sources. Furthermore, its Black Body coefficient is relatively small and the clock transition taking place between P and S states is not affected by quadrupole shifts. In principle, laser cooling could be done through the transition $^1S_0 - {}^1P_0$ at 159 nm, which is a strong transition. However, this cooling wavelength falls in the deep vacuum UV as in the case of Al^+ creating difficulties in frequency multiplication to reach such a short wavelength. Instead the transition $^1S_0 - {}^3P_1$ at 231 nm is used. It is a weaker

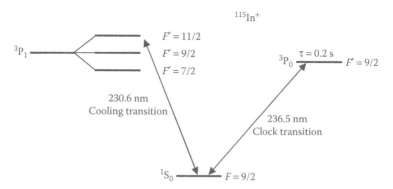

FIGURE 4.10   Lower manifolds of energy levels of the $^{115}In^+$ ion of interest in implementing an optical frequency standard. The nuclear spin of ^{115}In is 9/2.

transition and its line width is 360 kHz, a condition that requires good stability from the laser used. The clock transition is the $^1S_0 - ^3P_0$ transition at 236.5 nm with a lifetime of 0.2 s giving a natural line width of ~1 Hz. Preliminary experiments have been done in trapping a single laser-cooled ^{115}In$^+$ ion in a miniature rf trap in order to determine the characteristics of the ion in the implementation of an optical frequency standard (Sherman et al. 2005). Following further investigation, an actual standard was constructed (Liu et al. 2007). The system used a so-called miniature Paul–Straubel rf trap, composed essentially of a ring electrode of the order of 1 mm and compensation electrodes situated some 1 cm from the ring. The rf trapping voltage used in such a trap has a frequency of about 1 MHz at a voltage of the order of 700 V (Champenois et al. 2001). The clock frequency was determined against a Cs clock as follows:

$$v(^{115}\text{In}^+) = 1,267,402,452,900,967(63), (5 \times 10^{-14})\,\text{Hz}$$

The uncertainty originates from statistical errors in measurements and a systematic uncertainty of the frequency of the Cs clock used (Liu et al. 2007).

4.4.2.5 ^{137}Ba$^+$ ($I = 3/2$) and ^{138}Ba$^+$ ($I = 0$)

Two isotopes of Ba have been investigated early in proposals for implementation of an optical frequency standard. ^{137}Ba has a natural abundance of 11% while ^{138}Ba has an abundance of 72%. The lower energy level manifolds of interest of the ^{137}Ba$^+$ and ^{138}Ba$^+$ ions are shown in Figure 4.11. The nuclear spin of the ^{137}Ba$^+$ ion is $I = 3/2$ and the ground and excited states have hyperfine structures. Cooling is accomplished by means of the transition $6^2S_{1/2} - 6^2P_{1/2}$ with radiation at 493 nm.

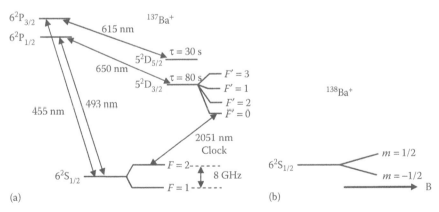

FIGURE 4.11 (a) Lower energy levels manifold of ^{137}Ba$^+$ of interest for implementing an optical frequency standard (Sherman, J.A., Trimble, W., Metz, S., Nagourney, W., and Fortson, N., Progress on indium and barium single ion optical frequency standards, arXiv:physics/0504013v2, 2005.); (b) ground state of ^{138}Ba$^+$ in a magnetic field. (With kind permission from Springer Science+Business Media: *Frequency Measurement and Control—Advanced Techniques and Future Trends*, Berlin, Germany, 2001, Madej, A. and Bernard, J.) It is recalled that the nuclear spin of ^{137}Ba is 3/2 and that of ^{138}Ba is 0.

The Ba ion was the first element on which quantum jumps and shelving were observed (Nagourney et al. 1986).

In the cooling process, optical pumping takes place and the preparation of the state of the ion needs some repumping with a well-tuned laser. There is also some leakage (branching) from the P state to the D states and repumping from those states must be done.

The $^2D_{3/2}$ state has a lifetime of 80 s. It is one of the longest lifetimes observed for D states in investigated ions. The quadrupole transition $6^2S_{1/2} - 5^2D_{3/2}$, at a wavelength of 2051 nm, has thus a very narrow natural linewidth giving a theoretical line Q greater than 10^{16}. It has been proposed as clock transition (Sherman et al. 2005). One advantage obtained in the use of that ion resides in the absence of quadrupole Stark shift for the clock transition. It is noted that the wavelength of that transition falls in the range of readily available laser diodes.

On the other hand, the even isotope $^{138}Ba^+$ has been studied extensively in early high-resolution spectroscopic measurements in the mid infrared (Madej et al. 1993). In particular, the dipole transition $5^2D_{5/2} - 5^2D_{3/2}$ at a frequency of 24 THz (12.48 μm) has possibly a limited resolution of 0.02 Hz. Its main interest at the time was that the frequency involved falls close to that of the NH_3 laser. It also made possible its measurement in terms of the Cs frequency by means of frequency chains existing at NRC, Canada, at the time (Whitford et al. 1994; Madej and Bernard 2001). The frequency of that transition was measured as follows:

$$\nu_{DD} = 24,012,048,317,170(440)\,Hz$$

It appears that the Ba^+ has attracted a lot of interest in the early days of development of optical frequency standards especially due to the long lifetime of the D state.

4.4.2.6 $^{171}Yb^+$ ($I = 1/2$), $^{172}Yb^+$ ($I = 0$), and $^{173}Yb^+$ ($I = 5/2$)

The natural abundance of the three isotopes listed is ^{171}Yb, 3.1%; $^{172}Yb^+$, 21.9%; and $^{173}Yb^+$, 16.2%. The small nuclear spin of $^{171}Yb^+$, similar to that of $^{199}Hg^+$, $I = 1/2$, has made this isotope more interesting than $^{173}Yb^+$ with its nuclear spin $I = 5/2$ and a larger number of levels in the ground state. The lower energy levels of interest of the isotope $^{171}Yb^+$ are shown in Figure 4.12.

The ion is characterized by the presence of three possible clock transitions at respective wavelengths, 411, 435, and 467 nm, respectively, as shown in the figure. The excited states associated with those transitions are characterized by their own lifetime and their possible branching with the $^2P_{1/2}$ state. The $^{173}Yb^+$ ion, due to its larger number of hyperfine levels, does not appear to offer advantages relative to the isotope $^{171}Yb^+$. Isotope $^{172}Yb^+$ has a similar level structure as the one shown but without hyperfine splitting of the states, since its nuclear spin is zero. It was studied extensively in early spectroscopic work for determination of the various transitions wavelength and lifetimes involved (Roberts et al. 1997). However, since the ground state depends on the magnetic field to first order, isotope $^{171}Yb^+$ is preferred having field-independent transitions in first order using $m_F = 0$ hyperfine sublevels.

In all Yb isotopes, cooling is accomplished by means of $^2S_{1/2} - ^2P_{1/2}$ transition at 369 nm. There is unfortunately decay from the $^2P_{1/2}$ state to the metastable $^2D_{3/2}$ state

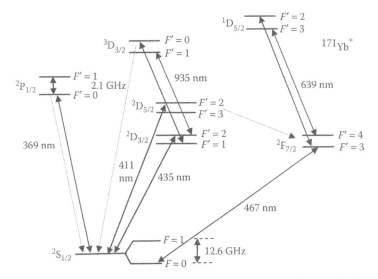

FIGURE 4.12 Lower manifolds of energy levels of ^{171}Yb+$(I = 1/2)$ of interest in implementing an optical frequency standard.

and repumping to the cooling cycle needs to be done. This can be done by means of radiation at 935 nm pumping the ion to state $^3D_{3/2}$, causing the ion to reintegrate the cooling cycle by decay. As in cases studied previously, transition $^2S_{1/2} - {}^2D_{5/2}$ was chosen as the clock transition. The lifetime of the $^2D_{5/2}$ state has been measured as 7.2 ms giving a natural width of 22 Hz and a line Q of 3×10^{13}. The short lifetime limits the length of the Rabi interrogation pulse and the observed line width. Coupling of the $^2D_{5/2}$ state to the long-live $^2F_{7/2}$ causes decay to that state which acts as a dark state. The branching ratio is 0.83. Consequently, repumping also needs to be done from that state by means of radiation at 638 nm. The frequency of the clock transition is as follows (Roberts et al. 1999):

$$\nu\left({}^{171}\text{Yb}^+, \left|S_{1/2}, F = 0\right\rangle \rightarrow \left|{}^2D_{5/2}, F' = 2\right\rangle\right) = 729,487,779,566(153)\,\text{kHz}(411\,\text{nm})$$

On the other hand, transition $^2S_{1/2} - {}^2D_{3/2}$ with a narrower line width can also be used as a clock transition. The lifetime of the $^2D_{3/2}$ state is 52 ms leading to a natural line width of 3.1 Hz. It was also extensively studied and its frequency was determined as follows (Sherstov et al. 2007):

$$\nu\left({}^{171}\text{Yb}^+, \left|S_{1/2}, F = 0\right\rangle \rightarrow \left|{}^2D_{3/2}, F' = 2\right\rangle\right) = 688,358,979,309,307.6(1.4)\,\text{Hz}(435\,\text{nm})$$

These transitions are quadrupole transitions. The limited lifetime of the D state involved limits the line Q and the length of the Rabi interrogation pulse. However, the octupole transition $\left|S_{1/2}, F = 0, m_F = 0\right\rangle \rightarrow \left|{}^2F_{7/2}, F' = 3, m_{F'} = 0\right\rangle$ offers several advantages and was also considered and studied as a possible clock transition. The lifetime of the F state is of the order of six years leading to a line Q in the

10^{20} range. A 120 ms interrogating Rabi pulse gave an observed line width of 6.6 Hz (Huntemann et al. 2012). Cooling is done as in the previous case in which quadrupole clock transitions are used. Pumping from ^2D states where decay takes place through branching is needed and accomplished by means of radiation at 935 nm. However, with the lifetime of the F level, it is necessary for the ion after excitation to the shelving F level to be returned to the cooling cycle. This is done by repumping by means of radiation at 639 nm clearing out F state. The measured frequency is as follows (Roberts et al. 2000):

$$\nu\left(^{171}Yb^+, \left|S_{1/2}, F=0\right\rangle \to \left|^2F_{7/2}, F'=3\right\rangle\right) = 642,116,785.3\,(0.7)\,MHz\,(467\,nm)$$

4.4.2.7 $^{198}Hg^+$ ($I = 0$) and $^{199}Hg^+$ ($I = 1/2$)

Along with Ba^+, Hg^+ was one of the first ions in which quantum jumps were observed. Hg^+ was also the ion with which important advances were made towards the proof of concept that single ion trapping in a Paul trap satisfied the Lamb–Dicke requirement for Doppler free spectroscopy, through the observation of secular motion sidebands at optical frequencies (Bergquist et al. 1987).

In the introduction of this section, outlining the basic concepts in implementing a single ion frequency standard, we have used as example the case of the $^{199}Hg^+$ ion. Its level structure was given in Figure 4.4 and an actual standard using that ion was described. Frequency comparison of the resulting standard against a Cs fountain by means of femtosecond laser frequency comb, to be described below, gave a frequency stability of $3.4 \times 10^{-13}\,\tau^{-1/2}$, dominated by the fountain frequency stability. The $^{199}Hg^+$ $S_{1/2} - D_{5/2}$ clock transition frequency was measured as follows:

$$\nu\left(Hg^+, \left|S_{1/2}, m_F=0\right\rangle \to \left|^2D_{5/2}, m_{F'}=0\right\rangle\right) = 1,064,721,609,899,144.94\,(97)\,Hz$$

The uncertainty in brackets corresponds to a fractional uncertainty of 9.1×10^{-16}, the greater part originating from the uncertainty on the Cs fountain (Oskay et al. 2006).

In early experiments such as those made for verifying the basic trap concepts relative to Lamb–Dicke regime, a $^{198}Hg^+$ ion was used. This element has a nuclear spin equal to zero. Although this makes the level structure simpler than with $^{199}Hg^+$, the energy levels are dependent of the magnetic field in first order and this adds a practical requirement regarding magnetic shielding and magnetic field evaluation. Interest was thus later concentrated on $^{199}Hg^+$ with its simple hyperfine structure ($I = 1/2$) although a similar approach using field-independent lines could be used to create a pseudo field-independent transition as in the case of ^{88}Sr.

4.4.3   SYSTEMATIC FREQUENCY SHIFTS IN SINGLE ION CLOCKS

4.4.3.1   Doppler Effect

The resonance frequency $\omega_{m\mu}$ of an atom, m and μ identifying two energy levels of that atom between which the clock transition takes place, is altered by the motion of that atom moving at speed v. The expression of the resulting frequency, made explicit in Appendix 2.A, is

$$\omega = \omega_{m\mu} + \mathbf{k_L} \cdot \mathbf{v}_\mu - \frac{1}{2}\frac{\omega_0 v^2}{c^2} + \frac{\hbar k_L^2}{2M} \tag{4.2}$$

where the second term is Doppler effect. The third term, called second-order Doppler effect, is caused by relativistic time dilation and the last term is the recoil effect when the atom absorbs or emits a photon. Let us recall some of the physics raised by that relation. In practice, in an ensemble of atoms, the velocity vector $\mathbf{v}$ is random and the resulting resonance frequency $\omega_{m\mu}$ is spread over a spectrum. For example, at room temperature, the hyperfine frequency of the ground state of the H atom (1.4 GHz) is spread over a range of the order of 12 kHz, a line width that would give a line Q of 1.2×10^5 not better than a poor quartz oscillator. Techniques were described to reduce this effect to a negligible value by means of storage in a region of space smaller than the wavelength of the interacting radiation or still keeping the atoms in a region of space where the phase of the radiation is constant, the atom essentially appearing at rest relative to the electromagnetic wave. This is the technique used in the H maser. On the other hand, in the fountain, for example, the ball of atoms traverses a cavity in which radiation is present in a standing wave with constant phase. In such a case, the line width is dictated by the time the atoms stay in interaction with the radiation (Rabi width) in the cavity. Upon passage in the same cavity a second time, narrow interference fringes are observed in the resonance signal detected.

However, we have shown that we could alter considerably the speed of the atoms by means of laser cooling. In the microwave, atomic speeds being reduced to the cm/s level, the spread in resonance frequency of an ensemble of atoms is considerably reduced. This property is exploited in the implementation of the fountain. However, at optical wavelengths, due to the short wavelength involved, cooling to the level of cm/s leaves a large spread in resonance frequency (~several kHz) because of the large wave vector $\mathbf{k_L}$ involved. Consequently in a MOT, with residual temperature in the 50 µK or so, speeds are reduced to the cm/s level and the optical line widths observed are still in the kHz range. Furthermore, at those high frequencies, a standard mechanical storage technique cannot be used. This is to be evaluated in the context of the fact that atomic lifetimes in metastable states are of the order of 1 s or more and leads to line widths of 1 Hz or less.

In the case of the single ion frequency standard approach, a most interesting phenomenon takes place. When the ion is cooled to the mK level, a limit reached in Doppler cooling, the remaining speed is of the order of m/s and consequently a rather large Doppler shift remains. However, the motion of the ion in the trap is periodic in the MHz range and its presence results in a modulation of the resonance phenomenon. As mentioned above, the effect was analyzed in an elegant experiment on ^{198}Hg, and well-resolved sidebands on the absorption spectrum were observed. In such a situation, the central peak is observed without shift and the recoil is absorbed by the whole structure as in Mossbauer effect in solids (Bergquist et al. 1987). Consequently, the ion is in a regime satisfying totally the Lamb–Dicke condition for trapping and no first-order frequency shift is present. In that condition, the second and last terms in Equation 4.2 disappear and the remaining perturbing term is second-order Doppler effect.

In the limit of Doppler cooling, the final temperature T_D reached is given by:

$$T_D = \frac{\hbar \Delta \omega_{1/2}}{2 k_B} \tag{4.3}$$

where:

$\Delta \omega_{1/2}$ is the line width of the transition used for cooling

One notices that the Doppler limit is higher for transitions connected to a fast decaying excited state giving a large spectral width $\Delta \omega_{1/2}$. For this reason, if a lower temperature is desired, one needs to use an extra step in the cooling process such as another cooling transition with a smaller line width. From Equation 4.3, the temperature limit of the ion, after a single step cooling, is of the order of a few millikelvins depending on the ion lifetime in the excited state. The second-order Doppler shift is then estimated from:

$$\frac{\Delta \nu_{D2}}{\nu}(\text{thermal}) = -\frac{3 k_B T_D}{2 M c^2} \tag{4.4}$$

This second-order Doppler shift is of the order of 10^{-17} for the usual temperatures reached.

The forced micromotion of the ion in the trap, however, can also have an influence on the frequency of the ion by means of an added contribution to the second-order Doppler effect. The size of the micromotion can be determined by means of a measurement on the scattering rate R_i when the monitor laser is tuned to the carrier ($i = 0$) or to the first sideband ($i = 1$). The ratio of those measurements provides information on the characteristics of the micromotion and calculation shows that the second-order Doppler shift due to that motion is given by (Berkeland et al. 1998):

$$\frac{\Delta \nu_{D2}}{\nu}(\text{micromotion}) = -\left(\frac{\Omega}{c k \cos \xi}\right)^2 \frac{R_1}{R_0} \tag{4.5}$$

where:

Ω is the driving frequency of the trap

ξ is the angle between the micromotion and the radiation wave vector $\mathbf{k}$

For example, if $R_1/R_0 = 0.1$, this shift, in the case of Hg$^+$, may be of the order of 10^{-17}. It is thus important to limit micromotion in order to not increase the thermal second-order Doppler shift, if higher accuracy is desired.

4.4.3.2   Zeeman Effect

The influence of magnetic fields on single ion optical clocks performance is less critical than in the microwave case using an ion cloud. One reason is that even in its secular and micro motion, the ion is rather well-localized, its motion being relatively small; the magnetic field can be made rather small and the resonance frequency of the ion is not influenced by magnetic field gradients. In the case where the ion chosen to implement a clock has a nuclear spin, the quadrupole clock transition can be chosen to take place between S and D levels with $m_F = 0$. In that case the clock frequency depends on the magnetic field only in second order and the frequency shift can be calculated. This was done in Volume 1 of *QPAFS* (1989) for the ground state hyperfine frequency of alkali atoms and selected odd isotope ions. The calculation was straightforward since in that case the clock transition is in the microwave range between levels within the S ground state and there are only two hyperfine levels in such a state. The present situation is different in that we are dealing with transitions between an S state and excited states such as D, P, and F states. Since most single ion optical standards have been implemented with S–D clock transitions, we will examine that particular case. The D state consists of a fine structure with levels $D_{3/2}$ and $D_{5/2}$. For the purpose of illustration, let us consider the case of the $^{43}Ca^+$ ion with a nuclear spin equal to 7/2. Its energy level structure of interest is shown in Figure 4.8. The $D_{3/2}$ and $D_{5/2}$ states contain respectively 4 and 6 hyperfine levels, as shown in Figure 4.13.

The goal is to select a clock transition that is least sensitive in second order to the applied magnetic field. The position of each m_F levels within the F states can be readily calculated. For the ground S state there are only two hyperfine levels and the calculation leads to the Breit–Rabi formula for the position of the m_F levels (*QPAFS* 1989). For the $m_F = 0$ levels, the equation can be approximated at low fields as:

$$\Delta E(F, m_F = 0) = -\frac{E_{HFS}}{2(2I+1)} \pm \frac{E_{HFS}}{2}\left(1 + \frac{x^2}{2}\right) \tag{4.6}$$

where x stands for:

$$x = \frac{g_J \mu_B B_0}{E_{HFS}} \tag{4.7}$$

In those equations, E_{HFS} is the hyperfine splitting of the state in question, μ_B is Bohr magneton, g_J is the electron spectroscopic splitting factor, and B_0 is the applied magnetic

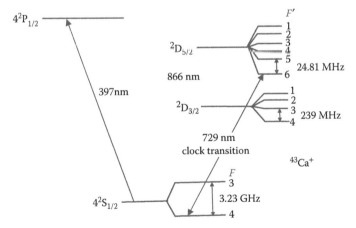

FIGURE 4.13 Illustration of hyperfine structure of ground state and ^2D state of $^{43}Ca^+$ for which $I = 7/2$.

induction. The nuclear spectroscopic splitting factor is neglected. For the $D_{5/2}$ state there are six hyperfine levels and they all enter into the perturbation calculation of the displacement of each level with $m_F = 0$. The result is approximated by the equation:

$$\Delta E\left(^2 D_{5/2}, F', m_F = 0\right) = \left(g_J \mu_B B_0\right)^2 \sum_{F' \neq F''} \frac{\left\langle IJF'm_{F'} | J_z | IJF''m_{F''} \right\rangle}{E_{HFS}^{F'-F''}} \tag{4.8}$$

where $E_{HFS}^{F'-F''}$ stands for the hyperfine splitting between state F' and F'' within the $D_{5/2}$ state. The total quadratic effect of the magnetic field effect can thus be written as:

$$\Delta v = \left(\beta_{D2} - \beta_{S0}\right) B_0^2 = \beta_{e,g} B_0^2 \tag{4.9}$$

where the coefficients β_i are calculated from the equation cited. It is readily observed, making explicit the definition of x in Equation 4.6 by means of Equation 4.7, that the displacement of the levels is an inverse function of the hyperfine splitting. Due to the large size of E_{HFS} for the ground S state compared to that of the D state, the contribution of the quadratic magnetic shift from the S ground state is much smaller than that from the D state and in some cases may be negligible.

The value of $\beta_{e,g}$ has been calculated by several authors and is given in Table 4.2 for selected ions. In the case of $^{43}Ca^+$, transition $F = 4$ to $F' = 6$ shown in Figures 4.8

TABLE 4.2

Quadratic Zeeman Coefficient of Selected Ion Transitions That Have Been Studied for Implementing an Optical Frequency Standard[a]

Ion	Clock Transition	Quadratic Zeeman Shift (mHz/μT²)	Reference
$^{27}Al^+$	$^1S_0 (F = 5/2, m_F = \pm 5/2) \rightarrow$ $^3P_0 (F = 5/2, m_F = \pm 5/2)$	0.072	Rosenband et al. 2008
$^{87}Sr^+$	$^2S_{1/2}, F = 5 \rightarrow {}^2D_{5/2}, F = 7$	6.4×10^3	Barwood et al. 2003
$^{88}Sr^+$	$^2S_{1/2} \rightarrow {}^2D_{5/2} (m_J = 1/2)$ $^2S_{1/2} \rightarrow {}^2D_{5/2} (m_J = 3/2)$	0.0056 0.0037	Madej et al. 2004
$^{40}Ca^+$	$^2S_{1/2} \rightarrow {}^2D_{5/2} (m_J = 1/2)$ $^2S_{1/2} \rightarrow {}^2D_{5/2} (m_J = 3/2)$	0.026 0.017	Chwalla et al. 2008 Madej et al. 2004
$^{43}Ca^+$	$^2S_{1/2} (F = 4) \rightarrow {}^2D_{5/2} (F' = 2)$ $^2S_{1/2} (F = 4) \rightarrow {}^2D_{5/2} (F' = 4)$ $^2S_{1/2} (F = 4) \rightarrow {}^2D_{5/2} (F' = 6)$	34.2×10^3 17.2×10^3 -8.99×10^3	Kajita et al. 2005
$^{171}Yb^+$ (oct)	$^2S_{1/2} (F = 0, m_F = 0) \rightarrow$ $^2F_{7/2} (F = 3, m_F = 0)$	-1.72	Hosaka et al. 2005
$^{171}Yb^+$ (quad)	$^2S_{1/2} (F = 0, m_F = 0) \rightarrow$ $^2D_{3/2} (F = 2, m_F = 0)$	52.1	Tamm et al. 2014
$^{199}Hg^+$	$^2S_{1/2} (F = 0) \rightarrow {}^2D_{5/2} (F = 2)$	-18.9	Oskay et al. 2006

[a] It should be noted that for even isotopes there is present a linear magnetic field shift that may be cancelled by the method described in the text.

and 4.13 satisfies quantum mechanical selection rules, $\Delta F = 0, \pm 1, \pm 2$, and has a smaller quadratic shift than the other transitions of the same ion. It is important in the implementation of a given optical ion frequency standard to select the appropriate transition to minimize the effect of the magnetic field and, in the case of $^{43}Ca^+$, that transition should be selected. In $^{87}Sr^+$, the transition listed is the least field-dependent one. In a field of 1 µT (10 mG), the fractional frequency shift of that transition is of the order of 10^{-15}. If the field can be controlled to 10%, its contribution to the uncertainty budget is then of the order of 10^{-17}.

In the case of even isotopes, having no nuclear spin, the position of the levels depends in first order on the magnetic field. We have described the technique used in creating a level corresponding to a pseudo-state with $m_F = m_J = 0$ independent of the magnetic field in first order, by averaging the frequency of two resonance lines corresponding to transitions involving levels with $m_F = m_J = \pm 1/2$. The effect of the magnetic field is cancelled automatically in first order. However, an exact calculation shows that the levels are also shifted in second order by the magnetic transition chosen. The shift is then also given by Equation 4.9. The values of the quadratic coefficient, $\beta_{e,g}$, are listed in Table 4.2 for ions of interest and the specific transitions listed.

In all cases, the magnetic field intensity must be known in order to make the appropriate correction to the measured frequency. It is usually measured by means of a first-order field-dependent transition. In the case of even isotopes without nuclear spin, this is done automatically since the clock transition frequency in zero field is determined by means of field-dependent transitions. In a typical example, as in the case of an optical standard based on a field-independent transition, a field of the order of 10 µT (100 mG) may be applied in a shielded environment. In the case of narrow interrogating laser radiation, this value is generally sufficient to resolve the individual m_F lines and the accuracy of determination of the transition is of the order of 1 mHz, corresponding to an accuracy of the order of a few in 10^{-18}.

4.4.3.3 Biases due to the Presence of Electric Fields

Electric fields affect the atomic energy levels through Stark effect. In early stages of development of atomic frequency standards, their accuracy was low and the effect of the possible presence of electric fields was not considered. The magnetic field was essentially the main calculable frequency bias. As stability and accuracy of clocks increased, the presence of such electric fields as in Paul traps was investigated to determine their importance. The effect of BBR was also investigated. It was soon found that they could no longer be neglected at the accuracy reached. We have recalled in Chapter 1 the calculation of the effect of an electric field on the hyperfine separation in the ground state of hydrogen and alkali atoms and selected ions. The calculation was done for the hyperfine structure within an S ground state. Such a state has spherical symmetry and the result showed that the shift of the hyperfine levels was proportional to the scalar polarizability of the atom or ion and to the square of the electric field present. In the case of optical frequency standards, the clock transition in many cases takes place between an S ground state and an excited D state. Due to the symmetry involved in the D state, the interaction depends on the orientation of the electric field and the quantization axis of the atom or ion, given normally by the applied magnetic field. The interaction takes a more complex form and the energy shift ΔW of state $\left| F, m_F \right\rangle$ is given by (Angel and Sandars 1968):

$$\Delta W(F, m_F) = -\frac{1}{2}\alpha_{sc}(F)E^2 - \frac{1}{4}\alpha_{ten}(F)\frac{3m_F^2 - F(F+1)}{F(2F+1)}E^2\left[3(\cos\theta)^2 - 1\right] \quad (4.10)$$

where:

  α_{sc} is the scalar polarizability
  α_{ten} is the so-called tensor polarizability reflecting the added symmetry of the excited state involved in the clock transition

The term $\cos\theta$ represents the ratio E_z/E, θ being thus the angle between the electric field and the quantization axis z set by the magnetic field in the system. In the case of an S state, only α_{sc} needs to be considered, which is what we have done in connection to the calculation of hyperfine frequency shift in the ground state of H, Cs, and some ions used in implementing a microwave frequency standard. The scalar polarizability can be calculated by means of perturbation theory extending summations over close connected states as done in *QPAFS*, Volume 1 (1989) for the S state. The tensor polarizability is calculated in a similar way but involves more complex terms. The Stark shift of the clock frequency corresponding to an S–D transition is thus given by:

$$\delta\nu_{st} = \frac{1}{h}\delta(\Delta W_S - \Delta W_D) \quad (4.11)$$

and is a function of the square of the electric field present.

It is important to establish the origin of electric fields and the size of the polarizability of the atom or ion considered. In practice, these fields can originate from several sources. These can be fields required to operate the trap or still stray static electric fields. On the other hand, the electric field associated by BBR is a low frequency field and its quadratic average can be used to evaluate its effect as we have done in the case of the microwave fountain.

In the case where the clock transition is an S–D transition, there may be a displacement of the D level due to the interaction between the quadrupole moment of the ion in that state and an electric field gradient that exists in the trap. We will examine these various effects and the size of the bias they introduce in the clock transition. Other electric fields created by the variety of lasers used for cooling and pumping can be considered as perturbations that we call light shifts and they will also be evaluated below.

4.4.3.3.1 Micro- and Secular-Motion Effects

In a Paul trap, the ion appears as if being trapped in a potential well. The ion potential energy is given by (we refer to Figure 1.31 and the condition that $r_o = \sqrt{2}\,z_o$):

$$E_p = \frac{1}{2}m\omega^2\left[(\bar{r})^2 + (\bar{z})^2\right] \quad (4.12)$$

where r and z are the average ion position in the trap and ω is the frequency of macro or secular motion of the ion within the trap given by:

$$\omega = \frac{eV_1}{m\Omega r_o^2} \qquad (4.13)$$

in which the terms V_1 and Ω were defined earlier and in Figure 4.5. This motion is modulated by the so-called micromotion at the driving frequency Ω of the trap. During its motion, the ion samples the electric field existing in the trap and the clock frequency is shifted by:

$$\delta v_S = \zeta_S \left\langle E(x_i,t)^2 \right\rangle \qquad (4.14)$$

where the coefficient ζ_S stands for the sensitivity of the frequency to the field, $\zeta_S = \partial v / \partial E^2$, and is proportional to the polarizability of the ion as made implicit by Equation 4.10. The term $\left\langle E(x_i,t)^2 \right\rangle$ stands for the time average of the squared field at the ion position x_i.

The motion forces the ion to sample the electric field in a small region close to the centre of the trap and consequently causes a frequency shift. Calculation shows that the corresponding Stark shift δv_S for the micromotion is given by (Berkeland et al. 1998):

$$\Delta v_S (\text{micromotion}) = 2\zeta_S \left(\frac{m\Omega^2}{qk\cos\theta} \right)^2 \frac{R_1}{R_0} \qquad (4.15)$$

where the symbols have the same meaning as defined previously. The secular motion contribution is also calculated and is given by (Madej et al. 2004):

$$\Delta v_S (\text{secular}) = \zeta_S \left(\frac{3m\Omega^2 KT_B}{q^2} \right) \qquad (4.16)$$

These shifts evaluated for example in the case of $^{88}\text{Sr}^+$ are of the order of 800 mHz for micromotion and 30 mHz for secular motion. It is thus clear that the micromotion should be reduced as much as possible in order to reduce that type of bias. This is done normally by means of added trim voltages between electrodes (Dubé et al. 2005). Table 4.3 summarizes some results obtained for selected ions of interest regarding Stark shifts and electric quadrupole shifts to be examined below.

4.4.3.3.2 BBR Shift

We have shown that above a given accuracy the effect of BBR on the frequency of an atomic transition needs to be considered. This is the case of the Cs fountain. Its study led to extensive work in order to determine coefficients involved in its exact determination. The shift is caused by the presence of the large spectrum of radiation that is off resonance. The electric field associated to that radiation causes a Stark shift. The same effect exists in the case of single ion optical clocks and a similar

TABLE 4.3

DC Stark Coefficient and Electric Quadrupole Shift as Reported by Authors for Selected Ions of Interest

Ion	Clock Transition	DC Stark Coefficient Hz/(V/m)^2	Electric Quadrupole Shift (Hz/V/m^2)	Reference
^{27}Al$^+$	$^1S_0 \rightarrow {}^3P_0$	-0.14×10^{-7}	0	Poli et al. 2014
^{40}Ca$^+$	$^2S_{1/2} \rightarrow {}^2D_{5/2}$	6×10^{-7}		Matsubara et al. 2005
^{88}Sr$^+$	$^2S_{1/2}$ (1/2, 1/2) $\rightarrow {}^2D_{5/2}(-1/2,$ $-1/2)$	-2.3×10^{-7}		Madej and Bernard 2001
^{115}In$^+$	$1S_0 - 3P_0$	0.5×10^{-7}	0	Poli et al. 2014
^{138}Ba$^+$	$6^2S_{1/2}$ ($m_S = 1/2$) $\rightarrow 5\,{}^2D_{5/2}$ ($m_D = 1/2$) $6S_{1/2}(F = 2) \rightarrow 5\,D_{3/2}(F = 0)$	$6,1(7) \times 10^{-7}$	0	Yu et al. 1994 Sherman et al. 2005
^{171}Yb$^+$ (oct)	$^2S_{1/2} \rightarrow {}^2F_{7/2}$	-1.44×10^{-7}	1.05×10^{-7}	Blythe 2004
^{171}Yb$^+$ (quad)	$^2S_{1/2} \rightarrow {}^2D_{3/2}$	-6×10^{-7}	60×10^{-7}	Poli et al. 2014
^{199}Hg$^+$	$^2S_{1/2} \rightarrow {}^2D_{5/2}$	-1.1×10^{-7}	-3.6×10^{-7}	Poli et al. 2014

analysis applies. We recall that the quadratic value of the electric field of BBR is calculated as:

$$\left\langle E^2\left(t\right)\right\rangle^{1/2} = 831.9\left(\frac{T}{300}\right)^2 \tag{4.17}$$

This electric field acts on the position of the energy levels through their polarizability and the frequency shift is given by Equation 4.11. However, the BBR is isotropic and only the scalar polarizability enters into play. We summarize in Table 4.4 the calculated relative shift at room temperature for some candidates for optical-ion clocks transitions. From the table it is observed that the temperature of the clock environment must be well-stabilized in order to achieve frequency stabilities in the desired range of 10^{-18}. There are ions that are rather interesting, such as Al$^+$, having very low BBR shift coefficients. On the other hand, we note that operation at liquid-nitrogen temperatures can reduce the BBR shift considerably, due to its T^4 dependence. This can be done readily in a miniature Paul trap. Consequently, even though the BBR shift appears to be important at the level of accuracy desired, it is not insurmountable and can be controlled to some extent.

4.4.3.3.3 Electric Quadrupole Frequency Shift

In optical frequency standards, the clock transition selected in many ions is that between an S state and a D state. In some cases, like in Yb$^+$, the transition chosen may be between an S state and an F state. An atom or ion in such states (F or D) possesses an electric quadrupole moment. In such a case, like in many other atomic systems, the atom or ion state is affected by the presence of an electric field in a particular way. It is found that

TABLE 4.4

Coefficient Characterizing the Effect of the Black Body Radiation on the Clock Frequency of Selected Ions of Interest

Ion	Relative BBR Shift ($\times 10^{-16}$) at 300 K	Reference
$^{27}Al^+$	−0.004256	Safronova et al. 2012
	−0.008(3) Hz	Rosenband et al. 2006
$^{43}Ca^+$	0.38	Arora et al. 2007
$^{88}Sr^+$	0.24	Jiang et al. 2009
$^{115}In^+$	−0.0173(17)	Safronova et al. 2012
$^{199}Hg^+$	−1.020(3)	Simmons et al. 2011
$^{201}Hg^+$	−0.994(3)	Simmons et al. 2011
$^{171}Yb^+$ octupole	−0.15(7)	Lea et al. 2006
	−0.16	Safronova et al. 2012
$^{171}Yb^+$ quad	−0.35(7)	Lea et al. 2006
	−0.37(5)	Tamm et al. 2007
	−0.35	Safronova et al. 2012

the atom electric quadrupole moment interacts with the gradient of an electric field that may be present. The atom or ion energy levels in such a state are affected by the interaction and are shifted. In microwave frequency standards, in which transitions are excited between hyperfine levels of an S ground state, no quadrupole moments are present and the clock frequency of ions in a Paul trap are not affected by an electric field gradient, although of course they are affected by Stark effect as outlined above.

The interaction energy W of a distribution of charge with an electric potential V both with axial symmetry can be calculated classically using a McLaurin's expansion and Laplace equation as (see Vanier 1960):

$$W = \frac{1}{4} eQ^+ \frac{\partial^2 V}{\partial z^2} \tag{4.18}$$

The term eQ^+ is the quadrupole moment and is a tensor. It stands for the integral:

$$eQ^+ = \int \rho'_n (3z'^2 - r^2) dv' \tag{4.19}$$

where:

The parameter ρ'_n stands for the charge density

dv' stands for an elementary volume

One observes that the interaction is a function of the gradient of the electric field, which is also a tensor. The energy is thus a product of two tensors and is a function of the orientation of the two axis of symmetry of the quadrupole moment and the electric field gradient.

In the case of a clock transition that takes place between an S state and a D state, the calculation thus consists in evaluating that interaction, taking care of the angle between the symmetry axis of the quadrupole moment and that of the electric field gradient using the appropriate wave functions and calculating their expectation value. As said, the S state is spherically symmetrical and no quadrupole moment is present. However, in D and F state the electronic distribution is such as to allow the presence of a quadrupole moment. It is noted first that the laboratory axis is given by the quantization axis provided by the presence of a magnetic field. The transformation to that frame involves tensors analysis and the energy is then given by:

$$W = \frac{1}{8} eQ \frac{\partial^2 V}{\partial z'^2} \left(3\cos^2 \theta - 1 \right) \qquad (4.20)$$

where:
   θ is the angle between the field gradient axis and the magnetic field

The splitting and shift of energy levels by the magnetic field are assumed to be large compared to the perturbation caused by the electric field gradient. The quadrupole interaction thus causes only a small shift of the energy level considered. The interaction is then evaluated for the state of the atom in question. This is done by means of a perturbation approach (Itano 2000). The result is approximately:

$$\langle \Psi | W | \Psi \rangle = \frac{-2A \left[3m_F^2 - F \left(F + 1 \right) \right] \left(\Psi \| Q_{op} \| \Psi \right)}{\left[\left(2F + 3 \right) \left(2F + 2 \right) \left(2F + 1 \right) 2F \left(2F - 1 \right) \right]^{1/2}} \left(3\cos^2 \theta - 1 \right) \quad (4.21)$$

where:
   A is a measure of the electric potential creating the electric field gradient expressed in V/cm^2
   $\left(\Psi \| Q_{op} \| \Psi \right)$ is the quantum mechanical evaluation of the quadrupole moment of the ion in the state in question

In practice A may be of the order of 10^3 V/cm^2 and the quadrupole moment for 5d electron evaluated from Equation 4.19 may be of the order of a few ea_o^2, where a_o is Bohr radius. The energy shift is then of the order of 1 Hz or a few parts in 10^{-15}. In view of the kind of accuracy aimed at, it is thus important to evaluate exactly that frequency bias.

Unfortunately the field gradient in the trap is not well-known being produced in some cases by so-called patch potentials produced on the surface of the electrodes. Consequently, in such a case, one is obliged to measure the shift in terms of known parameters and possibly extrapolate or interpolate the results. The effect of the shift on the clock frequency can be measured for various angles θ between the field gradient symmetry axis and the magnetic field applied. This orientation is thus an excellent parameter. Furthermore, it turns out that if three orthogonal directions θ are used and the average of these measurements is made, the quadrupole shift cancels out (Itano 2000). This technique has been used extensively (Blythe et al. 2003; Tamm et al. 2007). The technique, however, relies on the perfect orthogonality of the three directions chosen for orienting the magnetic field. This is difficult to achieve in practice. Another technique was also proposed (Dubé et al. 2005). It consists in realizing that the quadrupole shift is zero if $3m_F^2$ is equal to $F(F + 1)$ in Equation 4.21. These authors made measurements with transitions to various m levels in the excited state of ^{88}Sr$^+$, for various orientations of the magnetic field and interpolation of the results showed that a cancellation of the quadrupole shift could be made to an accuracy of a few in 10^{-18}. It thus appears that with that technique the quadrupole shift is no longer a major concern in the inaccuracy budget.

4.4.3.3.4 Other Frequency Shifts and Present State of the Art

The frequency biases examined are particular to trapped single ion frequency standards. There are, however, several other frequency biases that can affect their frequency and cause frequency instability. Those are common to many types of frequency standards and may include for example, gravitational shift, light shifts, and collisions with background atoms. These biases have been examined in Chapters 1 and 3 for various standards. Similar techniques as those described in those cases can be used to evaluate those shifts and can be adapted to particular circumstances encountered in implementing an optical single ion frequency standard.

It appears that the main contributors to uncertainty, even after careful evaluation, remain the Stark shift due to ion motion, the BBR shift and in the case of an S–D or S–F clock transition, the electric quadrupole shift. However, this last contribution can be avoided by using ions with clock transitions other than S–D or S–F. For example, a clock transition S – P$_0$ as in ^{27}Al$^+$ does not have an electric quadrupole shift. Such a clock is characterized by an accuracy of 8.6×10^{-18}. Two such clocks were compared and their frequency difference was measured as 1.8×10^{-17} (Chou et al. 2010).

As a guide, we give in Table 4.5 a summary of the order of magnitude of frequency shifts generally encountered in single ion frequency standards in the case of three selected ions: ^{27}Al$^+$, ^{88}Sr$^+$, and ^{199}Hg$^+$ ions. Table 4.6 summarizes the accuracy reached in some selected single ion optical frequency standards. As is readily seen in that table, it appears that the technique has reached a level of implementation that allows unsurpassed accuracy in a space of relatively small dimensions.

4.5   OPTICAL LATTICE NEUTRAL ATOMS CLOCK

4.5.1   THE CONCEPT

The concept of ion and atom trapping was introduced in Chapter 2. We have seen in Chapter 3 that the trapping of ions in a Paul trap could be used to implement microwave frequency standards with rather interesting properties regarding accuracy, frequency stability as well as size. The characteristics of those standards regarding frequency stability, as we have seen, are limited mainly by the width of the clock transition that can be used, leading to line Q's of the order of 10^{10} at best. We recall that the line Q has a direct effect on frequency stability and plays a major role in accuracy since larger line Q makes easier the determination of resonance line centre. Earlier, in this chapter we have shown that the same trapping technique could be used to implement optical frequency standards. The clock transition being at a much higher frequency, this property being coupled to a narrow resonance line made possible very large line Q's ($\sim 10^{12}$ and higher). Furthermore, the use of a single ion in such a trap made possible the observation of these high Q transitions in an environment that is close to free space and provided hope for implementation of atomic frequency standards with still

TABLE 4.5

Summary of the Size of the Shifts Encountered in the Implementation of a Single Ion Optical Frequency Standard with Three Selected Ions: ^{27}Al$^+$, ^{88}Sr$^+$, and ^{199}Hg$^+$

Type of Bias or Perturbation	Ion	Clock Shift (10^{-18})	Typical Uncertainty in Determination (10^{-18})	Reference
BBR shift	^{27}Al$^+$	−9	3	Chou et al. 2010
	^{88}Sr$^+$	0.05	22	Madej et al. 2012
Cooling laser Stark shift	^{27}Al$^+$	−3.6	1.5	Chou et al. 2010
Quad. Zeeman shift	^{27}Al$^+$	−1,079.9	0.7	Chou et al. 2010
	^{199}Hg$^+$	−1,130	5	Lorini et al. 2008
	^{88}Sr$^+$		0.002	Madej et al. 2012
Excess micromotion	^{27}Al$^+$	−9	6	Chou et al. 2010
	^{199}Hg$^+$	−4	4	Lorini et al. 2008
	^{88}Sr$^+$		1	Madej et al. 2012
Secular motion	^{27}Al$^+$	−16.3	5	Chou et al. 2010
	^{199}Hg$^+$	−3	3	Lorini et al. 2008
Clock laser Stark shift	^{27}Al$^+$	0	0.2	Chou et al. 2010
Background gas collisions	^{27}Al$^+$	0	0.5	Chou et al. 2010
	^{199}Hg$^+$	0	4	Lorini et al. 2008
Total shift	^{27}Al$^+$	−1,117.8	8.6	Chou et al. 2010
	^{199}Hg$^+$	−1,137	19	Lorini et al. 2008
	^{88}Sr$^+$		22	Madej et al. 2012

Note: The shift and uncertainty are given for three selected ions as typical for ion implementation of an optical frequency standard.

greater frequency stability and accuracy. However, as is readily realized by the reader in examining the various tables given above, several systematic frequency shifts or biases are still present in the implementation of those single ion standards and the measurement accuracy of those biases is limited. Furthermore, the fact that a single ion is used limits frequency stability due mainly to a limited S/N. Finally, its interaction with various stray electric fields, possibly present in the structure, limits its accuracy.

The idea of using a large number of neutral atoms in a trapping environment, which would have weak interaction with that environment always appeared as a challenge that could not be overcome. One main impediment was also the possible interaction between those atoms themselves. However, with the advent of laser cooling and trapping, it was realized that if the proper technique was used, those challenges could be overcome. In particular, the use of a trap implemented by means of optical standing waves, as described in Chapter 2, appeared to offer a very privileged situation. A relatively large number of atoms can be stored individually or in small numbers in microtraps for long periods without interacting much with each other. Those

TABLE 4.6
Level of Accuracy Reached with Selected Single Ion Standards

Ion	Total Frequency Bias (10^{-16})	Uncertainty in Determination (10^{-16})	Reference
$^{199}Hg^+$	−11.37	0.19	Lorini et al. 2008
$^{27}Al^+$	−11.178	0.086	Chou et al. 2010
$^{171}Yb^+$ quad	69.68	6.13	Godun et al. 2014
	4.26	1.1	Tamm et al. 2014
$^{171}Yb^+$ octupole	1.41	0.71	Huntemann et al. 2013
$^{88}Sr^+$		0.23	Dubé et al. 2013
$^{40}Ca^+$	47.4	6.5	Gao 2013
	24.5		Chwalla 2009

atoms are in a low energy state in a shallow well, thus essentially at rest and obeying Lamb–Dicke criteria for cancelling Doppler effect. Such a trap is called an *optical lattice* whose cells dimension is half the wavelength of the radiation used to create the standing wave. As shown earlier, the trapping is due to an interaction between the atoms and the radiation field with little perturbation on the energy levels of the atom aside from a so-called light shift that can be cancelled by a proper choice of lattice wavelength called the *magic wavelength*, a technique that will be outlined below.

Such an ideal structure can be created as in Figure 4.14 in which a standing wave is produced by means of a frequency-stable laser radiation field reflected on itself by a mirror.

Small potential wells of length $\lambda_m/2$ are created. The figure shows some of those potential wells in the central region of the system where the radiation beams have a width w of the order of the waist of the interfering laser beams. Not all wells contain atoms and although not made explicit in the figure wells may contain several atoms. The lattice shown is one-dimensional (1D) created by a laser beam reflected on itself. However, 2D and 3D lattices may be created using two or three orthogonal laser beams. The structure shown is horizontal, but may also be oriented vertically. A number of 10^4 to 10^5 atoms may be stored in a few thousand traps in a very small space. Those atoms are first cooled to very low temperature in the microkelvin range or less in a MOT superimposed on the lattice trapping field. The cooling lasers and the MOT magnetic field are turned off and, upon appropriate conditions related to wells' depth and atom temperature, a large proportion of the MOT atoms are trapped in the optical lattice. They occupy low energy levels n in the individual potential wells, depending on their residual temperature after cooling.

A typical optical lattice frequency standard is shown in Figure 4.15 and consists essentially of a source of atoms, a MOT for cooling and trapping an ensemble of atoms of the type chosen in the region of the optical lattice and a laser generally of the Ti sapphire type for creating the optical lattice. Frequency stabilized lasers are used to effectuate the cooling and the interrogation of the atomic ensemble at the clock transition frequency as in the case of single ion clocks described earlier.

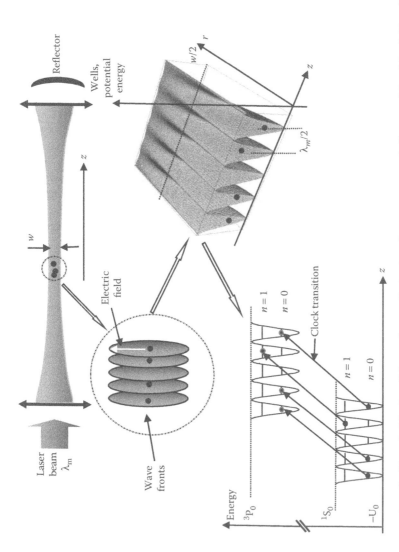

FIGURE 4.14 Representation of a 1D optical lattice. Coherent laser radiation interferes to create periodic potentials of depth U_o acting as small traps for atoms. The atoms are confined to a small region smaller than the clock transition wavelength. The laser radiation forming the lattice has a wavelength λ_m chosen such as to introduce an equal frequency shift of the S and P state involved in the clock transition. The number n characterizes the vibration states of atoms in a well. w is the laser beam waist.

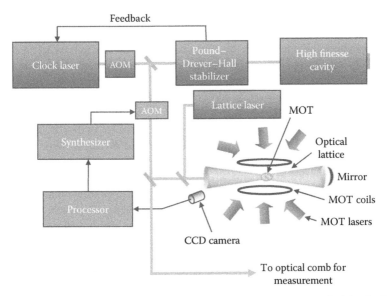

FIGURE 4.15 Schematic diagram illustrating the main components required in the implementation of an optical lattice frequency standard. The processor is also used for controlling the sequence of pulses for cooling and interrogating the atomic ensemble trapped in the optical lattice.

The operation of such a clock may be described as follows. To illustrate the process, we will use the level manifolds shown in Figure 4.16. This energy level structure is similar to that of Sr that we will examine in more detail later. The atomic ensemble is first cooled in the MOT superimposed on the lattice. The first-stage cooling is done by means of the transition $|{}^1S_0\rangle \rightarrow |{}^1P_1\rangle$. The MOT generally traps atoms with a velocity smaller than 10 m/s. It is thus best to feed the MOT with a 2D MOT with

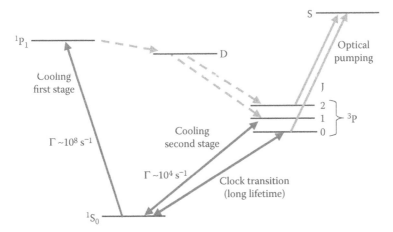

FIGURE 4.16 Lower energy levels manifold of a typical atom used in a lattice clock shown to illustrate the functioning of such a clock in the optical range. The clock operates on an S–P transition.

low velocity atoms to increase the number of atoms trapped. This approach is also advantageous when alkali-earth atoms are used since their vapour pressure is low. It is recalled that there exists a limit in the lowest temperature that can be reached in laser cooling. This is the Doppler limit that is a function of the decay rate of the excited state (see Chapter 2). This limit using the cycling transition mentioned is normally in the mK range. It is possible, however, in certain cases, to improve cooling by means of an added step, using another transition that has a narrower line width, for example, the $|^1S_0\rangle \rightarrow |3P_1\rangle$ illustrated in the figure. In that case temperatures in the μK range can be reached. After the MOT is established and a temperature in the μK range or lower is reached, the cooling lasers and the magnetic field are turned off and a certain amount of atoms remain trapped in the optical lattice. The atoms are trapped in the shallow wells at very low temperature in a Lamb–Dicke regime. The excitation of the clock transition is done in a similar way as in the case of single ion optical clock described earlier, applying a pulse of radiation at the clock transition, which is characterized by a very narrow width. The excitation causes atom shelving. The length of that pulse determines the width of the resonance line that is observed. Detection of fluorescence at the cooling wavelength is done to establish whether shelving has taken place or not, thus detecting resonance. Atoms sometime decay from the P level to other states and are lost for the clock transition. Optical pumping may be applied to drive the atom back to the cooling process. Optical pumping may also be done from the shelving level in order to return the atoms to the ground state for another cycle. All light pulses are applied in sequences in order to avoid light shifts during detection as illustrated in Figure 4.17. The whole process is very similar to that used in the single ion clock described earlier. We will now examine some basic properties of such an optical lattice.

4.5.1.1 Trapping Characteristics

The lattice is generally characterized by a potential experienced by the atom as (see Chapter 2):

$$U(r,z) = U_0 \exp\left\{-2\left[\frac{r}{w(z)}\right]^2 \cos^2(k_L z)\right\} \qquad (4.22)$$

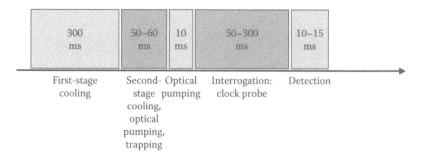

FIGURE 4.17   Typical sequence of operations that can be used in the cooling, trapping, interrogation, and detection of the optical transition in an optical lattice frequency standard in the optical range. (Adapted from Poli, N. et al., Optical atomic clocks, arXiv:1401.2378v1, 2014.)

where U_o is given by:

$$U_o = -\alpha\left(\omega_L\right)E_L^2 \tag{4.23}$$

in which E_L is the electric field of the interfering wave and $\alpha(\omega_L)$ is the polarizability of the atom at the lattice frequency ω_L in level μ and is given by:

$$\alpha_\mu\left(\omega_L\right) = \frac{\langle\mu|er\cdot e_\lambda|i\rangle}{\omega_{i\mu} - \omega_L} \tag{4.24}$$

It is readily observed that depending on the tuning of the laser, the polarizability may change sign. The force on the atoms being the gradient of electric field creating the potential well, the resultant energy may be minimum for a maximum or a minimum of the electric field. In Figure 4.17, we have chosen the case where the energy is a minimum for a maximum of the field. In practice, the laser power is limited and the wells, depth is shallow. It may be of the order of 100 μK and requires efficient cooling well below that temperature for trapping a sufficient number of atoms.

4.5.1.2 Atom Recoil

At the bottom of the wells, the potential is very nearly harmonic and the spacing between the levels may be written approximately as (Lemonde 2009; Derevianko and Katori 2011):

$$\hbar\omega_{har} = \frac{2\pi\hbar}{\lambda_L}\left(\frac{2|U_o|}{M}\right)^{1/2} \tag{4.25}$$

where:
  M is the mass of the atom trapped in the well

The interrogation at the clock frequency ω_c imparts a kick to the atom corresponding to a momentum change $\Delta p = \hbar\omega_c/c$. If the corresponding recoil energy $E_r = p^2/2M$ of the atom is smaller than the spacing $\hbar\omega_{har}$ of the wells' vibration states, the atom stays in the same vibration state n upon a clock transition. Consequently, the clock frequency is not altered by the residual motion of the atom at the low residual temperature. We can interpret this phenomenon classically. The atom oscillates in the well at the frequency ω_{har} and the frequency of the incoming wave at ω_c appears modulated by that oscillation. If the atom oscillation amplitude is small, analysis then shows that the modulation gives rise to a signal resonant with a central carrier and with two small sidebands separated from the carrier by the vibration frequency. Detection of the resonance at the central carrier is not affected then by the motion of the atom inside the well: this is the Lamb–Dicke regime. The situation is similar to that we have described in the case of ion trapping in a Paul trap. In the present case also, the recoil is absorbed by the whole structure since the atom does not change vibrational state within the well upon absorbing a photon.

4.5.1.3 Atom Localization

The optical lattice as we said in Chapter 2 is similar to a solid-state crystal. However, the wells are separated by several hundreds of nm while in a crystal the spacing of the cells is of the order of 0.1 nm. In an optical lattice, there is little interaction between atoms trapped in different wells. However, as mentioned above, the wells are very shallow and resonant tunnelling can take place with several atoms being in the same well. The construction of a vertical optical lattice may have an advantage over a horizontal one in that regards, since the successive wells are at different height z. The energy of the states between wells changes by mgz, thus inhibiting resonance between wells at different heights z.

4.5.1.4 Magic Wavelength

In order to trap a large number of atoms, the depth of the wells may be made as large as 100 times the recoil energy E_r of the atoms in first-stage cooling. Upon second-stage cooling, the depth of the wells can be reduced to a value of the order of $10\ E_r$ reducing at the same time the frequency shift caused by the lattice electric field. It is recalled that the energy levels of the atom trapped in such an optical well are shifted by the presence of those electric fields. Actually, the equivalent frequency shifts may be of the order of 10^4 Hz and such frequency shifts would inhibit the use of the approach for implementing a high quality frequency standard if some means did not exist to minimize them. We have analyzed in Chapter 2 the displacement of an energy level δE_i in the presence of an electric field E_L. A simple second-order analysis leads to the expression:

$$\delta E_i = \alpha_i\left(\omega_L\right)\left(\frac{E_L}{2}\right)^2 \tag{4.26}$$

where:

$\alpha_i(\omega_L)$ is the dynamic polarizability of the atom in level i as given by Equation 4.24

We note that the effect is essentially a light shift that was studied extensively in *QPAFS*, Volume 1 (1989). The clock transition takes place between two levels, S and P, and it is reasonable to ask about the possibility that both levels be shifted by the same amount causing the shifts of both levels to cancel each other. The difference in polarizability of the atom in the two states is:

$$\Delta\alpha\left(\omega_L\right) = \alpha_P\left(\omega_L\right) - \alpha_S\left(\omega_L\right) \tag{4.27}$$

causing a differential frequency shift of the two levels creating a light shift given by:

$$\Delta\nu_{LS} = \frac{1}{h}\left[\delta E_P\left(\omega_L\right) - \delta E_S\left(\omega_L\right)\right] \tag{4.28}$$

If both polarizability are equal for a given wavelength, a zero light shift of the clock transition results. In practice, the lattice laser is well-detuned from resonance with a given transition and the polarizability is evaluated using a summation over all levels connected to the states S and P of the clock transition:

$$\alpha_\mu\left(\omega_L\right) = \sum_{i\neq\mu}\frac{\langle\mu|er\cdot e_\lambda|i\rangle}{\omega_{i\mu} - \omega_L} \tag{4.29}$$

The resulting shift of each state varies differently with the lattice frequency ω_L. It is then a question of finding the correct frequency for which the shift is the same for each state. This analysis as presented is elementary. A more complete approach consists in introducing concepts such as the tensor polarizability that makes shifts dependent on relative direction of polarization e_λ, and magnetic field and considers higher order terms in E^4, for example. The present analysis provides, nevertheless, the essential of the physics involved. Calculations have been done for S and P states of atoms having promising characteristics and graphs have been obtained for various atoms in which the energy shifts of those states were equal for given laser wavelength (see, e.g., Katori et al. 2003; Derevianko et al. 2009; Lemonde 2009). Several graphs describing the behaviour of various atoms in such a field have been calculated. A typical sketch of such a graph describing the general behaviour of an atom such as Sr is shown in Figure 4.18. The resonances present in the expression of the polarizability of each level are made explicit.

Approximate values of the so-called magic wavelength are obtained and are used as a guide in their determination by means of accurate measurements. Those values will be given below.

4.5.1.5 Clock Transition

The clock transition in the few types of atoms chosen for the implementation of a lattice clock is a $|{}^1S_0\rangle \rightarrow |{}^3P_0\rangle$. This is a strictly forbidden transition from quantum mechanical rules. However, the presence of a nuclear spin, as in odd isotopes introduces a slight admixture of the upper states in the ${}^3P$ manifold resulting in a finite transition probability. In Sr with a nuclear spin $I = 9/2$ the decay rate of the ${}^3P_0$ state is calculated as 6.7×10^{-3} s^{-1}. In even isotopes with $I = 0$, the application of a small magnitude magnetic field also causes an admixture of other states causing also a finite transition probability. However, as is readily observed that transition probability is rather small and a narrow resonance with a high line Q is observed at the clock frequency.

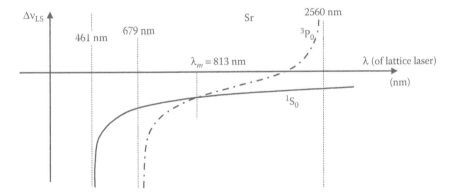

FIGURE 4.18 Sketch of the variation of the light shift of both levels entering in the clock transition as a function of the lattice laser wavelength in the case of the strontium atom. (Data from Lemonde 2009.)

4.5.2 TYPE OF ATOMS USED IN OPTICAL LATTICE CLOCKS

A few atoms have been proposed and used to implement an optical lattice clock. They are Sr, Hg, and Yb. Ca and Mg have also been proposed, but main efforts have been concentrated on the first three listed. We will describe the energy level structure of those atoms and outline the choice of frequencies made for laser cooling, trapping, and clock transitions.

4.5.2.1 Strontium Atom

Initial theoretical analysis on the use of Sr in a lattice clock configuration was done by Katori (2002) and Katori et al. (2003). The proposal was followed shortly after by first demonstrations based on optical lattice spectroscopic studies using transitions $^1S_0 - ^3P_1$ and $^1S_0 - ^3P_0$ (Takamoto and Katori 2003) as clock transitions. Sr has four stable isotopes that can be used for implementing an optical clock. The isotopes ^{88}Sr ($I = 0$) and ^{87}Sr ($I = 9/2$) are those that have raised interest. Their natural abundance is 82.6% and 7%, respectively.

The energy levels of interest for implementing a lattice clock with that ion are shown in Figure 4.19. Sr possesses two transitions from the $^1S_0$ ground state that are suitable for laser cooling: the broad transition $^1S_0 \rightarrow {}^1P_1$ (32 MHz) allows the formation of a MOT containing more than 10^9 atoms at a temperature of the order of 1 mK. Extra cooling by means of the narrow $^1S_0 \rightarrow {}^3P_1$ transition (7.5 kHz width) allows the reaching of a quantum-limited minimum temperature of ~250 nK. The alkaline earth-like level structure of the triplet and singlet manifolds possesses sufficient complexity to support the use of a specific *magic wavelength* inhibiting light shifts of the transition $^1S_0$ to $^3P_0$, both levels being shifted by the same amount. A magic wavelength of 813 nm, as shown in Figure 4.18, is found and is generated from a power oscillator.

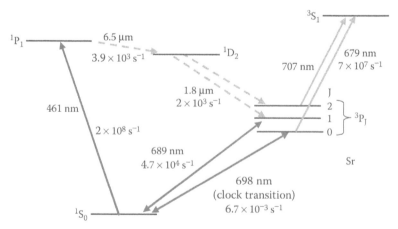

FIGURE 4.19 Lower energy levels of interest of Sr showing the important transitions for the implementation of an optical lattice frequency standard (hyperfine structure for the fermion is omitted). The decay rates shown from the excited states provide information on the line width of the transitions involved.

As was mentioned above, the clock transition $^1S_0 \rightarrow {}^3P_0$ is normally highly forbidden. However, in the case of the odd isotope, ^{87}Sr, hyperfine interactions causes a small admixture of P states and the transition is allowed with a line width in the mHz range. It is thus highly advantageous as a clock transition (Katori 2011). In general, the laser used for exciting that transition has a width a the order of 1 Hz and consequently the observed width is not limited by the clock transition itself but rather by the laser used or the length of the Rabi pulse used in the excitation.

The Sr system is very convenient in that all of the lasers for cooling and spectroscopy can be derived from inexpensive diode laser sources, as opposed to some other optical lattice clocks to be described below, Hg, for example, which requires laser wavelengths that are more challenging to generate. In the case of Sr, the first-stage cooling at 461 nm may be produced by means of a so-called master oscillator power amplifier (MOPA) diode system by means of frequency doubling. Second-stage cooling and clock transition wavelength at 689 and 698 nm, respectively, are available from existing laser diodes.

4.5.2.2 Mercury Atom

Mercury has two fermionic and five bosonic isotopes. Their natural abundance is in the 10%–30% except for ^{196}Hg whose natural abundance is an order of magnitude smaller. Both fermions (^{199}Hg, ^{201}Hg) have a weakly allowed intercombination line ($^1S_0 - {}^3P_0$) with a ~100 mHz natural line width at 265 nm (Bigeon 1967) and an accessible *magic wavelength*, at which the trapping dipole field does not disturb the clock transition (V. Pal'chikov 2004, pers. comm.; Ovsiannikov et al. 2007). At room temperature, Hg has a vapour pressure of 0.3 Pa and no oven is needed to produce the vapour pressure required for creating a MOT of sufficient density. In fact, a vapour pressure well-suited for a magneto-optical trap is obtained near −40°C, a temperature which is easily reached in vacuum with Peltier thermoelectric coolers.

Neutral mercury has an alkaline-earth like electronic structure similar to that of Sr. The lower energy level structure relevant to the implementation of an optical lattice clock is shown in Figure 4.20.

In comparison to Sr, neutral Hg has a cooling transition $\left| {}^1S_0 \right\rangle \rightarrow \left| {}^3P_1 \right\rangle$ better adapted for single-stage laser-cooling at a wavelength of 253.7 nm. Optical pumping by means of the $^3S_1$ is expected to be required (Mejri 2011). The clock transition sensitivity to room-temperature BBR is more than 10 times lower than that of Sr, which is another advantage. For these reasons, Hg appears to offer great advantages over other proposed atoms in the implementation of such an optical frequency standard and we will describe in some detail a particular approach such as the one used at SYRTE, Paris, France.

4.5.2.2.1 Vacuum Chambers

Laser-cooling and trapping of Hg is done in two steps:

- The first section of the system acts as a source of Hg vapour. Since Hg has an exceptionally high vapour pressure of 0.3 Pa at room temperature, the source of Hg is cooled down to reduce and control its pressure. A drop of Hg is placed in a small copper bowl glued to the cold surface of a two-stage

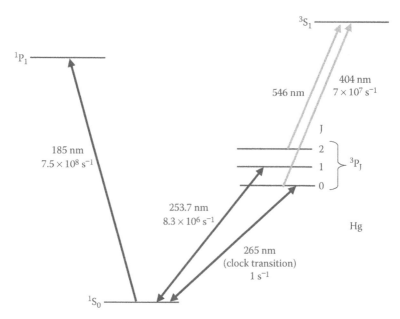

FIGURE 4.20 Lower energy level structure of Hg of interest for the implementation of an optical lattice clock (hyperfine structure for fermions is omitted). Vacuum wavelengths are shown.

Peltier thermoelectric element placed inside the vacuum chamber. The hot side of the thermoelectric cooler is attached to a water-cooled block of copper whose temperature is maintained near 10°C. A getter takes care of pumping impurities from the released Hg.

- The second section acts as the main chamber where the atoms are laser-cooled and trapped in a 3D MOT to decrease temperature and increase density. They are ultimately transferred to the optical lattice trap and then probed for detection of the clock transition.

4.5.2.2.2 MOT of Hg

A typical MOT only captures atoms from the tail end of the room temperature atomic velocity distribution. In order to have as many trapped atoms as possible and a long MOT lifetime, a 2D-MOT is used for pre-selecting slower atoms. The MOT is produced by the intersection of three orthogonal pairs of retroreflected $\sigma^+ - \sigma^-$ polarized UV laser beams with a diameter of ~15 mm. Anti-Helmholtz coils generate the magnetic quadrupole field with a gradient of 0.10 T/m (10 G/cm) along the axis of the coil pair at the MOT centre. The cooling laser radiation at 253.7 nm is frequency stabilized by means of saturated absorption in a 1-mm-long quartz cell containing Hg vapour at ~0.25 Pa at room temperature. A residual reflection of UV light from the 507.4 nm resonant frequency-doubling system is used for frequency stabilization and tuning purposes. This auxiliary beam is frequency shifted by two acousto-optic modulators. This allows frequency tuning of the main UV beam with respect to the Hg saturated absorption cell resonance. The source for the cooling light is a quadrupled Yb-doped thin disk laser. The laser delivers 7 W of CW single frequency laser light at 1 015 nm. This light is frequency doubled in a commercial cavity-enhanced second harmonic generation (SHG) unit, producing ~3.5 W at a wavelength of 507.4 nm. The second SHG of ~70 mW at the cooling wavelength of 253.7 nm UV light is performed with a home-made system similar to that described by Berkeland et al. (1997).

4.5.2.2.3   Clock Interrogation Laser

The clock laser is a quadrupled Yb-doped fibre laser at 1062 nm. In order to obtain an interrogating radiation with a frequency stability sufficient to observe the clock transition with a one Hertz resolution, the fibre laser is locked to an ultra-stable cavity as the one described earlier in Chapter 2. The cavity is placed inside double vacuum chambers and double heat shields connected to external radiators. It is also placed on a passive anti-vibration platform encapsulated in a soundproof box. The finesse of the FP cavity is ~914,800 as measured by the ringing technique. A comparison of two such systems showed a relative instability of no more than 8×10^{-16} over a one second averaging period.

4.5.2.2.4   Implementation of the Dipole Lattice at the Magic Wavelength

Feasibility of the optical lattice clock scheme relies on the existence of a so-called trapping magic wavelength for which the two states $^1S_0$ and $^3P_0$ are shifted equally. For Hg, such a magic wavelength exists between two strongly allowed transitions, $6 \ ^3P_0$ to $7 \ ^3S_1$ at 405 nm and $6 \ ^3P_0$ to $6 \ ^3D_1$ at 297 nm (V. Pal'chikov 2004, pers. comm.). A recent calculation gives that magic wavelength as 363 nm (Ye and Wang 2008). This radiation is generated by means of a continuous-wave (CW) titanium sapphire laser delivering about 900 mW at 724 nm. That laser light is frequency doubled using a Brewster-cut LiB_3O_5 crystal in a bow-tie resonant cavity using the Hänsch-Couillaud locking scheme (Hänsch and Couillaud 1980). About 200 mW of 363 nm lattice light is produced. The cooled atoms are then confined in the vertically oriented optical lattice, which is loaded from the single-stage MOT just described.

At the time of writing, a clock based on this atom and described in general terms previously is under investigation at SYRTE, Paris. The determined magic frequency was found to be 8826.8546 $\pm$ 0.0024 THz (λ_M = 362.5697 $\pm$ 0.0011 nm). The fre quency of the clock transition was measured as 1,128,575,290,808,162.0 $\pm$ 6.4 (sys.) $\pm$ 0.3 Hz with a fractional uncertainty of 5.7×10^{-15}. A fractional frequency sta-bility of $5.4 \times 10^{-15} \ \tau^{-1/2}$ has been reported in its early development and it is claimed that its accuracy should be in the 10^{-17} range (McFerran et al. 2012).

4.5.2.3   Ytterbium Atom

Another excellent candidate for an optical lattice clock is the Yb atom (Porsev et al. 2004; Hong et al. 2005; Barber et al. 2006). Yb isotopes exist as two fermions and two bosons (mass from 171 to 173) with natural abundance ranging from 3% to 22%.

The lower energy level structure of Yb relevant to the implementation of an optical lattice clock is shown in Figure 4.21. The Yb clock transition $^1S_0 \rightarrow {}^3P_0$ ($\lambda = 578.4$ nm) has a line width of ~10 mHz due to hyperfine state mixing of the $^3P_0$ sublevels.

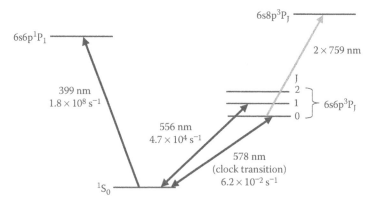

FIGURE 4.21 Lower energy levels of Ytterbium of interest in implementing an optical lattice frequency standard in the optical domain.

The magic wavelength for the clock transition is $^1S_0 \rightarrow {}^3P_0$ is 759 nm and may be created using a titanium sapphire laser (Barber et al. 2006; Poli et al. 2008).

Laser cooling of Yb is obtained by means of a double-stage MOT first, using the dipole transition $^1S_0 \rightarrow {}^1P_1$ at 399 nm (natural line width of 28.9 MHz), by means of InGaN diode lasers. Secondary cooling wavelength at 556 nm, using the intercombination line $^1S_0 \rightarrow {}^1P_1$ at 556 nm, may be obtained using a frequency doubled fibre laser emitting radiation at 1 112 nm (Hoyt et al. 2005). The trapping efficiency of Yb in the optical lattice using that scheme is claimed to be about 10%.

4.5.2.4 Magnesium Atom

Mg is another proposed candidate as the heart of an optical frequency standard due to its low sensitivity to BBR, which currently limits the accuracy of optical clocks. The lower energy levels manifolds of interest are shown in Figure 4.22.

At this time, some preliminary work has been done at the Institut für Quantenoptik, Leibniz Universitat Hannover in trapping Mg in an optical lattice as a first step towards an optical lattice-based frequency standard (Ertmer et al. 2010).

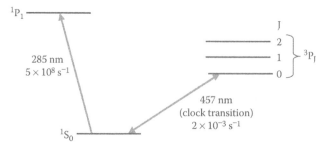

FIGURE 4.22 Lower manifolds of energy levels of interest of magnesium in implementing an optical lattice frequency standard in the optical range.

4.5.2.5 Calcium Atom

Ca is also a possible candidate for implementing an optical lattice clock in the optical range. We have shown above that it was used early to implement an optical clock by means of a freely expanding MOT. The energy level structure was given in Figure 4.3 and, in principle, it could be used for the implementation of a lattice clock.

4.5.3 IMPORTANT FREQUENCY BIASES

Optical clocks using an optical lattice as storage technique are affected as all other types of optical standards by several biases that have to be evaluated accurately. Several of those biases have been discussed earlier in connection to the description of single ion optical frequency standards. We recall the most important ones and adapt the explanations to the present situation. We also give a short analysis of those particular to the optical lattice trap.

4.5.3.1 Zeeman Effect

As in the case of the ion clocks described earlier, the magnetic field shift of the optical clock transition is a function of the dependence of each of the levels involved in the transition. We will examine in particular the case of the odd isotope of strontium, ^{87}Sr. The two states of the clock transitions have an angular momentum originating solely from the nuclear spin $I = 9/2$. If states $m_F = 9/2$ are chosen for each levels of the transition, there should not be any field dependence in first order since both states would be displaced equally by the magnetic field. However, the presence of the other P states introduces a small admixture into the P_0 state and there results a small field dependence in first and second order in the field. To simplify matters, it is possible, as was done earlier in the case of ion optical frequency standards to express the sensitivity of the clock frequency to the magnetic field by means of the difference in Landé factors g of the two states in question, the P state being corrected for the admixture of states. The effect on the transition is to shift the frequency by:

$$\Delta v = \alpha_{e,g} B m_I + \beta_{e,g} B^2 \qquad (4.30)$$

where:

$\alpha_{e,g}$ and $\beta_{e,g}$ are respectively the first and second-order coefficients of the dependence of the combined clock states, e for excited and g for ground states

m_I is the magnetic quantum number of the transition chosen

The linear coefficient, in the case of odd isotopes, depends of course on the particular transition or levels chosen. Those coefficients are given in Table 4.7 for selected atoms of interest.

In the case of even isotopes, the nuclear spin is zero and there is no linear frequency shift. There remains a quadratic frequency shift, which is given in Table 4.8.

As an example, we may mention the case of Sr which has been studied extensively. In the case of the odd isotope ^{87}Sr, with a nuclear spin I equal to 9/2, it is possible to

TABLE 4.7

Linear Zeeman Coefficient of the $1S_0 - 3P_0$

Isotope	Nuclear Spin I	Linear Shift $\alpha'_{e,g}$ (Hz/μT)	Reference
^{87}Sr	9/2	−1.10	Boyd et al. 2007
^{171}Yb	1/2	−4.1	Porsev et al. 2004
^{173}Yb	5/2	+1.1	Porsev et al. 2004
^{199}Hg	1/2	+6.6	Hachisu et al. 2008
^{201}Hg	3/2	−2.5	Hachisu et al. 2008

TABLE 4.8

Quadratic Zeeman Shift Coefficient of Clock Transition for Selected Atoms

Atom	$\beta'_{e,g}$ (MHz/T^2)	References
^{24}Mg	−217	Taichenachev et al. 2006
^{40}Ca	−83.5	Taichenachev et al. 2006
^{87}Sr	−23.3	Boyd et al. 2006
		Baillard et al. 2007
^{88}Sr	−23.3	Akatsuka et al. 2010
^{171}Yb	−6.2	Taichenachev et al. 2006
^{174}Yb		Taichenachev et al. 2006
^{199}Hg	−2.44	Hachisu et al. 2008

excite, using appropriate light polarization, transitions $\left|S_0, m_I = 9/2\right\rangle \rightarrow \left|P_0, m_I = 9/2\right\rangle$ as well as $\left|S_0, m_I = -9/2\right\rangle \rightarrow \left|P_0, m_I = -9/2\right\rangle$ (Le Targat et al. 2013). In such a case, a measurement of the resonance frequencies for each transition indicates the value of the magnetic field and the average of the frequencies measured for the two transitions taking care of second-order effect corresponds to the clock frequency in zero magnetic field. Several measurements are then done on both sides of the clock transition as in the case of ions described earlier and the average of these measurements is used to construct numerically an error signal.

4.5.3.2 BBR Shift

When the desired frequency accuracy exceeds 10^{-16}, the BBR shift is very important as was shown in the case of the fountain and optical ion clocks. In fact, in the case of Sr for example, the BBR shift is of the order of 2.3 Hz or 5.3×10^{-15}. As was shown, the shift can be evaluated theoretically from a knowledge of the polarizability of the atom used and integration over a broad spectrum of excited states of that atom. Attempts have been made to measure its actual value by changing temperature. One main problem is to know exactly the temperature of the environment in which the optical lattice is embedded. Inhomogeneities as large as 1 K may exist between various parts of the structure and it is difficult to assign an exact temperature to the radiation in which the atomic ensemble is embedded. In a typical case using Sr, a conservative accuracy in the determination of the BBR may be about 30 mHz or 7.5×10^{-17} (Le Targat et al. 2013). Proposals have been made in order to improve its determination by means of a technique in which the atomic ensemble or lattice is made to slip physically by a short distance, say 50 mm in a short time, where measurements of the dc Stark shift would be made. A second approach would be interrogation of the atomic ensemble in a cryogenic environment where the BBR shift is reduced to the order of a few $\times 10^{-18}$ (Middelmann et al. 2010) as was described in the case of the Cs fountain. Values of the BBR shifts evaluated for several atoms are given in Table 4.9. It appears from such a table that Hg has a net advantage over other atoms for implementing an optical frequency standard.

4.5.3.3 Lattice Light Shift

We have analyzed in Section 4.5.1 and in Chapter 2 the effect on the clock transition of the applied radiation field that creates the optical lattice. The elementary analysis done showed that the clock frequency is shifted by the interaction of the atoms with the relatively strong field present. The shift depends on the frequency of the laser field ω_L and its intensity E_L^2, and is different for the two energy levels of the clock transition. Fortunately, it is possible to find a wavelength, qualified as *magic*, for which both levels are shifted by the same amount. More advanced analysis, however, shows that it also depends on the relative direction of the lattice polarization and the system quantization axis generally defined by the orientation of the applied magnetic field. This effect makes the clock frequency sensitive to the stability of the lattice polarization fluctuations. The shift can be evaluated experimentally as a function of several parameters characterizing the lattice standing wave and adjustments can then be made to minimize it. This magic wavelength is given in Table 4.10 for various atoms of interest.

TABLE 4.9
Fractional Shift due to Black Body Radiation at 300 K for Several Selected Atoms

Atom	Transition	Fractional Shift due to BBR at 300K $\delta\nu_{BBR}/\nu_0$	Reference
Sr	$^1S_0 \rightarrow {}^3P_0$	-5.5×10^{-15}	Porsev and Derevianko 2006
Ca	$^1S_0 \rightarrow {}^3P_1$	-2.6×10^{-15}	Porsev and Derevianko 2006
Yb	$^1S_0 \rightarrow {}^3P_0$	-2.4×10^{-15}	Porsev and Derevianko 2006
Mg	$^1S_0 \rightarrow {}^3P_0$	-3.9×10^{-16}	Porsev and Derevianko 2006
Hg	$^1S_0 \rightarrow {}^3P_0$	-1.6×10^{-16}	Hachisu et al. 2008
Rb	5s $(F = 2 \rightarrow F = 1)$	-1.25×10^{-14}	Safronova et al. 2010
Cs	6s $(F = 4 \rightarrow F = 3)$	-1.7×10^{-14}	Simon et al. 1998

TABLE 4.10

Magic Wavelength for Various Atoms of Interest That Can Be Used to Implement an Optical Frequency Standard in the Optical Lattice Approach

Atom	Magic Wavelength (Frequency) (Experimental Results)	Magic Wavelength (Theoretical Results) (nm)	Reference (Experimental Results)	Reference (Theoretical Calculation)
^{24}Mg		466		Derevianko et al. 2009
^{40}Ca		739		Derevianko et al. 2009
^{87}Sr	385,554.718(5) GHz	813	Westergaard et al. 2011	
^{88}Sr	368,554.58(28) GHz		Akatsuka et al. 2010	
^{171}Yb	394,798.329(10) GHz	759	Lemke et al. 2009	
^{174}Yb	394,799.475(35) GHz		Barber et al. 2008	
^{199}Hg	362.53(0.21) nm	362	Yi et al. 2011	Derevianko et al. 2009

4.5.3.4 Other Shifts

The three shifts just examined are the most important ones encountered in the practical realization of an optical frequency standard based on the optical lattice approach. There are, however, several other shifts or biases that, although of smaller amplitude, can contribute to the inaccuracy of such a frequency standard.

4.5.3.4.1 Collision Shifts

Atoms in a potential well may collide and cause a frequency shift in the clock transition. We have studied extensively the effect of atomic collisions on the clock frequency in the case of microwave standards and have found that they may be important particularly in the case of the Cs fountain. It was found that the Rb atom was probably a better choice since in that case the frequency shift was smaller by a factor of about 30. In an optical lattice with atoms extracted from a MOT, there may be present 10^9 atoms/cm^3. The atoms are captured in the lattice at very low temperature, in the μK range, a temperature that depends on the potential well depth. The atoms are also in various vibration states n within a given well. There may be several atoms per sites and they may interact depending on their statistical properties. Due to the low temperature involved, S wave scattering is predominant in collisions as in the case of the fountain. Furthermore, stored atoms may be of the fermionic or bosonic type and their interaction is rather different. For example, in the case of fermionic atoms such as ^{87}Sr with a nuclear spin $I = 9/2$, due to the exclusion principle the effect of collisions is suppressed when the atoms are polarized. This is not the case for bosonic atoms. In the case of partial polarization of fermionic atoms, collisions may take place between atoms in different states. A similar situation may exist for atoms being excited by the clock interrogation radiation. The intensity of the applied field may vary slightly in space causing a different level of excitation and providing a basis for differentiating between atoms that are no longer in similar states. Collisions may then take place causing a frequency shift. Actual clock frequency measurements may be made with different densities in the original MOT, causing different numbers of atoms per site. It is then found that the effect may be of the order of few 10^{-17} or less (Le Targat et al. 2013).

4.5.3.4.2 Stark Shift

The laser tuned to the clock transition appears to the other possible transitions as a detuned radiation field. It may in principle cause a light shift. However, this laser field is weak, the radiation intensity being a few nW. Its contribution may be of the order of a few in 10^{-17}.

4.5.3.4.3 Motion Effect

The waist of the lattice may be of the order of 100 μm or less and the depth of the wells may be of the order of a few hundred times the recoil energy, E_r. With second step cooling, the temperature reaches the μK range and the atoms may be confined to lower internal levels of small quantum numbers. Strong confinement of the atoms in those lattice potential wells introduces a Lamb–Dicke regime and the motional effects are dramatically reduced. The effect on frequency is evaluated to be of the order of a few 10^{-17} or less. Even a possible slipping of the lattice causing a sliding of the atoms at a small velocity has been considered and its effect has also been evaluated to be less than a few in 10^{-17}.

4.5.4 Frequency Stability of an Optical Lattice Clock

The frequency stability of such clocks is given theoretically by Equation 4.1. The advantage of using optical lattice storage with large number of atoms is thus evident from the expression compared to single ion clocks using the Paul trap approach. Recent results have confirmed this prediction. For example, the frequency stability of an optical lattice Sr clock was determined as $3.3 \times 10^{-16}\,\tau^{-1/2}$ by means of a comparison to a close copy of the same clock (Bloom et al. 2014). This gives a frequency stability of 6×10^{-18} for an averaging time of 3000 s making possible measurements in a relatively short period with the corresponding resolution.

4.5.5 Practical Realizations

Several laboratories have constructed optical lattice clocks. They are built essentially along the general lines shown in Figures 4.14 and 4.15. One implementation of interest is that of using the atom ^{87}Sr (Le Targat 2007; Katori 2011). Two of those optical clocks using a 1D vertical lattice were constructed at SYRTE, Paris, France. They were compared in frequency directly and were also compared to the Cs and Rb fountains maintained in operation on a quasi continuous basis by means of a frequency comb to be described below (Le Targat et al. 2013). The properties of one of the two ^{87}Sr optical lattice clock realized at SYRTE, Paris, France, are given in Table 4.11. It is readily observed that the most important shifts are as mentioned

TABLE 4.11

Frequency Shifts and Uncertainty in Their Determination for One of the ^{87}Sr Lattice Clocks Implemented at SYRTE, Paris, France, during the Last Decade

Physical Perturbation	Size (mHz)	Uncertainty (10^{-17})
Quadratic Zeeman	846	2.1
Residual lattice light shift, first order	−21	1.2
Residual lattice light shift, second order	0	0.7
BBR shift	2,310	
1. Temperature uncertainty		7.5
2. Uncertainty in sensitivity		0.5
Density shift	−10	4.6
Line pulling	0	4.7
Light shift due to probe	0	0.15
Total	3,125	10.3

Source: Le Targat et al., 2013.

above, the Zeeman shift, the BBR and the lattice light shift. It appears that the total uncertainty in the frequency determination of the clock in terms of the Cs standard is about 1×10^{-16} and is essentially originating from the difficulty in determining exactly the BBR shift.

Another more recent development has been realized also using ^{87}Sr in a lattice clock (Bloom et al. 2014). Extreme care was taken in many aspects of the evaluation of the main frequency biases, particularly the BBR shift. This was done using a very careful stabilization of the temperature of the ensemble through the insertion of the entire clock in a black box and an evaluation with calibrated sensors of the internal temperature with an accuracy of 27 mK at the site of the atoms. In one of the two Sr standards constructed it is claimed that the BBR shift could be evaluated with an accuracy of the order of 5×10^{-18}, a factor of 10 better than that reported above. Several other improvements in the determination of various shifts lead to a total claimed uncertainty of 6.4×10^{-18} in the determination of the clock frequency. This result leads to the hope of reaching an accuracy of 10^{-18} in frequency accuracy of an optical lattice clock in the near future.

Yb whose energy levels structure was given above was also used to implement an optical lattice clock with rather interesting properties. It is constructed along the same line of thought as the Sr clock described above (Hinkley et al. 2013). The atoms are laser cooled in a MOT on the transition $^1S_0 - {}^1P_1$ ($\Gamma = 1.8 \times 10^8$ s^{-1}) at 399 nm, followed by a second-stage cooling on the transition $^1S_0 - {}^3P_1$ ($\Gamma = 4.7 \times 10^4$ s^{-1}) at 356 nm. The narrow width of this second transition allows the realization of temperatures in the tens of μK range. About 1000 wells of the 1D lattice can be loaded with tens of thousands of Yb atoms. The residual temperature of the atoms trapped in optical wells is about 10 μK. The advantage of Yb resides in a BBR shift about twice as small as that of Sr. Furthermore, the value of the nuclear spin, $I = 1/2$,

TABLE 4.12

Characteristics of Selected Implementations of Optical Lattice Clocks in the Optical Domain

Atom	Clock Transition Frequency (Hz)	Transition	Accuracy	Frequency Stability	References
^{24}Mg	$655,659,923,839.6(1.6) \times 10^3$	$^1S_0 \rightarrow {}^3P_1$	2.5×10^{-12}	3×10^{-13} @ 1s	Friebe et al. 2008
^{40}Ca	$456,986,240,494,135.8$	$^1S_0 \rightarrow {}^3P_1$	7.5×10^{-15}	4×10^{-15} @ 1s	Oates et al. 2006
^{87}Sr	$429,228,004,229,873.10$	$^1S_0 \rightarrow {}^3P_0$	1.5×10^{-16}	$3 \times 10^{-15}\tau^{-1/2}$	Le Targat et al. 2013
^{88}Sr		$^1S_0 \rightarrow {}^3P_0$		$2 \times 10^{-16}\,\tau^{-1/2}$	Akatsuka et al. 2012
^{88}Sr	$438,828,957,494(10) \times 10^3$	$^1S_0 \rightarrow {}^3P_1$			Ferrari et al. 2003
^{171}Yb	$518,295,836,590,865.2(0.7)$	$^1S_0 \rightarrow {}^3P_0$	3.4×10^{-16}	$3 \times 10^{-16}\tau^{-1/2}$	Hinkley et al. 2013
^{174}Yb	$518,294,025,309,217.8(0.9)$	$^1S_0 \rightarrow {}^3P_0$	1.5×10^{-15}	2.5×10^{-16}@100 s	Lemke et al. 2009; Barber et al. 2006
^{199}Hg	$1,128,575,290,808,162.0 \pm 6.4$ (sys.) ± 0.3 (stat.)	$^1S_0 \rightarrow {}^3P_0$	5.7×10^{-15}	5.4×10^{-15} @1s	McFerran 2012

reduces the number of hyperfine levels and makes state preparation easier. The system constructed along these general lines showed a frequency stability of $3.2 \times 10^{-16}\,\tau^{-1/2}$ approximately the same as a Sr clock (see also Smart 2014).

It should be mentioned that the use of the atom Hg can lead to a reduction of the contribution of the BBR and improve accuracy. As shown in Table 4.12, the BBR shift of Hg is a factor 20 smaller than that for Sr. Consequently, the use of Hg would permit a reduction of the shift to be associated with this bias.

4.6   FREQUENCY MEASUREMENT OF OPTICAL CLOCKS

Since the early development of optical frequency standards, measurement of their frequency has been an important goal. The frequency difference of two optical frequency standards implemented with the same atom can be measured by means of a frequency mixing of the two signals in a fast response sensor. However, the absolute measurement of their frequency in terms of the SI unit, the second, defined by means of the ground state hyperfine frequency, 9.2 GHz, of the Cs atom is a major challenge. We have described in *QPAFS*, Volume 2 (1989), several methods that have been developed in the past. These consisted in the use of a certain number of lasers compared to each other by means of heterodyne techniques, either frequency- or phase-locked, with various steps in frequency reaching the infrared and finally the microwave range. The number of lasers required is rather large and such arrangements could readily fill a room with instrumentation that was very sensitive to environmental conditions and could in general be operated continuously only for rather limited periods. This approach was very complex and allowed only a very limited number of frequencies to be measured.

4.6.1 OPTICAL COMB

Fortunately, a new approach was developed that made possible a measurement of optical frequencies over a broad range in term of the Cs frequency essentially in a single step. The system, called *optical comb* is illustrated in Figure 4.23 and works in the following way (Hansch 2006; Hall 2006). It uses the basic idea that the radiation of a laser emitting very short pulses of radiation is characterized by a spectrum of Fourier frequencies distributed over a large range. There exist such lasers that emit extremely short pulses of radiation and are called *femtosecond lasers*. In such lasers, the light may be in the visible range and the short pulses repetition rate may be in the rf or microwave range from 100 MHz to 1 GHz. If we examine the emitted spectrum of such a laser by means of a Fourier analysis, we find that it consists of lines spaced in frequency by the repetition rate. These lines are very sharp. They are called *teeth* and in principle should appear at a frequency nf_r where n is an integer representing a particular tooth and f_r is the repetition rate of the pulses. Unfortunately, this repetition rate is not synchronized to the phase of the optical radiation. In other words, the phase of the optical radiation slips relative to the pulse maximum from pulse to pulse. If that phase slippage is $\Delta\phi$ from pulse to pulse, the net result is a bias f_0 of the whole optical frequency spectrum. As made explicit in the Figure 4.23 the frequency of a given tooth is then given by:

$$f_n = f_0 + nf_r \qquad (4.31)$$

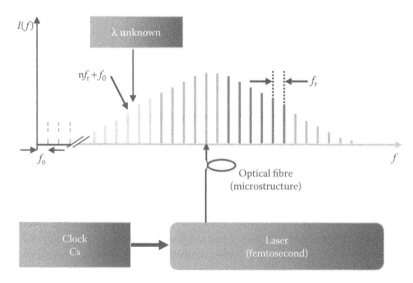

FIGURE 4.23 Conceptual diagram of the optical comb that makes possible measurement of an optical frequency directly in terms of the Cs frequency that acts a primary standard.

In order for the teeth to have stable frequencies, it is thus required to stabilize both f_0 and f_r. The frequency f_r may be stabilized directly by measuring the repetition rate and comparing the result to a stable microwave standard derived from a primary standard. A H maser may be used taking advantage of its low phase noise. The error signal developed can then be used as feedback on the laser mirrors that control that repetition rate. The frequency f_0, on the other hand can be measured through an heterodyne technique as follows. First, the frequency of the low frequency end of the output radiation is doubled by means of a non-linear device. The spectrum obtained is then beat against the high end of the original spectrum. The difference in frequency results in a beat note at frequency f_d equal to f_0 as shown by the following equation:

$$f_d = 2\left(f_0 + n_o f_r\right) - \left(f_0 + 2n_o f\right) = f_0 \qquad (4.32)$$

Consequently, an error signal can again be developed by comparing that frequency to a stable microwave frequency of the same quality as the one used above for stabilizing the repetition rate. That error signal can be used to stabilize the laser frequency through laser property on which the laser frequency is sensitive. Consequently, the laser spectrum is totally stabilized in frequency and phase. Any unknown laser frequencies in the range of the laser spectrum can then be measured by mixing that unknown laser signal to the comb spectrum and identifying the actual tooth that provides the observed beat. This comb technique is now used at large for measuring unknown optical frequencies.

4.6.2 CLOCK FREQUENCIES AND FREQUENCY STABILITIES REALIZED

The comb just described has been used extensively for determining the frequency of the clock transition of several atoms that have been used for implementing a lattice clock in the optical domain, as measured against primary Cs standards. The results are given in Table 4.12 with other characteristics, such as frequency stability and accuracy determined from an evaluation of various biases.

An accuracy, such as that shown in Table 4.12 is rather outstanding and from the discussion on the evaluation of the biases realized it allows hope of even doing better. As will be made explicit in the conclusion, this kind of accuracy can be used already to test the basic validity of assumptions imbedded in the formulation of some fundamental physical principles.

5 Summary, Conclusion, and Reflections

In the history of mankind, time has always played a crucial role either in explorations or quest for knowledge. Without a precise measurement of time, such activities as basic research, earth navigation and space exploration would hardly be feasible at the level of precision we know at the time of writing. Time plays a basic role in the universe's physical laws that we have uncovered and particularly in relativity. Our understanding of those physical laws is based on fundamental constants, which, as their name would indicate, have been believed to be invariable with time. With the precision now reached in measuring frequency and time we can now challenge this concept.

Technology based on precision timing has invaded our way of life, and even our daily activities require precise time/frequency standards: cell phones, transportation, and telecommunication systems, which are now part of our lives, all operate on precision time transfer. The second is the base unit of time in the SI system and it is also used to define other SI units such as length through the speed of light. Time is thus now at the heart of precise length measurements in industry; as we will see, time may play a unique role as being the single unit required in determining all basic units necessary to implement a complete measurement system.

Time has acquired this importance because better clocks have been invented and implemented over the years. In the early days of navigation, it was realized that if time was better known on a reference site and on the travelling vessel, better positioning on the surface of the planet would be accomplished resulting in increased security and improved commercial activities. In the nineteenth century, extensive development took place in mechanical clocks reaching stability in the 10 s range over a time required for a full accurate traversal of the Atlantic Ocean. In the early twentieth century, the quartz clock was developed, which increased the stability of time keeping by orders of magnitude and opened the door to new applications in such fields as communication, time keeping, and navigation. The advent of atomic clocks based on internal properties of atoms improved considerably the precision of all those applications and opened the door to new ones.

The basic physics behind the operation of those so-called atomic clocks and their state-of-the-art development at the end of the 1980s were described in *QPAFS*, Volumes 1 and 2. The present text is aimed at a description of the outstanding advances that were accomplished in that field since that time. The clocks themselves can be categorized under various criteria, such as nature, uses, as well as qualities. For example, these criteria may be:

Basic characteristic: Primary or secondary
Quality: Frequency stability and accuracy

Intent: Laboratory or industrial (commercial)
Physics principles: Passive or active; microwave or optical; ions or neutral atoms
Use: Earth surface, space
Dimensions: Volume and weight

That categorization may be helpful in deciding which type of clock is best suited for a given application. The intent of the present text is more oriented at outlining the basic physics involved in the implementation of such clocks; we have put more emphasis on such physics and the results obtained regarding frequency stability and accuracy. From that text, however, and the outline made relative on their physical construction, it is relatively easy to situate the clocks in those categories and decide from their properties on possible applications based on the state of the art of their development. We will not, however, go into the details of applications of such frequency standards. For an extensive survey of such applications, the reader is referred to Riehle (2004).

5.1 ACCURACY AND FREQUENCY STABILITY

Relative to accuracy and stability, it is best to represent these two characteristics by means of graphs that illustrate where we now stand. Accuracy refers to a clock's capability to represent the definition of the unit that it represents, the second, in the SI. The accuracy of a frequency standard used as a primary clock, on the other hand, can be qualified as our ability to evaluate all biases that can affect its frequency. Presently, the second is still defined in terms of the hyperfine frequency of the ground state of Cs. There are limits to the precision we can evaluate those biases, particularly in the microwave range, and we have found that we can do better in this regard in using optical frequency standards. Consequently, at first sight, it would appear that we could redefine the second in terms of an optical frequency with an atom other than Cs. This, however, is an important decision that needs to go through a heavy relative evaluation and bureaucratic process and it appears better for the moment to establish some rules by which these new standards can be used to implement the definition of the second. This can be done by means of knowledge of their frequency in terms of the Cs frequency, although we know that they are in principle more accurate. The accuracy of the measurements would then be limited to that of the Cs clock and a progress would result only if a new definition is adopted.

It is interesting to visualize the progress that has been made on atomic clocks accuracy since their first implementation. The evolution of that accuracy with time is presented in Figure 5.1 for selected standards using various technologies.

On the other hand, frequency stability, the important property that determines the precision of a frequency standard and plays an important role in determining its characteristics, has also been improving continuously with time. Actually it is basic for improvement in accuracy since without stability a frequency standard could not be evaluated regarding accuracy. Much work has been done in order to improve frequency stability in the short and medium term in order to reduce the time required to determine accuracy of a given standard. Present state-of-the-art frequency stability of selected standards described earlier in this text is given in Figure 5.2.

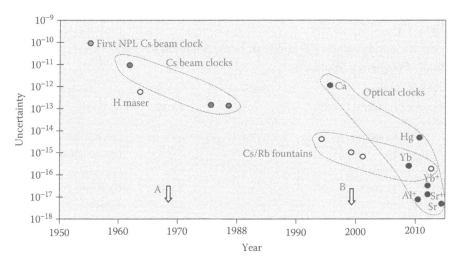

FIGURE 5.1 Evolution of accuracy of atomic frequency standards as a function of time. The graph makes evident steps accomplished with the change in technology. The two arrows A and B identify respectively the date of the introduction of the new definition of the second in terms of atomic properties and the invention of the optical comb making possible a direct connection between optical and microwave frequency standards.

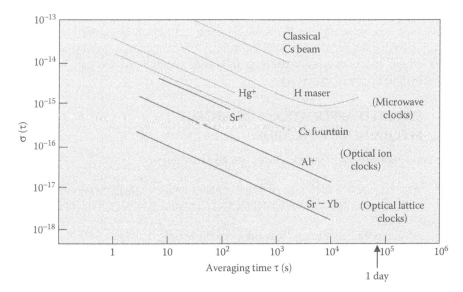

FIGURE 5.2 Present state-of-the-art frequency stability of various selected atomic frequency standards.

TABLE 5.1

Accuracy and Frequency Stability of Some Selected Recently Developed Atomic Frequency Standards

Frequency Standard		Stability	Accuracy	Reference
Microwave clocks	H maser	8.0×10^{-14} @ 1 s 2×10^{-16}/day	5×10^{-13}	www.T4Science.com
	Cs fountain	$1.6 \times 10^{-14} \tau^{-1/2}$	4.4×10^{-16}	Guena et al. 2014
	Hg^+	$5 \times 10^{-14} \tau^{-1/2}$		Burt et al. 2008
Optical clocks	$^{27}Al^+$	$2.8 \times 10^{-15} \tau^{-1/2}$	8.6×10^{-18}	Chou et al. 2010
	Yb^+ (oct)		7.1×10^{-17}	Huntemann et al. 2012
	$^{88}Sr^+$		2.2×10^{-17}	Madej et al. 2012
	^{171}Yb	$3 \times 10^{-16} \tau^{-1/2}$	3.4×10^{-16}	Lemke et al. 2009
	^{87}Sr	$3.4 \times 10^{-16} \tau^{-1/2}$	6.4×10^{-18}	Bloom et al. 2014
	^{40}Ca	2×10^{-16} @ 2,000 s	7.5×10^{-15}	Wilpers et al. 2007
	^{199}Hg	5.4×10^{-15} @ 1 s	5.7×10^{-15}	McFerran et al. 2012
Compact clocks	CPT Rb clock	$3 \times 10^{-11} \tau^{-1/2}$		Vanier 2005
	CPT Cs cell clock	$3.2 \times 10^{-13} \tau^{-1/2}$		Guerandel et al. 2014
	PHM	7×10^{-15} @ 10,000 s		Belloni et al. 2009
	Compact microwave mercury clock	$1 - 2 \times 10^{-13}$ @ 1 s		Prestage et al. 2007
	HORACE	$1 \times 10^{-13} \tau^{-1/2}$		Esnault et al. 2008
	Thermal beam CsIII	2.7×10^{-13} @ 10,000 s		www.symmetricom.com

We complete those figures by means of Table 5.1, which gives the actual frequency stability and accuracy of selected standards with appropriate references. It is observed that in less than a few decades, frequency stability and accuracy has been raised by four orders of magnitude.

5.2 SELECTED APPLICATIONS OF ATOMIC FREQUENCY STANDARDS

The main applications of atomic clocks can be classified in:

- Metrology, timescales (e.g., SI units, primary and secondary standard)
- Scientific research, instrumentation (e.g., gravity, relativity, fundamental constants)
- Telecommunications (e.g., networks synchronization)
- Radio astronomy (e.g., very long base line interferometry [VLBI], astronomy)
- Navigation and positioning (e.g., satellites systems, geodesy)

Several of those applications were outlined in Volume 2 of *QPAFS* (1989) (see also Maleki and Prestage 2005). We will examine briefly some of them in the light of the new developments described in the present text.

5.2.1 THE SI: TOWARDS A REDEFINITION OF THE SECOND

Time is the basic quantity in the SI, whose representative unit, the second, was the first to be defined in terms of an atomic property, the hyperfine frequency of the Cs atom in its ground state. Time is also the quantity that can be measured with the greatest accuracy due to the extraordinary development that has taken place in the field of atomic and laser physics.

The General Conference on Weights and Measures (Conférence Générale des Poids et Mesures CGPM 1967) decided in 1967 that "the second is the duration of 9,192,631,770 periods of the radiation corresponding to the transition between the two hyperfine levels of the ground state of the caesium 133 atom". This definition was qualified in 1997 by declaring that it refers to a Cs atom at rest at a thermodynamic temperature of 0 K.

In recent years, a large number of atomic species seem to be equally qualified to act as the basis of a new definition. That definition would be based on an optical transition of a particular atom selected from a group that shows promises when used in the implementation of a clock. In fact, the Conférence Internationale des Poids et Mesures (CIPM 2012) has established a List of Recommended Transitions, some of which are recognized as Secondary Representation of the Second (SRS) in the SI. The first SRS recognized in 2004 was Rb^{87} in the microwave range since its hyperfine frequency was less affected by atomic collisions and its frequency could be determined with better accuracy (Bize et al. 1999; see also Guena et al. 2014). On the other hand, due to development of accurate optical ion clocks using ions such as Hg^+, Al^+, Yb^+, Sr^+, Ca^+, as well as accurate optical lattice clocks using Sr, Yb, and Hg, several other optical transitions were recognized by the CIPM as SRS (see Table 5.2).

The second, one of the seven base units of the SI system is the unit that is realized with the greatest accuracy and stability. A logical approach is therefore to try to connect, using the laws of physics and fundamental constants, base SI units to the second. First the meter is directly connected to the second by means of the speed of light. The speed of light is now defined exactly as:

$$c = 299,792,458 \text{(m/s)} \tag{5.1}$$

a definition that gives an exact value to the electric constant equal to:

$$\varepsilon_0 = \frac{1}{\mu_0 c^2} = 8.854,187,817...\times 10^{-12} \text{(F/m)} \tag{5.2}$$

since the magnetic constant μ_0 is defined as

$$\mu_0 = 4\pi \times 10^{-7} \text{(N/A}^2\text{)} \tag{5.3}$$

On the other hand, it is possible to connect directly the volt to the second by means of Josephson effect, which is observed in a junction made of two superconductors separated by a thin dielectric film. The effect is observed as steps in the response of the junction exposed to radiation of frequency ν. Those steps are observed in the I-V curve of the junction obeying exactly the law:

$$V = n \, K_J \nu \tag{5.4}$$

TABLE 5.2

Characteristics (Transition and Frequency) of Selected Optical Frequency Standards That Can Be Used as Representation of the Definition of the Second

Ion/Atom	Nonperturbed Transition	Frequency (Hz)
^{27}Al$^+$	$3s^2\ ^1S_0 - 3s3p\ ^3P_0$	1,121,015,393,207,857.3
^{199}Hg$^+$	$5d^{10}6s\ ^2S_{1/2} - 5d^96s^2\ ^2D_{5/2}$	1,064,721,609,899,145.3
^{199}Hg	$6s^2\ ^1S_0 - 6s6p\ ^3P_0$	1,128,575,290,808,162
^{171}Yb$^+$(octupole)	$6s\ ^2S_{1/2} - 4f^{13}6s^2\ ^2F_{7/2}$	642,121,496,772,645.6
^{171}Yb$^+$(quadrupole)	$6s\ ^2S_{1/2}\ (F = 0,\ m_F = 0) - 5d\ ^2D_{3/2}\ (F = 2,\ m_F = 0)$	688,358,979,309,307.1
^{88}Sr$^+$	$5s\ ^2S_{1/2} - 4d\ ^2D_{5/2}$	444,779,044,095,485.3
^{40}Ca$^+$	$4s\ ^2S_{1/2} - 3d\ ^2D5_{/2}$	411,042,129,776,395
^{1}H	$1S - 2S$	1,233,030,706,593,518
^{87}Sr	$5s^2\ ^1S_0 - 5s5p\ ^3P_0$	429,228,004,229,873.4
^{171}Yb	$6s^2\ ^1S_0 - 6s6p\ ^3P_0$	518,295,836,590,865.0
^{87}Rb	$^1S_0,\ F = 1,\ m_F = 0 - {}^1S_0,\ F = 2,\ m_F = 0$	6,834,682,610.904312

Source: BIPM. 2012. Bureau International des Poids et Mesures, Consultative Committee for Time and Frequency (CCTF). Report of the 19th meeting to the International Committee for Weights and Measures 101e session du CIPM – Annexe 8 203, September 13–14, 2012. http://www.bipm.org/en/publications/mep.html.

where K_J is Josephson constant given by:

$$K_J = \frac{h}{2e} = 483,597.870(11)\ (\text{Hz/V}) \tag{5.5}$$

where:
   h is Planck's constant
   e the electronic charge

Furthermore, it is possible to construct a capacitor whose capacity C is exactly known in terms of its dimension (Thompson and Lampard theorem 1956) and the electric constant ε_0:

$$C_0 = \frac{\varepsilon_0 \ln 2}{\pi}\ (\text{F/m}) \tag{5.6}$$

Consequently, using these concepts, it is possible to imagine a system of units in which all base quantities length, time, ampere, and mass are connected together by simple relations which rely on fundamental constants as definition of their representing respective units. In that system, mass is obtained by means of a so-called watt balance as represented in Figure 5.3 by the box called force measurement, which is essentially a balance that measures the force developed by a

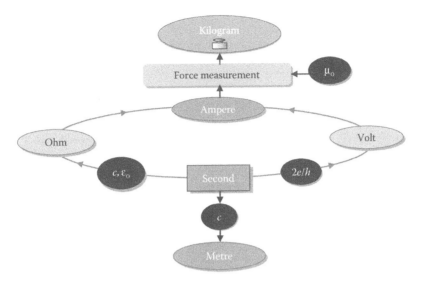

FIGURE 5.3 Schematic diagram making explicit a conceptual system of units based on fundamental constants and making the second as the central unit of measurement.

magnetic field induction B on a wire carrying a current I. The constant involved is μ_o defined earlier.

The consistency of the whole system can then be verified by means of the quantum Hall effect (von Klitzing et al. 1980) that leads to the implementation of an ohm standard in terms of the constant $R_K = h/e^2$. As of 2014, this system has not yet been implemented because of the limitation of accuracy of the watt balance. Accuracy better than a part in 10^9 is required. The kg is still defined as the mass of the Pt prototype that is maintained at the Bureau International des Poids et Mesures in Sèvres, France.

5.2.2 TESTS OF FUNDAMENTAL PHYSICAL LAWS

5.2.2.1 Fundamental Constants

The laws of physics are based on fundamental constants. Gravity relies on the gravitational constant G, electromagnetism relies on the electric and magnetic constants ε_o an μ_o as well as on the speed of light c, while quantum mechanics relies on such constants as, Plank's constant h and the mass and charge of the electron, m_e and e. In practice, so-called constants such as Rydberg constant $R_\infty = \alpha^2 m_e c/2h$, where α is the fine structure constant $\alpha = e^2/2\varepsilon_o hc$, play an important role in the description of energy level structure of atoms. Determination of those constants is, in some cases, made with the greatest accuracy by means of frequency measurements derived from atomic frequency standards techniques (see, e.g., De Beauvoir et al. 2000). A basic question is then raised as to the *constancy* with time of those physical constants. Hyperfine frequencies depend on the atom chosen and are functions of those constants (Prestage et al. 1995; Flambaum and Berengut 2009; Peik 2010). Similarly, optical frequencies depend on those constants in a manner different from microwave hyperfine

TABLE 5.3

Measurement of the Variation of the Ratio of Various Atomic Clocks Frequency with Time[a]

v_1/v_2	d/dt ln (v_1/v_2) $(10^{-16}/\text{yr})$	Reference
Rb/Cs	-1.36 ± 0.91	Guéna et al. 2012
H(1S – 2S)/Cs	-32 ± 63	Fischer et al. 2004
Yb$^+$/Cs	-4.9 ± 4.1	Tamm et al. 2009
Hg$^+$/Cs	3.7 ± 3.9	Fortier et al. 2007
Sr/Cs	-10 ± 18	Blatt et al. 2008
Al$^+$/Hg$^+$	-0.53 ± 0.79	Rosenband et al. 2008

Source: Bize, S. *Optical Frequency Standards and Applications*, Cours de 3ième cycle du Programme Doctoral en Physique de la Conférence Universitaire de Suisse Occidentale, 2014.

[a] Results have been obtained for hyperfine and optical clocks.

frequencies. Actually, the ratio of a hyperfine frequency to an atomic resonance optical frequency depends on the product $\alpha^2 g_1(m_e/m_p)$ (Bize 2001). Consequently, the advent of optical standards has provided a possibility of testing the constancy of those parameters, particularly α, over time by simply operating various clocks over long periods of time and comparing their frequencies. The exercise also allows a relation to astrophysical measurements (see Karshenboim and Peik 2008). Such a comparison has been done with several clocks and some of the results are summarized in Table 5.3.

The result is reported in terms of fluctuation with time according to the relation:

$$\frac{1}{R}\frac{dR}{dt} = \frac{d}{dt}\ln R \tag{5.7}$$

where:

R is a frequency ratio

These measurements provide a limit to the variation of the frequency of various clocks relative to each other and thus provide a limit to the variation of fundamental constants with time. From the table it appears that for the period of measurement no variation has yet been observed.

5.2.2.2 Time Dilation and Gravitational Red Shift

The twin paradox has always raised interest since its introduction by Einstein in the early 1900s. Experiments in the 1970s with clocks on airplanes around the earth have shown without ambiguity that time maintained by clocks in motion relative to clocks at rest did not flow at the same rate (Hafele and Keating 1972). Their results are summarized in Table 5.4.

Such an experiment was reproduced on a shorter flight by members of the National Physical Laboratory, UK, in 2005 on a flight from London to Washington and the results were found in agreement with theory to 1%.

TABLE 5.4

Results Obtained by Hafele and Keating (1972) on Time Dilation Measurements by Means of a Comparison of 4 Cs Atomic Clocks Travelling on Air Planes around the Earth Compared to Other Clocks Maintained at Rest Relative to Them on the Surface of the Earth

Δt(ns)	East	West
Calculated	-40 ± 23	275 ± 21
Measured	-59 ± 10	273 ± 7

Note: The results have been corrected for the gravitational red shift examined in this table.

On the other hand, it is known that a clock rate varies with gravitational potential. Near the surface of the earth, the fractional variation with height h is given by the expression:

$$\frac{\Delta \nu}{\nu} \cong \frac{gh}{c^2} \qquad (5.8)$$

where:
  g is the acceleration of earth's gravity

The change is about 10^{-16} per metre elevation.

The effect has been verified in the 1970s by means of an H maser sent in space by a rocket at an altitude of about 10,000 km (Vessot and Levine 1979; Vessot et al. 1980). The agreement with theory was of the order of 0.01%. Similar experiments done with clocks on airplane at various altitudes also agreed with theory within ~1% (Alley 1979, 1981).

It should be mentioned that those two phenomena, time dilation and gravitational red shift, have an important effect on navigation systems such as the GPS. Clocks in satellites in such a system must be corrected for their speed and altitude in their orbits by 38 μs/day of which 45 μs is due to gravitational potential and −7 μs due to speed.

With the advent of optical clocks, it is readily observed that with stabilities reaching the 10^{-18} over averaging times of the order of 10,000 s, it is now possible to measure altitude effectively to the level of centimetres (Rosenband et al. 2008).

5.2.2.3 Fundamental Physics in Space

With the advent of more stable atomic clocks, the subject of revisiting selected physics basic experiments in space has attracted interest. In particular, the International Space Station (ISS) offers a unique environment in microgravity that is rather appropriate for such a mission. For example, the European Space Agency (ESA) has approved a project called Atomic Clock Ensemble in Space (ACES) for testing some of the physics laws to a higher precision (Cacciapuoti and Salomon 2009). In particular, ACES will be composed of a set of clocks such as an active H maser for space (SHM) and a laser-cooled Cs beam clock (PHARAO) of the design described earlier in the text. Scientific objectives are in particular a measurement of the gravitational red shift with a 25-fold improvement over the experiment carried by Vessot and Levine (1979) with an H maser, a better test of the isotropy of the speed of light and a search for a possible drift of the fine structure constant α. Furthermore, with the time transfer equipment, ACES will allow synchronization of timescales of distant ground laboratories with 30 ps accuracy and frequency comparisons at the 10^{-16} accuracy level.

5.2.3 CLOCKS FOR ASTRONOMY AND EARTH SCIENCE

5.2.3.1 VLBI and Geodesy

The VLBI is a very powerful technique using antennas to observe stellar objects with an outstanding resolution. The antennas use atomic clocks for synchronization. The H maser has been the preferred frequency standard for this purpose due to its high reliability and its medium-term frequency stability. This technique can also be employed in geodesy in order to measure the distance between the antennas with precision of the order of millimetres. This type of measurement can also be used to follow, for example, the dynamics of tectonic plates. Furthermore, due to the dependence of clock frequency on gravitational potential, as predicted by general relativity, atomic clocks of high accuracy can be used in principle to improve gravitational field knowledge. This makes possible the definition of a *chronometric geoid* as the surface where precise clocks run with the same speed and the surface is nearest to mean sea level (see, e.g., Delva and Lodewyck 2013).

5.2.3.2 Deep Space Network

Frequency standards capable of operating simultaneously for a long term with high frequency stability are necessary for synchronization and navigation of the satellites and space probes. The Deep Space Network (DSN) is a worldwide network of large antennas and communication facilities, located in California, Spain, and Australia, that supports interplanetary spacecraft missions, performs radio and radar astronomy observations. The space network includes DSN which is part of the NASA Jet Propulsion Laboratory, European Space Operations Centre, Russian Deep Space Network, Chinese Deep Space Network, Indian Deep Space Network, and Usuda Deep Space Centre in Japan.

5.2.3.3 Earth Clocks Network

In the past, synchronization of clocks was done generally by means of transportable clocks such as a Cs beam frequency standards of the classical design. The accuracy of the technique relied entirely on the frequency stability of the clocks which was not obvious due to changing environmental conditions during travel. That alone considerably limited the accuracy of time synchronization between distant sites. It thus appeared logical to develop systems in which timing information could be transmitted by electromagnetic radiation. Since transmission is altered by atmospheric conditions, it also appeared logical to develop transmission networks that would not suffer from such an influence. The advent of lasers and transmission by optical fibres appears to correctly address the question and to offer a solution to the problem of transmission.

Such networks between the clocks using coherent optical fibre link is under development in several countries. In France, for example, a long link, in dark channel (fibre shared with Internet) of 540 km between Syrte, Paris, and Laboratoire de Physique des Lasers (LPL) was demonstrated. This network including bidirectional optical amplifiers and a signal regeneration station has demonstrated stability of 4×10^{-15} at 1 s and better than 10^{-19} at one day, in a measurement bandwidth of 10 Hz. No perturbation of internet traffic on the same fibre was observed. SYRTE is also developing free-space optical links for ground-to-space or space-to-space transfer and as a potential method for intercontinental comparison of ground-based standards. A laser system stabilized to a fibre spool delay line has been developed, and will be used to test coherent transmission to a corner-cube reflector on a low earth orbit satellite.

A range of networks using various techniques such as optical fibres has also been implemented in several countries allowing time synchronization as well as frequency comparisons between distant locations. These include Germany, the United Kingdom, Italy, Japan, and China. Accuracies vary with the network developed but in some, reached the range of 10^{-17} to 10^{-18} level. This activity in connection to time transfer makes evident the importance that those countries attach to the field.

5.2.3.4 Navigation Systems

Positioning on earth's surface has always been a great challenge from the early days of navigation across oceans by boats to modern days of transit around the globe by means of fast airplanes. Knowing time accurately was essential to navigating by means of astronomical observation. However, there was always a severe limit in positioning stars, by means of sextants, for example, and it appeared early that the use of frequency-stable electromagnetic radiation providing accurate timing would offer, by means of triangulation, both accuracy and reliability. It is recalled that timing signals carried by means of electromagnetic radiation stable to 1 ns provide a localization ability of 30 cm. Consequently, triangulation with frequency-stable and synchronized radio signals can in principle provide localization to that accuracy, limited mainly by the stability of the transmitted signals.

Several systems were developed early using the stability of atomic clocks controlling radio signals in transmission of low frequencies in the vicinity of the earth's surface. Such an approach was used in systems based on very low frequency (VLF) transmitters and in systems such as DECCA, OMEGA, and LORAN C. However, those systems had somewhat limited range or low accuracy. More recently, systems using satellites as transmitters of frequency-stable signals were developed with improved accuracy in positioning. One such system is the GPS. It consists of a constellation of 24 satellites, each of which contains redundant atomic clocks, including in its early stage Cs beam clocks as well as Rb optically pumped standards of the type described in the present text. The satellites are in Earth orbits of about 20,000 km thus orbiting Earth a few times a day. GPS was first designed as a US military support system for precise dissemination of time, velocity, and location anywhere on the earth (Parkinson 1996). Today, the spread of GPS in applications ranges from positioning in navigation with airplanes and ships to travelling by automobiles, automated location of vehicles in serious accidents by emergency crews, search and rescue functions, automated farming and many other domains requiring accurate positioning. A similar system was developed in the 1990s in Russia and called GLONASS. Other systems of a similar nature are being developed by other countries such as Galileo in Europe, BeiDou in China, and IRNSS in India.

5.3   LAST REFLECTIONS

In the two volumes of *QPAFS* (1989), we described the state of the art of atomic clocks development at the end of the 1980s. At that time, the best clocks were H masers for short- and medium-term frequency stability at the level of a few in 10^{-13} for 1 s averaging time improving to better than 10^{-15} at about 10,000 s averaging time. Its accuracy was at best 10^{-12} due to the presence of the wall shift, which is not reproducible to better than about 10%. Laboratory Cs beam frequency standards were more reliable in the long term and offered accuracy of the order of 10^{-14}. At that time, although great efforts were dedicated to solve the problems that were believed to be the cause of those limits, it did not seem, in the light of the progress made, that important gains were possible within a reasonable time simply by improving the technology that was used. However, with the advent of solid-state lasers and accomplishments in the field of atomic physics such as laser cooling, a substantial jump was made in the early 1990s with the development of the Cs and Rb fountains. Immediately, improvements by an order of magnitude in frequency stability as well as in accuracy followed. Then, the implementation of the optical standards with another improvement in frequency stability and accuracy of nearly two orders of magnitude followed. We are now talking of frequency stability in the 10^{-15} to 10^{-16} range at an averaging time of 1 s and accuracy of nearly 10^{-18}. This is an improvement of nearly four orders of magnitude in a time lapse of 25 years.

Can we expect an improvement of that magnitude in the next few decades? At the end of the 1980s we did not even think of much possible improvements in frequency stability and accuracy and we were more interested in volume, size, and transportability of the standards. Furthermore, although developments in optical frequency standards with higher qualities, such as high line Q and relative independence of environmental fluctuations were in progress, it appeared that the connection of their frequency to the microwave range in order to make them available for measurements other than length was not a simple task. That connection was limited essentially to very specialized laboratories equipped with stacks of interconnected lasers integrated in complex systems difficult to maintain in operation for long periods. The invention of the optical comb changed the landscape totally. The frequency of optical frequency standards could be divided readily to the microwave range in a manageable and relatively small system. Its characteristics could be transposed and could be used in the so-called electronic medium and even become a clock to measure time.

It appears that the kind of improvements in clock implementation, which we have observed recently and described above, relied on parallel developments in connected fields, in particular atomic physics. As far as optical frequency standards are concerned, we can certainly improve slightly the situation by improving the present standards by means of developments at shorter wavelength. However, with some of those standards we are already in the ultraviolet and the use of optics at those short wavelengths is a challenge. Furthermore the gain would not be as substantial as the one made in going from microwave to optical frequencies. It appears that to make a real step one would have to go to nuclear radiation (Peik et al. 2009). However, this is a totally different technology and the techniques to be used would certainly be totally different from those used in the field of optical and microwave frequency standards

There is one thing assured, however. Our quest for better frequency standards, regarding frequency stability, and accuracy will continue. It is driven by the passion of knowledge and the will to implement better instruments that would open doors to new applications. The predominant hope is that with better standards we will be able to challenge some of our beliefs, regarding basic concepts, the stability of fundamental constant, for example, as well as to challenge some of the validity of fundamental laws of physics that have been accepted as representing the true dynamics of the universe (see, e.g., Vanier 2011). Time is a quantity we live by and is one of the most fundamental concepts in physics, which appears to regulate the evolution of the universe. However, we really do not know what it is and we do not understand entirely its properties. The theory of relativity, special and general, makes it part of coordinate systems as a fourth dimension and arrives at the conclusion that its flow depends on the dynamics of the reference frame we live in. Quantum mechanics deals with particles, quanta and discontinuous phenomena. However, time in those theories stays continuous. Is there a minimum value by which time can be divided? Was time created with space at the Big Bang which gave birth to the universe we live in? It appears that in the photon frame of reference, time does not exist. The photon, however, travels in space and is affected by space expansion: its wavelength changes on a continuous basis with the universe expansion, but it does not age. This is really a different world than the one we live in. In our world, time is a master that we cannot control, but, most amazingly, it is the quantity that we measure best and it appears that we will continue to be interested in doing it better.

References

CHAPTER 1

Abashev Y.G., Elkin G.A., and Pushkin S.B. 1983. Primary time and frequency standards, *Measurement Techniques*, USSR, **26**: 996.

Abashev Y.G., Elkin G.A., and Pushkin S.B. 1987. The main characteristics and the results of long-term comparisons between primary caesium beam standards and ensemble of hydrogen clocks of the national time standard. *IEEE T. Instrum. Meas.* **IM-36**: 627.

Affolderbach C., Droz F., and Mileti G. 2006. Experimental demonstration of a compact and high-performance laser-pumped rubidium gas cell atomic frequency standard. *IEEE T. Instrum. Meas.* **55**: 429.

Ashby N., Heavner T.P., Jeffferts S.R., Parker T.E., Radnaev A.G., and Dudin Y.O. 2007. Testing local position invariance with four caesium-fountain primary frequency standards and four NIST hydrogen masers *Phys. Rev. Lett.* **98**: 070802.

Audoin C. 1981. Fast cavity auto-tuning system for hydrogen maser. *Rev. Phys. Appl.* **16**: 125.

Audoin C. 1982. Addendum: fast cavity auto-tuning system for hydrogen maser. *Rev. Phys. Appl.* **17**: 273.

Audoin C. 1992. Caesium beam frequency standards: classical and optically pumped. *Metrologia* **29**: 113.

Audoin C., Jardino M., Cutler L.S., and Lacey R. 1978. Frequency offset due to spectral impurities in caesium-beam standards *IEEE T. Instrum. Meas.* **IM-27**: 325.

Audoin C., Lesage P., and Mungall A.G. 1974. Second-order Doppler and cavity phase dependent frequency shifts in atomic beam frequency standards *IEEE T. Instrum. Meas.* **IM-23**: 501.

Barwood G.P., Bell A.S., Gill P., and Klein H.A. 1988. Trapped Yb$^+$ as a potential optical frequency standard. In *Proceedings of the 4th Symposium on Frequency Standards and Metrology*, A. DeMarchi Ed. (Springer-Verlag, Berlin, Germany) 451.

Bauch A. 2005. The PTB primary clocks CS1 and CS2. *Metrologia* **42**: S43.

Bauch A., Dorenwendt K., Fischer B., Heindorff T., Müller E.K., and Schröder R. 1987. CS2: the PTB's new primary clock. *IEEE T. Instrum. Meas.* **IM-36**: 613.

Bauch A., Fischer B., Heindorff T., Hetzel P., Petit G., Schröder R., and Wolf P. 2000a. Comparisons of the PTB primary clocks with TAI in 1999. *Metrologia* **37**: 683.

Bauch A., Fischer B., Heindorff T., and Schröder R. 1993. The new PTB primary caesium clocks. *IEEE T. Instrum. Meas.* **IM-42**: 444.

Bauch A., Fischer B., Heindorff T., and Schröder R. 1998. Performance of the PTB reconstructed primary clock CS1 and an estimate of its current uncertainty. *Metrologia* **35**: 829.

Bauch A., Fischer B., Heindorff T., and Schröder R. 1999a. Some results and an estimate of the current uncertainty of the PTB's reconstructed primary clock CS1. *IEEE T. Instrum. Meas.* **48**: 508.

Bauch A., Fischer B., Heindorff T., and Schröder R. 1999b. Recent results of PTB's primary clock Cs1. In *Proceedings of the Joint European Forum on Time and Frequency/IEEE International Frequency Control Symposium* 43.

Bauch A., Fischer B., Heindorff T., and Schröder R. 2000b. Recent results of PTB's primary clock CS1. *IEEE Trans. Ultrason. Ferroelectr. Freq. Control* **47**: 443.

Bauch A., Heindorff T., Schröder R., and Fischer B. 1996. The PTB primary clock CS3: type B evaluation of its standard uncertainty. *Metrologia* **33**: 249.

Bauch A. and Schröder R. 1997. Experimental verification of the shift of the cesium hyperfine transition frequency due to blackbody radiation. *Phys. Rev. Lett.* **78** (4): 622.

Bauch A., Schröder R., and Weyers S. 2003. Discussion of the uncertainty budget and of long term comparison of PTB's primary frequency standards CS1, CS2 and CSF1. In *Proceedings of the Joint Meeting of the IEEE International Frequency Control Symposium/European Forum on Time and Frequency* 191.

Beard R., Golding W., and White J. 2002. Design factors for atomic clocks in space. In *Proceedings of the IEEE International Frequency Control Symposium* 483.

Becker G. 1976. Recent progress in primary Cs beam frequency standards at the PTB. *IEEE T. Instrum. Meas.* **IM-25**: 458.

Belloni M., Gioia M., Beretta S., Droz F., Mosset P., Wang Q., Rochat P., Resti A., Waller P., and Ostillio A. 2010. Space passive hydrogen maser-performances, lifetime data and GIVE-B related telemetries. In *Proceedings of the European Forum on Time and Frequency*.

Belloni M., Gioia M., Beretta S., Waller P., Droz F., Mosset P., and Busca G. 2011. Space mini passive hydrogen maser—a compact passive hydrogen maser for space applications. In *Proceedings of the Joint International Frequency Control Symposium and European Forum on Time and Frequency* 906.

Belyaev A.A. and Savin V.A. 1987. Calculation and analysis of frequency properties of special axisymmetric resonators of hydrogen quantum frequency discriminators. *Izmeritel'naya Tekhnica 2* (Translation, Plenum Publishing Co., New York) 29.

Bernier L.G. 1994. Preliminary design and breadboarding of a compact space qualified hydrogen maser based on a sapphire loaded microwave cavity. In *Proceedings of the European Forum on Time and Frequency* 965.

Bernier L.G. and Busca G. 1990. Some results on the line Q degradation in hydrogen masers. In *Proceedings of the European Forum on Time and Frequency* 713.

Berthoud P., Pavlenko I., Wang Q., and Schweda H. 2003. The engineering model of the space passive hydrogen maser for the European global navigation satellite system Galileo. In *Proceedings of the Joint Meeting of the European Forum on Time and Frequency/IEEE International Frequency Control Symposium* 90.

Blatt R., Casdorff R., Enders V., Neuhauser W., and Toschek P.E. 1988. New frequency standards based on Yb+. In *Proceedings of the 4th Symposium on Frequency Standards and Metrology*, A. De Marchi Ed. (Springer-Verlag, Berlin, Germany) 306.

Blatt R., Schnatz H., and Werth G. 1983. Precise determination of the ^{171}Yb+ ground state hyperfine separation. *Z. Phys. A* **312**: 143.

Bloch F. and Siegert A. 1940. Magnetic resonance for nonrotating fields. *Phys. Rev.* **57**: 522.

Boulanger J.S. 1986. New method for the determination of velocity distribution in caesium beam clocks. *Metrologia* **23**: 37.

Boussert B., Théobald G., Cérez P., and de Clercq E. 1998. Frequency performances of a miniature optically pumped caesium beam frequency standard. *IEEE Trans. Ultrason. Ferroelectr. Freq. Control* **45**: 728.

Burt E.A., Diener W.A., and Tjoelker R.L. 2007. Improvements to JPL's compensated multi-pole linear ion trap standard and long-term measurements at the 10^{-16} level. In *Proceedings of the Joint IEEE International Frequency Control Symposium/European Forum on Time and Frequency* 1041.

Burt E.A., Taghavi-Larigani S., Prestage J.D., and Tjoelker R.L. 2008a. Stability evaluation of systematic effects in a compensated multi-pole mercury trapped ion frequency standard. In *Proceedings of the IEEE International Frequency Control Symposium* 371.

Burt E.A., Taghavi-Larigani S., Prestage J.D., and Tjoelker R.L. 2008b. A compensated multi-pole mercury trapped ion frequency standard and stability evaluation of systematic effects. In *Proceedings of the 7th Symposium on Frequency Standards and Metrology*, L. Maleki Ed. (World Scientific, Singapore) 321.

Busca G., Bernier L.G., Silvestrin P., Feltham S., Gaygorov B.A., and Tatarenkov V.M. 1993. Hydrogen maser clocks in space for solid-earth research and time-transfer applications: experiment overview and evaluation of Russian miniature sapphire loaded cavity. In *Proceedings of the Annual Precise Time and Time Interval Applications and Planning Meeting* 467.

Busca G. and Brandenberger H. 1979. Passive H-Maser. In *Proceedings of the Frequency Control Symposium* 563.

Busca G., Frelchoz C., Wang Q., Merino M.R., Hugentobler U., Dach R., Dudle G., Graglia G., Luingo F., Rochat P., Droz F., Mosset P., Emma F., and Hahn J. 2003. Space clocks for navigation satellites. In *Proceedings of the Joint European Forum on Time and Frequency/IEEE International Frequency Control Symposium* 172.

Camparo J. 2005. Does the light shift drive frequency aging in the rubidium atomic clock? *IEEE Trans. Ultrason. Ferroelectr. Freq. Control* **52**: 1075.

Camparo J.C., Hagerman J.O., and McClelland T.A. 2012. Long-term behaviour of rubidium clocks in space. In *Proceedings of the European Forum on Time and Frequency* 501.

Camparo J.C., Klimcak C.M., and Herbulock S.J. 2005. Frequency equilibration in the vapor-cell atomic clock. *IEEE T. Instrum. Meas.* **54**: 1873.

CGPM, *Conférence générale des poids et measures*. 1967–1968. Resolution 1 of the 13th CGPM and 1968 News from the International Bureau of Weights and Measures. *Metrologia* **4** (1): 41.

Chung S.K., Prestage J.D., Tjoelker R.L., and Maleki L. 2004. Buffer gas experiments in mercury (Hg^+) ion clock. In *Proceedings of the IEEE International Frequency Control Symposium* 130.

Cline R.W., Smith D.A., Greytak T.J., and Kleppner D. 1980. Magnetic confinement of spin-polarized atomic hydrogen. *Phys. Rev. Lett.* **45** (2): 117.

Collin R.E. 1991. *Field Theory of Guided Waves*, 2nd edition (IEEE Press, New York).

Crampton S.B., Jones K.M., Nunes G., and Souza S.P. 1984. Hydrogen maser oscillation at 10 K. In *Proceedings of the Annual Precise Time and Time Interval Applications and Planning Meeting* 339.

Cutler L.S., Flory C.A., Giffard R.P., and De Marchi A. 1991. Frequency pulling by hyperfine σ transitions in caesium beam atomic frequency standards. *J. App. Phys.* **69**: 2780.

Cutler L.S., Giffard R.P., and McGuire M.D. 1983. Mercury-199 trapped ion frequency standard. In *Proceedings of the Annual Symposium on Frequency Control* 32.

Cutler L.S., Giffard R.P., Wheeler P.J., and Winkler G.M.R. 1987. Initial operational experience with a mercury ion storage frequency standard. In *Proceedings of the Annual Symposium on Frequency Control* 12.

Cyr N., Têtu M., and Breton M. 1993. All-optical microwave frequency standard: a proposal. *IEEE T. Instrum. Meas.* **42**: 640.

Daams H. 1974. Corrections for second-order Doppler shift and cavity phase error in caesium atomic beam frequency standards. *IEEE T. Instrum. Meas.* **IM-23**: 509.

Dehmelt H.G. 1967. Radiofrequency spectroscopy of stored ions: I storage. *Adv. At. Mol. Phys.* **3**: 53.

De Marchi A. 1986. A novel cavity design for minimization of distributed phase shift in atomic beam frequency standards. In *Proceedings of the Annual Symposium on Frequency Control* 441.

De Marchi A. 1987. Rabi pulling and long-term stability in caesium beam frequency standards. *IEEE Trans. Ultrason. Ferroelectr. Freq. Control* **34**: 598.

De Marchi A., Rovera G.D., and Premoli A. 1984. Pulling by neighbouring transitions and its effects on the performance of caesium-beam frequency standards. *Metrologia* **20**: 37.

De Marchi A., Shirley J., Glaze D.J., and Drullinger R. 1988. A new cavity configuration for caesium beam primary frequency standards. *IEEE T. Instrum. Meas.* **37**: 185.

Demidov N., Vorontsov V., Belyaev A., and Blinov I. 2012. Studies of a short and long-term stability of an active hydrogen maser with stand alone cavity auto tuning. In *Proceedings of the European Forum on Time and Frequency* 488.

Demidov N.A., Pstukhov A.V., and Uljanov A.A. 1999. Progress in the development of IEM Kvartz passive hydrogen masers. In *Proceedings of the Annual Precise Time and Time Interval Applications and Planning Meeting* 579.

Dicke R.H. 1953. The effect of collisions upon the Doppler width of spectral lines. *Phys. Rev.* **89**: 472.

Dicke R.H. 1954. Coherence in spontaneous radiation processes. *Phys. Rev.* **93**: 99.

Droz F., Mosset T., Barmaverain G., Rochat P., Wang Q., Belloni M., Mattioni L., Schmidt U., Pike T., Emma F., and Waller P. 2006. The on-board Galileo clocks: rubidium standard and passive hydrogen maser current status and performance. In *Proceedings of the European Forum on Time and Frequency* 420.

Droz F., Mosset T., Wang Q., Rochat P., Belloni M., Gioia M., Resti A., and Waller P. 2009. Space passive hydrogen maser—performances and lifetime data. In *Proceedings of the European Forum on Time and Frequency* 393.

Dubé P., Madej A.A., Bernard J.E., Marmet L., Boulanger J.-S., and Cundy S. 2005. Electric quadrupole shift cancellation in single-ion optical frequency standards. *Phys. Rev. Lett.* **95**: 033001.

Essen L. and Parry. VI. 1955. An atomic standard of frequency and time interval: a caesium resonator. *Nature* **176**: 280.

Fischer B. 2001. Frequency pulling by hyperfine sigma-transitions in the conventional laboratory frequency standards of the PTB. *Metrologia* **38**: 115.

Fisk P.T.H., Lawn M.A., and Coles C. 1993. Progress at CSIRO Australia towards a microwave frequency standard based on trapped, laser-cooled $^{171}Yb^+$ ions. In *Proceedings of the IEEE International Frequency Control Symposium* 139.

Flambaum V.V. and Berengut J.C. 2008. Variation of fundamental constants from the Big Bang to atomic clocks: theory and observations. In *Proceedings of the 7th Symposium of Frequency Standards and Metrology*, L. Maleki Ed. (World Scientific, Singapore) 3.

Forman P. 1985. Atomichron®: the atomic clock from concept to commercial product. *Proc. IEEE* **73**: 1181.

Fronzcis W. and Hyde J.S. 1982. The loop-gap resonator: a new microwave lumped circuit ESR sample structure. *J. Mag. Res.* **47**: 515.

Gaigerov B. and Elkin G. 1968. *Cavity Autotuning in Hydrogen Frequency Reference. Measurement Equipment* No 6.

Garvey R.M. 1982. 4 Caesium beam frequency standard with microprocessor control. In *Proceedings of the Annual Symposium on Frequency Control* 236.

Gaygorov B.A., Rusin F.S., and Sysoev V.P. 1991. Portable atomic clock on the basis of an active hydrogen maser, "sapphir." In *Proceedings of the European Forum on Time and Frequency* 293.

Gill P., Webster S.A., Huang G., Hosaka K., Stannard A., Lea S.N., Godun R.M., King S.A., Walton B.R., and Margolis H.S. 2008. A trapped $^{171}Yb^+$ ion optical frequency standard based on the $S_{1/2} - F_{7/2}$ transition. In *Proceedings of the 7th Symposium on Frequency Standards and Metrology*, L. Maleki Ed. (World Scientific, Singapore) 250.

Godone A., Levi F., and Vanier J. 1999. Coherent microwave emission in caesium under coherent population trapping. *Phys. Rev. A* **59**: R12.

Godone A., Micalizio S., Calosso C.E., and Levi F. 2006. The pulsed rubidium clock. *IEEE Trans. Ultrason. Ferroelectr. Freq. Control* **53**: 525.

Godone A., Micalizio M., and Levi F. 2004. Pulsed optically pumped frequency standard. *Phys. Rev. A* **70**: 023409.

Goldenberg H.M., Kleppner D., and Ramsey N.F. 1960. Atomic hydrogen maser. *Phys. Rev. Lett.* **8**: 361.

Goldenberg H.M., Kleppner D., and Ramsey N.F. 1961. Atomic beam resonance experiments with stored beams. *Phys. Rev.* **123**: 530.

Goujon D., Rochat P., Mosset P., Boving D., Perri A., Rochat J., Ramanan N, Simonet D., Vernez X., Froidevaux S., and Perruchoud G. 2010. Development of the space active hydrogen maser for the ACES mission. In *Proceedings of the European Forum on Time and Frequency* 17-02.

Hardy W.N. and Morrow M. 1981. Prospects for low temperature H masers using liquid helium coated walls. *J. Phys. Colloq.* **42** (Suppl.): C8 171.

Hardy W.N. and Whitehead L.A. 1981. Split ring resonator for use in magnetic resonance from 200–2000 MHz. *Rev. Sci. Instrum.* **52**: 213.

Hartnett J.G., Tobar M.E., Stanwix P., Morikawa T., Cros D., and Piquet O. 2004. Cavity designs for a space hydrogen maser. *IEEE International Ultrasonics, Ferroelectronics and Frequency Control Symposium Joint 50th Anniversary Conference* 608.

Harrach R.J. 1966. Some accuracy limiting effects in an atomic beam frequency standards. In *Proceedings of the Annual Symposium on Frequency Control* 424.

Harrach R.J. 1967. Radiation-field-dependent frequency shifts of atomic beam resonances. *J. App. Phys.* **18**: 1808.

Hasegawa A., Fukuda K., Kajita M., Ito H., Kumagai M., Hosokawa M., Kotake N., and Morikawa T. 2004. Accuracy evaluation of optically pumped primary frequency standard CRL-O1. *Metrologia* **41**: 257.

Hellwig H., Jarvis S. Jr., Halford D., and Bell H.E. 1973. Evaluation and operation of atomic beam tube frequency standards using time domain velocity selection modulation. *Metrologia* **9**: 107.

Howe D.A., Walls F.L., Bell H.E., and Hellwig H. 1979. A small, passively operated hydrogen maser. In *Proceedings of the Annual Symposium on Frequency Control* 554.

Hu J., Xia B., Xie Y., Wang Q., Zhong D., An S., Mei G., and Xia B. 2007. A subminiature microwave cavity for rubidium atomic frequency standards. In *Proceedings of the Joint European Forum on Time and Frequency/International Frequency Control Symposium* 599.

Huang X., Xia B., Zhong D., An S., Zhu X., and Mei G. 2001. A microwave cavity with low temperature coefficient for a passive rubidium frequency standard. In *Proceedings of the IEEE International Frequency Control Symposium* 105.

Itano W.N., Lewis L.L., and Wineland D.J. 1981. Shift of $^2S_{1/2}$ hyperfine splitting due to black-body radiation and its influence on frequency standards *J. Phys. Colloq.* **42**: C8 283.

Ito H., Hosokawa M., Umezu J., Morikawa T., Takahei K., Uehara M., Mori K., and Tsidu M. 2002. Development and preliminary performance evaluation of a spaceborne hydrogen maser. In *Proceedings of the Asia Pacific Workshop on Time and Frequency* 103.

Ito H., Morikawa T., Ishida H., Hama S., Kimura K., Yokota S., Matori S., Numata Y., Kitayama M., and Takahei K. 2004. Development of a spaceborne hydrogen maser atomic clock for quasi-zenith satellites. In *Proceedings of the Annual Precise Time and Time Interval Applications and Planning Meeting* 423.

Jardino M., Desaintfuscien M., Barillet R., Viennet J., Petit P., and Audoin C. 1981. Frequency stability of a mercury ion frequency standard. *Appl. Phys.* **24**: 107.

Jornod A., Goujon D., Gritti D., and Bernier L.G. 2003. The 35 kg space active hydrogen maser (SHM-35). In *Proceedings of the Joint IEEE International Frequency Control Symposium/European Forum on Time and Frequency* 82.

King S.A., Godun R.M., Webster S.A., Margolis H.S., Johnson L.A.M., Szymaniec K., Baird P.E.G., and Gill P. 2012. Absolute frequency measurement of the $^2S_{1/2}$–$^2F_{7/2}$ electric octupole transition in a single ion of $^{171}Yb^+$ with 10^{-15} fractional uncertainty. *New J. Phys.* **14**: 013045.

Kleppner D., Berg H.C., Crampton S.B., Ramsey N.F., Vessot R.F.C., Peters H.E., and Vanier J. 1965. Hydrogen-maser principles techniques. *Phys. Rev. A* **138**: 972.

Koga Y., Nakadan Y., and Yoda J. 1981. The caesium beam frequency standard NRLM-01. *J. Phys. Colloq.* **42**: C8 247.

Koyama Y., Matsuura H., Atsumi K., Nakajima Y., and Chiba K. 1995. An ultra-miniature rubidium frequency standard with two-cell scheme. In *Proceedings of the IEEE International Frequency Control Symposium* 33.

Koyama Y., Matsuura H., Atsumi K., Nakamuta K., Sakai M., and Maruyama I. 2000. An ultra-miniature rubidium frequency standard. In *Proceedings of the IEEE/EIA International Frequency Control Symposium* 394.

Kramer G. 1973. Bestimmung der Geschwindigkeitsverteilung des Cs-Atomstrahls im Frequenznormal CS1 mittels Fourieranalyse der Resonanzkurve. *PTB-jahresbericht* 134.

Lee H.S., Kwon T.Y., Kang H.S., Park Y.H., Oh C.H., Park S.E., Cho H., and Minogin V.G. 2003. Comparison of the Rabi and Ramsey pulling in an optically pumped caesium-beam standard. *Metrologia* **40**: 224.

Lesage P., Audoin C., and Têtu M. 1979. Amplitude noise in passively and actively operated masers. In *Proceedings of the Annual Symposium on Frequency Control* 515.

Levi F., Godone A., Novero C., and Vanier J. 1997. On the use of a modulated laser for hyper-frne frequency excitation in passive atomic frequency standards. In *Proceedings of the European Forum on Time and Frequency* 216.

Lewis L.L., Walls F.L., and Glaze D.J. 1981. Design considerations and performance of NBS-6, the NBS primary frequency standard. *J. Phys. Colloq.* C8 241.

Lin C., Liu T., Zhai Z., Zhang W., Lu J., Peng J., and Wang Q. 2001. Miniature passive hydrogen maser at shanghai observatory. In *Proceedings of the IEEE International Frequency Control Symposium* 89.

Majorana E. 1932. Oriented atoms in variable magnetic field. *Nuovo Cimento* **9**: 43.

Makdissi A. and de Clercq E. 2001. Evaluation of the accuracy of the optically pumped caesium beam primary frequency standard of BNM-LPTF. *Metrologia* **38**: 409.

Mandache C., Bastin T., Nizet J., and Leonard D. 2012. Development of an active hydrogen maser in Belgium – first results. In *Proceedings of the European Forum on Time and Frequency* 290.

Mandache C., Nizet J., Leonard D., and Bastin T. 2012. On the hydrogen maser oscillation threshold. *Appl. Phys. B: Lasers Opt.* **107**: 675.

Markowitz W., Hall R.G., Essen L., and Parry J.V.L. 1958. Frequency of caesium in terms of ephemeris time. *Phys. Rev. Lett.* **1**: 105.

Mattioni L., Berthoud P., Pavlenko I., Schweda H., Wang Q., Rochat P., Droz F., Mosset P., and Ruedin H. 2002. The development of a passive hydrogen maser clock for the Galileo navigation system. In *Proceedings of the Annual Precise Time and Time Interval Applications and Planning Meeting* 161.

Mattison E.M., Blomberg E.L., Nystrom G.H., and Vessot R.F.C. 1979. Design, construction and testing of a small passive hydrogen maser. In *Proceedings of the Annual Precise Time and Time Interval Applications and Planning Meeting* 549.

Mattison E.M., Vessot R.F.C., and Levine M. 1976. A study of hydrogen maser resonators and storage bulbs for use in ground and satellites masers. In *Proceedings of the Annual Precise Time and Time Interval Applications and Planning Meeting* 243.

McCoubrey A.O. 1996. History of atomic frequency standards: a trip through 20th century physics. In *Proceedings of the IEEE International Frequency Control Symposium* 1225.

Micalizio S., Calosso C.E., Godone A., and Levi F. 2012. Metrological characterization of the pulsed Rb clock with optical detection. *Metrologia* **49**: 425.

Micalizio S., Calosso C.E., Levi F., and Godone A. 2013. Ramsey-fringe shape in an alkali-metal vapor cell with buffer gas. *Phys. Rev. A* **88**: 033401.

Micalizio S., Godone A., Calonico D., Levi F., and Loroni L. 2004. Blackbody radiation shift of the Cs hyperfine transition frequency. *Phys. Rev. A* **69**: 053401.

Micalizio S., Godone A., Levi F., and Calosso C.E. 2008. The pulsed optically pumped clock: microwave and optical detection. In *Proceedings of the 7th Symposium on Frequency Standards and Metrology*, L. Maleki Ed. (World Scientific, Singapore) 343.

Micalizio S., Godone A., Levi F., and Vanier J. 2006. Spin-exchange frequency shift in alkali-metal-vapor cell frequency standards. *Phys. Rev. A* **73**: 033414; Correction *Phys. Rev. A* **74**: 059905(E).

Morikawa T., Umezu J., Takahei K., Uehara M., Mori K., and Tsuda M. 2000. Development of a small hydrogen maser with a sapphire loaded cavity for space applications. In *Proceedings of the Asian Workshop on Time and Frequency* 224.

Morris D. 1971. Hydrogen maser wall shift experiments at the National Research Council of Canada. In *Proceedings of the Annual Symposium on Frequency Control* 343.

Morris D. 1990. Report on special hydrogen maser workshop. In *Proceedings of the Annual Precise Time and Time Interval Applications and Planning Meeting* 349.

Morrow M., Jochemsen R., Berlinski A.J., and Hardy W.N. 1981. Zero-field hyperfine resonance of atomic hydrogen for $0.18 < \sim T < \sim 1$ K: The binding energy of H on Liquid ^{4}He. *Phys. Rev. Lett.* **46**: 195.

Münch A., Berkler M., Gerz Ch., Wilsdorf D., and Werth G. 1987. Precise ground-state hyperfine splitting in ^{173}Yb$^+$. *Phys. Rev. A* **35**: 4147.

Mungall A.G. 1971. The second order Doppler shift in caesium beam atomic frequency standards. *Metrologia* **7**: 49.

Mungall A.G., Bailey R., Daams H., Morris D., and Costain C.C. 1973. The new NRC 2.1 metre primary caesium beam frequency standard, Cs V. *Metrologia* **9**: 113.

Mungall A.G. and Costain C.C. 1977. NRC CsV primary clock performance. *Metrologia* **13**: 105.

Mungall A.G., Damms H., and Boulanger J.S. 1981. Design, construction, and performance of the NRCCs VI primary caesium clocks. *Metrologia* **17**: 123.

Nagakiri K., Shibuki M., Okazawa H., Umezu J., Ohta Y., and Saitoh H. 1987. Studies on the accurate evaluation of the RRL primary caesium beam frequency standard. *IEEE T. Instrum. Meas.* **IM-36**: 617.

Nagakiri K., Shibuki M., Uabe S., Hayashi Rm. and Saburi Y. 1981. Caesium beam frequency standard at the radio research laboratories. *J. Phys. Colloq.* C8 253.

Nakadan Y. and Koga Y. 1982. A squarewave F.M. servo system with a digital signal processing for caesium frequency standards. In *Proceedings of the Annual Symposium on Frequency Control* 223.

Nakadan Y. and Koga Y. 1985. Recent progress in Cs beam frequency standards at the NRLM. *IEEE T. Instrum. Meas.* **IM-34**: 133.

Park S.J., Manson P.J., Wouters M.J., Warrington R B., Lawn M.A., and Fisk P.T.H. 2007. ^{171}Yb$^+$ microwave frequency standard. In *Proceedings of the Joint Meeting of the International Frequency Control Symposium/European Forum on Time and Frequency* 613.

Paul W. Nobel lectures. 1990. Electromagnetic traps for charged and neutral particles. *Rev. Mod. Phys.* **62**: 531.

Paul W., Osberghaus O., and Fisher E. 1958. Ein Ionenkäfig. *Forsch*ungsber. *Wirtsch. Verkehrsministerium. Nordrhei-Westfallen* **415**.

Peters H., Owings B., Oakley T., and Beno L. 1987. Hydrogen masers for radio astronomy. In *Proceedings of the Annual Symposium on Frequency Control* 75.

Peters H.E. 1978a. *Atomic standard with reduced size and weight*, Patent 4,123,727.

Peters H.E. 1978b. *S*mall, very small, and extremely small hydrogen masers. In *Proceedings of the Annual Symposium on Frequency Control* 469.

Peters H.E. 1984. Design and performance of new hydrogen masers using cavity frequency switching servos. In *Proceedings of the Annual Symposium on Frequency Control* 420.

Peters H.E. and Washburn P.J. 1984. Atomic hydrogen maser active oscillator cavity and bulb design optimization. In *Proceedings of the Annual Precise Time and Time Interval Applications and Planning Meeting* 313.

Pound R.V. 1946. Electronic frequency stabilization of microwave oscillators. *Rev. Sci. Instrum.* **17**: 490.

Prestage D., Dick G.J., and Maleki L. 1987. JPL trapped ion frequency standard development. In *Proceedings of the Annual Symposium on Frequency Control* 20.

Prestage D., Dick G.J., and Maleki L. 1989. New ion trap for frequency standard. *J. Appl. Phys.* **66**: 1013.

Prestage J.D., Chung S., Le T., Lim L., and Maleki L. 2005. Liter sized ion clock with 10^{-15} stability. In *Proceedings of the Joint IEEE International Frequency Symposium and Precise Time and Time Interval Systems and Applications Meeting* 472.

Prestage J.D., Chung S., Thomson R., MacNeal P., and Thanh L. 2008. Small mercury microwave ion clock for navigation and radio-science. In *Proceedings of the 7th Symposium on Frequency Standards and Metrology*, L. Maleki Ed. (World Scientific, Singapore) 156.

Prestage J.D., Chung S.K., Lim L., and Matevosian A. 2007. Compact microwave mercury ion clock for deep-space applications. In *Proceedings of the Joint IEEE International Frequency Control Symposium/European Forum on Time and Frequency* 1113.

Prestage J.D., Janik G.R., Dick G.J., and Maleki L. 1990. Linear ion trap for second-order Doppler shift reduction in frequency standard applications. *IEEE Trans. Ultrason. Ferroelectron. Freq. Control* **37**: 535.

Prestage J.D., Tjoelker R.L., Dick G.J., and Maleki L. 1993. Improved linear ion trap physics package. In *Proceedings of the IEEE International Frequency Control Symposium* 144.

Prestage J.D., Tjoelker R.L., Dick G.J., and Maleki L. 1995. Progress report on the improved linear ion trap physics package. In *Proceedings of the IEEE International Frequency Control Symposium* 82.

QPAFS. 1989. *The quantum physics of atomic frequency standards*, Vols. 1 and 2, J. Vanier and C. Audoin Eds. (Adam Hilger, Bristol).

Rabian J. and Rochat P. 1988. Full digital-processing in a new commercial caesium standard. In *Proceedings of the European Forum on Time and Frequency* 461.

Ramsey N.F. 1950. A molecular beam resonance method with separated oscillating fields *Phys. Rev.* **78**: 695.

Ramsey N.F. 1956. *Molecular beams* (Oxford at the Clarendon Press, London).

Rinard G.A. and Eaton G.R. 2005. Loop gap resonator *Biological Magnetic Resonance*, S.S. Eaton, G.R. Eaton, and L.J. Berliner Eds. (Academic Plenum, New York) 19–52.

Rochat P., Doz F., Mosset P., Barmaverain G., Wang Q., Boving D., Mattioni L., Schmidt U., Pike T., and Emma F. 2007. The onboard galileo rubidium and passive maser, status & performance. In *Proceedings of the Annual Precise Time and Time Interval Applications and Planning Meeting* 26.

Sellars M.J., Fisk P., Lawn M.A., and Coles C. 1995. Further investigation of a prototype microwave frequency standard based on trapped $^{171}Yb^+$ ions. In *Proceedings of the IEEE International Frequency Control Symposium* 66.

Shirley J.H. 1997. Velocity distributions calculated from the Fourier transforms of Ramsey line shapes. *IEEE T. Instrum. Meas.* **46**: 117.

Shirley J.H., Lee W.D., and Drullinger R.E. 2001. Accuracy evaluation of the primary frequency standard NIST-7. *Metrologia* **38**: 427.

Shirley J.H., Lee W.D., Rovera G.D., and Drullinger R.E. 1995. Rabi pedestal shifts as a diagnostic tool in primary frequency standards. *IEEE T. Instrum. Meas.* **44**: 136.

Siegman A.E. 1964. *Microwave Solid State Masers* (McGraw-Hill, New York).

Siegman A.E. 1971. *An Introduction to Lasers and Masers* (McGraw-Hill, New York).

Silvera I.F. and Walraven J.T.H. 1980. Stabilization of atomic hydrogen at low temperature. *Phys. Rev. Lett.* **44**: 164.

Sing L.T., Viennet J., and Audoin C. 1990. Digital synchronous detector and frequency control loop for caesium beam frequency standard. *IEEE T. Instrum. Meas.* **39**: 428.

Sphicopoulos T. Thèse de doctorat. 1986. *Conception et analyse de cavités compacte pour étalons de fréquence atomique, contribution `a la représentation du champ intégral du champ électromagnétique* (École Polytechnique Fédérale de Lausanne, Switzerland).

Sphicopoulos T., Bernier L.G., and Gardiol F. 1984. Theoretical basis for the design of the radially stratified dielectric-loaded cavities used in miniaturized atomic frequency standards. *IEE Proc. H* **131** (2): 94.

Stefanucci C., Bandi T., Merli F., Pellaton M., Affolderbach C., Mileti G., and Skrivervik A.K. 2012. Compact microwave cavity for high performance Rubidium frequency standards. *Rev. Sci. Instr.* **83** (10): 104706.

Taghavi-Larigani S., Burt E.A., Lea S.N., Prestage J.D., and Tjoelker R.L. 2009. A new trapped ion clock based on ^{201}Hg$^+$. In *Proceedings of the Joint IEEE International Frequency Control Symposium/European Forum on Time and Frequency* 774.

Tamm Chr., Lipphardt B., Mehlstauber T.E., Okhapkin M., Sherstov I., Stein B., and Peik E. 2008. ^{171}Yb$^+$ single-ion optical frequency standards. In *Proceedings of the 7th Symposium on Frequency Standards and Metrology*, L. Maleki Ed. (World Scientific, Singapore) 235.

Tamm Chr., Schrier D., and Bauch A. 1995. Radiofrequency laser double-resonance spectroscopy of trapped Yb-171 ions and determination of line shifts of the ground-state hyperfine resonance *Appl. Phys. B: Lasers Opt.* **60** (1): 19.

Tanaka U., Bize S., Tanner C.E., Drullinger R.E., Diddams S.A., Hollberg L., Itano W.M., Wineland D.J., and Bergquist J.C. 2003. The ^{199}Hg$^+$ single ion optical clock: progress. *J. Phys. B At. Mol. Opt. Phys* **36**: 545.

Tjoelker R.L., Prestage J.D., and Maleki L. 1995. Record frequency stability with mercury in a linear ion trap. In *Proceedings of the 5th Symposium on Frequency Standards and Metrology*, J.C. Berquist Ed. (World Scientific, Singapore) 33.

Vanier J. 1968. Relaxation in Rubidium-87 and the Rubidium Maser. *Phys. Rev.* **168**: 129.

Vanier J. 1969. *Tuning of atomic masers by magnetic quenching using transverse magnetic fields*, Patent #3,435,369.

Vanier J. 2002. Atomic frequency standards: basic physics and impact on metrology. In *Proceedings of the International School of Physics*, T.J. Quinn, S. Leschiutta, and P. Tavella Eds. (IOS Press, Amsterdam, the Netherlands).

Vanier J. and Audoin C. 2005. The classical caesium beam frequency standard: fifty years later. *Metrologia* **42**: S31.

Vanier J., Blier R., Gingras D., and Paulin P. 1984. Hydrogen maser work at Laval University. *Acta Metrologica Sinica* **5**: 267.

Vanier J., Godone A., and Levi F. 1998. Coherent population trapping in caesium: dark lines and coherent microwave emission. *Phys. Rev. A* **58**: 2345.

Vanier J., Kunski R., Cyr N., Savard J.Y., and Têtu M. 1982. On hyperfine frequency shifts caused by buffer gases: application to the optically pumped passive rubidium frequency standard. *J. Appl. Phys.* **53**: 5387.

Vanier J. and Larouche R. 1978. A comparison of the wall shift of TFE and FEP teflon coatings in the hydrogen maser. *Metrologia* **14**: 31.

Vanier J., Larouche R., and Audoin C. 1975. The hydrogen maser wall shift problem. In *Proceedings of the Annual Symposium on Frequency Control* 371.

Vanier J. and Vessot R.F.C. 1964. Cavity tuning and pressure dependence of frequency in the hydrogen maser. *Appl. Phys. Lett.* **4**: 122.

Vanier J. and Vessot R.F.C. 1966. Relaxation in the level F = 1 of the ground state of hydrogen; application to the hydrogen maser. *IEEE J. Quant. Elect.* **QE2**: 391.

Vanier J. and Vessot R.F.C. 1970. H maser wall shift. *Metrologia* **6**: 52.

Vessot R.F.C. 2005. The atomic hydrogen maser oscillator. *Metrologia* **42**: S80.

Vessot R.F.C., Levine M.W., and Mattison E.M. 1977. Comparison of theoretical and observed maser stability limitation due to thermal noise and the prospect of improvement by low temperature operation. In *Proceedings of the Annual Precise Time and Time Interval Applications and Planning Meeting* 549.

Vessot R.F.C., Levine M., Mattison E.M., Blomberg E.L., Hoffman T.E., Nystrom G.U., Farrel B.F., Decher R., Eby P.B., Baugher C.R., Watts J.W., Teuber D.L., and Wills F.D. 1980. Test of relativistic gravitation with a space-borne hydrogen maser. *Phys. Rev. Lett.* **45**: 2081.

Vessot R.F.C., Mattison E.M., and Blomberg E.M. 1979. Research with a cold atomic hydrogen maser. In *Proceedings of the Annual Symposium on Frequency Control* 511.

Vessot R.F.C., Mattison E.M., Imbier E.A., Zhai Z.C., Klepczynski W.J., Wheeler P.G., Kubik A.J., and Winkler G.M.K. 1984. Performance data of U.S. Naval Observatory VLG-11 hydrogen masers since September 1983. In *Proceedings of the Annual Precise Time and Time Interval Applications and Planning Meeting* 375.

Vessot R.F.C., Mattison E.M., Walsworth R.L., and Silvera I.F. 1988. The cold hydrogen maser. In *Proceedings of the 4th Symposium on Frequency Standards and Metrology*, A. de Marchi Ed. (Springer-Verlag, Berlin, Germany) 88.

Vessot R.F.C., Mattison E.M., Walsworth R.L., Silvera I.F., Godfried H.P., and Agosta C.C. 1986. A hydrogen maser at temperatures below 1 K. In *Proceedings of the Annual Symposium on Frequency Control* 413.

Violetti M., Merli F., Zurcher J.F., Skrivervik A.K., Pellaton M., Affolderbach C., and Mileti G. 2014. The microloop-gap resonator: a novel miniaturized microwave cavity for Rubidium double-resonance atomic clocks. *IEEE Sensors J.* **14** (3): 194.

Violetti M., Pellaton M., Affolderbach C., Merli F., Zurcher J.F., Mileti G., and Skrivervik A.K. 2012. New miniaturized microwave cavity for Rubidium atomic clocks. In *Proceedings of the IEEE Sensors Conference* 315.

VREMYA-CH. 2012. Hydrogen maser frequency and time standard model VCH-1003A. Brief description and maintenance instruction.

Vuylsteke A.A. 1960. *Elements of maser theory* (Van Norstrand, Princeton, NJ).

Waller P., Gonzalez F., Binda S., Rodriguez D., Tobias G., Cernigliaro A., Sesia I., and Tavella P. 2010. Long-term performance analysis of GIOVE clocks. In *Proceedings of the Annual Precise Time and Time Interval Applications and Planning Meeting* 171.

Walls F.L. 1987. Characteristics and performance of miniature NBS passive hydrogen masers. *IEEE T. Instrum. Meas.* **IM-36** (2): 585.

Walls F.L. and Howe D.A. 1978. A passive hydrogen maser frequency standard. In *Proceedings of the Annual Symposium on Frequency Control* 492.

Wallsworth R.L., Silvera I.F., Godfried H.P., Agosta C.C., Vessot R.F.C., and Mattison E.M. 1986. Hydrogen maser at temperatures below 1 K. *Phys. Rev. A* **34**: 2550.

Wang H.T.M. 1980. An oscillating compact hydrogen maser. In *Proceedings of the Annual Symposium on Frequency Control* 364.

Wang H.T.M. 1989. Subcompact hydrogen maser atomic clocks. *Proc. IEEE* **77** (7): 982.

Wang H.T.M., Lewis J.B., and Crampton S.B. 1979. Compact cavity for hydrogen frequency standard. In *Proceedings of the Annual Symposium on Frequency Control* 543.

Wang Q., Mosset P., Droz F., and Rochat P. 2013. Lifetime of space passive hydrogen maser. In *Proceedings of the Joint UFFC, EFTF and PFM Symposium* 973.

Wang Q., Mosset P., Droz F., Rochat P., and Busca G. 2006. Verification and optimization of the physics parameters of the onboard Galileo passive hydrogen maser. In *Proceedings of the Annual Precise Time and Time Interval System and Applications Meeting* 1.

Wang Q., Zhai Z., Zhang W., and Lin C. 2000. An experimental study for the compact hydrogen maser with a TE111 septum cavity. *IEEE Trans Utrason. Ferroelectr. Freq. Control* **47**: 197.

Warrington R.B., Fisk P., Wouters M.J., Lawn M.A., and Coles C. 1999. The CSIRO trapped ^{171}Yb$^+$ ion clock: improved accuracy through laser-cooled operation. In *Proceedings of the Joint Meeting of the European Forum on Time and Frequency/IEEE International Frequency Control Symposium* 125.

Weber C., Duerrenberger M., and Schweda H. 2007. Principle of pulsed-interrogation automatic cavity tuning for the ACES hydrogen maser. In *Proceedings of the Joint Meeting of the European Forum on Time and Frequency/IEEE International Frequency Control Symposium* 71.

Webster S.A., Taylor P., Roberts M., Barwood G.P., Blythe P., and Gill P. 2001. A frequency standard using the $^2S_{1/2} - {^2F_{7/2}}$ octupole transition in ^{171}Yb$^+$. In *Proceedings of the 6th Symposium on Frequency Standards and Metrology*, P. Gill Ed. (World Scientific, Singapore) 115.

Wineland D.J., Itano W.M., and Bergquist J.C. 1981. Proposed stored ^{201}Hg$^+$ ion frequency standard. In *Proceedings of the Annual Symposium on Frequency Control* 602.

Wittke J.P. and Dicke R.H. 1956. Determination of the hyperfine splitting in the ground state of atomic hydrogen. *Phys. Rev.* **103**: 620.

Xia B., Zhong D., An S., and Mei G. 2006. Characteristics of a novel kind of miniaturized cavity-cell assembly for rubidium frequency standards. *IEEE T. Instrum. Meas.* **55**: 1000.

Xiaoren Y. 1981. Works on chinese primary frequency standards. *J. Phys. Colloq.* C8 257.

Yahyabey N., Lesage P., and Audoin C. 1989. Studies of dielectrically loaded cavities for small size hydrogen masers. *IEEE T. Instrum. Meas.* **38**: 74.

CHAPTER 2

Acernese F., Amico P., Alshourbagy M., and 127 colleagues. 2006. The virgo status. *Classical Quant. Grav.* **23**: S635.

Adams C.S. and Riis E. 1997. Laser cooling and trapping of neutral atoms. *Prog. Quant. Elec.* **21**: 1.

Affolderbach C. and Mileti G. 2003. A compact, frequency stabilized laser head for optical pumping in space Rb clocks. In *Proceedings of the Joint IEEE International Frequency Control Symposium/European Forum on Time and Frequency*, IEEE, 109.

Affolderbach C. and Mileti G. 2005. A compact laser head with high-frequency stability for Rb atomic clocks and optical instrumentation. *Rev. Sci. Instrum.* **76**: 073108.

Affolderbach C., Mileti G., Slavov D., Andreeva C., and Cartaleva S. 2004. Comparison of simple and compact *Doppler* and *sub-Doppler* laser frequency stabilisation schemes. In *Proceedings of the European Forum on Time and Frequency* 84.

Alzetta G., Gozzini A., Moi M., and Orriols G. 1976. An experimental method for observation of Rf transitions and laser beat resonances in oriented Na vapor. *Il Nuovo Cimento B* **36**: 5.

Arimondo E. 1996. Coherent population trapping in laser spectroscopy. In *Progress in Optics*, E. Wolf Ed. (Elsevier, Amsterdam, the Netherlands) 257.

Aspect A., Arimondo E., Kaizer R., Vansteenkiste N., and Cohen-Tannoudji C. 1989. Laser cooling below the one-photon recoil energy by velocity-selective coherent population trapping: theoretical analysis. *J. Opt. Soc. Am. B* **6**: 2112.

Aucouturier E. 1997. *Nouvelle source d'atomes froids pour l'horloge atomique*. Thèse Université de Paris XI, Orsay, France.

Avila G., Giordano V., Candelier V., deClercq E., Theobald G., and Cerez P. 1987. State selection in a cesium beam by laser-diode optical pumping. *Phys. Rev. A* **36**: 3719.

Baillard X., Gauguet A., Bize S., Lemonde P., Laurent Ph., Clairon A., and Rosenbusch P. 2006. Interference-filter-stabilized external-cavity diode lasers. *Optics Communications* **266**: 609.

Balykin V., Letokhov V., and Mushin V. 1979. Observation of free cooling sodium atoms in a resonant laser field with a frequency scan. *JEPT Let.* **29**: 560.

Barwood G.P., Gill P., and Rowley W.R.C. 1988. A simple rubidium-stabilised laser diode for interferometric applications. *J. Phys. E: Sci. Instrum.* **21**: 966.

Barwood G.P., Gill P., and Rowley W.R.C. 1991. Frequency measurements on optically narrowed Rb stabilised laser diodes at 780 nm and 795 nm. *Appl. Phys. B* **53**: 142.

Batelaan H., Padua S., Yang D.H., Xie C., Gupta R., and Metcalf H. 1994. Slowing of 85rb atoms with isotropic light. *Phys. Rev. A* **49**: 2780.

Bell W.E. and Bloom A.L. 1961. Optically driven spin precession. *Phys. Rev. Lett.* **6**: 280.

Berkeland D.J., Miller J.D., Bergquist J.C., Itano W.N., and Wineland D.J. 1998. Laser-cooled mercury ion trap frequency standard. *Phys. Rev. Lett.* **80**: 2089.

Beverini N., Maccioni E., Marsili P., Ruffini A., and Serrentino F. 2001. Frequency stabilization of a diode laser on the Cs D2 resonance line by Zeeman effect in a vapor cell. *Appl. Phys. B* **73**: 133.

Camposeo A., Piombini A., Cervelli F., Tantussi F., Fuso F., and Arimondo E. 2001. A cold cesium atomic beam produced out of pyramidal funnel. *Opt. Comm.* **200**: 231.

Chapelet F. 2008. *Fontaine atomique double de césium et de rubidium avec une exactitude de quelques 10^{-16} et applications.* Thèse de l'Universite Paris XI.

Cheiney P., Carraz O., Bartoszek-Bober D., Faure S., Vermerasch F., Fabre C.M., Gattobigio G.L., Lahaye T., Guéry-Odelin D., and Mathevet R. 2011. Zeeman slower design with permanent magnets in a Halbach configuration. *Rev. Sci. Instrum.* **82**: 063115.

Chow W.W. and Koch S.W. 2011. *Semiconductor–Laser Fundamentals* (Springer, New York).

Chu S., Hollberg L., Bjorkholm J.E., Cable A., and Ashkin A. 1985. Three-dimensional viscous confinement and cooling of atoms by resonance radiation pressure. *Phys. Rev. Lett.* **55**: 48.

Cohen-Tannoudji C., Dupont-Roc J., and Grynberg G. 1988. *Processus d'interaction entre photons et atomes* (Editions du CNRS, Paris, France).

Cohen-Tannoudji C. and Guéry-Odelin D. 2011. *Advances in Atomic Physics* (World Scientific, Singapore).

Cohen-Tannoudji C. and Phillips W.D. 1990. New mechanisms for laser cooling. *Physics Today* **33**.

Cyr N., Têtu M., and Breton M. 1993. All-optical microwave frequency standard: a proposal. *IEEE Trans. Instrum. Meas.* **42**: 640.

Dahmani B., Hollberg L., and Drullinger R. 1987. Frequency stabilization of semiconductor lasers by resonant optical feedback. *Opt. Lett.* **12**: 876.

Dalibard J. 1986. *Le rôle des fluctuations dans la dynamique d'un atome couple au champ électromagnétique.* Thèse, Un. de Paris VI France.

Dalibard J. and Cohen-Tannoudji C. 1989. Laser cooling below the Doppler limit by polarization gradients: simple theoretical models. *J. Opt. Soc. Am. B* **6**: 2023.

Danzmann K. and Rudiger A. 2003. LISA technology—concept, status, prospects. *Classical and Quantum Gravity* **20**: S1.

Dawkins S.T., Chicireanu R., Petersen M., Millo J., Magalhães D.V., Mandache C., Le Coq Y., and Bize S. 2010. An ultra-stable referenced interrogation system in the deep ultraviolet for a mercury optical lattice clock. *Appl. Phys. B* **99**: 41.

de Labachelerie M. 1988. *Principales caractéristiques des lasers à semiconducteurs à cavité étendue: application à l'amélioration des proprietés spectrales des diodes laser.* Thèse Université Orsay.

de Labachelerie M. and Cerez P. 1985. An 850 nm semiconductor laser tunable over 300 A range. *Opt. Commun.* **55**: 174.

de Labachelerie de M., Latrasse C., Kemssu P., and Cerez P. 1992. The frequency control of laser diodes. *J. Phys. III France* **2**: 1557.

Derevianko A. and Katori H. 2011. Colloquium physics of optical lattice clocks. *Rev. Mod. Phys.* **83**: 331.

Dicke R.H. 1953. The effect of collisions upon the doppler width of spectral lines. *Phys. Rev.* **89**: 472.

Dieckmann K., Spreeuw R.J.C., Weidemuller M., and Walraven J.T.M. 1998. Two-dimensional magneto-optical trap as a source of slow atoms. *Phys. Rev. A* **58**: 3891.

Drever R., Hall J., Kowalski F., Hough J., Ford G., Munley A., and Ward H. 1983. Laser phase and frequency stabilization using an optical resonator. *Appl. Phys. B* **31**: 97.

Esnault F.X. 2009. *Étude des performances ultimes d'une horloge compacte a atomes froids: optimisation de la stabilité court terme*, Thèse Université de Paris.

Esnault F.X., Perrin S., Holleville D., Guerandel S., Dimarcq N., and Delporte J. 2008. Reaching a few $10^{-13}\, \tau^{-1/2}$ stability level with the compact cold atom clock Horace. In *Proceedings of the IEEE International Frequency Control Symposium*, IEEE, 381.

Fleming M.W. and Mooradian A. 1981. Spectral characteristics of external-cavity controlled semiconductor lasers. *IEEE J. Quant. Electr.* **QE-17**: 44.

Forman P. 1985. Atomichron®: the atomic clock from concept to commercial product. *Proc. IEEE* **73**: 1181.

Frisch R. 1933. Experimenteller Nachweis des Einsteinschen Strahlungsriickstolies. *Z. Phys.* **86**: 42.

Godone A., Levi F., Micalizio S., and Vanier J. 2000. Theory of the coherent population trapping maser: a strong-field self-consistent approach. *Phys. Rev. A* **62**: 053402.

Godone A., Levi F., Micalizio S., and Vanier J. 2002. Line shape of dark line and maser emission profile in CPT. *Eur. Phys. J. D* **18**: 5.

Godone A., Levi F., and Vanier J. 1999. Coherent microwave emission in cesium under coherent population trapping. *Phys. Rev. A* **59**: R12.

Goldberg L., Taylor H.F., Dandridge A., Weller J.F., and Miles R.O. 1982. Spectral characteristics of semiconductor lasers with optical feedback. *IEEE J. Quant. Electr.* **QE-18**: 555.

Gray H.R., Whitley R.M., and Stroud C.R. Jr. 1978. Coherent trapping of atomic populations. *Opt. Lett.* **3**: 218.

Guillemot C., Vareille Ch., Valentin C., and Dimarcq N. 1997. 3D cooling of cesium atoms with isotropic light. In *Proceedings of the European Forum on Time and Frequency* 156.

Guillot E., Pottie P.E., Valentin C., Petit P., and Dimarcq N. 1999. HORACE: atomic clock with cooled atoms in a cell. In *Proceedings of the Joint Meeting of the European Forum on Time and Frequency/IEEE International Frequency Control Symposium*, IEEE, 81.

Halbach K. 1980. Design of permanent multipole magnets with oriented rare earth cobalt materials. *Nucl. Instrum. Methods* **169**: 1.

Hansch T. and Shawlow A. 1975. Cooling of gases by laser radiation. *Opt. Commun.* **13**: 68.

Happer W. and Mather B.S. 1967. Effective operator formalism in optical pumping. *Phys. Rev.* **163**: 12.

Harris S.E. 1997. Electromagnetically induced transparency. *Physics Today* **50**: 36.

Hashimoto M. and Ohtsu M. 1987. Experiments on a semiconductor laser pumped Rb atomic clock. *IEEE J. Quant. Electr.* **23**: 446.

Hess H.F. 1986. Evaporative cooling of magnetically trapped and compressed spin-polarized hydrogen. *Phys. Rev. B* **34**: 3476.

Hill I.R., Ovchinnikov Y.B., Bridge L., Curtis E.A., Donnellan S., and Gill P. 2012. A simple, configurable, permanent magnet Zeeman slower for Sr. In *Proceedings of the European Forum on Time and Frequency* 545.

Jiang H., Kefelian F., Crane S., Lopez O., Lours M., Millo J., Holleville D., Lemonde P., Chardonnet C., Klein A.A., and Santarelli G. 2008. Long-distance frequency transfer over an urban fiber link using optical phase stabilization. *J. Opt. Soc. Am. B* **25**: 2029.

Kanada T. and Nawata K. 1979. Single-mode operation of a modulated laser diode with a short external cavity. *Opt. Commun.* **31**: 81.

Kasapi A., Jain M., Yin G.Y., and Harris S.E. 1995. Electromagnetically induced transparency: propagation dynamics. *Phys. Rev. Lett.* **74**: 2447.

Kasevich M. and Chu S. 1992. Laser cooling below a photon recoil with three-level atoms. *Phys. Rev. Lett.* **69**: 1741.

Kastler A. 1950. Quelques suggestions concernant la production optique et la détection optique d'une inégalité de population des niveaux de quantification spatiale des atomes. Application à l'expérience de Stern et Gerlach et à la résonance magnétique. *J. Phys.* **II**: 255.

Katori H. 2001. Spectroscopy of strontium atoms in the Lamb-Dicke confinement. In *Proceedings of the 6th Symposium on Frequency Standards and Metrology*, P. Gill Ed. (World Scientific: Singapore) 323.

Ketterle W., Martin A., Joffe M., and Pritchard D. 1992. Slowing and cooling of atoms in isotropic light. *Phys. Rev. Lett.* **69**: 2483.

Kozuma M., Kourogi M., and Ohtsu M. 1992. Frequency stabilization, linewidth reduction and fine detuning of a semiconductor laser using velocity selective optical pumping of atomic resonance line. *Appl. Phys. Lett.* **61**: 1895.

Laurent P., Clairon A., and Breant C. 1989. Frequency noise analysis of optically self–locked diode lasers. *IEEE J. Quant. Electr.* **QE-25**: 1131.

Letokhov V.S. 1968. Doppler line narrowing in a standing line wave. *JEPT Lett.* **7**: 272.

Letokhov V.S. 2007. *Laser Control of Atoms and Molecules* (Oxford University Press, Oxford).

Letokhov V.S., Minogin V.G., and Pavlik B.D. 1976. Cooling and trapping of atoms and molecules by a resonant laser field. *Opt. Commun.* **19**: 72.

Lett P.D., Watts R.N., Westbrook C.I., and Phillips W.D. 1988. Observation of atoms laser cooled below the Doppler limit. *Phys. Rev. Lett.* **61**: 169.

Levi F., Godone A., Micalizio S., Calosso C., Detoma E., Morsaniga P., and Zanello R. 2002. CPT maser clock evaluation for galileo. In *Proceedings of the Precise Time and Time Interval Systems and Applications Meeting*, ION Publications, 139.

Levi F., Godone A., Micalizio S., and Vanier J. 1999. On the use of Λ transitions in atomic frequency standard. In *Proceedings Precise Time and Time Interval Systems and Applications Planning Meeting*, ION Publications, 216.

Levi F., Godone A., Novero C., and Vanier J. 1997. On the use of a modulated laser for hyperfine frequency excitation in passive atomic frequency standards. In *Proceedings of the European Forum on Time and Frequency* 216.

Li H. and Telle H.R. 1989. Efficient frequency noise reduction of GaAlAs semiconductor lasers by optical feedback from an external high-finesse resonator. *IEEE J. Quant. Electr.* **QE-25**: 257.

Lu Z.T., Corwin K.L., Renn M.J., Anderson M.H., Cornell E.A., and Wieman C.E. 1996. Low-velocity intense source of atoms from a magneto-optical trap. *Phys. Rev. Lett.* **77**: 3331.

Ludlow A.D., Huang X., Notcutt M., Zanon-Willette T., Foreman S.M., Boyd M.M., Blatt S., and Ye J. 2007. Compact, thermal-noise-limited optical cavity for diode laser stabilization at 1×10^{-15}. *Opt. Lett.* **32**: 641.

Ludlow A.D., Zelevinsky T., Campbell G.K., Blatt S., Boyd M.M., de Miranda M.H.G., Martin M.J., Thomsen J.W., Foreman S.M., Jun Ye., Fortier T.M., Stalnaker J.E., Diddams S.A., Le Coq Y., Barber Z.W., Poli N., Lemke N.D., Beck K.M., and Oates C.W. 2008. Sr Lattice clock at 1×10^{-16} fractional uncertainty by remote optical evaluation with a Ca clock. *Science* **319**: 1805.

Mandache C., Petersen M., Magalhaes D., Acef O., Clairon A., and Bize S. 2008. Towards an optical lattice clock based on neutral mercury. *Rom. Rep. Phys.* **60**: 581.

Mejri S. 2012. *Horloge à réseau optique de mercure: détermination de la longueur d'onde magique*. Thèse Université Paris VI France.

Metcalf J.H. and van der Straten P. 1999. *Laser Cooling and Trapping* (Springer, New York).

Mileti G. 1995. *Etude du pompage optique par laser et par lampe spectrale dans les horloges à vapeur de rubidium*. Thèse, Université de Neuchâtel, Suisse.

Millo J., Magalhães D.V., Mandache C., Le Coq Y., English E.M.L., Westergaard P.G., Lodewyck J., Bize S., Lemonde P., and Santarelli G. 2009. Ultrastable lasers based on vibration insensitive cavities. *Phys. Rev. A* **79**: 053829.

Monroe C., Swann W., Robinson H., and Wieman C. 1990. Very cold trapped atoms in a vapor cell. *Phys. Rev. Lett.* **65**: 1571.

Moulton P.F. 1986. Spectroscopic and laser characteristics of Ti:Al₂O₃. *J. Opt. Soc. B* 3: 125.

Müller H., Stanwix P.L., Tobar M.E., Ivanov E., Wolf P., Herrmann S., Senger A., Kovalchuk E., and Peters A. 2007. Tests of relativity by complementary rotating Michelson-Morley experiments. *Phys. Rev. Lett.* **99**: 050401.

Nagel A., Graf L., Naumov A., Mariotti E., Biancalana V., Meschede D., and Wynands R. 1998. Experimental realization of coherent dark-state magnetometers. *Europhys. Lett.* **44**: 31.

Nazarova T., Riehle F., and Sterr U. 2006. Vibration-insensitive reference cavity for an ultra-narrow-linewidth laser. *Appl. Phys. B: Lasers Opt.* **83**: 531.

Numata K., Kemery A., and Camp J. 2004. Thermal-noise limit in the frequency stabilization of lasers with rigid cavities. *Phys. Rev. Lett.* **93**: 250602.

Ohtsu M., Hashimoto M., and Hidetaka O. 1985. A highly stabilized semiconductor laser and its application to optically pumped Rb atomic clock. In *Proceedings of the Annual Symposium on Frequency Control*, IEEE, 43.

Ohtsu M., Nakagawa K., Kourogi M., and Wang W. 1993. Frequency control of semiconductor lasers. *J. Appl. Phys.* **73**: R1.

Orriols G. 1979. Nonabsorption resonances by nonlinear coherent effects in a three-level system. *Il Nuovo Cimento B* **53**: 1.

Ovchinnikov Y.B. 2007. A Zeeman slower based on magnetic dipoles. *Opt. Commun.* **276**: 261.

Ovchinnikov Y.B. 2008. A permanent Zeeman slower for Sr atomic clock. *Eur. Phys. J. Spec. Top.* **163**: 95.

Ovchinnikov Y.B. 2012. Longitudinal Zeeman slowers based on permanent magnetic dipoles. *Opt. Commun.* **285**: 1175.

Ovsiannikov V.D., Pal'chikov V.G., Taichenachev A.V., Yudin V.I., Katori H., and Takamoto M. 2007. Magic-wave-induced S01-P03 transition in even isotopes of alkaline-earth-metal-like atoms. *Phys. Rev. A* **75**: 020501.

Petersen M., Magalhaes D., Mandache C., Acef O., Clairon A., and Bize S. 2007. Towards an optical lattice clock based on neutral mercury. In *Proceedings of the Joint IEEE International Frequency Control Symposium/European Forum on Time and Frequency,* IEEE, 649.

Phillips W.D. 1992. Laser manipulation of atoms and ions. In *Proceedings of the International School of Physics,* Enrico Fermi Course CXVIII ed. E. Arimondo, W.D. Phillips, and F. Strumia Ed. (North Holland, Amsterdam) 239.

Phillips W.D. 1998. Laser cooling and trapping of neutral atoms. *Rev. Mod. Phys.* **70**; 3.

Phillips W.D. and Metcalf H. 1982. Laser deceleration of an atomic beam. *Phys. Rev. Lett.* **48**: 596.

Phillips W.D. and Prodan J.V. 1983. *Laser Cooled and Trapped Atoms* (NBS, Washington, DC, Spec. Pub. 653) 137.

Pottie P.E. 2004. *Étude du refroidissement laser en cellule: contribution au développement d'une horloge atomique miniature à ¹³³Cs.* Thèse, Un de Paris VI, France.

Prodan J.V., Migdall A., Phillips W.D., So I., Metcalf H., and Dalibard J. 1985. Stopping atoms with laser light. *Phys. Rev. Lett.* **54**: 992.

Prodan J.V., Philips W.D., and Metcaf H. 1982. Laser production of a very slow monoenergetic atomic beam. *Phys. Rev. Lett.* **49**: 1149.

QPAFS. 1989. *The Quantum Physics of Atomic Frequency Standards*, Vols. 1 and 2, J. Vanier and C. Audoin Ed. (Adam Hilger ed., Bristol).

Raab E.L., Prentiss M., Cable A., Chu S., and Pritchard D.E. 1987. Trapping of neutral sodium atoms with radiation pressure. *Phys. Rev. Lett.* **23**: 2631.

Renzoni F., Lindner A., and Arimondo E. 1999. Coherent population trapping in open system: a coupled/noncoupked state analysis. *Phys. Rev. A* **60**: 450.

Riis E., Weiss D.S., Moler K.A., and Chu S. 1990. Atom funnel for the production of a slow, high-density atomic beam. *Phys. Rev. Lett.* **64**: 1658.

Rosenband T., Hume D.B., Schmidt P.O., Chou C.W., Brusch A., Lorini L., Oskay W.H., Drullinger R.E., Fortier T.M., Stalnaker J.E., Diddams S.A., Swann W.C., Newbury N.R., Itano W.M., Wineland D.J., and Bergquist J.C. 2008. Frequency ratio of Al$^+$ and Hg$^+$ single-ion optical clocks; Metrology at the 17th decimal place. *Science* **319**: 1808.

Saito S., Nilsson O., and Yamamoto Y. 1982. Semiconductor laser stabilization by external optical feedback. *IEEE J. Quant. Electr.* **QE-18**: 961.

Schawlow A.L. and Townes C.H. 1958. Infrared and optical masers. *Phys. Rev.* **112** (6): 1940.

Schiff L.I. 1968. *Quantum Mechanics* (McGraw-Hill, New York).

Schoser J., Batar A., Low R., Schweikhard V., Grabowski A., Ovchinnikov Yu.B., and Pfau T. 2002. Intense source of cold Rb atoms from a pure two-dimensional magneto-optical trap. *Phys. Rev. A* **66**: 023410.

Scully M.O. and Fleischhauer M. 1992. High-sensitivity magnetometer based on index-enhanced media. *Phys. Rev. Lett.* **69**: 1360.

Scully M.O. and Zubairy M.S. 1999. *Quantum Optics* (Cambridge University Press: London).

Slowe C., Vernac L., and Nau L.V. 2005. High flux source of cold rubidium atoms. *Rev. Sci. Instrum.* **75**: 103101.

Sortais Y. 2001. *Construction d'une fontaine double a atomes froids de ^{87}Rb et ^{133}Cs: étude des effets dépendant de nombre d'atomes dans une fontaine.* Thèse Université de Paris VI, France.

Sortais Y., Bize S., Abgrall M., Zhang S., Nicolas C., Mandache C., Lemonde P., Laurent P., Santarelli G., Dimarcq N., Petit P., Clairon A., Mann A., Luiten A., Chang S., and Salomon C. 2001. Cold atomic clocks. *Phys. Scripta T* **95**: 50.

Srinivasan R. 1999. Laser cooling and trapping of ions and atoms. *Current Sci.* **76**: 2.

Steck D. 2003, 2008. *Cesium D Line Data, Rubidium 87 D Line Data.* http://steck.us/alkalidata.

Taylor C.T., Notcutt M., and Blair D.G. 1995. Cryogenic, all-sapphire Fabry–Perot optical frequency reference *Rev. Sci. Instrum.* **66**: 955.

Thomas J.E., Hemmer P.R., Ezekiel S., Leiby C.C., Picard H., and Willis R. 1982. Observation of Ramsey fringes using a stimulated, resonance Raman transition in a sodium atomic beam. *Phys. Rev. Lett.* **48**: 867.

Tsuchida H., Ohtsu M., and Tako T. 1982. Frequency stabilization of AlGaAs semiconductor laser based on the ^{85}Rb D$_2$ line. *Jpn. J. Appl. Phys.* **21**: L561.

Vanier J. 1969. Optical pumping as a relaxation process. *Can. J. Phys.* **47**: 1461.

Vanier J. 2005. Atomic clocks based on coherent population trapping: a review. *Appl. Phys. B* **81**: 421.

Vanier J. and Audoin C. 1989. *The Quantum Physics of Atomic Frequency Standards* (Adam Hilger, editor, Bristol).

Vanier J., Godone A., and Levi F. 1998. Coherent population trapping in cesium: dark lines and coherent microwave emission. *Phys. Rev. A* **58**: 2345.

Vanier J., Godone A., Levi F., and Micalizio S. 2003a. Atomic clocks based on coherent population trapping: basic theoretical models and frequency stability. In *Proceedings of the Joint IEEE International Frequency Control Symposium/European Forum on Time and Frequency*, IEEE, 2.

Vanier J., Levine M., Janssen D., and Delaney M. 2003b. Contrast and linewidth of the coherent population trapping transmission hyperfine resonance line in ^{87}Rb: effect of optical pumping. *Phys Rev. A* **67**: 06581.

Vanier J., Levine M., Janssen D., and Delaney M. 2003c. The coherent population trapping pasive frequency standard. *IEEE Trans. Instrum. Meas.* **52**: 258.

Vanier J., Levine M., Janssen D., and Delaney M. 2003d. On the use of intensity optical pumping and coherent population trapping techniques in the implementation of atomic frequency standards. *IEEE Trans. Instrum. Meas.* **52**: 822.

Vanier J., Levine M., Kendig S., Janssen D., Everson C., and Delaney M. 2004. Practical realization of a passive coherent population trapping frequency standard. In *Proceedings of the IEEE Ultrasonics, Ferroelectrics, and Frequency Control Joint 50th Anniversary Conference*, IEEE, 92.

Vanier J., Levine M., Kendig S., Janssen D., Everson C., and Delaney M. 2005. Practical realization of a passive coherent population trapping frequency standards. *IEEE Trans. Instrum. Meas.* **54**: 2531.

Vanier J. and Mandache C. 2007. The passive optically pumped Rb frequency standard: The laser approach. *Appl. Phys. B: Lasers Opt.* **87**: 665.

Waldman S.J. 2006. Status of LIGO at the start of the fifth science run. *Classical Quant. Grav.* **23** (19): S653.

Wallace C.D., Dinneen T.P., Tan K.-Y.N., Grove T.T., and Gould P.L. 1992. Isotopic difference in trap loss collisions of laser cooled rubidium atoms. *Phys. Rev. Lett.* **69**: 897.

Webster S.A., Oxborrow M., Pugla S., Millo J., and Gill P. 2008. Thermal-noise limited optical cavity. *Phys. Rev. A* **77** (3): 033847.

Williams P.A., Swann W.C., and Newbury N.R. 2008. High-stability transfer of an optical frequency over long fiber-optic links. *J. Opt. Soc. Am. B* **25**: 1284.

Wineland D. and Dehmelt H. 1975. Proposed $10^{14}\Delta v < v$ laser fluoresence spectroscopy on TV~ Mono-Ion Oscillator IIT. *Bull. Am. Phys. Soc.* **20**: 637.

Wineland D. and Itano W.M. 1979. Laser cooling of atoms. *Phys. Rev.* **20**: 1521.

Wynands R. and Nagel A. 1999. Precision spectroscopy with coherent dark states. *Appl. Phys. B* **68**.

Zanon T., Guérandel S., de Clercq E., Dimarcq N., and Clairon A. 2004. Observation of Ramsey fringes with optical CPT pulses. In *Proceedings of the Joint IEEE International Frequency Control Symposium/European Forum on Time and Frequency*, IEEE, 29.

Zanon-Willette T., de Clercq E., and Arimondo E. 2011. Ultrahigh-resolution spectroscopy with atomic or molecular dark resonances: exact steady-state line shapes and asymptotic profiles in the adiabatic pulse regime. *Phys. Rev. A* **84**: 062502.

CHAPTER 3

Abgrall M. 2003. *Évaluation des Performances de la Fontaine Atomique PHARAO. Participation à l'étude de l'horloge Spatiale PHARAO*. PhD Thesis of the Universite ParisVI, France.

Affolderbach C., Andreeva C., Cartaleva S., Karaulanov T., Mileti G., and Slavov D. 2005. Light shift suppression in laser optically-pumped vapour-cell atomic frequency standards. *Appl. Phys. B* **80**: 841.

Affolderbach C., Droz F., and Mileti G. 2006. Experimental demonstration of a compact and high-performance laser-pumped rubidium gas cell atomic frequency standard. *IEEE Trans. Instrum. Meas.* **55**: 429.

Affolderbach C. and Mileti G. 2003a. Development of new Rb clocks in observatoire de Neuchâtel. In *Proceedings of the Annual Precise Time and Time Interval Applications and Planning Meeting*, ION Publication, 489.

Affolderbach C. and Mileti G. 2003b. A compact, frequency stabilized laser head for optical pumping in space Rb clocks. In *Proceedings of the Joint IEEE International Frequency Control Symposium/European Forum on Time and Frequency*, IEEE, 109.

Affolderbach C., Mileti G., Andreeva C., Slavov D., Karaulanov T., and Cataleva S. 2003. Reducing light-shift effects in optically-pumped gas-cell atomic frequency standards. In *Proceedings of the Joint IEEE International Frequency Control Symposium/European Forum on Time and Frequency*, IEEE, 27.

Affolderbach C., Mileti G., Slavov D., Andreeva C., and Cartaleva S. 2004. Comparison of simple and compact "Doppler" and "Sub-Doppler" laser frequency stabilisation schemes. In *Proceedings of the European Forum on Time and Frequency*, ION Publication, 84.

Alekseev E.I., Bazarov Y.N., and Telegin G.I. 1975. Light shift in a quantum frequency measure with pulse optical-pumping and optical detection in Ramsey pattern of 0-0 transition in atoms of Rb[87]. *Radiotekhnika i elektronika* **20** (4): 777.

Anderson N. 1961. Oscillations of a plasma in a static magnetic field. *Proc. Phys. Soc.* **77** (5): 971.

Angstmann E.J., Dzuba V.A., and Flambaum V.V. 2006. Frequency shift of the cesium clock transition due to blackbody radiation. *Phys. Rev. Lett.* **97**: 040802.

Arditi M. and Carver T.R. 1964. Atomic clock using microwave pulse-coherent techniques. *IEEE Trans. Instrum. Meas.* **IM-13**: 146.

Arditi M. and Cerez P. 1972. Hyperfine structure separation of the ground state of [87]Rb measured with an optically pumped atomic beam. *IEEE Trans. Instrum. Meas.* **IM-21**: 391.

Arditi M. and Picqué J.-L. 1975. Precision measurements of light shifts induced by a narrow-band GaAs laser in the 0-0 [133]Cs hyperfine transition. *J. Phys. B: At. Mol. Phys.* **8**: L331.

Arditi M. and Picqué J.L. 1980. A cesium beam atomic clock using laser optical pumping. Preliminary tests. *J. Phys. Lett.* **41**: L379.

Audoin C., Candelier V., and Dimarcq N. 1991. A limit to the frequency stability of passive frequency standards due to an intermodulation effect. *IEEE Trans. Instrum. Meas.* **40**: 121.

Audoin C., Giordano V., Dimarcq N., Cerez P., Petit P., and Theobald G. 1994. Properties of an optically pumped cesium beam frequency standard with $\Phi=\pi$ between the two oscillatory fields. *IEEE Trans. Instrum. Meas.* **43** (4): 515.

Audoin C., Santarelli G., Makdissi A., and Clairon A. 1998. Properties of an oscillator slaved to a periodically interrogated atomic resonator. *IEEE Trans. Ultrason. Ferroelectr. Freq. Control* **45**: 877.

Avila G., Giordano V., Candelier V., de Clercq E., Theobald G., and Cerez P. 1987. State selection in a cesium beam by laser-diode optical pumping. *Phys. Rev. A* **36**: 3719.

Bablewski I., Coffer J., Driskell T., and Camparo J. 2011. Progress in the development of a simple laser-pumped, vapor cell clock. In *Proceedings of the 43rd Annual Precise Time and Time Interval System and Applications Meeting*, ION Publication, 483.

Bandi T., Affolderbach C., and Mileti G., 2012. Laser-pumped paraffin-coated cell rubidium frequency standard. *J. Appl. Phys.* **111**: 124906.

Bandi T., Affolderbach Chr., Stefanucci C., Merli F., Skrivervik A., and Mileti G. 2014. Compact high-performance continuous-wave double-resonance Rubidium standard with $1.4 \times 10^{-13}\, \tau^{-1/2}$ stability. *IEEE Trans. Ultrason. Ferroelectr. Freq. Control* **61**: 1769.

Barrat J.P. and Cohen-Tannoudji C. 1961. Élargissement et déplacement des raies de résonance magnétique causés par une excitation optique. *J. Phys. Rad.* **22**: 443.

Bauch A., Fischer B., Heindorff T., and Schröder R. 1998. Performance of the PTB reconstructed primary clock CS1 and an estimate of its current uncertainty. *Metrologia* **35**: 829.

Bauch A., Heindorff T., Schröder R., and Fischer B. 1996. The PTB primary clock CS3: type B evaluation of its standard uncertainty. *Metrologia* **33**: 249.

Bauch A. and Schröder R. 1997. Experimental verification of the shift of the Cesium hyperfine transition frequency due to blackbody radiation. *Phys. Rev. Lett.* **78**: 622.

Bauch A., Schröder R., and Weyers S. 2003. Discussion of the uncertainty budget and of long term comparison of PTB's primary frequency standards CS1, CS2 and CSF1. In *Proceedings of the Joint IEEE International Frequency Control Symposium/European Forum on Time and Frequency*, IEEE, 191.

Becker W., Blatt R., and Werth G. 1981. Precise determination of ^{135}Ba$^+$ and ^{137}Ba$^+$ hyperfine htructure. *J. Phys Col.* **C8** (Suppl. 12): 339.

Beloy K., Safronova U.I., and Derevianko A. 2006. High-accuracy calculation of the blackbody radiation shift in the ^{133}Cs primary frequency standard. *Phys. Rev. Lett.* **97**: 040801.

Berthoud P., Fretel E., Joyet A., Dudle G., and Thomann P. 1998. Toward a primary frequency standard based on a continuous fountain of laser-cooled cesium atoms. *IEEE Trans. Instrum. Meas.* **48** (2): 516.

Berthoud P., Joyet A., Thomann P., and Dudle G. 1998. A frequency standard based on a continuous beam of laser cooled atoms. In *Proceedings of the IEEE Confernce on Precision Electromagnetic Measurements*, IEEE, 128.

Beverini N., Ortolano M., Costanzo G.A., De Marchi A., Maccioni E., Marsili P., Ruffini A., Periale F., and Barychev V. 2001. Cs cell atomic clock optically pumped by a diode laser. *Laser Phys.* **11**: 1110.

Bhaskar N.D. 1995. Potential for improving the Rubidium frequency standard with a novel optical pumping scheme using diode lasers. *IEEE Trans. Ultrason. Ferroelectr. Freq. Control* **42**: 15.

BIPM. 2006. Brochure sur le SI: Le Système International d'Unités (8ième edition) Section 2.1.1.3.

BIPM. 2012. Bureau International des Poids et Mesures, Consultative Committee for Time and Frequency. In *Report of the 19th meeting to the International Committee for Weights and Measures* 101e session du CIPM – Annexe 8 203, September 13–14, 2012. http://www.bipm.org/en/publications/mep.html.

Bize S. 2001. *Tests fondamentaux a l'aide d'horloges a atomes froids de rubidium et de cesium.* PhD dissertation, Department of Physics, Universite Pierre et Marie Curie, Paris VI, France.

Bize S., Laurent Ph., Abgrall M., Marion H., Maksimovic I., Cacciapuoti L., Grünert J., Vian C., Pereira dos Santos F., and Rosenbusch P. 2004. Advances in atomic fountains. *C. R. Phys.* **5**: 829.

Bize S., Laurent Ph., Rosenbusch P., Guena J., Rovera D., Abgrall M., Santarelli G., Lemonde P., Chapelet F., Wolf P., Mandache C., Luiten A., Tobar M., Salomon Ch., and Clairon A. 2009. Réalisation and diffusion of the second at LNE-SYRTE. *Revue Française de Métrologie* **n°18**.

Bize S., Sortais Y., Mandache C., Clairon A., and Salomon C. 2001. Cavity frequency pulling in cold atom fountains. *IEEE Trans. Instrum. Meas.* **50** (2): 503.

Blatt R. and Werth G. 1982. Precision determination of the ground-state hyperfine splitting in ^{137}Ba$^+$ using the ion-storage technique. *Phys. Rev. A* **25**: 1476.

Bollinger J.J., Heinzeb D.J., Itano W.M., Gilbert S.L., and Wineland D.J. 1991. A 303-MHz frequency standard based on trapped Be$^+$ ions. *IEEE Trans. Instrum. Meas.* **40**: 126.

Bollinger J.J., Itano W.M., and Wineland D.J. 1983a. Laser cooled ^{9}Be$^+$ accurate clock. In *Proceedings of the Annual Symposium on Frequency Control*, IEEE, 37.

Bollinger J.J., Prestage J.D., Itano W.M., and Wineland D.J. 1985. Laser-cooled atomic frequency standard. *Phys. Rev. Lett.* **54**: 1000.

Bollinger J.J., Wineland D.J., Itano W.M., and Wells J.S. 1983b. Precision measurements of laser cooled ^{9}Be$^+$ ion. *Laser Spectroscopy VI*, H.P. Weber and W. Luthy Eds. (Springer-Verlag, Berlin, Germany) TN 132.

Bouchiat M.A. 1965. *Etude par Pompage Optique de la Relaxation d'atomes de Rubidium* (Publications scientifiques et techniques du Ministère de l'air, Paris, France).

Boudot R., Liu X., Abbé P., Chutani R., Passily N., Galliou S., Gorecki C., and Giordado V. 2012. A high-performance frequency stability compact CPT clock based on a CXs-Ne microcell. *IEEE Trans. Ultrason. Ferroelectr. Freq. Control* **59**: 2584.

Busca G., Brousseau R., and Vanier J. 1975. Long-term frequency stability of the Rb87 maser. *IEEE Trans. Instrum. Meas.* **IM-24**: 291.

Camparo J.C. 1996. Reducing the light-shift in the diode laser pumped rubidium atomic clock. In *Proceedings of the IEEE International Frequency Control Symposium*, IEEE, 988.

Camparo J.C. 1998a. Atomic stabilization of electromagnetic field strength using Rabi resonances. *Phys. Rev. Lett.* **80**: 222.

Camparo J.C. 1998b. Conversion of laser phase noise to amplitude noise in an optically thick vapor. *J. Opt. Soc. Am. B* **15**: 1177.

Camparo J.C. 2000. Report No. TR-96(8555)-2 25430. The Aerospace Corporation, El Segundo, CA.

Camparo J.C. 2005. Does the light shift drive frequency aging in the rubidium atomic clock? *IEEE Trans. Ultrason. Ferroelectr. Freq. Control* **52**: 1075.

Camparo J.C. and Buell W.F. 1997. Laser PM to AM conversion in atomic vapors and short term clock stability. In *Proceedings of the IEEE International Frequency Control Symposium*, IEEE, 253.

Camparo J.C. and Coffer J.G. 1999. Conversion of laser phase noise to amplitude noise in a resonant atomic vapor: the role of laser linewidth. *Phys. Rev. A* **59**: 728.

Camparo J.C., Coffer J.G., and Townsend J.J. 2004. Reducing PM-to-AM conversion and the light-shift in laser-pumped, vapor-cell atomic clocks. In *Proceedings of the IEEE International Frequency Control Symposium, IEEE*, 134.

Camparo J.C., Coffer J., and Townsend J. 2005. Laser-pumped atomic clock exploiting pressure-broadened optical transitions. *J. Opt. Soc. Am. B* **22**: 521.

Camparo J.C. and Delcamp S.B. 1995. Optical-pumping with laser-induced-fluorescence. *Opt. Commun.* **120**: 257.

Camparo J.C. and Frueholz R.P. 1985. Linewidths of the 0–0 hyperfine transition in optically pumped alkali-metal vapors. *Phys. Rev. A* **31**: 144.

Camparo J.C. and Frueholz R.P. 1989. A three-dimensional model of the gas cell atomic frequency standard. *IEEE Trans. Ultrason. Ferroelectr. Freq. Control* **36**: 185.

Camparo J.C., Frueholz R.P., and Volk C.H. 1983. Inhomogeneous light shift in alkali-metal atoms. *Phys. Rev. A* **27**: 1914.

Camparo J.C., Klimcak C.M., and Herbulock S.J. 2005. Frequency equilibration in the vapor-cell atomic clock. *IEEE Trans. Instrum. Meas.* **IM-54**: 1873.

Castagna N., Guéna J., Plimmer M.D., and Thomann P. 2006. A novel simplified two-dimensional magneto-optical trap as an intense source of slow caesium atoms. *Eur. Phys. J. Appl. Phys.* **34**: 21.

CCTF. 2004. *Report of the 16th meeting to the International Committee for Weights and Measures.* Halford: Un. Laval Kramer. Digest of abstracts, not published. April 1–2, 2004.

Cérez P., Théobald G., Giordano V., Dimarcq N., and de Labachelerie M. 1991. Comparison of pumping a cesium beam tube with D1 and D2 lines. *IEEE Trans. Instrum. Meas.* **40**: 137.

Chantry P.J., Liberman I.I., Verbanets W.R., Petronio C.E., Cather R.F., and Partlow W.D. 1996. Miniature laser-pumped cesium cell atomic clock oscillator. In *Proceedings of the IEEE International Frequency Control Symposium*, IEEE, 1002.

Chantry P.J., McAvoy B.R., Zomp J.M., and Liberman I. 1992. Towards a miniature laser-pumped cesium cell frequency standard. In *Proceedings of the IEEE International Frequency Control Symposium* 114.

Chapelet F. 2008. *Fontaine Atomique Double de Cesium et de Rubidium avec une Exactitude de quelques 10^{-16} et Applications.* PhD dissertation, Department of Physics, Universite de Paris XI, France.

Clairon A., Ghezali S., Santarelli G., Laurent P., Simon E., Lea S., Bouhara M., Weyers S., and Szymaniec K. 1996. The LPTF preliminary accuracy evaluation of cesium fountain frequency standard. In *European Forum on Time and Frequency*, IEEE, 219.

Clairon A., Laurent Ph., Santarelli G., Ghezali S., Lea S.N., and Baoura M. 1995. A cesium fountain frequency standard: recent results. *IEEE Trans. Instrum. Meas.* **44**: 128.

Clairon A., Salomon Ch., Guelatti S., and Phillips W.D. 1991. Ramsey resonances in a Zacharias fountain. *Europhys. Lett.* **16**: 165.

Coffer J.G., Anderson M., and Camparo J.C. 2002. Collisional dephasing and the reduction of laser phase-noise to amplitude-noise conversion in a resonant atomic vapor. *Phys. Rev. A* **65**: 033807.

Coffer J.G. and Camparo J.C. 1998. Diode laser linewidth and phase noise to intensity noise conversion in the gas-cell atomic clock. In *Proceedings of the IEEE International Frequency Control Symposium*, IEEE, 52.

Coffer J.G. and Camparo J.C. 2000. Atomic stabilization of field intensity using Rabi resonances. *Phys. Rev. A* **62**: 013812.

Coffer J.G., Sickmiller B., and Camparo J.C. 2004. Cavity-Q aging observed via an atomic-candle signa. *IEEE Trans. Ultrason. Ferroelectr. Freq. Control* **51**: 139.

Cohen-Tannoudji C., Dupont-Roc J., and Grynberg G. 1988. *Processus d'nteraction Entre photons et Atomes* (InterEditions/Editions du CNRS, Paris, France).

Collin R.E. 1991. *Field Theory of Guided Waves* 2nd edition (IEEE, New York).

Cozijn F.M.J., Biesheuvel J., Flores A.S., Ubachs W., Blume G., Wicht A., Paschke K., Erbert G., and Koelemeij J.C.J. 2013. Laser cooling of beryllium ions using a frequency-doubled 626 nm diode laser. *Opt. Lett.* **38** (13): 2370.

Cyr N., Tetu M., and Breton M. 1993. All-optical microwave frequency standard: a proposal. *IEEE Trans. Instrum. Meas.* **IM-42**: 640.

Danet J.-M., Lours M., Guérandel S., and de Clercq E. 2014. Dick effect in a pulsed atomic clock using coherent population trapping. *IEEE Trans. Ultrason. Ferroelectr. Freq. Control* **61**: 567.

Davidovits P. and Novick R. 1966. The optically pumped rubidium maser. *Proc. IEEE* **54**: 155.

de Clercq E., Clairon A., Dahmani B., Gérard A., and Aynié P. 1989. Design of an optically pumped Cs laboratory frequency standard. In *Proceedings of the 4th Symposium on Frequency Standards and Metrology*, A. De Marchi Ed. (Springer-Verlag, Berlin, Germany) 120.

Dehmelt H.G. 1967. Radiofrequency spectroscopy of stored ions: I Storage. *Adv. At. Mol. Phys.* **3**: 53.

De Marchi A., Lo Presti L., and Rovera G.D. 1998. Square wave frequency modulation as a mean to reduce the effect of local oscillator instabilities in alkali vapor passive frequency standards. In *Proceedings of the IEEE International Frequency Control Symposium*, IEEE, 104.

De Marchi A., Shirley J., Glaze D.J., and Drullinger R. 1988. A new cavity configuration for cesium beam primary frequency standards. *IEEE Trans. Instrum. Meas.* **37**: 185.

Deng J. 2000. Light shift compensation in a Rb gas cell frequency standard with two-laser-pumping. In *Proceedings of the IEEE International Frequency Control Symposium*, IEEE, 659.

Deng J., Liu J., An S., Tan Y., and Zhu X. 1994. Light shift measurements in a diode laser-pumped ^{87}Rb maser. *IEEE Trans. Instrum. Meas.* **IM-43**: 549.

Deng J.Q., De Marchi A., Walls F., and Drullinger R.E. 1998. Frequency stability of cell-based passive frequency standards: reducing the effects of local-oscillator PM noise. In *Proceedings of the IEEE International Frequency Control Symposium*, IEEE, 95.

Deng J.Q., Mileti G., Jennings D.A., Drullinger R.E., and Walls F. 1997. Improving the short-term stability of laser pumped Rb clocks by reducing the effects of the interrogation oscillator. In *Proceedings of the IEEE International Frequency Control Symposium*, IEEE, 438.

Devenoges L., Bernier L.G., Morel J., Di Domenico G., Jallageas A., Petersen M., and Thomann P. 2013. Design and realization of a low phase gradient microwave cavity for a continuous atomic fountain clock. In *Proceedings of the Joint European Forum on Time and Frequency/IEEE International Frequency Control Symposium*, IEEE, 235.

Devenoges L., Stefanov A., Joyet A., Thomann P., and Di Domenico G. 2012. Improvement of the frequency stability below the dick limit with a continuous atomic fountain clock. *IEEE Trans. Ultrason. Ferroelectr. Freq. Control* **59**: 0885.

Dick G.J. 1987. Local oscillator induced instabilities in trapped ion frequency standards. In *Proceedings of the Annual Precise Time and Time Interval Applications and Planning Metting*, ION Publication, 133.

Dicke R.H. 1953. The effect of collisions upon the Doppler width of spectral lines. *Phys. Rev.* **89**: 472.

Di Domenico G., Castagna N., Mileti G., Thomann P., Taichenachev A.V., and Yudin V.I. 2004. Laser collimation of a continuous beam of cold atoms using Zeeman-shift degenerate-Raman-sideband cooling. *Phys. Rev. A* **69**: 063403.

Di Domenico G., Devenoges L., Dumas C., and Thomann P. 2010. Combined quantum-state preparation and laser cooling of a continuous beam of cold atoms. *Phys. Rev. A* **82**: 053417.

Di Domenico G., Devenoges L., Joyet A., Stefanov A., and Thomann P. 2011a. Uncertainty evaluation of the continuous caesium fountain frequency standard FOCS-2. In *Proceedings of the European Forum on Time and Frequency* 51.

Di Domenico G., Devenoges L., Stefanov A., Joyet A., and Thomann P. 2011b. Fourier analysis of Ramsey fringes observed in a continuous atomic fountain for in situ magnetometry. *Eur. Phys. J. Appl. Phys.* **56**: 11001.

Dimarcq N., Giordano V., Cérez P., and Théobald G. 1993. Analysis of the noise sources in an optically pumped cesium beam resonator. *IEEE Trans. Instrum. Meas.* **IM-42**: 11.

Dudle G., Joyet A., Berthoud P., Mileti G., and Thomann P. 2001. First results with a cold cesium continuous fountain resonator. *IEEE Trans. Instrum. Meas.* **50**: 510.

Dudle G., Mileti G., Jolivet A., Fretel E., Berthoud P., and Thomann P. 2000. An alternative cold caesium frequency standard: the continuous fountain. *IEEE Trans. Ultrason. Ferroelectr. Freq. Control* **47**: 438.

English T.C., Jechart E., and Kwon T.M. 1978. Elimination of the light shift in rubidium gas cell frequency standards using pulsed optical pumping. In *Proceedings of the Annual Precise Time and Time Interval Applications and Planning Meeting*, SAO/NASA ADS Physics, 147.

Ertmer W., Blatt R., Hall J.L., and Zhu M. 1985. Laser manipulation of atomic beam velocities: demonstration of stopped atoms and velocity reversal. *Phys. Rev. Lett.* **54**: 996.

Esnault F.X., Blanshan E., Ivanov E.N., Scholten R.E., Kitching J., and Donley E.A. 2013. Cold-atom double coherent population trapping clock. *Phys. Rev.* **88**: 042120.

Esnault F.X., Holleville D., Rossetto N., Guerandel S., and Dimarcq N. 2010. High-stability compact clock based on isotropic laser cooling. *Phys. Rev. A* **82**: 033436-1.

Esnault F.X., Rossetto N., Holleville D., Delporte J., and Dimarcq N. 2011. Horace a compact cold atomic clock for Galileo. *Adv. Space Res.* **47**: 854.

Fertig C., Rees J.I., and Gibble K. 2001. A juggling Rb fountain clock and a direct measurement of population differences. In *Proceedings of the IEEE International Frequency Control Symposium, IEEE*, 18.

Forman P. 1985. Atomichron®: The atomic clock from concept to commercial product. *Proc. IEEE* **73**: 1181.

Füzesi F., Jornod A., Thomann P., Plimmer M., Dudle G., Moser R., Sache L., and Bleuler H. 2007. An electrostatic glass actuator for ultra-high vacuum: a rotating light trap for continuous beams of laser-cooled atoms. *Rev. Sci. Instrum.* **78**: 102109.

Ghezali S., Laurent Ph., Lea S., and Clairon A. 1996. An experimental study of the spin exchange frequency shift in a laser cooled caesium fountain. *Europhys. Lett.* **36**: 25.

Gibble K. and Chu S. 1993. A laser cooled Cs frequency standard and a measurement of the frequency shift due to ultra-cold collisions. *Phys. Rev. Lett.* **70**: 1771.

Gibbs H.M. 1965. *Total Spin-Exchange Cross Sections For Alkali Atoms From Optical Pumping Experiments*. PhD thesis, Lawrence Berkeley National Laboratory, University of California, Berkeley, CA.

Godone A., Calonico D., Levi F., Micalizio S., and Calosso C. 2005. Stark-shift measurement of the $^2S_{1/2}$, F = 3 → F = 4 hyperfine transition of ^{133}Cs. *Phys. Rev. A* **71**: 063401.

Godone A., Levi F., and Micalizio S. 2002a. *Coherent Population Trapping Maser* (CLUT, Torino, Italy).

Godone A., Levi F., and Micalizio S. 2002b. Propagation and density effects in the coherent-population-trapping maser. *Phys. Rev. A* **65**: 033802.

Godone A., Levi F., Micalizio S., and Calosso C., 2004b. Coherent-population-trapping maser: noise spectrum and frequency stability. *Phys. Rev. A* **70**: 012508.

Godone A., Levi F., Micalizio S., and Vanier J. 2000. Theory of the coherent population trapping maser: a strong-field self-consistent approach. *Phys. Rev. A* **62**: 053402.

Godone A., Levi F., and Vanier J. 1999. Coherent microwave emission in cesium under coherent population trapping. *Phys. Rev. A* **59**: R12.

Godone A., Micalizio S., Calosso C., and Levi F. 2006a. The pulsed rubidium clock. *IEEE Trans. Ultrason. Ferroelectr. Freq. Control* **53**: 525.

Godone A., Micalizio S., and Levi F. 2004a. Pulsed optically pumped frequency standard. *Phys. Rev. A* **70**: 023409.

Godone A., Micalizio S., Levi F., and Calosso C. 2006b. Physics characterization and frequency stability of the pulsed rubidium maser. *Phys. Rev. A* **74**: 043401.

Greenhall C.A. 1998. A derivation of the long-term degradation of a pulsed atomic frequency standard from a control-loop model. *IEEE Trans. Ultrason. Ferroelectr. Freq. Control* **45**: 895.

Guena J., Abgrall M., Clairon A., and Bize S. 2014. Contributing to TAI with a secondary representation of the SI second. *Metrologia* **51**: 108.

Guena J., Abgrall M., Rovera D., Laurent Ph., Chupin B., Lours M., Santarelli G., Rosenbusch P., Tobar M.E., Li R., Gibble K., Clairon A., and Bize S. 2012. Progress in atomic fountains at LNE-SYRTE. *IEEE Trans. Ultrason. Ferroelectr. Freq. Control* **59** (3): 391.

Guéna J., Li R., Gibble K., Bize S., and Clairon A. 2011. Evaluation of Doppler shifts to improve the accuracy of primary atomic fountain clocks. *Phys. Rev. Lett.* **106** (3): 130801.

Guillemot C., Petit P., Forget S., Valentin C., and Dimarcq N. 1998. A simple configuration of clock using cold atoms. In *Proceedings of the European Forum on Time and Frequency* 55.

Hagimoto K., Koga Y., Ikegami T. 2008. Re-evaluation of the optically pumped cesium frequency standard NRLM-4 with an H-Bend ring. *Cavity IEEE Trans. Instrum. Meas.* **57** (10) 2212.

Halford D. 1971. *Proceedings of the Frequency Standards and Metrology Seminar*, 413 (unpublished), Laval University Quebec, Canada.

Happer W. 1972. Optical pumping. *Rev. Mod. Phys.* **44**: 169.

Hasegawa A., Fukuda K., Kajita M., Ito H., Kumagai M., Hosokawa M., Kotake N., and Morikawa T. 2004. Accuracy evaluation of optically pumped primary frequency standard CRL-O1. *Metrologia* **41**: 257.

Hashimoto H., Ohtsu M., and Furuta H. 1987. Ultra-sensitive frequency discrimination—in a diode laser pumped ^{87}Rb atomic clock. In *Proceedings of the Annual Symposium on Frequency Control*, IEEE, 25.

Hashimoto M. and Ohtsu M. 1987. Experiments on a semiconductor laser pumped rubidium atomic clock. *IEEE J. Quantum Electronics* **23**: 446.

Hashimoto M. and Ohtsu M. 1989. Modulation transfer and optical Stark effect in a rubidium atomic clock pumped by a semiconductor laser. *J. Opt. Soc. Am. B* **6**: 1777.

Hashimoto M. and Ohtsu M. 1990. A novel method to compensate for the effect of light shift in a rubidium atomic clock pumped by a semiconductor laser. *IEEE Trans. Instrum. Meas.* **IM-39**: 458.

Heavner T.P., Donley E.A., Levi F., Costanzo G., Parker T.E., Shirley J.H., Ashby N., Barlow S., and Jefferts S.R. 2014. First accuracy evaluation of NIST-F2. *Metrologia* **51**: 174.

Heavner T.P., Parker T.E., Shirley J.H., Donley L., Jefferts S.R., Levi F., Calonico D., Calosso C., Costanzo G., and Mongino B. 2011. Comparing room temperature and cryogenic cesium fountains. In *Proceedings of the Joint IEEE International Frequency Control Symposium/European Forum on Time and Frequency*, IEEE, 1.

Hemmer P.R., Ezekiel S., and Leiby C.C. Jr. 1983. Stabilization of a microwave oscillator using a resonance Raman transition in a sodium beam. *Opt. Lett.* **8**: 440.

Hemmer P.R., Ontai G.P., and Ezekiel S. 1986. Precision studies of stimulated resonance Raman interactions in an atomic beam. *J. Opt. Soc. Am. B* **3**: 219.

Hemmer P.R., Ontai G.P., Rosenberg A., and Ezekiel S. 1985. Performance of laser-induced resonance Raman clock. In *Proceedings of the Annual Symposium on Frequency Control*, IEEE, 88.

Hemmer P.R., Shahriar M.S., Lamela-Rivera H., Smith S.P., Bernacki B.E., and Ezekiel S. 1993. Semiconductor laser excitation of Ramsey fringes by using a Raman transition in a cesium atomic beam. *J. Opt. Soc. Am. B* **10**: 1326.

Hemmer P.R., Shahriar M.S., Natoli V.D., and Ezekiel S. 1989. AC Stark shifts in a two-zone Raman interaction. *J. Opt. Soc. Am. B* **6**: 1519.

Huang X., Xia B., Zhong D., An S., Zhu X., and Mei G. 2001. A microwave cavity with low temperature coefficient for passive rubidium frequency standards. In *Proceedings of the International Frequency Control Symposium*, IEEE, 105.

Itano W.M., Bergquist J.C., Bollinger J.J., Gilligan J.M., Heinzen D.J., Moore F.L., Raizen M.G., and Wineland D.J. 1993. Quantum projection noise: population fluctuations in two-level systems. *Phys. Rev. A* **47**: 3554.

Itano W.M., Lewis L., and Wineland D. 1982. Shift of 2S1/2 hyperfine splittings due to black-body radiation. *Phys. Rev. A* **25**: 1233.

Itano W.M. and Wineland D.J. 1982. Laser cooling of ions stored in harmonic and Penning traps. *Phys. Rev. A* **25**: 35.

Jau Y.Y., Miron E., Post A.B., Kuzma N.N., and Happer W. 2004. Push-pull optical pumping of pure superposition states. *Phys. Rev. Lett.* **93**: 160802.

Jau Y.Y., Post A.B., Kuzma N.N., Braun A.M., Romalis M.V., and Happer W. 2003. The physics of miniature atomic clocks: 0–0 versus "end" resonances. In *Proceedings of the Joint IEEE International Frequency Control Symposium/European Forum on Time and Frequency*, IEEE, 33.

Jau Y.Y., Post A.B., Kuzma N.N., Braun A.M., Romalis M.V., and Happer W. 2004. Intense, narrow atomic-clock resonances. *Phys. Rev. Lett.* **92**: 110801.

Jefferts S.R., Heavner T., Donley E., Shirley J., and Parker T.E. 2003. Second generation cesium fountain primary frequency standards at NIST. In *Proceedings of the Joint IEEE International Frequency Control Symposium/European Forum on Time and Frequency*, IEEE, 1084.

Jefferts S.R., Heavner T.P., Parker T.E., Shirley J.H., Donley E.A., Ashby N., Levi F., Calonico D., and Costanzo G.H. 2014. High-accuracy measurement of the blackbody radiation frequency shift of the ground-state hyperfine transition in ^{133}Cs. *Phys. Rev. Lett.* **112**: 050801.

Jefferts S.R., Shirley J., Ashby N., Heavner T., Donley E., and Levi F. 2005. On the power dependence of extraneous microwave fields in atomic frequency standards. In *Proceedings of the IEEE International Frequency Control Symposium*, IEEE, 6.

Joindot I. 1982. Measurements of relative intensity noise (RIN) in semiconductor lasers. *J. Phys. III France* **2**: 1591.

Joyet A. 2003. *Aspects Metrologiques d'une Fontaine Continue a Atoms Froids*. PhD Thesis, Universite de Neuchâtel, Switzerland.

Joyet A., Mileti G., Dudle G., and Thomann P. 2001. Theoretical study of the Dick effect in a continuously operated Ramsey resonator. *IEEE Trans. Instrum. Meas.* **50**: 150.

Jun J.W., Lee H.S., Kwon T.Y., and Minogin V.G. 2001. Light shift in an optically pumped caesium-beam frequency standard. *Metrologia* **38**: 221.

Kargapoltsev S.V., Kitching J., Hollberg L., Taichenachev A.V., Velichansky V.L., and Yudin V.I. 2004. High-contrast dark resonance in $\sigma_+ - \sigma_-$ optical field. *Laser Phys. Lett.* **1**: 495.

Kasevich M.A., Riis E., and Chu S. 1989. RF spectroscopy in an atomic fountain. *Phys. Rev. Lett.* **63**: 612.

Kim J.H. and Cho D. 2000. Stimulated Raman clock transition without a differential ac Stark shift. *J. Korean Phys. Soc.* **37**: 744.

Kitching J., Knappe S., Vukicecic N., Hollberg L., Wynands R., and Weidmann W. 2000. A microwave frequency reference based on VCSEL-driven dark line resonances in Cs vapor. *IEEE Trans. Instrum. Meas.* **49**: 1313.

Kitching J., Robinson H.D., and Hollberg L. 2001. Compact microwave frequency reference based on coherent population trapping. In *Proceedings of the 6th Symposium on Frequency Standards and Metrology*, P. Gill Ed. (World Scientific, London) 167.

Knappe S., Schwindt P., Shah V., Hollberg L., Kitching J., Liew L., and Moreland J. 2004. Microfabricated atomic frequency references. In *Proceedings of the IEEE International Ultrasonics, Ferroelectrics, and Frequency Control Joint 50th Anniversary Conference*, IEEE, 87.

Kokkelmans S.J.J.M.F., Verhaar B.J., Gibble K., and Heinzen D.J. 1997. Predictions for laser-cooled Rb clocks. *Phys. Rev. A* **56**: R4389.

Kol'chenko A.P., Rautian S.G., and Sokolovskii R.I. 1969. Interaction of an atom with a strong electromagnetic field with the recoil effect taken into consideration. *Sov. Phys. JETP* **28** (5): 986.

Kozlova O., Danet J.-M., Guérandel S., and De Clercq E. 2014. Limitations of long-term stability in a coherent population trapping Cs clock. *IEEE Trans. Instrum. Meas.* **63**: 1863.

Kramer G. 1974. Noise in passive frequency standards. *IEEE Conference on Precision Electromagnetic Measurements* **113**: 157.

Larson D.J., Bergquist J.C., Bollinger J.J., Itano W.M., and Wineland D.J. 1986. Sympathetic cooling of trapped ions: a laser-cooled two-species nonneutral ion plasma. *Phys. Rev. Lett.* **57**: 70.

Laurent Ph., Abgrall M., Jentsch Ch., Lemonde P., Santarelli G., Clairon A., Maksimovic J., Bize S., Salomon Ch., Blonde D., Vega J.F., Grosjean O., Picard F., Saccoccio M., Chaubet M., Ladiette N., Guillet L., Zenone I., Delaroche Ch., and Sirmain Ch. 2006. Design of the cold atom PHARAO space clock and initial test results. *Appl. Phys. B* **84**: 683.

Lee H.S., Kwon T.Y., Park S.E., and Choi S.K. 2004. Research on cesium atomic clocks at the Korea Research Institute of Standards and Science. *J. Korean Phys. Soc.* **45**: 256.

Lee H.S., Park S.E., Kwon T.Y., Yang S.H., and Cho H. 2001. Toward a cesium frequency standard based on a continuous slow atomic beam: preliminary results. *IEEE Trans. Instrum. Meas.* **50**: 531.

Lee T., Das T.P., and Sternheimer R.M. 1975. Perturbation theory for the Stark effect in the hyperfine structure of alkali-metal atoms. *Phys. Rev. A* **11** (6) 1784.

Legere R., and Gibble K. 1998. Quantum scattering in a juggling atomic fountain. *Phys. Rev. Lett.* **81**: 5780.

Lemonde P., Santarelli G., Laurent Ph., Pereira Dos Santos F., Clairon A., and Salomon C. 1998. The sensitivity function: a new tool for the evaluation of frequency shifts in atomic spectroscopy. In *Proceedings of the IEEE International Frequency Control Symposium*, IEEE, 110.

Leo P.J., Julienne P.S., Mies F.H., and Williams C.J. 2000. Collisional frequency shifts in ^{133}Cs fountain clocks. *Phys.Rev. Lett.* **86**: 3743.

Levi F. 1995. *Campione Atomicco di Frecquenza a Vapori di Rb con Pompagio Ottico Mediante Laser a Semiconduttore*. Thesis, Politecnico di Torino, Turin, Italy.

Levi F., Calonico D., Calosso C.E., Godone A., Micalizio S., Costanzo G.A., Mongino B., Jefferts S.R., Heavner T.P., and Donley E.A. 2009. The cryogenic fountain ITCsF2. In *Proceedings of the Joint Meeting of the European Forum on Time and Frequency/IEEE International Frequency Control Symposium*, IEEE, 769.

Levi F., Calosso C., Calonico D., Lorini L., Bertacco E.K., Godone A., and Costanzo G.A. 2014. Accuracy evaluation of ITCsF2; a nitrogen cooled caesium fountain. *Metrologia* 51: 270.

Levi F., Godone A., and Lorini L. 2001. Reduction of the cold collisions frequency shift in a multiple velocity fountain: a new proposal. *IEEE Trans. Ultrason. Ferroelectr. Freq. Control* **48**: 847.

Levi F., Godone A., Novero C., and Vanier J. 1997. On the use of a modulated laser on hyperfine frequency excitation in passive atomic frequency standards. In *Proceedings of the European Forum on Time and Frequency* 216.

Levi F., Godone A., and Vanier J. 2000. The light shift effect in the coherent population trapping cesium maser. *IEEE Trans. Instrum. Meas.* **47**: 466.

Levi F., Godone A., Vanier J., Micalizio S., and Modugno G. 2000. Line-shape of dark line and maser emission profile in CPT. *Eur. Phys. J. D* **12**: 53.

Levi F., Lorini L., Calonico D., and Godone A. 2004. IEN-CsF1 accuracy evaluation and two-way frequency comparison. *IEEE Trans. Ultrason. Ferroelectr. Freq. Control* **51**: 216.

Levi F., Novero C., Godone A., and Brida G. 1997. Analysis of the light shift effect in the ^{87}Rb frequency standard. *IEEE Trans. Instrum. Meas.* **46**: 126.

Lewis L.L. and Feldman M. 1981. Optical pumping by lasers in atomic frequency standards. In *Proceedings of the Annual Symposium on Frequency Control*, IEEE, 612.

Li R. and Gibble K. 2004. Phase variations in microwave cavities for atomic clocks. *Metrologia* **41**: 376.

Li R. and Gibble K. 2010. Evaluating and minimizing distributed cavity phase errors in atomic clocks. *Metrologia* **47** (5): 534.

Li R., Gibble K., and Szymaniec K. 2011. Improved accuracy of the NPL-CsF2 primary frequency standard: evaluation of distributed cavity phase and microwave lensing frequency shifts. *Metrologia* **48**: 283.

Lindvall T., Merimaa M., Tittonen I., and Ikonen E. 2001. All-optical atomic clock based on dark states of Rb. In *Proceedings of the 6th Symposium on Frequency Standards and Metrology*, P. Gill Ed. (World Scientific, London) 183.

Liu X., Mérolla J.-M., Guérandel S., de Clercq E., and Boudot R. 2013b. Ramsey spectroscopy of high-contrast CPT resonances with push-pull optical pumping in Cs vapor. *Opt. Express* **21**: 12451.

Liu X., Mérolla J.-M., Guérandel S., Gorecki C., de Clercq E., and Boudot R. 2013a. Coherent-population-trapping resonances in buffer-gas-filled Cs-vapor cells with push-pull optical pumping. *Phys. Rev. A* **87**: 013416.

Luiten A., Mann A.G., and Blair D.G. 1994. Cryogenic sapphire microwave resonator-oscillator with exceptional stability. *Electron. Lett.* **30**: 417.

Lutwak R., Raseed M., Varghese M., Tepolt G., LeBlanc J., Mescher M., Serkland D.K., Geib K.M., and Peake G.M. 2009. The chip scale atomic clock. In *Proceedings of the 7th Symposium Frequency Standards and Metrology*, ed. L. Maleki, 454, World Scientific, Singapore.

Makdissi A., Berthet J.P., and de Clercq E. 1997. Improvement of the short term stability of the LPTF cesium beam frequency standard. In *Proceedings of the European Forum on Time and Frequency* 564.

Makdissi A., Berthet J.P., and de Clercq E. 2000. Phase and light shift determination in an optically pumped cesium beam frequency standard. *IEEE Trans. Ultrason. Ferroelectr. Freq. Control.* **47**: 461.

Makdissi A. and de Clercq E. 2001. Evaluation of the accuracy of the optically pumped caesium beam primary frequency standard of BNM-LPTF. *Metrologia* **38**: 409.

Mandache C. 2006. *Rapport de stage* (Mairie de Paris, Paris, France) (unpublished).

Marion H. 2005. *Contrôle des Collisions Froides du Césium, Tests de la Variation de la Constante de Structure Fine à l'aide d'une Fontaine Atomique Double Rubidium-Césium*. Thèse, de l'Université Paris VI, SYRTE.

Mathur B.S., Tang H., and Happer W. 1968. Light shifts in the alkali atoms. *Phys. Rev.* **171**: 11.

Matsuda J., Yamaguchi S., and Suzuki M. 1990. Measurement of the characteristics of the optical-microwave double-resonance effect of the ^{87}Rb D$_1$line for application as an atomic frequency standard. *IEEE J. Quantum Electron.* **26**: 9.

McGuyer B.H., Jau Y.Y., and Happer W. 2009. Simple method of light-shift suppression in optical pumping systems. *Appl. Phys. Lett.* **94**: 251110.

Meekhof D.M., Jefferts S.R., Stepanovíc M., and Parker T.E. 2001. Accuracy evaluation of a cesium fountain primary frequency standard at NIST. *IEEE Trans. Instrum. Meas.* **50** (2): 507.

Merimaa M., Lindvall T., Tittonen I., and Ikonen E. 2003. All-optical atomic clock based on coherent population trapping in ^{85}Rb. *J. Opt. Soc. Am. B* **20**: 273.

Micalizio S., Calosso C.E., Godone A., and Levi F. 2012b. Metrological characterization of the pulsed Rb clock with optical detection. *Metrologia* **49**: 1.

Micalizio S., Calosso C.E., Godone A., and Levi F. 2012c. Cell-related effects in the pulsed optically pumped frequency standard. In *Proceedings of the European Forum on Time and Frequency* 74.

Micalizio S., Godone A., Calonico D., Levi F., and Lorini L. 2004. Black-body radiation shift of the 133Cs hyperfine transition frequency. *Phys. Rev. A* **69** (5): 053401.

Micalizio S., Godone A., Calosso C.E., Levi F., Affolderbach C., and Gruet F. 2012a. Pulsed optically pumped Rubidium clock with high frequency–stability performance. *IEEE Trans. Ultrason. Ferroelectr. Freq. Control* **59**: 457.

Micalizio S., Godone A., Levi F., and Calosso C.E. 2009. Pulsed optically pumped ^{87}Rb vapor cell frequency standard: a multilevel approach. *Phys. Rev. A* **79**: 013403.

Micalizio S., Godone A., Levi F., and Vanier J. 2006. Spin-exchange frequency shift in alkali-metal-vapor cell frequency standards. *Phys. Rev. A* **73**: 033414.

Michaud A., Tremblay P., and Têtu M. 1990. Experimental study of the laser diode pumped rubidium maser. In *IEEE CPEM Digest on Precision Electromagnetic Measurements*, IEEE, 155.

Michaud A., Tremblay P., and Têtu M. 1991. Experimental study of the laser diode pumped rubidium maser. *IEEE Trans. Instrum. Meas.* **40**: 170.

Mileti G. 1995. *Etude du Pompage Optique par Laser et par Lampe Spectrale dans les Horloges a Vapeur de Rubidium.* Thesis, Université de Neuchâtel, Switzerland.

Mileti G., Deng J.Q., Walls F., Jennings D.A., and Drullinger R.E. 1998. Laser-pumped rubidium requency standards: new analysis and progress. *IEEE J. Quantum Electron.* **34**: 233.

Mileti G., Deng J.Q., Walls F., Low J.P., and Drullinger R.E. 1996. Recent progress in laser-pumped rubidium gas-cell frequency standards. In *Proceedings of the IEEE International Frequency Control Symposium*, IEEE, 1066.

Mileti G., Rüedi I., and Schweda M. 1992. Line Inhomogeneity effects and power shift in miniaturized Rubidium frequency standard. In *Proceedings of the European Forum on Time and Frequency*, IEEE, 515.

Mileti G. and Thomann P. 1995. Study of the S/N performance of passive atomic clocks using a laser pumped vapour. In *Proceedings of the European Forum on Time and Frequency*, Neuchâtel, Switzerland, 271.

Millo J., Abgrall M., Lours M., English E.M.L., Jiang H., Guéna J., Clairon A., Tobar M.E., Bize S., Le Coq Y., and Santarelli G. 2009. Ultralow noise microwave generation with fiber-based optical frequency comb and application to atomic fountain clock. *Appl. Phys. Lett.* **94**: 141105.

Missout G. and Vanier J. 1975. Pressure and temperature coefficients of the more commonly used buffer gases in Rubidium vapor frequency standards. *IEEE Trans. Instrum. Meas.* **IM-24**: 180.

Mowat J.R. 1972. Stark effect in alkali-metal ground-state hyperfine structure. *Phys. Rev. A* **5**: 1059.

Müller S.T. 2010. *Padrão de Frequencia Compacto*. Thesis, Instituto de Fisico São Carlos, Brazil.

Müller S.T., Magalhães D.V., Alves R.F., and Bagnato V.S. 2011. Compact frequency standard based on an intra-cavity sample cold cesium stoms. *J. Opt. Soc. Am. B* **28** (11): 2592.

Mungall A.G. 1983. Comment on Observation of Ramsey fringes using a stimulated, resonance Raman transition in sodium atomic beam. *Phys. Rev. Letters* **50**: 548.

Ohtsu M., Hashimoto M., and Hidetaka O. 1985. A highly stabilized semiconductor laser and its application to optically pumped Rb atomic clock. In *Proceedings of the Annual Symposium on Frequency Control*, IEEE, 43.

Ohuchi Y., Suga H., Fujita M., Suzuki T., Uchino M., Takahei K., Tsuda M., and Saburi Y. 2000. A high-stability laser-pumped Cs gas-cell frequency standard. In *Proceedings of the IEEE International Frequency Control Symposium*, IEEE, 651.

Orriols G. 1979. Nonabsorption resonances by nonlinear coherent effects in a three-level system. *Il Nuovo Cimento B* **53**: 1.

Ortolano F., Beverini N., and De Marchi A. 2000. A dynamic analysis of the LO noise transfer mechanism in a Rb-cell frequency standard. *IEEE Trans. Ultrason. Ferroelectr. Freq. Control* **47**: 471.

Park S.J., Manson P.J., Wouters M.J., Warrington R.B., Lawn M.A., and Fisk P.T.H. 2007. ^{171}Yb$^+$ microwave frequency standard. In *Proceedings of the European Forum on Time and Frequency* 613.

Pereira dos Santos F., Marion H., Abgrall M., Zhang S., Sortais Y., Bize S., Maksimovic L., Calonico D., Grunert J., Mandache C., Vian C., Rosenbuch P., Lemonde P., Santarelli G., Laurent P., Clairon A., and Salomon C. 2003. ^{87}Rb and ^{133}Cs laser cooled clocks: testing the stability of fundamental constants. In *Proceedings of the Joint Meeting of the European Forum on Time and Frequency/IEEE International Frequency Control Symposium*, IEEE, 55.

Phillips W.D. and Metcalf H.J. 1982. Laser deceleration of an atomic beam. *Phys. Rev. Lett.* **48**: 596.

Picqué J.L. 1974. Hyperfine optical pumping of cesium with a CW GaAs laser. *IEEE J. Quant. Electron.* **10** (12): 892.

Prestage J.D., Chung S., Le T., Lim L., and Maleki L. 2005. Liter sized ion clock with 10^{-15} stability. In *Proceedings of the Joint IEEE International Frequency Symposium and Precise Time and Time Interval Systems and Applications Meeting*, IEEE, 472.

QPAFS, Vanier J., and Audoin C. 1989. *The Quantum Physics of Atomic Frequency Standards*, Vols. 1 and 2 (Adam Hilger, Bristol).

Ramsey N.F. 1956. *Molecular beams* (Oxford at the Clarendon Press, Oxford).

Risley A. and Busca G. 1978. Effect of line inhomogeneity on the frequency of passive Rb^{87} frequency standards. In *Proceedings of the Annual Symposium on Frequency Control*, IEEE, 506.

Risley A., Jarvis S., and Vanier J. 1980. The dependence of frequency upon microwave power of wall-coated and buffer–gas-filled gas cell Rb^{87} frequency standards. *J. Appl. Phys.* **51**: 4571.

Robinson H. and Johnson C. 1982. Narrow ^{87}Rb hyperfine-structure resonances in an evacuated wall-coated cell. *Appl. Phys. Lett.* **40**: 771.

Rosenbusch P., Zhang S., and Clairon A. 2007. Blackbody radiation shift in primary frequency standards. In *Proceedings of the Joint IEEE International Frequency Control Symposium/ European Forum on Time and Frequency*, IEEE, 1060.

Ruffieux R., Berthoud P., Haldimann M., Lecomte S., Hermann V., Gazard M., Barilet R., Guérandel E., De Clercq E., and Audoin C. 2009. Optically pumped space cesium clock for galileo: results of the breadboard. In *Proceedings of the 7th Symphosium Frequency Standands and Metrology*, Ed. L. Maleki (World Scientific: Singapore) 184.

Saburi Y., Koga Y., Kinugawa S., Imamura T., Suga H., and Ohuchi Y. 1994. Short-term stability of laser-pumped rubidium gas cell frequency standard. *IEEE Electron. Lett.* **30**: 633.

Sagna N., Mandache C., and Thomann P. 1992. Noise measurement in single-mode GaAl As diode lasers. In *Proceedings of the European Forum on Time and Frequency* 521.

Santarelli G., Audoin C., Makdissi A., Laurent P., Dick G.J., and Clairon C. 1998. Frequency stability degradation of an oscillator slaved to a periodically interrogated atomic resonator. *IEEE Trans. Ultrason. Ferroelectr. Freq. Control* **45**: 887.

Santarelli G., Governatori G., Chambon D., Lours M., Rosenbusch P., Guena J., Chapelet F., Bize S., Tobar M., Laurent Ph., Potier T., and Clairon A. 2009. Switching atomic fountain clock microwave interrogation signal and high-resolution phase measurements. *IEEE Trans. Ultrason. Ferroelectr. Freq. Control* **56** (7): 1319.

Santarelli G., Laurent Ph., Lemonde P., Clairon A., Mann A.G., Chang S., Luiten A.N., and Salomon C. 1999. Quantum projection noise in an atomic fountain: A high stability cesium frequency standard. *Phys. Rev. Lett.* **82**: 4619.

Sarosy E.B., Johnson W.A., Karuza S.K., and Voit F.J. 1992. Measuring frequency changes due to microwave power variations as a function of C-field setting in a rubidium frequency standard. In *Proceedings of the Annual Precise Time and Time Interval Applications and Planning Meeting*, ION Publication, 229.

Schwindt P.D.D., Olsson R., Wojciechowski K., Serkland D., Statom T., Partner H., Biedermann G., Fang L., Casias A., and Manginell R. 2009. Micro ion frequency standard. In *Proceedings of the Annual Precise Time and Time Interval Applications and Planning Meeting*, ION Publication, 509.

Shah V., Gerginov V., Schwindt P.D.D., Knappe S., Hollberg L., and Kitching J. 2006. Continuous light-shift correction in modulated coherent population trapping clocks. *Appl. Phys. Lett.* **89**: 151124.

Shahriar M.S. and Hemmer P.R. 1990. Direct excitation of microwave–spin dressed state using a laser-excited resonance Raman interaction. *Phys. Rev. Lett.* **65**: 1865.

Shahriar M.S., Hemmer P.R., Kalz D.P., Lee A., and Prentiss M.G. 1997. Dark-state-based three-element vector model for the stimulated Raman interaction. *Phys. Rev. A* **55**: 2272.

Shirley J.H., Lee W.D., and Drullinger R.E. 2001. Accuracy evaluation of the primary frequency standard NIST-7. *Metrologia* **38**: 427.

Simon E., Laurent Ph., and Clairon A. 1998. Measurement of the Stark shift of the Cs hyperfine splitting in an atomic fountain. *Phys. Rev. A* **57**: 436.

Singh G., DiLavore P., and Alley C.D. 1971. GaAs-laser-induced population inversion in the ground-state hyperfine levels of Cs^{133}. *IEEE J. Quantum Elect.* **QE7**: 196.

Sortais Y. 2001. *Construction d'une Fontaine Double à Atomes Froids de 87Rb et 133Cs; étude des Effets Dépendant du nombre d'atomes dans une Fontaine*. PhD Thesis, Université Paris VI, France.

Steane A. 1997. The ion trap quantum information processor. *Appl. Phys B* **64**: 623.

Szekely C., Walls F., Lowe J.P., Drullinger R.E., and Novick A. 1994. Reducing local oscillator phase noise limitations on the frequency stability of passive frequency standards: tests of a new concept. *IEEE Trans. Ultrason. Ferroelectr. Freq. Control* **41**: 518.

Szymaniec K., Chałupczak W., Tiesinga E., Williams C.J., Weyers S., and Wynands R. 2007. Cancellation of the collisional frequency shift in caesium fountain clocks. *Phys. Rev. Lett.* **98**: 153002.

Szymaniec K., Chałupczak W., Whibberley P.B., Lea S.N., and Henderson D. 2005. Evaluation of the primary frequency standard NPL-CsF1. *Metrologia* **42**: 49.

Szymaniec K., Lea S.N., and Liu K. 2014. An evaluation of the frequency shift caused by collisions with background gas in the primary frequency standard NPL-CsF2. *IEEE Trans. Ultrason. Feroelectr. Freq. Control* **61** (1): 203.

Taghavi-Larigani S., Burt E.A., Lea S.N., Prestage J.D., and Tjoelker R.L. 2009. A new trapped ion clock based on $^{201}Hg^+$. In *Proceedings of the IEEE International Frequency Control Symposium*, IEEE, 774.

Taichenachev A.V., Tumaikin A.M., Yudin V.I., Stähler M., Wynands R., Kitching J., and Hollberg L. 2004. Nonlinear-resonance line shapes: dependence on the transverse intensity distribution of a light beam. *Phys. Rev. A* **69**: 024501.

Taichenachev A.V., Yudin V.I., Velichansky V.L., Kargapoltsev S.V., Wynands R., Kitching J., and Hollberg L. 2004. High-contrast dark resonances on the D_1 line of alkali metals in the field of counterpropagating waves. *JEPT Lett.* **80**: 236.

Têtu M., Busca G., and Vanier J. 1973. Short-term frequency stability of the Rb87 maser. *IEEE Trans. Instrum. Meas.* **22**: 250.

Thomas J.E., Hemmer P.R., Ezekiel S., Leiby J.C.C., Picard H., and Willis R. 1982. Observation of Ramsey fringes using a stimulated, resonance Raman transition in a sodium atomic beam. *Phys. Rev. Lett.* **48**: 867.

Tiesinga E., Verhaar B.J., Stoof H.T.C., and van Bragt D. 1992. Spin-exchange frequency shift in a caesium atomic fountain. *Phys. Rev. A* **45**: R2671.

Tsuchida H. and Tako T. 1983. Relation between frequency and intensity stabilities in AlGaAs semiconductor lasers. *Jpn. J. Appl. Phys.* **22**: 1152.

Vanier J. 1968. Relaxation in rubidium-87 and the rubidium maser. *Phys. Rev.* **168**: 129.

Vanier J. 1969. Optical pumping as a relaxation process. *Can. J. Phys.* **47**: 1461.

Vanier J. 2002. Coherent population trapping for the realization of a small, stable, atomic clock. In *Proceedings of the IEEE International Frequency Control Symposium*, IEEE, 424.

Vanier J. 2005. Atomic clocks based on coherent population trapping: a review. *Appl. Phys. B* **81**: 421.

Vanier J. and Audoin C. 1989. *The Quantum Physics of Atomic Frequency Standards* (Adam Hilger, Bristol).

Vanier J. and Audoin C. 2005. The classical caesium beam frequency standard: fifty years later. *Metrologia* **42** (3): S31.

Vanier J. and Bernier L.G. 1981. On the signal-to-noise ratio and short-term stability of passive rubidium frequency standards. *IEEE Trans. Instrum. Meas.* **IM-30**: 277.

Vanier J., Godone A., and Levi F. 1998. Coherent population trapping in cesium: dark lines and coherent microwave emission. *Phys. Rev. A* **58**: 234.

Vanier J., Godone A., and Levi F. 1999. Coherent microwave emission in coherent population trapping: origin of the energy and of the quadratic light shift. In *Proceedings of the Joint Meeting of the European Forum on Time and Frequency/IEEE International Frequency Control Symposium*, IEEE, 96.

Vanier J., Kunski R., Brisson A., and Paulin P. 1981. Progress and prospects in rubidium frequency standards. *J. Phys. Coll.* **42** (C8), Suppl. 12: 139.

Vanier J., Kunski R., Cyr N., Savard J.Y., and Têtu M. 1982a. On hyperfine frequency shifts caused by buffer gases: application to the optically pumped passive rubidium frequency standard. *J. Appl. Phys.* **53**: 5387.

Vanier J., Kunski R., Paulin P., Têtu M., and Cyr N. 1982b. On the light shift in optical pumping of rubidium 87: the techniques of "separated" and "integrated" hyperfine filtering. *Can. J. Phys.* **60**: 1396.

Vanier J., Levine M., Janssen D., and Delaney M. 2001. Coherent population trapping and intensity optical pumping: on their use in atomic frequency standards. In *Proceedings of the 6th Symposium on Frequency Standards and Metrology*, P. Gil Ed. (World Scientific, London) 155.

Vanier J., Levine M., Janssen D., and Delaney M. 2003a. Contrast and linewidth of the coherent population trapping transmission hyperfine resonance line in ⁸⁷Rb: effect of optical pumping. *Phys Rev. A* **67**: 065801.

Vanier J., Levine M., Janssen D., and Delaney M. 2003b. On the use of intensity optical pumping and coherent population trapping techniques in the implementation of atomic frequency standards. *IEEE Trans. Instrum. Meas.* **52**: 822.

Vanier J., Levine M., Kendig S., Janssen D., Everson C., and Delaney M. 2004. Practical real-ization of a passive coherent population trapping frequency standard. In *Proceedings of the IEEE International Ultrasonics, Ferroelectrics, and Frequency Control Joint 50th Anniversary Conference*, IEEE, 92.

Vanier J., Levine M., Kendig S., Janssen D., Everson C., and Delaney M. 2005. Practical realization of a passive coherent population trapping frequency standard. *IEEE Trans. Instrum. Meas.* **54** (6): 2531.

Vanier J., Simard J.F., and Boulanger J.S. 1974. Relaxation and frequency shifts in the ground state of Rb^{85}. *Phys. Rev. A* **9**: 1031.

Vanier J. and Strumia F. 1976. Theory of the optically pumped cesium maser. *Can. J. Phys.* **54**: 2355.

Vanier J., Têtu M., and Bernier L.G. 1979. Transfer of frequency stability from an atomic frequency reference to a quartz-crystal oscillator. *IEEE Trans. Instrum. Meas.* **28**: 188.

Vian C., Rosenbusch P., Marion H., Bize S., Cacciapuoti L., Zhang S., Abgrall M., Chambon D., Maksimovic I., Laurent Ph., Santarelli G., Clairon A., Luiten A., Tobar M., and Salomon C. 2005. BNM-SYRTE fountains: recent results. *IEEE Trans. Instrum. Meas.* **54** (2): 833.

Warrington R.B., Fisk P.T.H., Wouters M.J., and Lawn M.A. 2002. A microwave frequency standard based on laser-cooled $^{171}Yb^+$ ions. In *Proceedings of the 6th Symposium Frequency Standards and Metrology*, P. Gil Ed. (World Scientific, London) 297.

Weyers R.S.S. and Wynands R. 2006. Effects of microwave leakage in caesium clocks: theo-retical and experimental results. In *Proceedings of the European Forum on Time and Frequency* 173.

Weyers S., Gerginov V., Nemitz N., Li R., and Gibble K. 2012. Distributed cavity phase fre-quency shifts of the caesium fountain PTB-CSF2. *Metrologia* **49**: 82.

Weyers S., Lipphardt B., and Schnatz H. 2009. Reaching the quantum limit in a fountain clock using a microwave oscillator phase locked to an ultra stable laser. *Phys. Rev. A* **79**: 031803(R).

Wineland D.J. and Itano W.M. 1981. Spectroscopy of a single Mg^+ ion. *Phys. Lett. A* **82**: 75.

Wolf P., Bize S., Clairon A., Landragin A., Laurent Ph., Lemonde P., and Borde Ch.J. 2001. Recoil effects in microwave atomic frequency standards: an update. In *Proceedings of the 6th Symposium on Frequency Standards and Metrology*, P. Gil Ed. (World Scientific, London) 593.

Wynands R. and Weyers S. 2005. Atomic fountain clocks. *Metrologia* **42**: S64.

Xiao Y., Novikova I., Phillips D.F., and Walsworth R.L. 2006. Diffusion-induced Ramsey nar-rowing. *Phys. Rev. Lett.* **96**: 043601.

Yamagushi S., Matsuda I., and Suzuki M. 1992. Dependence of the frequency shift of the optical-microwave double resonance signal in the Cs D_2 line on the pumping frequency and power of a GaAs semiconductor laser. *IEEE J. Quantum Electron.* **28**: 2551.

Zanon T., Guerandel S., de Clercq E., Dimarcq N., and Clairon A. 2003. Coherent population trapping with cold atoms. In *Proceedings of the Joint IEEE International Frequency Control Symposium/European Forum on Time and Frequency*, IEEE, 49.

Zanon T., Guerandel S., de Clercq E., Holleville D., Dimarcq N., and Clairon A. 2005a. High contrast Ramsey fringes with coherent-population-trapping pulses in a double lambdaa-tomic system. *Phys. Rev. Lett.* **94**: 193002–1.

Zanon T., Tremine S., Guerandel S., de Clercq E., Holleville D., Dimarcq N., and Clairon A. 2005b. Observation of Raman-Ramsey fringes with optical CPT pulses. *IEEE Trans. Instrum. Meas.* **54**: 776.

Zhang J.W., Wang S.G., Miao K., Wang Z.B., and Wang L.J. 2014. Toward a transportable micro-wave frequency standard based on laser cooled $^{113}Cd^+$ ions. *Appl. Phys. B* **114** (1–2): 183.

Zhang J.W., Wang Z.B., Wang S.G., Miao K., Wang B., and Wang L.J. 2012. High-resolution laser microwave double-resonance spectroscopy of hyperfine splitting of trapped $^{113}Cd^+$ and $^{111}Cd^+$ ions. *Phys. Rev. A* **86** (2): 022523.

Zhang S. 2004. *Déplacement de Fréquence du au Rayonnement du corps noir dans une Fontaine Atomique a Césium et Amélioration des Performances de l'horloge*, PhD dissertation, Department of Physics, Universite Pierre et Marie Curie, Paris VI, France.

Zhu, M. and Cutler L. 2000. Theoretical and experimental study of light shift in a CPT-based Rb vapor cell frequency standard. In *Proceedings of the Annual Precise Time and Time Interval Systems and Applications Meeting*, ION Publication, 311.

Zhu M., Cutler L.S., Berberian J.E., DeNatale J.F., Stupar P.A., and Tsai C. 2004. Narrow linewidth CPT signal in small vapor cells for chip scale atomic clocks. In *Proceedings of the IEEE International Ultrasonics, Ferroelectrics, and Frequency Control Joint 50th Anniversary Conference*, IEEE, 100.

CHAPTER 4

Akatsuka T., Takamoto M., and Katori H. 2010. Three-dimensional optical lattice clock with bosonic ^{88}Sr atoms. *Phys. Rev. A* **81**: 023402.

Angel J.R.P. and Sandars P.G.H. 1968. The hyperfine structure Stark effect. I. Theory. *Proc. Royal Soc. A* **305**: 125.

Arbes F., Gudjons T., Kurth F., Werth G., Marin F., and Inguscion M. 1993. Lifetime measurements of the ^{3}D$_{3/2}$ and ^{3}D$_{5/2}$ metastable states in Ca II. *Z. Phys. D: Atoms Mol. Clusters* **25**: 295.

Arora B., Safronova M.S., and Clark Ch.W. 2007. Blackbody radiation shift in a ^{43}Ca$^+$ ion optical frequency standard. *Phys. Rev. A* **76**: 064501.

Baillard X., Fouch M., Le Targat R., Westergaard P.G., Lecallier A., Le Coq Y., Rovera G.D., Bize S., and Lemonde P. 2007. Accuracy evaluation of an optical lattice clock with bosonic atoms. *Opt. Lett.* **32**: 1812.

Barber Z.W., Hoyt C.W., Oates C.W., Hollberg L., Taichenachev A.V., and Yudin V.I. 2006. Direct excitation of the forbidden clock transition in neutral ^{174}Yb atoms confined to an optical lattice. *Phys. Rev. Lett.* **96**: 083002.

Barber Z.W., Stalnaker J.E., Lemke N.D., Poli N., Oates C.W., Fortier T.M., Diddams S.A., Hollberg L., and Hoyt C.W. 2008. Optical lattice induced light shifts in an Yb atomic clock. *Phys. Rev. Lett.* **100**: 103002.

Barwood G.P., Gao K., Gill P., Huang G., and Klein H.A. 2001. Development of optical frequency standards based upon the ^{2}S$_{1/2}$ – ^{2}D$_{5/2}$ Transition in ^{88}Sr$^+$ and ^{87}Sr$^+$. *IEEE Trans. Instrum. Meas.* **50**: 543.

Barwood G.P., Gao K., Gill P., Huang G., and Klein H.A. 2003. Observation of the hyperfine structure of the ^{2}S$_{1/2}$ – ^{2}D$_{5/2}$ transition in ^{87}Sr$^+$. *Phys. Rev. A* **67**: 013402.

Barwood G.P., Gill P., Huang G., and Klein H.A. 2012. Characterization of a ^{88}Sr$^+$ optical frequency standard at 445 THz by two-trap comparison. In *Proceedings of the Conference Precision Electromagnetic Measurements*, IEEE, 270.

Bergquist J.C., Itano W.M., and Wineland D.J. 1987. Recoilless optical absorption and Doppler sidebands of a single trapped ion. *Phys. Rev. A* **36**: 428.

Bergquist J.C., Wineland D.J., Itano W.M., Hemmati H., Daniel H.-U., and Leuchs G. 1985. Energy and radiative lifetime of the 6d^9 6s^2 ^{2}D$_{5/2}$ state in Hg II by Doppelr-free two photon laser spectroscopy. *Phys. Rev. Lett.* **55** (15): 1567.

Berkeland D.J., Cruz F., and Bergquist J. C. 1997. Sum-frequency generation of continuous-wave light at 194 nm. *Appl. Opt.* **36**: 4159.

Berkeland D.J., Miller J.D., Bergquist J.C., Itano W.M., and Wineland D.J. 1998. Minimization of ion micromotion in a Paul trap. *J. Appl. Phys.* **83** (10): 5025.

Bernard J.E., Marmet L., and Madej A.A. 1998. A laser frequency lock referenced to a single trapped ion. *Opt. Commun.* **150**: 170.

Bigeon M. 1967. Probabilite de transition de la raie 6 ^{1}S$_0$ – 6 ^{3}P$_0$ du mercure. *J. Phys.* **28**: 51.

Binnewies T., Wilpers G., Sterr U., Riehle F., Helmcke J., Mehlstäubler T.E., Rasel E.M., and Ertmer W. 2001. Doppler cooling and trapping on forbidden transitions. *Phys. Rev. Lett.* **87**: 123002.

Blythe P.J. 2004. *Optical Frequency Measurement and Ground State Cooling of Single Trapped Yb+ Ions.* PhD Thesis, University of London.

Blythe P.J., Webster S.A., Hosaka K., and Gill P. 2003. Systematic frequency shifts of the 467 nm electric octupole transition in ^{171}Yb+. *J. Phys. B: At. Mol. Opt. Phys.* **36**: 981.

Bloom B.J., Nicholson T.L., Williams J.R., Campbell S.L., Bishof M., Zhang X., Zhang W., Bromley S.L., and Ye J. 2014. An optical lattice clock with accuracy and stability at the 10^{-18} level. *Nature* **506**: 71.

Boshier M.G., Barwood G.P., Huang G., and Klein H.A. 2000. Polarisation-dependent optical pumping for interrogation of magnetic-field-independent "clock" transition in laser cooled trapped ^{87}Sr+. *Appl. Phys. B: Lasers Opt.* **71**: 51.

Boyd M.M., Zelevinsky T., Ludlow A.D., Blatt S., Zanon-Willette T., Foreman S.M., and Ye J. 2007. Nuclear spin effects in optical lattice clocks. *Phys. Rev. A* **76**: 022510.

Boyd M.M., Zelevinsky T., Ludlow A.D., Foreman S.M., Blatt S., Ido T., and Ye J. 2006. Optical atomic coherence at the 1-second time scale. *Science* **314**: 1430.

Castin Y., Wallis H., and Dalibard J. 1989. Limit of Doppler cooling. *J. Opt. Soc. Am. B* **6** (11): 2047.

Champenois C., Houssin M., Lisowski C., Knoop M., Hagel G., Vedel M., and Vedel F. 2004. Evaluation of the ultimate performances of a Ca+ single-ion frequency standard. *Phys. Rev. A* **33**: 298.

Champenois C., Knoop M., Herbane M., Houssin M., Kaing T., Vedel M., and Vedel F. 2001. Characterization of a miniature Paul-Straubel trap. *Eur. Phys. J. D* **15**: 105.

Chou C.W., Hume D.B., Koelemeij J.C.J., Wineland D.J., and Rosenband T. 2010. Frequency comparison of two high-accuracy Al+ optical clocks. *Phys. Rev. Lett.* **104**: 070802.

Chwalla M. 2009. *Precision spectroscopy with ^{40}Ca+ ions in a Paul trap.* PhD Thesis, Leopold-Franzens-Universitat, Innsbruck, Austria.

Chwalla M., Benhelm J., and Kim K. 2009. Absolute frequency measurement of the ^{40}Ca+ $4s2S_{1/2} - 3d^2D_{5/2}$ clock transition. *Phys. Rev. Lett.* **102**: 023002.

Chwalla M., Benhelm J., Kim K., Kirchmair G., Monz T., Riebe M., Schindler P., Villar A.S., Hansel W., Roos C.F., Blatt R., Abgrall M., Santarelli G., Rovera G.D., and Laurent Ph. 2008. Absolute frequency measurement of the ^{40}Ca+ $4s\ ^2S_{1/2} - 3d\ ^2D_{5/2}$ clock transition. arXiv:0806.1414v1(physics.atom-ph).

Courtillot I., Quessada A., Kovacich R.P., Brusch A., Kolker D., Zondy J.J., Rovera G.D., and Lemonde P. 2003. A clock transition for a future optical frequency standard with trapped atoms. arXiv:physiscs/0303023.

Curtis E.A., Oates C.W., and Hollberg L. 2001. Quenched narrow-line laser cooling of ^{40}Ca to near the photon recoil limit. *Phys. Rev. A* **64**: 031403(R).

Degenhardt C., Stoehr H., Lisdat C., Wilpers G., Schnatz H., Lipphardt B., Nazarova T., Pottie P.E., Sterr U., Helmcke J., and Riehle F. 2005. Calcium optical frequency standard with ultracold atoms: approaching 10–15 relative uncertainty. *Phys. Rev. A* **72**: 062111.

Dehmelt H. 1973. Proposed 10^{14} Dv < v laser fluorescence spectroscopy on Tl+ mono-ion oscillator. *Bull. Am. Phys. Soc.* **18**: 1521.

Dehmelt H. 1975. Proposed 10^{14} Dv < v laser fluorescence spectroscopy on Tl+ mono-ion oscillator II (spontaneous quantum jumps). *Bull. Am. Phys. Soc.* **20**: 60.

Deng K., Xu Z.T., Yuan W.H., Qin C.B., Deng A., Sun Y., Zhang J., Lu Z.H., and Luo J. 2013. Progress report of an ^{27}Al+ ion optical clock. In *Proceedings of the European Frequency and Time Forum & International Frequency Control Symposium*, IEEE, 383.

Derevianko A. and Katori H. 2011. Colloquium: physics of optical lattice clocks. *Rev. Mod. Phys.* **83**: 331.

Derevianko A., Obreshkov B., and Dzuba V.A. 2009. Mapping out atom-wall interaction with atomic clocks. *Phys. Rev. Lett.* **103**: 133201.

Dubé P., Madej A.A., Bernard J.E., Marmet L., Boulanger J.S., and Cundy S. 2005. Electric Quadrupole shift cancellation in single-ion optical frequency standards. *Phys. Rev. Lett.* **95**: 033001.

Dubé P., Madej A.A., Zhou Z., and Bernard J.E. 2013. Evaluation of systematic shifts of the $^{88}Sr^+$ single-ion optical frequency standard at the 10^{-17} level. *Phys. Rev. A* **87**: 023806.

Ertmer W., Friebe J., Riedmann M., Pape A., Wbbena T., Kulosa A., Amairi S., Kelkar H., Rasel E.-M., Terra O., Grosche G., Lipphardt B., Feldmann T., Schnatz H., and Predehl K. 2010. Towards a lattice based neutral magnesium optical frequency standard. In *FOMO Conference Frontiers of Matter Wave Optics*.

Ferrari G., Cancio P., Drullinger R., Giusfredi G., Poli N., Prevedelli M., Toninelli C., and Tino G.M. 2003. Precision frequency measurement of visible intercombination lines of strontium. *Phys. Rev. Lett.* **91**: 243002.

Friebe J., Pape A., Riedmann M., Moldenhauer K., Mehlstäubler T., Rehbein N., Lisdat Ch.E, Rasel E.M., and Ertmer W. 2008. Absolute frequency measurement of the magnesium intercombination transition $^1S_0 \rightarrow {}^3P_1$. *Phys. Rev. A* **78**: 033830.

Gao K. 2013. Optical frequency standard based on a single $^{40}Ca^+$. *Chin. Sci. Bull.* **58**: 853.

Gill P., Barwood G.P., Huang G., Klein H.A., Blythe P.J., Hosaka K., Thompson R.C., Webster S.A., Lea S.N., and Margolis H.S. 2004. Trapped ion optical frequency standards. *Phys. Scripta. T* **112**: 63.

Godun R.M., Nisbet-Jones P.B.R., Jones J.M, King S.A., Johnson L.A.M., Margolis H.S., Szymaniec K., Lea S.N., Bongs K., and Gill P. 2014. Frequency ratio of two optical clock transitions in $^{171}Yb^+$ and constraints on the time-variation of fundamental constants. arXiv:1407.0164 v1(physics.atom-ph).

Hachisu H., Miyagishi K., Porsev S.G., Derevianko A., Ovsiannikov V.D., Pal'chikov V.G., Takamoto M., and Katori H. 2008. Trapping of neutral mercury atoms and prospects for optical lattice clocks. *Phys. Rev. Lett.* **100**: 053001.

Hall J.L. 2006. Nobel lecture: defining and measuring optical frequencies. *Rev. of Mod. Phys.* **78**: 1279.

Hänsch T. and Couillaud B. 1980. Laser frequency stabilization by polarization spectroscopy of a reflecting reference cavity. *Opt. Commun.* **35** (3): 441.

Hansch T.W. 2006. Noble lecture: passion for precision. *Rev. of Mod. Phys.* **78**: 1297.

Hinkley N., Sherman J.A., Phillips N.B., Schioppa M., Lemke N.D., Beloy K., Pizzocaro M., Oates C.W., and Ludlow A.D. 2013. An atomic clock with 10^{-18} instability. *Science* 341 (6151): 1215–1218. doi:10.1126/science.1240420.

Hong T., Cramer C., Nagourney W., and Fortson E.N. 2005. Optical clocks based on ultranarrow three-photon resonances in alkaline earth atoms. *Phys. Rev. Lett.* **94**: 050801.

Hosaka K., Webster S.A., Blythe P.J., Stannard A., Beaton D., Margolis H.S., Lea S.N., and Gill P. 2005. Anoptical frequency standard based on the electric octupole transition in $^{171}Yb^+$. *IEEE Trans. Instrum. Meas.* **54** (2): 759.

Hoyt C.W., Barber Z.W., Oates C.W., Fortier T.M., Diddams S.A., and Hollberg L. 2005. Observation and absolute frequency measurements of the $^1S_0 \rightarrow {}^3P_0$ optical clock transition in neutral ytterbium. *Phys. Rev. Lett.* **95**: 083003.

Huntemann N., Okhapkin M., Lipphardt B., Weyers S., Tamm Chr., and Peik E. 2012. High-accuracy optical clock based on the octupole transition in $^{171}Yb^+$. *Phys. Rev. Lett.* **108**: 090801.

Itano W.M. 2000. External-field shift of the $^{199}Hg^+$ optical frequency standard. *Res. Nat. Inst. Stand. Technol.* **105**: 829.

Jiang D., Arora B., Safronova M.S., and Clark Ch.W. 2009. Blackbody-radiation shift in a $^{88}Sr^+$ ion optical frequency standard. *J. Phys. B: At. Mol. Opt. Phys.* **42**: 154020.

Kajita M., Li Y., Matsubara K., Hayasaka K., and Hosokawa M. 2005. Prospect of optical frequency standard based on a ^{43}Ca$^+$ ion. *Phys. Rev. A* **72**: 043404.

Katori H. 2002. Spectroscopy of strontium atoms in the Lamb-Dicke confinement. In *Proceedings of the 6th Symposium on Frequency Standards and Metrology*, P. Gill Ed. (World Scientific, Singapore) 323.

Katori H. 2011. Optical lattice clocks and quantum metrology. *Nature Photon.* **45** (5): 201.

Katori H., Takamoto M., Pal'chikov V. G., and Ovsiannikov V.D. 2003. Ultrastable optical clock with neutral atoms in an engineered ligth shift trap. *Phys. Rev. Lett.* **91**: 173005.

Keupp J., Douillet A., Mehlstäubler T.E., Rehbein N., Rasel E.M., and Ertmer W. 2005. A high-resolution Ramsey-Bordé spectrometer for optical clocks based on cold Mg atoms. *EPJ D* **36** (3): 289.

Larson D.J., Bergquist J.C., Bollinger J.J., Itano W.M., and Wineland D.J. 1986. Sympathetic cooling of trapped ions: a laser-cooled two-species nonneutral ion plasma. *Phys. Rev. Lett.* **57**: 70.

Lea S.N., Webste R.S.A., and Barwood G.P. 2006. Polarisabilities and blackbody shifts in Sr$^+$ and Yb$^+$. In *Proceedings of the 20th European Frequency and Time Forum*.

Lemke N.D., Ludlow A.D., Barber Z.W., Fortier T.M., Diddams S.A., Jiang Y., Jefferts S.R., Heavner T.P., Parker T.E., and Oates C.W. 2009. Spin-1/2 optical lattice clock. *Phys. Rev. Lett.* **103**: 063001.

Lemonde P. 2009. Optical lattice clock. *Eur. Phys. J. Spec. Top.* **172**: 81.

Le Targat R. 2007. *Horloge a Réseau Optique au Strontium: une 2eme Génération d'horloges a Atomes Froids*. Thèse de L'EDITE DE PARIS.

Le Targat R., Lorini L., LeCoq Y., Zawada M., Guéna J., Abgrall M., Gurov M., Rosenbusch P., Rovera D.G., Nagórny B., Gartman R., Westergaard P.G., Tobar M.E., Lours M., Santarelli G., Clairon A., Bize S., Laurent P., Lemonde P., and Lodewyck J. 2013. Experimental realization of an optical second with strontium lattice clocks. *Nature Communications* **4**: 2109.

Letchumann V., Gill P., Riis E., and Sinclair A.G. 2004. Optical Ramsey spectroscopy of a single trapped ^{88}Sr$^+$ ion. *Phys. Rev. A* **70**: 033419.

Liu T., Wang Y.H. Elman V., Stejskal A., Zhao Y.N., Zhang J., Lu Z.H., Wang L.J., Dumke R., Becker T., and Walther H. 2007. Progress toward a single indium ion optical frequency standard. In *Proceedings of the IEEE International Frequency Control Symposium, 2007 Joint with the 21st European Frequency and Time Forum*, IEEE, 407.

Lorini L., Ashby N., Brusch A., Diddams S., Drullinger R., Eason E., Fortier T., Hastings P., Heavner T., Hume D., Itano W., Jefferts S., Newbury N., Parker T., Rosenband T., Stalnaker J., Swann W., Wineland D., and Bergquist J. 2008. Recent atomic clock comparisons at NIST. *Eur. Phys. J. Spec. Top.* **163**: 19.

Madej A. and Bernard J. 2001. *Frequency Measurement and Control—Advanced Techniques and Future Trends* (Springer-Verlag, Berlin, Germany) 28.

Madej A.A., Bernard J.E., Dubé P., Marmet L., and Windeler R.S. 2004. Absolute frequency of the ^{88}Sr $5s^2s^{1/2} - 4d^2D^{5/2}$ reference transition at 445 THz and evaluation of systematic shifts. *Phys. Rev. A* **70**: 012507.

Madej A.A., Dubé P., Zhou Z., Bernard J.E., and Gertsvolf M. 2012. ^{88}Sr$^+$ 445-THz single-ion reference at the 10–17 level via control and cancellation of systematic uncertainties and its measurement against the SI second. *Phys. Rev. Lett.* **109**: 203002.

Madej A.A., Siemsen K.J., Sankey J.D., Clark R.F., Vanier J. 1993. High-resolution spectroscopy and frequency measurement of the mid infrared 5d2D3/2 – 5d 2D5/2 transition of a single laser-cooled bariumion. *IEEE Trans. Instrum. Meas.* **42**: 2234.

Margolis H.S., Barwood G.P., Hosaka K., Klein H.A., Lea S.N., Stannard A., Walton B.R., Webster S.A., Gill P. 2006. Optical frequency standards and clocks based on single trapped ions. In *Proceedings of the Conference CLEO/QELS*, Optical Society of America, 1.

Margolis H.S., Barwood G.P., Huang G., Klein H.A., Lea S.N., Szymaniec K., and Gill P. 2004. Hertz-level measurement of the optical clock frequency in a single ^{88}Sr$^+$ ion. *Science* **306**: 1355.

Matsubara K., Hachisu H., Li Y., Nagano S., Locke C., Nogami A., Kajita M., Hayasaka K., Ido T., and Hosokawa M. 2012. Direct comparison of a Ca$^+$ single-ion clock against a Sr lattice clock to verify the absolute frequency measurement. *Opt. Express* 20: 22034.

Matsubara K., Hayasaka K., Li Y., Ito H., Nagano S., Kajita M., and Hosokawa M. 2008. Frequency measurement of the optical clock transition of ^{40}Ca$^+$ ions with an uncertainty of 10^{-14}. *Appl. Phys. Express* **1** (6): 067011.

Matsubara K., Toyoda K., Li Y., Tanaka U., Uetake S., Hayasaka K., Urabe S., and Hosokawa M. 2004. Study for a ^{43}Ca$^+$ optical frequency standard. In *Proceedings of the Conference Digest Precision Electromagnetic Measurements*, IEEE, 430.

McFerran J., Magalhães D.V., Mandache C., Millo J., Zhang W., LeCoq Y., Santarelli G., and Bize S. 2012. Laser locking to the ^{199}Hg ^{1}S$_0$ − ^{3}P$_0$ clock transition with 5.4 × 10^{-15} τ$^{-1/2}$ fractional frequency instability. *Opt. Lett.* **37** (17): 3477.

McFerran J.J., Yi L., Mejri S., Di Manno S., Zhang W., Guena J., Le Coq Y., and Bize S. 2012. Neutral atom frequency reference in the deep ultraviolet with a fractional uncertainty = 5.7 × 10^{-15}. *Phys. Rev. Lett.* **108**: 183004.

Mejri S., McFerran J.J., Yi L., Le Coq Y., and Bize S. 2011. Ultraviolet laser spectroscopy of neutral mercury in a one-dimensional optical lattice. *Phys. Rev. A* **84**: 032507.

Middelmann Th., Lisdat Ch., Falke St., Vellore Winfred J.S.R., Riehle F., and Sterr U. 2010. Tackling the blackbody shift in a strontium optical lattice clock. arXiv:1009.2017v1 (physics.atom-ph).

Nägerl H.C., Bechter W., Eschner J., Schmidt-Kaler Fr., and Blatt R. 1998. On strings for quantum gates. *Appl. Phys. B: Lasers Opt.* **66**: 603.

Nagourney W., Sandberg J., and Dehmelt H. 1986. Shelved optical electron amplifier: observation of quantum jumps. *Phys. Rev. Lett.* **56**: 2797.

Oates C.W., Bondu F., Fox R.W., and Hollberg L. 1999. A diode-laser optical frequency standard based on laser-cooled Ca atoms: sub-kilohertz spectroscopy by optical shelving detection. *Eur. Phys. J. D* **7**: 449.

Oates C.W., Hoyt C.W., LeCoq Y., Barber Z.W., Fortier T.M., Stalnaker J.E., Diddams S.A., Hollberg L. 2006. Stability measurements of the Ca and Yb optical frequency standards. In *Proceedings of the IEEE International Frequency Control Symposium and Exposition, IEEE*, 74.

Oskay W.H., Diddams S.A., Donley E.A., Fortier T.M., Heavner T.P., Hollberg L., Itano W.M., Jefferts S.R., Delaney M.J., Kim K., Levi F., Parker T.E., and Bergquist J.C. 2006. Single atom optical clock with high accuracy. *Phys. Rev. Lett.* **97** (2): 020801.

Ovsiannikov V.D., Pal'chikov V.G., Taichenachev A.V., Yudin V.I., Katori H., and Takamoto M. 2007. Magic-wave-induced ^{1}S$_0$ − ^{3}P$_0$ transition in even isotopes of alkaline-earth-like atoms. *Phys. Rev. A* **75**: 020501(R).

Peik E., Lipphardt B., Schnatz H., Sherstov I., Stein B., Tamm Chr., Weyers S., and Wynand R. 2007. ^{171}Yb$^+$ single-ion optical frequency standards and search for variations of the fine structure constant. In *Proceedings of the Quantum-Atom Optics Downunder* (Wollongong, Australia).

Petersen M., Chicireanu R., Dawkins S.T., Magalhães D.V., Mandache C., LeCoq Y., Clairon A., and Bize S. 2008. Doppler-free spectroscopy of the ^{1}S$_0$ − ^{3}P$_0$ optical clock transition in laser-cooled fermionic isotopes of neutral mercury. *Phys. Rev. Lett.* **101**: 183004.

Poli N., Barber Z.W., Lemke N.D., Oates C.W., Ma L.S., Stalnaker J.E., Fortier T.M., Diddams S.A., Hollberg L., Bergquist J.C., Brusch A., Jefferts S., Heavner T., and Parker T. 2008. Frequency evaluation of the doubly forbidden ^{1}S$_0$ → ^{3}P$_0$ transition in bosonic ^{174}Yb. *Phys. Rev. A* **77**: 050501(R).

Poli N., Oates C.W., Gill P., and Tino G.M. 2014. Optical atomic clocks. arXiv:1401.2378v1.

Porsev S.G., and Derevianko A. 2006. Multipolar theory of blackbody radiation shift of atomic energy levels and its implications for optical lattice clocks. *Phys. Rev. A* **74**: 020502.

Porsev S.G., Derevianko A., and Fortson E.N. 2004. Possibility of an optical clock using the $6^1S_0 - 6^3P_0$ transition in ^{171,173}Yb atoms held in an optical lattice. *Phys. Rev. A* **69**: 021403.

QPAFS. 1989. *The Quantum Physics of Atomic Frequency Standards,* Vols. 1 and 2, J. Vanier and C. Audoin, Eds. (Adam Hilger ed., Bristol).

Roberts M., Taylor P., Barwood G.P., Gill P., Klein H.A., and Rowley W.R.C. 1997. Observation of an electric octupole transition in a single ion. *Phys. Rev. Lett.* **78**: 1876.

Roberts M., Taylor P., Barwood G.P., Rowley W.R.C., and Gill P. 2000. Observation of the $^2S_{1/2} - ^2F_{7/2}$ electric octupole transition in a single ^{171}Yb$^+$ ion. *Phys. Rev. A* **62**: 020501(R).

Roberts M., Taylor P., Gateva-Kostova S.V., Clarke R.B.M., Rowley W.R.C., and Gill P. 1999. Measurementofthe $^2S_{1/2} - ^2D_{5/2}$ clocktransitioninasingle ^{171}Yb$^+$ ion. *Phys. Rev. A* **60**: 2867.

Rosenband T., Hume D.B., Schmidt P.O., Chou C.W., Brusch A., Lorini L., Oskay W.H., Drullinger R.E., Fortier T.M., Stalnaker J.E., Diddams S.A., Swann W.C., Newbury N.R., Itano W.M., Wineland D.J., and Bergquist J.C. 2008. Frequency ratio of Al$^+$ and Hg$^+$ single-ion optical clocks; Metrology at the 17th decimal place. *Science* **319** (5871): 1808.

Rosenband T., Itano W.M., Schmidt P.O., Hume D.B., Koelemeij J.C.J., Bergquist J.C., and Wineland D.J. 2006. Blackbody radiation shift of the ^{27}Al$^+$ $^1S_0 - ^3P_0$ transition. arXiv:physics/0611125v2.

Rosenband T., Schmidt P.O., Hume D.B., Itano W.M., Fortier T.M., Stalnaker J.E., Kim K., Diddams S.A., Koelemeij J.C.J, Bergquist J.C., and Wineland D.J. 2007. Observation of the $^1S_0 - ^3P_0$ clock transition in ^{27}Al$^+$. *Phys. Rev. Lett.* **98**: 220801.

Safronova M.A., Jiang D., and Safronova U.I. 2010. Blackbody radiation shift in ^{87}Rb frequency standard. *Phys. Rev. A* **82**: 022510.

Safronova M.S., Kozlov M.G., and Clark C.W. 2012. Black body radiation shifts in optical atomic clocks. *IEEE Trans. Ultrason. Ferroelectr. Freq. Control* **59**: 439.

Schmidt P.O., Rosenband T., Langer C., Itano W.M., Bergquist J.C., and Wineland D.J. 2005. Spectroscopy using quantum logic. *Science* 309 (5735): 749.

Sherman J.A., Trimble W., Metz S., Nagourney W., and Fortson N. 2005. Progress on indium and barium single ion optical frequency standards. arXiv:physics/0504013v2.

Sherstov I., Tamm C., Stein B., Lipphardt B., Schnatz H., Wynands R., Weyers S., Schneider T., and Peik E. 2007. ^{171}Yb$^+$ single-ion optical frequency standards. In *IEEE Conference Publications on Frequency Control Symposium, Joint with the 21st European Frequency and Time Forum, IEEE,* 405.

Simmons M., Safronova U.I., and Safronova M.S. 2011. Blackbody radiation shift, multipole polarizabilities, oscillator strengths, lifetimes, hyperfine constants, and excitation energies in Hg. *Phys. Rev. A* **84**: 052510.

Simon E., Laurent P., and Clairon A. 1998. Measurement of the stark shift of the Cs hyperfine splitting in an atomic fountain. *Phys. Rev. A* **57**: 436.

Smart A.G. 2014. Optical-lattice clock sets new standard for timekeeping. *Phys. Today* **67** (3): 12.

Taichenachev A.V., Yudin V.I., Oates C.W., Hoyt C.W., Barber Z.W., and Hollberg L. 2006. Magnetic field-induced spectroscopy of forbidden optical transitions with application to lattice-based optical atomic clocks. *Phys. Rev. Lett.* **96**: 083001.

Takamoto M., and Katori H. 2003. Spectroscopy of the $^1S_0 - ^3P_0$ clock transition of ^{87}Sr in an optical lattice. *Phys. Rev. Lett.* **91**: 223001.

Tamm Chr., Huntemann N., Lipphardt B., Gerginov V., Nemitz N., Kazda M., Weyers S., and Paik E. 2014. Cs-based frequency measurement using cross-linked optical and microwave oscillators. *Phys. Rev. A* **89**: 023820.

Tamm Chr., Lipphardt B., Schnatz H., Wynands R., Weyers S., Schneider T., and Peik E. 2007. ^{171}Yb$^+$ single-ion optical frequency standard at 688 THz. *IEEE Trans. Instrum. Meas.* **56**: 601.

Tanaka U., Bize S., Tanner C.E., Drullinger R.E., Diddams S.A., Hollberg L., Itano W.M., Wineland D.J., and Bergquist J.C. 2003. The ^{199}Hg$^+$ single ion optical clock: progress. *J. Phys. B: At. Mol. Opt. Phys.* **36**: 545.

Telle H.R., Steynmeyer G., Dunlop A.E., Stenger J., Sutter D.H., and Keller U. 1999. Carrier-envelope offset phase control: a novel concept for absolute optical frequency measurements and ultrashort pulse generation. *Appl. Phys. B: Lasers Opt.* **69**: 327.

Vanier J. 1960. Temperature dependence of a pure nuclear quadrupole resonance frequency in KClO$_3$. *Can. J. Phys.* **38**: 1397.

Vanier J. 2011. *The Universe, a Challenge to the Mind* (Imperial College Press, World Scientific, Singapore).

Vanier J. and Audoin C. 1989. *Th Quantum Physics of Atomic Frequency Standards* (Adam Hilger, Bristol).

Vogel K.R., Dinneen T.P., Gallangher A., and Hall J.L. 1999. Narrow-line Doppler cooling of strontium to recoil limit. *IEEE Trans. Instrum. Meas.* **48** (2): 618.

von Zanthier J., Becker T., Eichenseer M., Nevsky A.Y., Schwedes C., Peik E., Walther H., Holzwarth R., Reichert J., Udem T., Hänsch T.W., Pokasov P.V., Skvortsov M.N., and Bagayev S.N. 2000. Absolute frequency measurement of the In$^+$ clock transition with a mode-locked laser. *Opt. Lett.* **25**: 1729.

Wallin A., Fordell Th., Lindvall Th., and Merimaa M. 2013. Progress towards a ^{88}Sr$^+$ ion clock at MIKES. In *XXXIII Finnish URSI Convention on Radio Science and SMARAD Seminar 2013*.

Wang Y.H., Liu T., Dumke R., Stejskal A., Zhao Y.N., Zhang J., Lu Z.H., Wang L.J., Becker Th., and Walther H. 2007. Absolute frequency measurement of ^{115}In$^+$ clock transition. In *Conference Paper International Quantum Electronics Conference Munich, Germany*.

Westergaard P.G., Lodewyck J., Lorini L., Lecallier A., Burt E.A., Zawada M., Millo J., and Lemonde P. 2011. Lattice-induced frequency shifts in Sr optical lattice clocks at the 10^{-17} level. *Phys. Rev.Lett.* **106**: 210801.

Whitford B.G., Siemsen K.J., Madej A.A., and Sankey J.D. 1994. Absolute-frequency measurement of the narrow-linewidth 24-THz D-D transition of a single laser-cooled barium ion. *Opt. Lett.* **19** (5): 356.

Wilpers G., Oates C., and Hollberg L. 2006. Improved uncertainty budget for optical frequency measurements with microkelvin neutral atoms: results for a high-stability ^{40}Ca optical frequency standard. *Appl. Phys. B: Lasers Opt.* **85**: 31.

Ye A. and Wang G. 2008. Dipole polarizabilities of ns2 ^{1}S0 and nsnp^{3}P0 states and relevant magic wavelengths of group-IIB atoms. *Phys. Rev. A* **78**: 014502.

Yi L., Mejri S., McFerran J.J., LeCoq Y., and Bize S. 2011. Optical lattice trapping of ^{199}Hg and determination of the magic wavelength for the ultraviolet ^{1}S$_0$? ^{3}P$_0$ clock transition. *Phys. Rev. Lett.* **106**: 073005.

Yu N., Dehmelt H., and Nagourney W. 1992. The 3^1S$_0 \rightarrow 3^3$P$_0$ transition in the aluminum isotope ion ^{26}Al$^+$: a potentially superior passive laser frequency standard and spectrum analyzer. *Proc. Natl. Acad. Sci. USA* **89**: 7289.

Yu N., Zhao X., Dehmelt H., and Nagourney W. 1994. Stark shift of a single barium ion and potential application to zero-point confinement in a rf trap. *Phys. Rev. A* **50**: 2738.

CHAPTER 5

Alley C.O. 1979. Relativity and clocks. In *Proceedings of the Annual Symposium on Frequency Control*, IEEE, 4.

Alley C.O. 1981. Introduction to some fundamental concepts of general relativity and to the irrequired use in some modern time keeping systems. In *Proceedings of the Annual Precise Time and Time Interval Applications and Planning Meeting*, IEEE, 687.

Belloni M., Battisti A., Cosentino A., Sapia A., Borella A., Micalizio S., Godone A., Levi F., Calosso C., Zuliani L., Longo F., and Donati M. 2009. A space rubidium pulsed optical pumped clock—current status, results, and future activities. In *Proceedings of the Annual Precise Time and Time Interval Meeting*, ION Publications, 519.

Bize S. 2001. *Tests Fondamentaux à l'aide d'horloges à Atomes Froids de Rubidium at de Cesium*. Thesis, Université de Paris VI, France.

Bize S., Sortais Y., Santos M.S., Mandache C., Clairon A., and Salomon C. 1999. High-accuracy measurement of the ^{87}Rb ground-state hyperfine splitting in an atomic fountain. *Europhys. Lett.* **45**: 558.

Blatt S., Ludlow A.D., Campbell G.K., Thomsen J.W., Zelevinsky T., Boyd M.M., Ye J., Baillard X., Fouché M., Le Targat R., Brusch A., Lemonde P., Takamoto M., Hong F.L., Katori H., and Flambaum V.V. 2008. New limits on coupling of fundamental constants to gravity using Sr^{87} optical lattice clocks. *Phys. Rev. Lett.* **100**: 140801.

Bloom B.J., Nicholson T.L., Williams J.R., Campbell S.L., Bishof M., Zhang X., Zhang W., Bromley S.L., and Ye J. 2014. An optical lattice clock with accuracy and stability at the 10^{-18} level. *Nature* **506**: 71.

Burt E.A., Diener W.A., and Tjoelker R.L. 2008. A compensated multi-pole linear ion trap mercury frequency standard for ultra-stable timekeeping. *IEEE Trans. Ultrason. Ferroelectr. Freq. Control* **55**.

Cacciapuoti L. and Salomon Ch. 2009. Space clocks and fundamental tests: the ACES experiment. *Eur. Phys. J. Spec. Top.* **172**: 57.

CGPM 1967. Conference Générale des Poids et Mesures. Résolution 1 of the 13th CGPM and 1968 news from the international bureau of weights and measures. *Metrologia* **4** (1): 41.

Chou C.W., Hume D.B., Rosenband T., and Wineland D.J. 2010. Optical clocks and relativity. *Science* **329**: 1630.

CIPM 2012. Consultative Committee for Time and frequency. *Report of the 19th Meeting to the International Committee for Weights and Measures*.

De Beauvoir B., Schwob C., Acef O., Jozefowski L., Hilico L., Nez F., Julien L., Clairon A., and Biraben F. 2000. Metrology of the hydrogen and deuterium atoms: determination of the Rydberg constant and Lamb shifts. *Eur. Phys. J. D* **12**: 61.

Delva P. and Lodewyck J. 2013. Atomic clocks: new prospects in metrology and Geodesy. *Acta Futura* **7**: 67; see also arXiv;1308.6766v1 (physics.atom-ph), August 29, 2013.

Esnault F.X., Perrin S., Holleville D., Guerandel S., Dimarcq N., and Delporte J. 2008. Reaching a few 10^{-13} $\tau^{-1/2}$ stability level with the compact cold atom clock Horace. In *Proceedings of the IEEE International Frequency Control Symposium*, IEEE, 381.

Fischer M., Kolachevsky N., Zimmermann M., Holzwarth R., Udem Th., Hänsch T.W., Abgrall M., Grünert J., Maksimovic I., Bize S., Marion H., Dos Santos F.P., Lemonde P., Santarelli G., Laurent P., Clairon A., Salomon C., Haas M., Jentschura U.D., and Keite C.H. 2004. New limits on the drift of fundamental constants from laboratory measurements. *Phys. Rev. Lett.* **92**: 230802.

Flambaum V.V. and Berengut J.C. 2009. Variation of fundamental constants from the big bang to atomic clocks: theory and observations. In *Proceedings of the 7th Symphosium Frequency Standards and Metrology*, L. Maleki Ed. (World Scientific, Singapore) 3.

Fortier T.M., Ashby N., Bergquist J.C., Delaney M.J., Diddams S.A., Heavner T.P., Hollberg L., Itano W.M., Jefferts S.R., Kim K., Levi F., Lorini L., Oskay W.H., Parker T.E., Shirley J., and Stalnaker J.E. 2007. Precision atomic spectroscopy for improved limits on variation of the fine structure constant and local position invariance. *Phys. Rev. Lett.* **98**: 070801.

Guena J., Abgrall M., Clairon A., and Bize S. 2014. Contributing to TAI with a secondary representation of the SI second. *Metrologia* **51**: 108.

Guéna J., Abgrall M., Rovera D., Rosenbusch P., Tobar M.E., Laurent Ph., Clairon A., and Bize S. 2012. Improved tests of local position invariance using Rb^{87} and Cs^{133} fountains. *Phys. Rev. Lett.* **109**: 080801.

Guerandel S., Danet J.M., Yun P., and de Clercq E. 2014. High performance compact atomic clock based on coherent population trapping. In *Proceedings of the IEEE International Frequency Control Symposium*, IEEE, 1.

Hafele J.C. and Keating R.E. 1972. Around-the-World atomic clocks: predicted relativistic time gains. *Science* 177 (4044): 166.

Huntemann N., Lipphardt B., Okhapkin M., Tamm C., Peik E., Taichenachev A.V., and Yudin V.I. 2012. Generalized Ramsey excitation scheme with suppressed light shift. *Phys. Rev. Lett.* **109**: 213002.

Karshenboim S.G. and Peik E. 2008. Astrophysics, atomic clocks and fundamental constants. *Eur. Phys. J. Spec. Top.* **163**: 1.

Lemke N.D., Ludlow A.D., Barber Z.W., Fortier T.M., Diddams S.A., Jiang Y., Jefferts S.R., Heavner T.P., Parker T.E., and Oates C.W. 2009. Spin-1/2 optical lattice clock. *Phys. Rev. Lett.* **103**: 063001.

Madej A.A., Dubé P., Zhou Z., Bernard J.E., and Gertsvolf M. 2012. ^{88}Sr$^+$ 445-THz single-ion reference at the 10–17 level via control and cancellation of systematic uncertainties and its measurement against the SI second. *Phys. Rev. Lett.* **109**: 203002.

Maleki L. and Prestage J. 2005. Applications of clocks and frequency standards: from the routine to tests of fundamental models. *Metrologia* **42**: S145.

McFerran J., Magalhães D.V., Mandache C., Millo J., Zhang W., LeCoq Y., Santarelli G., and Bize S. 2012. Laser locking to the ^{199}Hg $^1S_0 - {}^3P_0$ clock transition with $5.4 \times 10-15/\sqrt{\tau}$ fractional frequency instability. *Opt. Lett.* **37** (17): 3477.

Parkinson B.W. 1996. Introduction and heritage of NAVSTAR, the global positioning system. In *Global Positioning System: Theory and Applications*, B.W. Parkinson et al. Eds. (AIAA, New York) 3.

Peik E. 2010. Fundamental constants and units and the search for temporal variations. *Nuclear Physics* **203**: 18.

Peik E., Zimmermann K., Okhapkin M., and Tamm Chr. 2009. Prospects for a nuclear opticxal frequency standard based on Thorium-229. In *Proceedings of the 7th Symposium on Frequency Standards and Metrology*, L. Maleki Ed. (World Scientific, Singapore) 532.

Prestage J., Chung S.K., Lim L., and Matevosian A. 2007. Compact microwave mercury ion clock for deep-space applications. In *Proceedings of the Joint IEEE International Frequency Control Symposium/European Forum on Time and Frequency*, IEEE, 1113.

Prestage J.D., Tjoelker R.L., and Maleki L. 1995. Atomic clocks and variations of the fine structure constant. *Phys. Rev Lett.* **174**: 3511.

QPAFS. 1989. *The Quantum Physics of Atomic Frequency Standards*, Vols. 1 and 2, J. Vanier and C. Audoin Eds. (Adam Hilger ed., Bristol).

Riehle F. 2004. *Frequency Standards Basics and Applications* (Wiley-VCH, Weinheim, Germany).

Rosenband T., Hume D.B., Schmidt P.O., Chou C.W., Brusch A., Lorini L., Oskay W.H., Drullinger R.E., Fortier T.M., Stalnaker J.E., Diddams S.A., Swann W.C., Newbury N.R., Itano W.M., Wineland D.J., and Bergquist J.C. 2008. Frequency ratio of Al$^+$ and Hg$^+$ single-ion optical clocks; metrology at the 17th decimal place. *Science* **319**: 1808.

Tamm Chr., Weyers S., Lipphardt B., and Peik E. 2009. Stray-field-induced quadrupole shift and absolute frequency of the 688-THz ^{171}Yb$^+$ single-ion optical frequency standard. *Phys. Rev. A* **80**: 043403.

Thompson A.M. and Lampard D.G. 1956. A new theorem in electrostatics and its application to calculable standards of capacitance. *Nature* **177**: 888.

Vanier J. 2005. Atomic clocks based on coherent population trapping: a review. *Appl. Phys. B* **81**: 421.

Vanier J. 2011. *The Universe, a Challenge to the Mind* (Imperial College Press/World Scientific, London).

Vessot R.F.C. and Levine M.W. 1979. Gravity Probe A-A review of vessot and levine experiment GP-A project final report.

Vessot R.F.C., Levine M.W., Mattison E.M., Blomberg E.L., Hoffman T.E., Nystrom G.U., Farrel B.F., Decher R., Eby P.B., Baugher C.R., Watts J.W., Teuber D.L., and Wills F.D. 1980. Test of relativistic gravitation with a space-borne hydrogen maser. *Phys. Rev. Lett.* **45**: 2081.

von Klitzing K., Dorda G., and Pepper M. 1980. New method for high-accuracy determination of the fine-structure constant based on quantized hall resistance. *Phys. Rev. Lett.* **45**: 494.

Wilpers G., Oates C.W., Diddams S.A., Bartels A., Fortier T.M., Oskay W.H., Bergquist J.C., Jefferts S.R., Heavner T.P., Parker T.R., and Hollberg L. 2007. Absolute frequency measurement of the neutral [40]Ca optical frequency standard at 657 nm based on microkelvin atoms. *Metrologia* **44**: 146.

Index